MW00862377

Introduction to Probability

SECOND EDITION

Dimitri P. Bertsekas and John N. Tsitsiklis

Massachusetts Institute of Technology

WWW site for book information and orders

http://www.athenasc.com

Athena Scientific, Belmont, Massachusetts

Athena Scientific
Post Office Box 805
Nashua, NH 03061-0805
U.S.A.

Email: info@athenasc.com
WWW: http://www.athenasc.com

Cover Design: *Ann Gallager*

Publisher's Cataloging-in-Publication Data

Bertsekas, Dimitri P., Tsitsiklis, John N.
Introduction to Probability
Includes bibliographical references and index
1. Probabilities. 2. Stochastic Processes. I. Title.
QA273.B475 2008 519.2 – 21
Library of Congress Control Number: 2002092167

ISBN 978-1-886529-23-6

To the memory of

Pantelis Bertsekas and Nikos Tsitsiklis

Preface

Probability is common sense reduced to calculation

Laplace

This book is an outgrowth of our involvement in teaching an introductory probability course ("Probabilistic Systems Analysis") at the Massachusetts Institute of Technology.

The course is attended by a large number of students with diverse backgrounds, and a broad range of interests. They span the entire spectrum from freshmen to beginning graduate students, and from the engineering school to the school of management. Accordingly, we have tried to strike a balance between simplicity in exposition and sophistication in analytical reasoning. Our key aim has been to develop the ability to construct and analyze probabilistic models in a manner that combines intuitive understanding and mathematical precision.

In this spirit, some of the more mathematically rigorous analysis has been just sketched or intuitively explained in the text, so that complex proofs do not stand in the way of an otherwise simple exposition. At the same time, some of this analysis is developed (at the level of advanced calculus) in theoretical problems, that are included at the end of the corresponding chapter. Furthermore, some of the subtler mathematical issues are hinted at in footnotes addressed to the more attentive reader.

The book covers the fundamentals of probability theory (probabilistic models, discrete and continuous random variables, multiple random variables, and limit theorems), which are typically part of a first course on the subject. It also contains, in Chapters 4-6 a number of more advanced topics, from which an instructor can choose to match the goals of a particular course. In particular, in Chapter 4, we develop transforms, a more advanced view of conditioning, sums of random variables, least squares estimation, and the bivariate normal distribution. Furthermore, in Chapters 5 and 6, we provide a fairly detailed introduction to Bernoulli, Poisson, and Markov processes.

Our M.I.T. course covers all seven chapters in a single semester, with the exception of the material on the bivariate normal (Section 4.7), and on continuous-time Markov chains (Section 6.5). However, in an alternative course, the material on stochastic processes could be omitted, thereby allowing additional emphasis on foundational material, or coverage of other topics of the instructor's choice.

Our most notable omission in coverage is an introduction to statistics. While we develop all the basic elements of Bayesian statistics, in the form of Bayes' rule for discrete and continuous models, and least squares estimation, we do not enter the subjects of parameter estimation, or non-Bayesian hypothesis testing.

The problems that supplement the main text are divided in three categories:

(a) *Theoretical problems:* The theoretical problems (marked by *) constitute an important component of the text, and ensure that the mathematically oriented reader will find here a smooth development without major gaps. Their solutions are given in the text, but an ambitious reader may be able to solve many of them, especially in earlier chapters, before looking at the solutions.

(b) *Problems in the text:* Besides theoretical problems, the text contains several problems, of various levels of difficulty. These are representative of the problems that are usually covered in recitation and tutorial sessions at M.I.T., and are a primary mechanism through which many of our students learn the material. Our hope is that students elsewhere will attempt to solve these problems, and then refer to their solutions to calibrate and enhance their understanding of the material. The solutions are posted on the book's www site

http://www.athenasc.com/probbook.html

(c) *Supplementary problems:* There is a large (and growing) collection of additional problems, which is not included in the book, but is made available at the book's www site. Many of these problems have been assigned as homework or exam problems at M.I.T., and we expect that instructors elsewhere will use them for a similar purpose. While the statements of these additional problems are publicly accessible, the solutions are made available from the authors only to course instructors.

We would like to acknowledge our debt to several people who contributed in various ways to the book. Our writing project began when we assumed responsibility for a popular probability class at M.I.T. that our colleague Al Drake had taught for several decades. We were thus fortunate to start with an organization of the subject that had stood the test of time, a lively presentation of the various topics in Al's classic textbook, and a rich set of material that had been used in recitation sessions and for homework. We are thus indebted to Al Drake for providing a very favorable set of initial conditions.

We are thankful to the several colleagues who have either taught from the draft of the book at various universities or have read it, and have provided us with valuable feedback. In particular, we thank Ibrahim Abou Faycal, Gustavo de Veciana, Eugene Feinberg, Bob Gray, Muriel Médard, Jason Papastavrou, Ilya Pollak, David Tse, and Terry Wagner.

The teaching assistants for the M.I.T. class have been very helpful. They pointed out corrections to various drafts, they developed problems and solutions suitable for the class, and through their direct interaction with the student body, they provided a robust mechanism for calibrating the level of the material.

Reaching thousands of bright students at M.I.T. at an early stage in their studies was a great source of satisfaction for us. We thank them for their valuable feedback and for being patient while they were taught from a textbook-in-progress.

Last but not least, we are grateful to our families for their support throughout the course of this long project.

Dimitri P. Bertsekas, dimitrib@mit.edu
John N. Tsitsiklis, jnt@mit.edu

Cambridge, Mass., May 2002

Preface to the Second Edition

This is a substantial revision of the 1st edition, involving a reorganization of old material and the addition of new material. The length of the book has increased by about 25 percent. The main changes are the following:

(a) Two new chapters on statistical inference have been added, one on Bayesian and one on classical methods. Our philosophy has been to focus on the main concepts and to facilitate understanding of the main methodologies through some key examples.

(b) Chapters 3 and 4 have been revised, in part to accommodate the new material of the inference chapters and in part to streamline the presentation. Section 4.7 of the 1st edition (bivariate normal distribution) has been omitted from the new edition, but is available at the book's website.

(c) A number of new examples and end-of-chapter problems have been added.

The main objective of the new edition is to provide flexibility to instructors in their choice of material, and in particular to give them the option of including an introduction to statistical inference. Note that Chapters 6-7, and Chapters 8-9 are mutually independent, thus allowing for different paths through the book. Furthermore, Chapter 4 is not needed for Chapters 5-7, and only Sections 4.2-4.3 from Chapter 4 are needed for Chapters 8 and 9. Thus, some possible course offerings based on this book are:

(a) Probability and introduction to statistical inference: Chapters 1-3, Sections 4.2-4.3, Chapter 5, Chapters 8-9.

(b) Probability and introduction to stochastic processes: Chapters 1-3 and 5-7, with possibly a few sections from Chapter 4.

We would like to express our thanks to various colleagues who have contributed valuable comments on the material in the 1st edition and/or the organization of the material in the new chapters. Ed Coffman, Munther Dahleh, Vivek Goyal, Anant Sahai, David Tse, George Verghese, Alan Willsky, and John Wyatt have been very helpful in this regard. Finally, we thank Mengdi Wang for her help with figures and problems for the new chapters.

Dimitri P. Bertsekas, dimitrib@mit.edu
John N. Tsitsiklis, jnt@mit.edu

Cambridge, Mass., June 2008

Contents

1

Sample Space and Probability

Contents

"Probability" is a very useful concept, but can be interpreted in a number of ways. As an illustration, consider the following.

> A patient is admitted to the hospital and a potentially life-saving drug is administered. The following dialog takes place between the nurse and a concerned relative.
>
> RELATIVE: Nurse, what is the probability that the drug will work?
> NURSE: I hope it works, we'll know tomorrow.
> RELATIVE: Yes, but what is the probability that it will?
> NURSE: Each case is different, we have to wait.
> RELATIVE: But let's see, out of a hundred patients that are treated under similar conditions, how many times would you expect it to work?
> NURSE (somewhat annoyed): I told you, every person is different, for some it works, for some it doesn't.
> RELATIVE (insisting): Then tell me, if you had to bet whether it will work or not, which side of the bet would you take?
> NURSE (cheering up for a moment): I'd bet it will work.
> RELATIVE (somewhat relieved): OK, now, would you be willing to lose two dollars if it doesn't work, and gain one dollar if it does?
> NURSE (exasperated): What a sick thought! You are wasting my time!

In this conversation, the relative attempts to use the concept of probability to discuss an **uncertain** situation. The nurse's initial response indicates that the meaning of "probability" is not uniformly shared or understood, and the relative tries to make it more concrete. The first approach is to define probability in terms of **frequency of occurrence**, as a percentage of successes in a moderately large number of similar situations. Such an interpretation is often natural. For example, when we say that a perfectly manufactured coin lands on heads "with probability 50%," we typically mean "roughly half of the time." But the nurse may not be entirely wrong in refusing to discuss in such terms. What if this was an experimental drug that was administered for the very first time in this hospital or in the nurse's experience?

While there are many situations involving uncertainty in which the frequency interpretation is appropriate, there are other situations in which it is not. Consider, for example, a scholar who asserts that the Iliad and the Odyssey were composed by the same person, with probability 90%. Such an assertion conveys some information, but not in terms of frequencies, since the subject is a one-time event. Rather, it is an expression of the scholar's **subjective belief**. One might think that subjective beliefs are not interesting, at least from a mathematical or scientific point of view. On the other hand, people often have to make choices in the presence of uncertainty, and a systematic way of making use of their beliefs is a prerequisite for successful, or at least consistent, decision making.

In fact, the choices and actions of a rational person can reveal a lot about the inner-held subjective probabilities, even if the person does not make conscious use of probabilistic reasoning. Indeed, the last part of the earlier dialog was an attempt to infer the nurse's beliefs in an indirect manner. Since the nurse was willing to accept a one-for-one bet that the drug would work, we may infer that the probability of success was judged to be at least 50%. Had the nurse accepted the last proposed bet (two-for-one), this would have indicated a success probability of at least 2/3.

Rather than dwelling further on philosophical issues about the appropriateness of probabilistic reasoning, we will simply take it as a given that the theory of probability is useful in a broad variety of contexts, including some where the assumed probabilities only reflect subjective beliefs. There is a large body of successful applications in science, engineering, medicine, management, etc., and on the basis of this empirical evidence, probability theory is an extremely useful tool.

Our main objective in this book is to develop the art of describing uncertainty in terms of probabilistic models, as well as the skill of probabilistic reasoning. The first step, which is the subject of this chapter, is to describe the generic structure of such models and their basic properties. The models we consider assign probabilities to collections (sets) of possible outcomes. For this reason, we must begin with a short review of set theory.

1.1 SETS

Probability makes extensive use of set operations, so let us introduce at the outset the relevant notation and terminology.

A **set** is a collection of objects, which are the **elements** of the set. If S is a set and x is an element of S, we write $x \in S$. If x is not an element of S, we write $x \notin S$. A set can have no elements, in which case it is called the **empty set**, denoted by $\emptyset$.

Sets can be specified in a variety of ways. If S contains a finite number of elements, say $x_1, x_2, \ldots, x_n$, we write it as a list of the elements, in braces:

$$S = \{x_1, x_2, \ldots, x_n\}.$$

For example, the set of possible outcomes of a die roll is $\{1, 2, 3, 4, 5, 6\}$, and the set of possible outcomes of a coin toss is $\{H, T\}$, where H stands for "heads" and T stands for "tails."

If S contains infinitely many elements $x_1, x_2, \ldots$, which can be enumerated in a list (so that there are as many elements as there are positive integers) we write

$$S = \{x_1, x_2, \ldots\},$$

and we say that S is **countably infinite**. For example, the set of even integers can be written as $\{0, 2, -2, 4, -4, \ldots\}$, and is countably infinite.

Alternatively, we can consider the set of all x that have a certain property P, and denote it by

$$\{x \,|\, x \text{ satisfies } P\}.$$

(The symbol "$|$" is to be read as "such that.") For example, the set of even integers can be written as $\{k \,|\, k/2 \text{ is integer}\}$. Similarly, the set of all scalars x in the interval $[0, 1]$ can be written as $\{x \,|\, 0 \leq x \leq 1\}$. Note that the elements x of the latter set take a continuous range of values, and cannot be written down in a list (a proof is sketched in the end-of-chapter problems); such a set is said to be **uncountable**.

If every element of a set S is also an element of a set T, we say that S is a **subset** of T, and we write $S \subset T$ or $T \supset S$. If $S \subset T$ and $T \subset S$, the two sets are **equal**, and we write $S = T$. It is also expedient to introduce a **universal set**, denoted by Ω, which contains all objects that could conceivably be of interest in a particular context. Having specified the context in terms of a universal set Ω, we only consider sets S that are subsets of Ω.

Set Operations

The **complement** of a set S, with respect to the universe Ω, is the set $\{x \in \Omega \,|\, x \notin S\}$ of all elements of Ω that do not belong to S, and is denoted by S^c. Note that $\Omega^c = \emptyset$.

The **union** of two sets S and T is the set of all elements that belong to S or T (or both), and is denoted by $S \cup T$. The **intersection** of two sets S and T is the set of all elements that belong to both S and T, and is denoted by $S \cap T$. Thus,

$$S \cup T = \{x \,|\, x \in S \text{ or } x \in T\},$$

and

$$S \cap T = \{x \,|\, x \in S \text{ and } x \in T\}.$$

In some cases, we will have to consider the union or the intersection of several, even infinitely many sets, defined in the obvious way. For example, if for every positive integer n, we are given a set S_n, then

$$\bigcup_{n=1}^{\infty} S_n = S_1 \cup S_2 \cup \cdots = \{x \,|\, x \in S_n \text{ for some } n\},$$

and

$$\bigcap_{n=1}^{\infty} S_n = S_1 \cap S_2 \cap \cdots = \{x \,|\, x \in S_n \text{ for all } n\}.$$

Two sets are said to be disjoint if their intersection is empty. More generally, several sets are said to be **disjoint** if no two of them have a common element. A collection of sets is said to be a **partition** of a set S if the sets in the collection are disjoint and their union is S.

If x and y are two objects, we use (x, y) to denote the **ordered pair** of x and y. The set of scalars (real numbers) is denoted by $\Re$; the set of pairs (or triplets) of scalars, i.e., the two-dimensional plane (or three-dimensional space, respectively) is denoted by $\Re^2$ (or $\Re^3$, respectively).

Sets and the associated operations are easy to visualize in terms of **Venn diagrams**, as illustrated in Fig. 1.1.

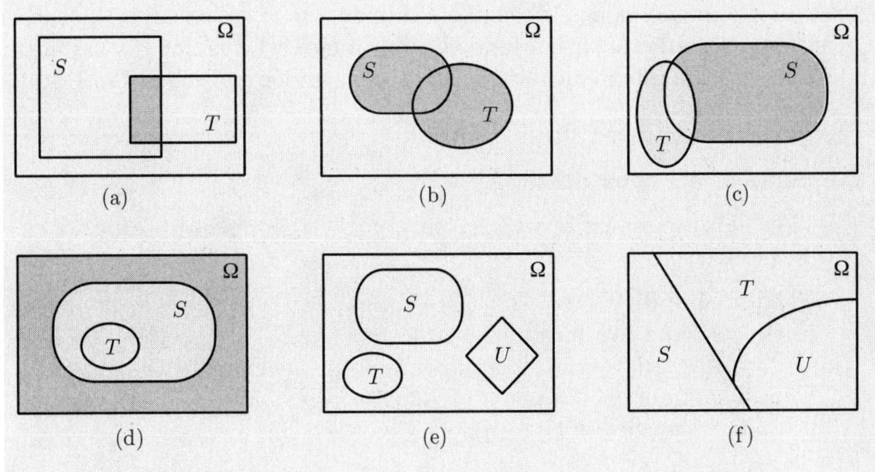

Figure 1.1: Examples of Venn diagrams. (a) The shaded region is $S \cap T$. (b) The shaded region is $S \cup T$. (c) The shaded region is $S \cap T^c$. (d) Here, $T \subset S$. The shaded region is the complement of S. (e) The sets S, T, and U are disjoint. (f) The sets S, T, and U form a partition of the set Ω.

The Algebra of Sets

Set operations have several properties, which are elementary consequences of the definitions. Some examples are:

$$\begin{aligned} S \cup T &= T \cup S, & S \cup (T \cup U) &= (S \cup T) \cup U, \\ S \cap (T \cup U) &= (S \cap T) \cup (S \cap U), & S \cup (T \cap U) &= (S \cup T) \cap (S \cup U), \\ (S^c)^c &= S, & S \cap S^c &= \varnothing, \\ S \cup \Omega &= \Omega, & S \cap \Omega &= S. \end{aligned}$$

Two particularly useful properties are given by **De Morgan's laws** which state that

$$\left(\bigcup_n S_n\right)^c = \bigcap_n S_n^c, \qquad \left(\bigcap_n S_n\right)^c = \bigcup_n S_n^c.$$

To establish the first law, suppose that $x \in (\cup_n S_n)^c$. Then, $x \notin \cup_n S_n$, which implies that for every n, we have $x \notin S_n$. Thus, x belongs to the complement of every S_n, and $x \in \cap_n S_n^c$. This shows that $(\cup_n S_n)^c \subset \cap_n S_n^c$. The converse inclusion is established by reversing the above argument, and the first law follows. The argument for the second law is similar.

1.2 PROBABILISTIC MODELS

A probabilistic model is a mathematical description of an uncertain situation. It must be in accordance with a fundamental framework that we discuss in this section. Its two main ingredients are listed below and are visualized in Fig. 1.2.

Elements of a Probabilistic Model

- The **sample space** Ω, which is the set of all possible **outcomes** of an experiment.
- The **probability law**, which assigns to a set A of possible outcomes (also called an **event**) a nonnegative number $\mathbf{P}(A)$ (called the **probability** of A) that encodes our knowledge or belief about the collective "likelihood" of the elements of A. The probability law must satisfy certain properties to be introduced shortly.

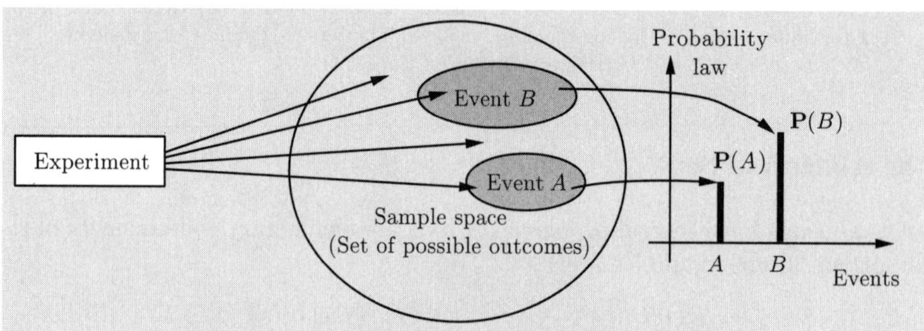

Figure 1.2: The main ingredients of a probabilistic model.

Sample Spaces and Events

Every probabilistic model involves an underlying process, called the **experiment**, that will produce exactly one out of several possible **outcomes**. The set of all possible outcomes is called the **sample space** of the experiment, and is denoted by Ω. A subset of the sample space, that is, a collection of possible outcomes, is called an **event**.† There is no restriction on what constitutes an experiment. For example, it could be a single toss of a coin, or three tosses, or an infinite sequence of tosses. However, it is important to note that in our formulation of a probabilistic model, there is only one experiment. So, three tosses of a coin constitute a single experiment, rather than three experiments.

The sample space of an experiment may consist of a finite or an infinite number of possible outcomes. Finite sample spaces are conceptually and mathematically simpler. Still, sample spaces with an infinite number of elements are quite common. As an example, consider throwing a dart on a square target and viewing the point of impact as the outcome.

Choosing an Appropriate Sample Space

Regardless of their number, different elements of the sample space should be distinct and **mutually exclusive**, so that when the experiment is carried out there is a unique outcome. For example, the sample space associated with the roll of a die cannot contain "1 or 3" as a possible outcome and also "1 or 4" as another possible outcome. If it did, we would not be able to assign a unique outcome when the roll is a 1.

A given physical situation may be modeled in several different ways, depending on the kind of questions that we are interested in. Generally, the sample space chosen for a probabilistic model must be **collectively exhaustive**, in the sense that no matter what happens in the experiment, we always obtain an outcome that has been included in the sample space. In addition, the sample space should have enough detail to distinguish between all outcomes of interest to the modeler, while avoiding irrelevant details.

Example 1.1. Consider two alternative games, both involving ten successive coin tosses:

Game 1: We receive \$1 each time a head comes up.

Game 2: We receive \$1 for every coin toss, up to and including the first time a head comes up. Then, we receive \$2 for every coin toss, up to the second time a head comes up. More generally, the dollar amount per toss is doubled each time a head comes up.

---

† Any collection of possible outcomes, including the entire sample space Ω and its complement, the empty set $\emptyset$, may qualify as an event. Strictly speaking, however, some sets have to be excluded. In particular, when dealing with probabilistic models involving an uncountably infinite sample space, there are certain unusual subsets for which one cannot associate meaningful probabilities. This is an intricate technical issue, involving the mathematics of measure theory. Fortunately, such pathological subsets do not arise in the problems considered in this text or in practice, and the issue can be safely ignored.

In game 1, it is only the total number of heads in the ten-toss sequence that matters, while in game 2, the order of heads and tails is also important. Thus, in a probabilistic model for game 1, we can work with a sample space consisting of eleven possible outcomes, namely, $0, 1, \ldots, 10$. In game 2, a finer grain description of the experiment is called for, and it is more appropriate to let the sample space consist of every possible ten-long sequence of heads and tails.

Sequential Models

Many experiments have an inherently sequential character; for example, tossing a coin three times, observing the value of a stock on five successive days, or receiving eight successive digits at a communication receiver. It is then often useful to describe the experiment and the associated sample space by means of a **tree-based sequential description**, as in Fig. 1.3.

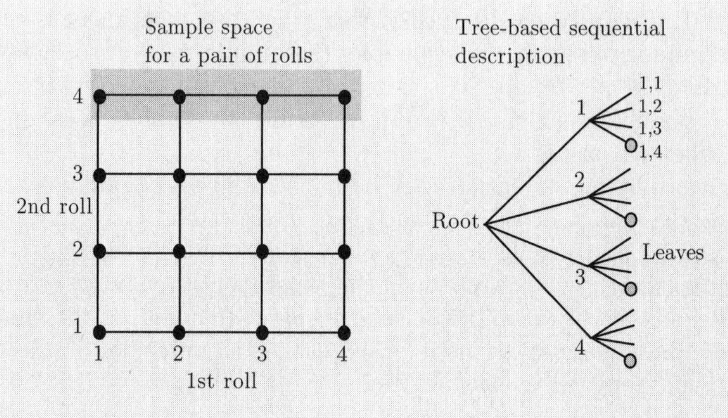

Figure 1.3: Two equivalent descriptions of the sample space of an experiment involving two rolls of a 4-sided die. The possible outcomes are all the ordered pairs of the form (i, j), where i is the result of the first roll, and j is the result of the second. These outcomes can be arranged in a 2-dimensional grid as in the figure on the left, or they can be described by the tree on the right, which reflects the sequential character of the experiment. Here, each possible outcome corresponds to a leaf of the tree and is associated with the unique path from the root to that leaf. The shaded area on the left is the event $\{(1,4),\ (2,4),\ (3,4),\ (4,4)\}$ that the result of the second roll is 4. That same event can be described by the set of leaves highlighted on the right. Note also that every node of the tree can be identified with an event, namely, the set of all leaves downstream from that node. For example, the node labeled by a 1 can be identified with the event $\{(1,1), (1,2), (1,3), (1,4)\}$ that the result of the first roll is 1.

Probability Laws

Suppose we have settled on the sample space Ω associated with an experiment. To complete the probabilistic model, we must now introduce a **probability law**.

Intuitively, this specifies the "likelihood" of any outcome, or of any set of possible outcomes (an event, as we have called it earlier). More precisely, the probability law assigns to every event A, a number $\mathbf{P}(A)$, called the **probability** of A, satisfying the following axioms.

Probability Axioms

1. **(Nonnegativity)** $\mathbf{P}(A) \geq 0$, for every event A.
2. **(Additivity)** If A and B are two disjoint events, then the probability of their union satisfies

$$\mathbf{P}(A \cup B) = \mathbf{P}(A) + \mathbf{P}(B).$$

   More generally, if the sample space has an infinite number of elements and $A_1, A_2, \ldots$ is a sequence of disjoint events, then the probability of their union satisfies

$$\mathbf{P}(A_1 \cup A_2 \cup \cdots) = \mathbf{P}(A_1) + \mathbf{P}(A_2) + \cdots.$$

3. **(Normalization)** The probability of the entire sample space Ω is equal to 1, that is, $\mathbf{P}(\Omega) = 1$.

In order to visualize a probability law, consider a unit of mass which is "spread" over the sample space. Then, $\mathbf{P}(A)$ is simply the total mass that was assigned collectively to the elements of A. In terms of this analogy, the additivity axiom becomes quite intuitive: the total mass in a sequence of disjoint events is the sum of their individual masses.

A more concrete interpretation of probabilities is in terms of relative frequencies: a statement such as $\mathbf{P}(A) = 2/3$ often represents a belief that event A will occur in about two thirds out of a large number of repetitions of the experiment. Such an interpretation, though not always appropriate, can sometimes facilitate our intuitive understanding. It will be revisited in Chapter 5, in our study of limit theorems.

There are many natural properties of a probability law, which have not been included in the above axioms for the simple reason that they can be **derived** from them. For example, note that the normalization and additivity axioms imply that

$$1 = \mathbf{P}(\Omega) = \mathbf{P}(\Omega \cup \emptyset) = \mathbf{P}(\Omega) + \mathbf{P}(\emptyset) = 1 + \mathbf{P}(\emptyset),$$

and this shows that the probability of the empty event is 0:

$$\mathbf{P}(\emptyset) = 0.$$

As another example, consider three disjoint events A_1, A_2, and A_3. We can use the additivity axiom for two disjoint events repeatedly, to obtain

$$\begin{aligned}\mathbf{P}(A_1 \cup A_2 \cup A_3) &= \mathbf{P}\big(A_1 \cup (A_2 \cup A_3)\big)\\ &= \mathbf{P}(A_1) + \mathbf{P}(A_2 \cup A_3)\\ &= \mathbf{P}(A_1) + \mathbf{P}(A_2) + \mathbf{P}(A_3).\end{aligned}$$

Proceeding similarly, we obtain that the probability of the union of finitely many disjoint events is always equal to the sum of the probabilities of these events. More such properties will be considered shortly.

Discrete Models

Here is an illustration of how to construct a probability law starting from some common sense assumptions about a model.

Example 1.2. Consider an experiment involving a single coin toss. There are two possible outcomes, heads (H) and tails (T). The sample space is $\Omega = \{H, T\}$, and the events are

$$\{H, T\},\ \{H\},\ \{T\},\ \emptyset.$$

If the coin is fair, i.e., if we believe that heads and tails are "equally likely," we should assign equal probabilities to the two possible outcomes and specify that $\mathbf{P}(\{H\}) = \mathbf{P}(\{T\}) = 0.5$. The additivity axiom implies that

$$\mathbf{P}\big(\{H, T\}\big) = \mathbf{P}\big(\{H\}\big) + \mathbf{P}\big(\{T\}\big) = 1,$$

which is consistent with the normalization axiom. Thus, the probability law is given by

$$\mathbf{P}\big(\{H, T\}\big) = 1, \qquad \mathbf{P}\big(\{H\}\big) = 0.5, \qquad \mathbf{P}\big(\{T\}\big) = 0.5, \qquad \mathbf{P}(\emptyset) = 0,$$

and satisfies all three axioms.

Consider another experiment involving three coin tosses. The outcome will now be a 3-long string of heads or tails. The sample space is

$$\Omega = \{HHH,\ HHT,\ HTH,\ HTT,\ THH,\ THT,\ TTH,\ TTT\}.$$

We assume that each possible outcome has the same probability of 1/8. Let us construct a probability law that satisfies the three axioms. Consider, as an example, the event

$$A = \{\text{exactly 2 heads occur}\} = \{HHT,\ HTH,\ THH\}.$$

Using additivity, the probability of A is the sum of the probabilities of its elements:

$$\begin{aligned}\mathbf{P}\big(\{HHT,\ HTH,\ THH\}\big) &= \mathbf{P}\big(\{HHT\}\big) + \mathbf{P}\big(\{HTH\}\big) + \mathbf{P}\big(\{THH\}\big)\\ &= \frac{1}{8} + \frac{1}{8} + \frac{1}{8}\\ &= \frac{3}{8}.\end{aligned}$$

Similarly, the probability of any event is equal to 1/8 times the number of possible outcomes contained in the event. This defines a probability law that satisfies the three axioms.

By using the additivity axiom and by generalizing the reasoning in the preceding example, we reach the following conclusion.

Discrete Probability Law

If the sample space consists of a finite number of possible outcomes, then the probability law is specified by the probabilities of the events that consist of a single element. In particular, the probability of any event $\{s_1, s_2, \ldots, s_n\}$ is the sum of the probabilities of its elements:

$$\mathbf{P}\big(\{s_1, s_2, \ldots, s_n\}\big) = \mathbf{P}(s_1) + \mathbf{P}(s_2) + \cdots + \mathbf{P}(s_n).$$

Note that we are using here the simpler notation $\mathbf{P}(s_i)$ to denote the probability of the event $\{s_i\}$, instead of the more precise $\mathbf{P}(\{s_i\})$. This convention will be used throughout the remainder of the book.

In the special case where the probabilities $\mathbf{P}(s_1), \ldots, \mathbf{P}(s_n)$ are all the same (by necessity equal to $1/n$, in view of the normalization axiom), we obtain the following.

Discrete Uniform Probability Law

If the sample space consists of n possible outcomes which are equally likely (i.e., all single-element events have the same probability), then the probability of any event A is given by

$$\mathbf{P}(A) = \frac{\text{number of elements of } A}{n}.$$

Let us provide a few more examples of sample spaces and probability laws.

Example 1.3. Consider the experiment of rolling a pair of 4-sided dice (cf. Fig. 1.4). We assume the dice are fair, and we interpret this assumption to mean that each of the sixteen possible outcomes [pairs (i, j), with $i, j = 1, 2, 3, 4$] has the same probability of 1/16. To calculate the probability of an event, we must count the number of elements of the event and divide by 16 (the total number of possible outcomes). Here are some event probabilities calculated in this way:

$$\mathbf{P}\big(\{\text{the sum of the rolls is even}\}\big) = 8/16 = 1/2,$$
$$\mathbf{P}\big(\{\text{the sum of the rolls is odd}\}\big) = 8/16 = 1/2,$$
$$\mathbf{P}\big(\{\text{the first roll is equal to the second}\}\big) = 4/16 = 1/4,$$
$$\mathbf{P}\big(\{\text{the first roll is larger than the second}\}\big) = 6/16 = 3/8,$$
$$\mathbf{P}\big(\{\text{at least one roll is equal to 4}\}\big) = 7/16.$$

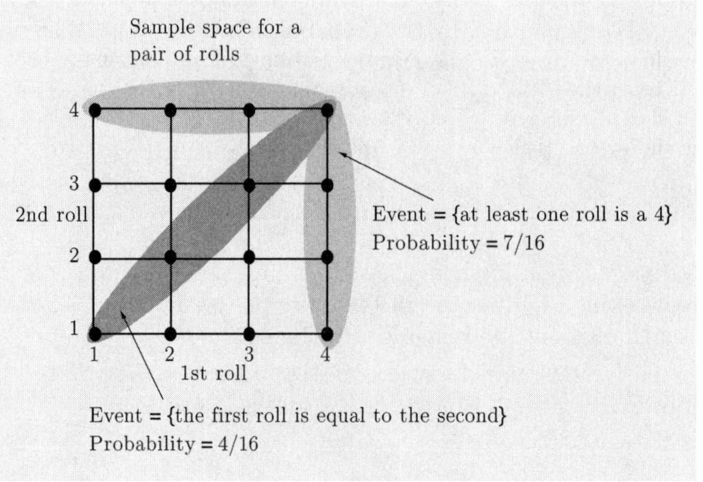

Figure 1.4: Various events in the experiment of rolling a pair of 4-sided dice, and their probabilities, calculated according to the discrete uniform law.

Continuous Models

Probabilistic models with continuous sample spaces differ from their discrete counterparts in that the probabilities of the single-element events may not be sufficient to characterize the probability law. This is illustrated in the following examples, which also indicate how to generalize the uniform probability law to the case of a continuous sample space.

Example 1.4. A wheel of fortune is continuously calibrated from 0 to 1, so the possible outcomes of an experiment consisting of a single spin are the numbers in the interval $\Omega = [0, 1]$. Assuming a fair wheel, it is appropriate to consider all outcomes equally likely, but what is the probability of the event consisting of a single element? It cannot be positive, because then, using the additivity axiom, it would follow that events with a sufficiently large number of elements would have probability larger than 1. Therefore, the probability of any event that consists of a single element must be 0.

In this example, it makes sense to assign probability $b - a$ to any subinterval $[a, b]$ of $[0, 1]$, and to calculate the probability of a more complicated set by evaluating its "length."† This assignment satisfies the three probability axioms and qualifies as a legitimate probability law.

Example 1.5. Romeo and Juliet have a date at a given time, and each will arrive at the meeting place with a delay between 0 and 1 hour, with all pairs of delays being equally likely. The first to arrive will wait for 15 minutes and will leave if the other has not yet arrived. What is the probability that they will meet?

Let us use as sample space the unit square, whose elements are the possible pairs of delays for the two of them. Our interpretation of "equally likely" pairs of delays is to let the probability of a subset of Ω be equal to its area. This probability law satisfies the three probability axioms. The event that Romeo and Juliet will meet is the shaded region in Fig. 1.5, and its probability is calculated to be 7/16.

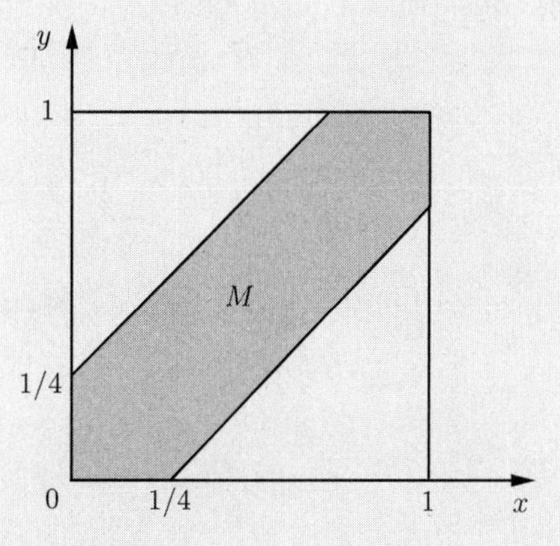

Figure 1.5: The event M that Romeo and Juliet will arrive within 15 minutes of each other (cf. Example 1.5) is

$$M = \left\{(x, y) \;\middle|\; |x - y| \leq 1/4,\ 0 \leq x \leq 1,\ 0 \leq y \leq 1\right\},$$

and is shaded in the figure. The area of M is 1 minus the area of the two unshaded triangles, or $1 - (3/4) \cdot (3/4) = 7/16$. Thus, the probability of meeting is 7/16.

---

† The "length" of a subset S of $[0, 1]$ is the integral $\int_S dt$, which is defined, for "nice" sets S, in the usual calculus sense. For unusual sets, this integral may not be well defined mathematically, but such issues belong to a more advanced treatment of the subject. Incidentally, the legitimacy of using length as a probability law hinges on the fact that the unit interval has an uncountably infinite number of elements. Indeed, if the unit interval had a countable number of elements, with each element having zero probability, the additivity axiom would imply that the whole interval has zero probability, which would contradict the normalization axiom.

Properties of Probability Laws

Probability laws have a number of properties, which can be deduced from the axioms. Some of them are summarized below.

Some Properties of Probability Laws

Consider a probability law, and let A, B, and C be events.

(a) If $A \subset B$, then $\mathbf{P}(A) \leq \mathbf{P}(B)$.

(b) $\mathbf{P}(A \cup B) = \mathbf{P}(A) + \mathbf{P}(B) - \mathbf{P}(A \cap B)$.

(c) $\mathbf{P}(A \cup B) \leq \mathbf{P}(A) + \mathbf{P}(B)$.

(d) $\mathbf{P}(A \cup B \cup C) = \mathbf{P}(A) + \mathbf{P}(A^c \cap B) + \mathbf{P}(A^c \cap B^c \cap C)$.

These properties, and other similar ones, can be visualized and verified graphically using Venn diagrams, as in Fig. 1.6. Note that property (c) can be generalized as follows:

$$\mathbf{P}(A_1 \cup A_2 \cup \cdots \cup A_n) \leq \sum_{i=1}^{n} \mathbf{P}(A_i).$$

To see this, we apply property (c) to the sets A_1 and $A_2 \cup \cdots \cup A_n$, to obtain

$$\mathbf{P}(A_1 \cup A_2 \cup \cdots \cup A_n) \leq \mathbf{P}(A_1) + \mathbf{P}(A_2 \cup \cdots \cup A_n).$$

We also apply property (c) to the sets A_2 and $A_3 \cup \cdots \cup A_n$, to obtain

$$\mathbf{P}(A_2 \cup \cdots \cup A_n) \leq \mathbf{P}(A_2) + \mathbf{P}(A_3 \cup \cdots \cup A_n).$$

We continue similarly, and finally add.

Models and Reality

The framework of probability theory can be used to analyze uncertainty in a wide variety of physical contexts. Typically, this involves two distinct stages.

(a) In the first stage, we construct a probabilistic model by specifying a probability law on a suitably defined sample space. There are no hard rules to guide this step, other than the requirement that the probability law conform to the three axioms. Reasonable people may disagree on which model best represents reality. In many cases, one may even want to use a somewhat "incorrect" model, if it is simpler than the "correct" one or allows for tractable calculations. This is consistent with common practice in science and engineering, where the choice of a model often involves a tradeoff between accuracy, simplicity, and tractability. Sometimes, a model is chosen on the basis of historical data or past outcomes of similar experiments, using statistical inference methods, which will be discussed in Chapters 8 and 9.

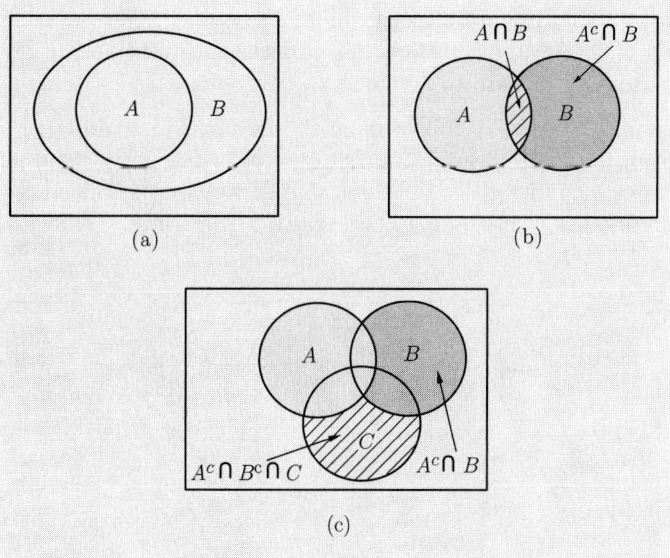

Figure 1.6: Visualization and verification of various properties of probability laws using Venn diagrams. If $A \subset B$, then B is the union of the two disjoint events A and $A^c \cap B$; see diagram (a). Therefore, by the additivity axiom, we have

$$\mathbf{P}(B) = \mathbf{P}(A) + \mathbf{P}(A^c \cap B) \geq \mathbf{P}(A),$$

where the inequality follows from the nonnegativity axiom, and verifies property (a).

From diagram (b), we can express the events $A \cup B$ and B as unions of disjoint events:

$$A \cup B = A \cup (A^c \cap B), \qquad B = (A \cap B) \cup (A^c \cap B).$$

Using the additivity axiom, we have

$$\mathbf{P}(A \cup B) = \mathbf{P}(A) + \mathbf{P}(A^c \cap B), \qquad \mathbf{P}(B) = \mathbf{P}(A \cap B) + \mathbf{P}(A^c \cap B).$$

Subtracting the second equality from the first and rearranging terms, we obtain $\mathbf{P}(A \cup B) = \mathbf{P}(A) + \mathbf{P}(B) - \mathbf{P}(A \cap B)$, verifying property (b). Using also the fact $\mathbf{P}(A \cap B) \geq 0$ (the nonnegativity axiom), we obtain $\mathbf{P}(A \cup B) \leq \mathbf{P}(A) + \mathbf{P}(B)$, verifying property (c).

From diagram (c), we see that the event $A \cup B \cup C$ can be expressed as a union of three disjoint events:

$$A \cup B \cup C = A \cup (A^c \cap B) \cup (A^c \cap B^c \cap C),$$

so property (d) follows as a consequence of the additivity axiom.

(b) In the second stage, we work within a fully specified probabilistic model and derive the probabilities of certain events, or deduce some interesting properties. While the first stage entails the often open-ended task of connecting the real world with mathematics, the second one is tightly regulated by the rules of ordinary logic and the axioms of probability. Difficulties may arise in the latter if some required calculations are complex, or if a probability law is specified in an indirect fashion. Even so, there is no room for ambiguity: all conceivable questions have precise answers and it is only a matter of developing the skill to arrive at them.

Probability theory is full of "paradoxes" in which different calculation methods seem to give different answers to the same question. Invariably though, these apparent inconsistencies turn out to reflect poorly specified or ambiguous probabilistic models. An example, **Bertrand's paradox**, is shown in Fig. 1.7.

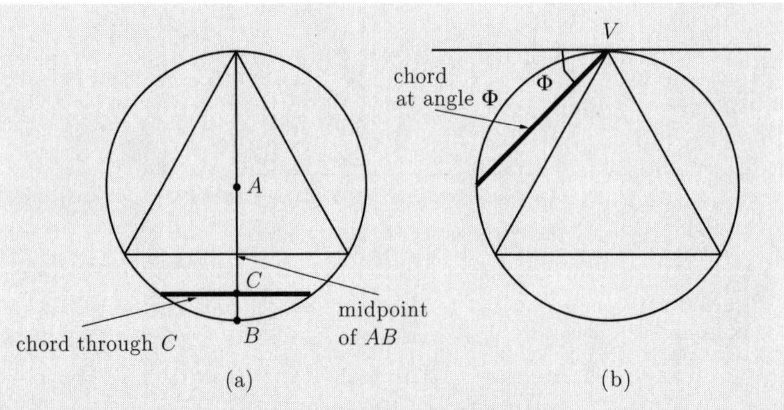

Figure 1.7: This example, presented by L. F. Bertrand in 1889, illustrates the need to specify unambiguously a probabilistic model. Consider a circle and an equilateral triangle inscribed in the circle. What is the probability that the length of a randomly chosen chord of the circle is greater than the side of the triangle? The answer here depends on the precise meaning of "randomly chosen." The two methods illustrated in parts (a) and (b) of the figure lead to contradictory results.

In (a), we take a radius of the circle, such as AB, and we choose a point C on that radius, with all points being equally likely. We then draw the chord through C that is orthogonal to AB. From elementary geometry, AB intersects the triangle at the midpoint of AB, so the probability that the length of the chord is greater than the side is $1/2$.

In (b), we take a point on the circle, such as the vertex V, we draw the tangent to the circle through V, and we draw a line through V that forms a random angle Φ with the tangent, with all angles being equally likely. We consider the chord obtained by the intersection of this line with the circle. From elementary geometry, the length of the chord is greater than the side of the triangle if Φ is between $\pi/3$ and $2\pi/3$. Since Φ takes values between 0 and π, the probability that the length of the chord is greater than the side is $1/3$.

A Brief History of Probability

- **B.C.E.** Games of chance were popular in ancient Greece and Rome, but no scientific development of the subject took place, possibly because the number system used by the Greeks did not facilitate algebraic calculations. The development of probability based on sound scientific analysis had to await the development of the modern arithmetic system by the Hindus and the Arabs in the second half of the first millennium, as well as the flood of scientific ideas generated by the Renaissance.

- **16th century.** Girolamo Cardano, a colorful and controversial Italian mathematician, publishes the first book describing correct methods for calculating probabilities in games of chance involving dice and cards.

- **17th century.** A correspondence between Fermat and Pascal touches upon several interesting probability questions and motivates further study in the field.

- **18th century.** Jacob Bernoulli studies repeated coin tossing and introduces the first law of large numbers, which lays a foundation for linking theoretical probability concepts and empirical fact. Several mathematicians, such as Daniel Bernoulli, Leibnitz, Bayes, and Lagrange, make important contributions to probability theory and its use in analyzing real-world phenomena. De Moivre introduces the normal distribution and proves the first form of the central limit theorem.

- **19th century.** Laplace publishes an influential book that establishes the importance of probability as a quantitative field and contains many original contributions, including a more general version of the central limit theorem. Legendre and Gauss apply probability to astronomical predictions, using the method of least squares, thus pointing the way to a vast range of applications. Poisson publishes an influential book with many original contributions, including the Poisson distribution. Chebyshev, and his students Markov and Lyapunov, study limit theorems and raise the standards of mathematical rigor in the field. Throughout this period, probability theory is largely viewed as a natural science, its primary goal being the explanation of physical phenomena. Consistently with this goal, probabilities are mainly interpreted as limits of relative frequencies in the context of repeatable experiments.

- **20th century.** Relative frequency is abandoned as the conceptual foundation of probability theory in favor of a now universally used axiomatic system, introduced by Kolmogorov. Similar to other branches of mathematics, the development of probability theory from the axioms relies only on logical correctness, regardless of its relevance to physical phenomena. Nonetheless, probability theory is used pervasively in science and engineering because of its ability to describe and interpret most types of uncertain phenomena in the real world.

1.3 CONDITIONAL PROBABILITY

Conditional probability provides us with a way to reason about the outcome of an experiment, based on **partial information**. Here are some examples of situations we have in mind:

(a) In an experiment involving two successive rolls of a die, you are told that the sum of the two rolls is 9. How likely is it that the first roll was a 6?

(b) In a word guessing game, the first letter of the word is a "t". What is the likelihood that the second letter is an "h"?

(c) How likely is it that a person has a certain disease given that a medical test was negative?

(d) A spot shows up on a radar screen. How likely is it to correspond to an aircraft?

In more precise terms, given an experiment, a corresponding sample space, and a probability law, suppose that we know that the outcome is within some given event B. We wish to quantify the likelihood that the outcome also belongs to some other given event A. We thus seek to construct a new probability law that takes into account the available knowledge: a probability law that for any event A, specifies the **conditional probability of A given B**, denoted by $\mathbf{P}(A \,|\, B)$.

We would like the conditional probabilities $\mathbf{P}(A \,|\, B)$ of different events A to constitute a legitimate probability law, which satisfies the probability axioms. The conditional probabilities should also be consistent with our intuition in important special cases, e.g., when all possible outcomes of the experiment are equally likely. For example, suppose that all six possible outcomes of a fair die roll are equally likely. If we are told that the outcome is even, we are left with only three possible outcomes, namely, 2, 4, and 6. These three outcomes were equally likely to start with, and so they should remain equally likely given the additional knowledge that the outcome was even. Thus, it is reasonable to let

$$\mathbf{P}(\text{the outcome is } 6 \,|\, \text{the outcome is even}) = \frac{1}{3}.$$

This argument suggests that an appropriate definition of conditional probability when all outcomes are equally likely, is given by

$$\mathbf{P}(A \,|\, B) = \frac{\text{number of elements of } A \cap B}{\text{number of elements of } B}.$$

Generalizing the argument, we introduce the following definition of conditional probability:

$$\mathbf{P}(A \,|\, B) = \frac{\mathbf{P}(A \cap B)}{\mathbf{P}(B)},$$

where we assume that $\mathbf{P}(B) > 0$; the conditional probability is undefined if the conditioning event has zero probability. In words, out of the total probability of the elements of B, $\mathbf{P}(A \mid B)$ is the fraction that is assigned to possible outcomes that also belong to A.

Conditional Probabilities Specify a Probability Law

For a fixed event B, it can be verified that the conditional probabilities $\mathbf{P}(A \mid B)$ form a legitimate probability law that satisfies the three axioms. Indeed, nonnegativity is clear. Furthermore,

$$\mathbf{P}(\Omega \mid B) = \frac{\mathbf{P}(\Omega \cap B)}{\mathbf{P}(B)} = \frac{\mathbf{P}(B)}{\mathbf{P}(B)} = 1,$$

and the normalization axiom is also satisfied. To verify the additivity axiom, we write for any two disjoint events A_1 and A_2,

$$\begin{aligned}
\mathbf{P}(A_1 \cup A_2 \mid B) &= \frac{\mathbf{P}\big((A_1 \cup A_2) \cap B\big)}{\mathbf{P}(B)} \\
&= \frac{\mathbf{P}((A_1 \cap B) \cup (A_2 \cap B))}{\mathbf{P}(B)} \\
&= \frac{\mathbf{P}(A_1 \cap B) + \mathbf{P}(A_2 \cap B)}{\mathbf{P}(B)} \\
&= \frac{\mathbf{P}(A_1 \cap B)}{\mathbf{P}(B)} + \frac{\mathbf{P}(A_2 \cap B)}{\mathbf{P}(B)} \\
&= \mathbf{P}(A_1 \mid B) + \mathbf{P}(A_2 \mid B),
\end{aligned}$$

where for the third equality, we used the fact that $A_1 \cap B$ and $A_2 \cap B$ are disjoint sets, and the additivity axiom for the (unconditional) probability law. The argument for a countable collection of disjoint sets is similar.

Since conditional probabilities constitute a legitimate probability law, all general properties of probability laws remain valid. For example, a fact such as $\mathbf{P}(A \cup C) \leq \mathbf{P}(A) + \mathbf{P}(C)$ translates to the new fact

$$\mathbf{P}(A \cup C \mid B) \leq \mathbf{P}(A \mid B) + \mathbf{P}(C \mid B).$$

Let us also note that since we have $\mathbf{P}(B \mid B) = \mathbf{P}(B)/\mathbf{P}(B) = 1$, all of the conditional probability is concentrated on B. Thus, we might as well discard all possible outcomes outside B and treat the conditional probabilities as a probability law defined on the new universe B.

Let us summarize the conclusions reached so far.

Properties of Conditional Probability

- The conditional probability of an event A, given an event B with $\mathbf{P}(B) > 0$, is defined by

$$\mathbf{P}(A \mid B) = \frac{\mathbf{P}(A \cap B)}{\mathbf{P}(B)},$$

and specifies a new (conditional) probability law on the same sample space Ω. In particular, all properties of probability laws remain valid for conditional probability laws.

- Conditional probabilities can also be viewed as a probability law on a new universe B, because all of the conditional probability is concentrated on B.

- If the possible outcomes are finitely many and equally likely, then

$$\mathbf{P}(A \mid B) = \frac{\text{number of elements of } A \cap B}{\text{number of elements of } B}.$$

Example 1.6. We toss a fair coin three successive times. We wish to find the conditional probability $\mathbf{P}(A \mid B)$ when A and B are the events

$$A = \{\text{more heads than tails come up}\}, \qquad B = \{\text{1st toss is a head}\}.$$

The sample space consists of eight sequences,

$$\Omega = \{HHH, HHT, HTH, HTT, THH, THT, TTH, TTT\},$$

which we assume to be equally likely. The event B consists of the four elements HHH, HHT, HTH, HTT, so its probability is

$$\mathbf{P}(B) = \frac{4}{8}.$$

The event $A \cap B$ consists of the three elements HHH, HHT, HTH, so its probability is

$$\mathbf{P}(A \cap B) = \frac{3}{8}.$$

Thus, the conditional probability $\mathbf{P}(A \mid B)$ is

$$\mathbf{P}(A \mid B) = \frac{\mathbf{P}(A \cap B)}{\mathbf{P}(B)} = \frac{3/8}{4/8} = \frac{3}{4}.$$

Because all possible outcomes are equally likely here, we can also compute $\mathbf{P}(A \mid B)$ using a shortcut. We can bypass the calculation of $\mathbf{P}(B)$ and $\mathbf{P}(A \cap B)$, and simply divide the number of elements shared by A and B (which is 3) with the number of elements of B (which is 4), to obtain the same result $3/4$.

Example 1.7. A fair 4-sided die is rolled twice and we assume that all sixteen possible outcomes are equally likely. Let X and Y be the result of the 1st and the 2nd roll, respectively. We wish to determine the conditional probability $\mathbf{P}(A \mid B)$, where

$$A = \{\max(X, Y) = m\}, \qquad B = \{\min(X, Y) = 2\},$$

and m takes each of the values 1, 2, 3, 4.

As in the preceding example, we can first determine the probabilities $\mathbf{P}(A \cap B)$ and $\mathbf{P}(B)$ by counting the number of elements of $A \cap B$ and B, respectively, and dividing by 16. Alternatively, we can directly divide the number of elements of $A \cap B$ with the number of elements of B; see Fig. 1.8.

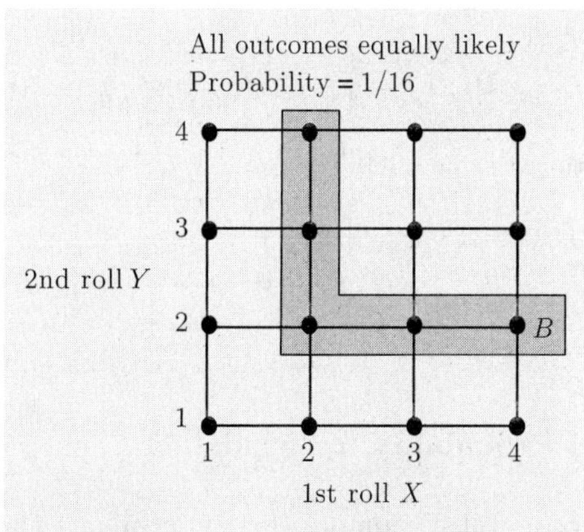

Figure 1.8: Sample space of an experiment involving two rolls of a 4-sided die. (cf. Example 1.7). The conditioning event $B = \{\min(X, Y) = 2\}$ consists of the 5-element shaded set. The set $A = \{\max(X, Y) = m\}$ shares with B two elements if $m = 3$ or $m = 4$, one element if $m = 2$, and no element if $m = 1$. Thus, we have

$$\mathbf{P}\big(\{\max(X, Y) = m\} \bigm| B\big) = \begin{cases} 2/5, & \text{if } m = 3 \text{ or } m = 4, \\ 1/5, & \text{if } m = 2, \\ 0, & \text{if } m = 1. \end{cases}$$

Example 1.8. A conservative design team, call it C, and an innovative design team, call it N, are asked to separately design a new product within a month. From past experience we know that:

(a) The probability that team C is successful is $2/3$.

(b) The probability that team N is successful is $1/2$.

(c) The probability that at least one team is successful is $3/4$.

Assuming that exactly one successful design is produced, what is the probability that it was designed by team N?

There are four possible outcomes here, corresponding to the four combinations of success and failure of the two teams:

SS: both succeed, FF: both fail,
SF: C succeeds, N fails, FS: C fails, N succeeds.

We were given that the probabilities of these outcomes satisfy

$$\mathbf{P}(SS)+\mathbf{P}(SF)=\frac{2}{3}, \quad \mathbf{P}(SS)+\mathbf{P}(FS)=\frac{1}{2}, \quad \mathbf{P}(SS)+\mathbf{P}(SF)+\mathbf{P}(FS)=\frac{3}{4}.$$

From these relations, together with the normalization equation

$$\mathbf{P}(SS)+\mathbf{P}(SF)+\mathbf{P}(FS)+\mathbf{P}(FF)=1,$$

we can obtain the probabilities of individual outcomes:

$$\mathbf{P}(SS)=\frac{5}{12}, \qquad \mathbf{P}(SF)=\frac{1}{4}, \qquad \mathbf{P}(FS)=\frac{1}{12}, \qquad \mathbf{P}(FF)=\frac{1}{4}.$$

The desired conditional probability is

$$\mathbf{P}\big(FS \mid \{SF, FS\}\big)=\frac{\frac{1}{12}}{\frac{1}{4}+\frac{1}{12}}=\frac{1}{4}.$$

Using Conditional Probability for Modeling

When constructing probabilistic models for experiments that have a sequential character, it is often natural and convenient to first specify conditional probabilities and then use them to determine unconditional probabilities. The rule $\mathbf{P}(A\cap B)=\mathbf{P}(B)\mathbf{P}(A \mid B)$, which is a restatement of the definition of conditional probability, is often helpful in this process.

Example 1.9. Radar Detection. If an aircraft is present in a certain area, a radar detects it and generates an alarm signal with probability 0.99. If an aircraft is not present, the radar generates a (false) alarm, with probability 0.10. We assume that an aircraft is present with probability 0.05. What is the probability of no aircraft presence and a false alarm? What is the probability of aircraft presence and no detection?

A sequential representation of the experiment is appropriate here, as shown in Fig. 1.9. Let A and B be the events

$$A=\{\text{an aircraft is present}\},$$

$$B=\{\text{the radar generates an alarm}\},$$

and consider also their complements

$$A^c = \{\text{an aircraft is not present}\},$$

$$B^c = \{\text{the radar does not generate an alarm}\}.$$

The given probabilities are recorded along the corresponding branches of the tree describing the sample space, as shown in Fig. 1.9. Each possible outcome corresponds to a leaf of the tree, and its probability is equal to the product of the probabilities associated with the branches in a path from the root to the corresponding leaf. The desired probabilities are

$$\mathbf{P}(\text{not present, false alarm}) = \mathbf{P}(A^c \cap B) = \mathbf{P}(A^c)\mathbf{P}(B \mid A^c) = 0.95 \cdot 0.10 = 0.095,$$

$$\mathbf{P}(\text{present, no detection}) = \mathbf{P}(A \cap B^c) = \mathbf{P}(A)\mathbf{P}(B^c \mid A) = 0.05 \cdot 0.01 = 0.0005.$$

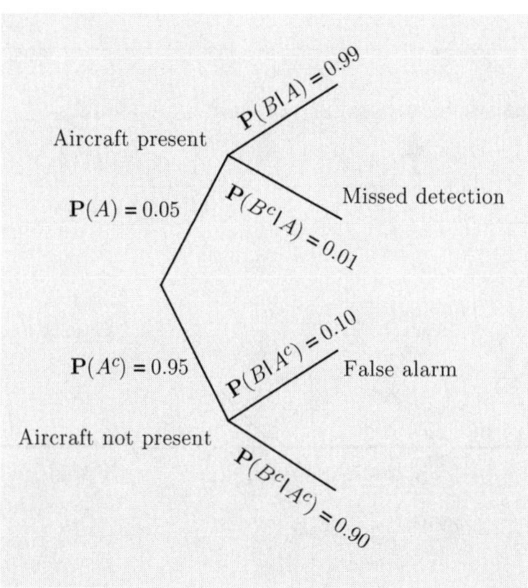

Figure 1.9: Sequential description of the experiment for the radar detection problem in Example 1.9.

Extending the preceding example, we have a general rule for calculating various probabilities in conjunction with a tree-based sequential description of an experiment. In particular:

(a) We set up the tree so that an event of interest is associated with a leaf. We view the occurrence of the event as a sequence of steps, namely, the traversals of the branches along the path from the root to the leaf.

(b) We record the conditional probabilities associated with the branches of the tree.

(c) We obtain the probability of a leaf by multiplying the probabilities recorded along the corresponding path of the tree.

In mathematical terms, we are dealing with an event A which occurs if and only if each one of several events $A_1, \ldots, A_n$ has occurred, i.e., $A = A_1 \cap A_2 \cap \cdots \cap A_n$. The occurrence of A is viewed as an occurrence of A_1, followed by the occurrence of A_2, then of A_3, etc., and it is visualized as a path with n branches, corresponding to the events $A_1, \ldots, A_n$. The probability of A is given by the following rule (see also Fig. 1.10).

Multiplication Rule

Assuming that all of the conditioning events have positive probability, we have

$$\mathbf{P}\big(\cap_{i=1}^n A_i\big) = \mathbf{P}(A_1)\mathbf{P}(A_2 \,|\, A_1)\mathbf{P}(A_3 \,|\, A_1 \cap A_2) \cdots \mathbf{P}\big(A_n \,|\, \cap_{i=1}^{n-1} A_i\big).$$

The multiplication rule can be verified by writing

$$\mathbf{P}\big(\cap_{i=1}^n A_i\big) = \mathbf{P}(A_1) \cdot \frac{\mathbf{P}(A_1 \cap A_2)}{\mathbf{P}(A_1)} \cdot \frac{\mathbf{P}(A_1 \cap A_2 \cap A_3)}{\mathbf{P}(A_1 \cap A_2)} \cdots \frac{\mathbf{P}\big(\cap_{i=1}^n A_i\big)}{\mathbf{P}\big(\cap_{i=1}^{n-1} A_i\big)},$$

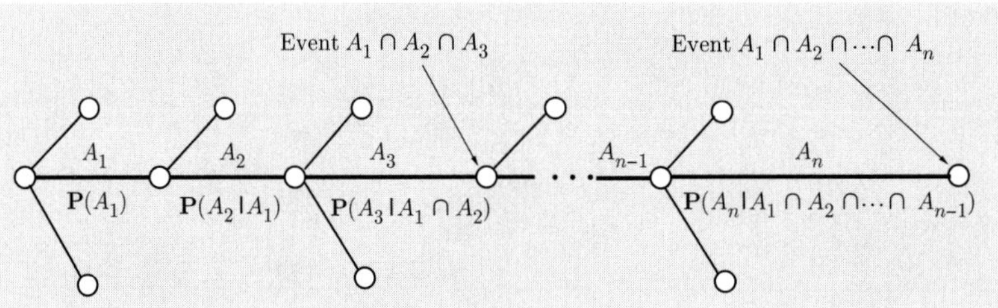

Figure 1.10: Visualization of the multiplication rule. The intersection event $A = A_1 \cap A_2 \cap \cdots \cap A_n$ is associated with a particular path on a tree that describes the experiment. We associate the branches of this path with the events $A_1, \ldots, A_n$, and we record next to the branches the corresponding conditional probabilities.

The final node of the path corresponds to the intersection event A, and its probability is obtained by multiplying the conditional probabilities recorded along the branches of the path

$$\mathbf{P}(A_1 \cap A_2 \cap \cdots \cap A_n) = \mathbf{P}(A_1)\mathbf{P}(A_2 \,|\, A_1) \cdots \mathbf{P}(A_n \,|\, A_1 \cap A_2 \cap \cdots \cap A_{n-1}).$$

Note that any intermediate node along the path also corresponds to some intersection event and its probability is obtained by multiplying the corresponding conditional probabilities up to that node. For example, the event $A_1 \cap A_2 \cap A_3$ corresponds to the node shown in the figure, and its probability is

$$\mathbf{P}(A_1 \cap A_2 \cap A_3) = \mathbf{P}(A_1)\mathbf{P}(A_2 \,|\, A_1)\mathbf{P}(A_3 \,|\, A_1 \cap A_2).$$

and by using the definition of conditional probability to rewrite the right-hand side above as

$$\mathbf{P}(A_1)\mathbf{P}(A_2 \,|\, A_1)\mathbf{P}(A_3 \,|\, A_1 \cap A_2) \cdots \mathbf{P}\big(A_n \,|\, \cap_{i=1}^{n-1} A_i\big).$$

For the case of just two events, A_1 and A_2, the multiplication rule is simply the definition of conditional probability.

Example 1.10. Three cards are drawn from an ordinary 52-card deck without replacement (drawn cards are not placed back in the deck). We wish to find the probability that none of the three cards is a heart. We assume that at each step, each one of the remaining cards is equally likely to be picked. By symmetry, this implies that every triplet of cards is equally likely to be drawn. A cumbersome approach, which we will not use, is to count the number of all card triplets that do not include a heart, and divide it with the number of all possible card triplets. Instead, we use a sequential description of the experiment in conjunction with the multiplication rule (cf. Fig. 1.11).

Define the events

$$A_i = \{\text{the } i\text{th card is not a heart}\}, \qquad i = 1, 2, 3.$$

We will calculate $\mathbf{P}(A_1 \cap A_2 \cap A_3)$, the probability that none of the three cards is a heart, using the multiplication rule

$$\mathbf{P}(A_1 \cap A_2 \cap A_3) = \mathbf{P}(A_1)\mathbf{P}(A_2 \,|\, A_1)\mathbf{P}(A_3 \,|\, A_1 \cap A_2).$$

We have

$$\mathbf{P}(A_1) = \frac{39}{52},$$

since there are 39 cards that are not hearts in the 52-card deck. Given that the first card is not a heart, we are left with 51 cards, 38 of which are not hearts, and

$$\mathbf{P}(A_2 \,|\, A_1) = \frac{38}{51}.$$

Finally, given that the first two cards drawn are not hearts, there are 37 cards which are not hearts in the remaining 50-card deck, and

$$\mathbf{P}(A_3 \,|\, A_1 \cap A_2) = \frac{37}{50}.$$

These probabilities are recorded along the corresponding branches of the tree describing the sample space, as shown in Fig. 1.11. The desired probability is now obtained by multiplying the probabilities recorded along the corresponding path of the tree:

$$\mathbf{P}(A_1 \cap A_2 \cap A_3) = \frac{39}{52} \cdot \frac{38}{51} \cdot \frac{37}{50}.$$

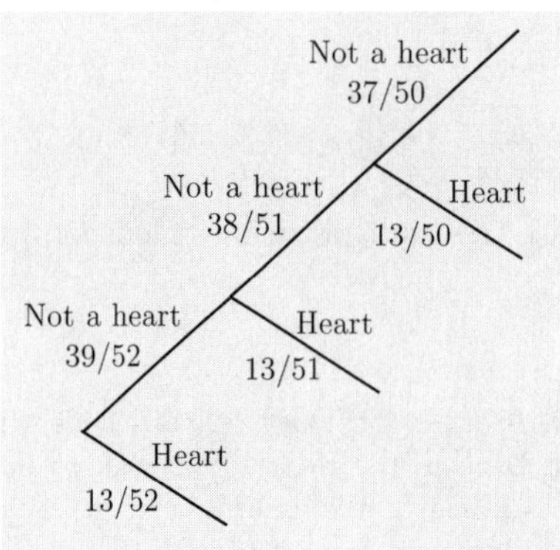

Figure 1.11: Sequential description of the experiment in the 3-card selection problem of Example 1.10.

Note that once the probabilities are recorded along the tree, the probability of several other events can be similarly calculated. For example,

$$\mathbf{P}(\text{1st is not a heart and 2nd is a heart}) = \frac{39}{52} \cdot \frac{13}{51},$$

$$\mathbf{P}(\text{1st and 2nd are not hearts, and 3rd is a heart}) = \frac{39}{52} \cdot \frac{38}{51} \cdot \frac{13}{50}.$$

Example 1.11. A class consisting of 4 graduate and 12 undergraduate students is randomly divided into 4 groups of 4. What is the probability that each group includes a graduate student? We interpret "randomly" to mean that given the assignment of some students to certain slots, any of the remaining students is equally likely to be assigned to any of the remaining slots. We then calculate the desired probability using the multiplication rule, based on the sequential description shown in Fig. 1.12. Let us denote the four graduate students by 1, 2, 3, 4, and consider the events

$$A_1 = \{\text{students 1 and 2 are in different groups}\},$$

$$A_2 = \{\text{students 1, 2, and 3 are in different groups}\},$$

$$A_3 = \{\text{students 1, 2, 3, and 4 are in different groups}\}.$$

We will calculate $\mathbf{P}(A_3)$ using the multiplication rule:

$$\mathbf{P}(A_3) = \mathbf{P}(A_1 \cap A_2 \cap A_3) = \mathbf{P}(A_1)\mathbf{P}(A_2 \,|\, A_1)\mathbf{P}(A_3 \,|\, A_1 \cap A_2).$$

We have

$$\mathbf{P}(A_1) = \frac{12}{15},$$

since there are 12 student slots in groups other than the one of student 1, and there are 15 student slots overall, excluding student 1. Similarly,

$$\mathbf{P}(A_2 \,|\, A_1) = \frac{8}{14},$$

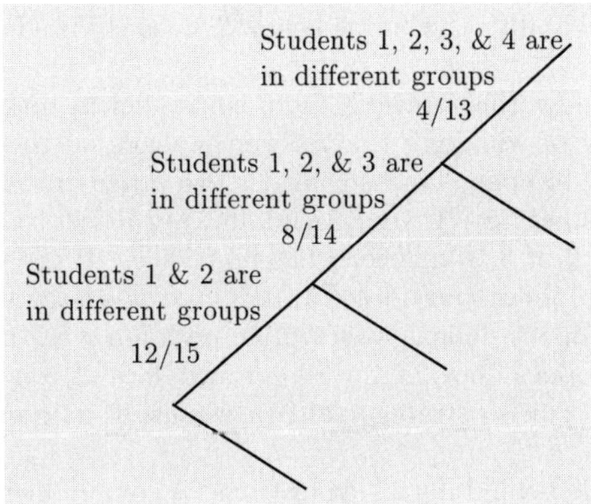

Figure 1.12: Sequential description of the experiment in the student problem of Example 1.11.

since there are 8 student slots in groups other than those of students 1 and 2, and there are 14 student slots, excluding students 1 and 2. Also,

$$\mathbf{P}(A_3 \mid A_1 \cap A_2) = \frac{4}{13},$$

since there are 4 student slots in groups other than those of students 1, 2, and 3, and there are 13 student slots, excluding students 1, 2, and 3. Thus, the desired probability is

$$\frac{12}{15} \cdot \frac{8}{14} \cdot \frac{4}{13},$$

and is obtained by multiplying the conditional probabilities along the corresponding path of the tree in Fig. 1.12.

Example 1.12. The Monty Hall Problem. This is a much discussed puzzle, based on an old American game show. You are told that a prize is equally likely to be found behind any one of three closed doors in front of you. You point to one of the doors. A friend opens for you one of the remaining two doors, after making sure that the prize is not behind it. At this point, you can stick to your initial choice, or switch to the other unopened door. You win the prize if it lies behind your final choice of a door. Consider the following strategies:

(a) Stick to your initial choice.

(b) Switch to the other unopened door.

(c) You first point to door 1. If door 2 is opened, you do not switch. If door 3 is opened, you switch.

Which is the best strategy? To answer the question, let us calculate the probability of winning under each of the three strategies.

Under the strategy of no switching, your initial choice will determine whether you win or not, and the probability of winning is 1/3. This is because the prize is equally likely to be behind each door.

Under the strategy of switching, if the prize is behind the initially chosen door (probability 1/3), you do not win. If it is not (probability 2/3), and given that another door without a prize has been opened for you, you will get to the winning door once you switch. Thus, the probability of winning is now 2/3, so (b) is a better strategy than (a).

Consider now strategy (c). Under this strategy, there is insufficient information for determining the probability of winning. The answer depends on the way that your friend chooses which door to open. Let us consider two possibilities.

Suppose that if the prize is behind door 1, your friend always chooses to open door 2. (If the prize is behind door 2 or 3, your friend has no choice.) If the prize is behind door 1, your friend opens door 2, you do not switch, and you win. If the prize is behind door 2, your friend opens door 3, you switch, and you win. If the prize is behind door 3, your friend opens door 2, you do not switch, and you lose. Thus, the probability of winning is 2/3, so strategy (c) in this case is as good as strategy (b).

Suppose now that if the prize is behind door 1, your friend is equally likely to open either door 2 or 3. If the prize is behind door 1 (probability 1/3), and if your friend opens door 2 (probability 1/2), you do not switch and you win (probability 1/6). But if your friend opens door 3, you switch and you lose. If the prize is behind door 2, your friend opens door 3, you switch, and you win (probability 1/3). If the prize is behind door 3, your friend opens door 2, you do not switch and you lose. Thus, the probability of winning is $1/6 + 1/3 = 1/2$, so strategy (c) in this case is inferior to strategy (b).

1.4 TOTAL PROBABILITY THEOREM AND BAYES' RULE

In this section, we explore some applications of conditional probability. We start with the following theorem, which is often useful for computing the probabilities of various events, using a "divide-and-conquer" approach.

Total Probability Theorem

Let $A_1, \ldots, A_n$ be disjoint events that form a partition of the sample space (each possible outcome is included in exactly one of the events $A_1, \ldots, A_n$) and assume that $\mathbf{P}(A_i) > 0$, for all i. Then, for any event B, we have

$$\begin{aligned}\mathbf{P}(B) &= \mathbf{P}(A_1 \cap B) + \cdots + \mathbf{P}(A_n \cap B) \\ &= \mathbf{P}(A_1)\mathbf{P}(B \mid A_1) + \cdots + \mathbf{P}(A_n)\mathbf{P}(B \mid A_n).\end{aligned}$$

The theorem is visualized and proved in Fig. 1.13. Intuitively, we are partitioning the sample space into a number of scenarios (events) A_i. Then, the probability that B occurs is a weighted average of its conditional probability under each scenario, where each scenario is weighted according to its (unconditional) probability. One of the uses of the theorem is to compute the probability of various events B for which the conditional probabilities $\mathbf{P}(B \mid A_i)$ are known or easy to derive. The key is to choose appropriately the partition $A_1, \ldots, A_n$, and this choice is often suggested by the problem structure. Here are some examples.

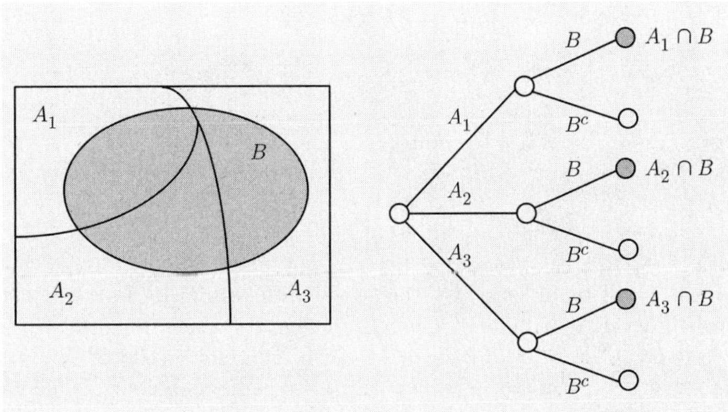

Figure 1.13: Visualization and verification of the total probability theorem. The events $A_1, \ldots, A_n$ form a partition of the sample space, so the event B can be decomposed into the disjoint union of its intersections $A_i \cap B$ with the sets A_i, i.e.,

$$B = (A_1 \cap B) \cup \cdots \cup (A_n \cap B).$$

Using the additivity axiom, it follows that

$$\mathbf{P}(B) = \mathbf{P}(A_1 \cap B) + \cdots + \mathbf{P}(A_n \cap B).$$

Since, by the definition of conditional probability, we have

$$\mathbf{P}(A_i \cap B) = \mathbf{P}(A_i)\mathbf{P}(B \,|\, A_i),$$

the preceding equality yields

$$\mathbf{P}(B) = \mathbf{P}(A_1)\mathbf{P}(B \,|\, A_1) + \cdots + \mathbf{P}(A_n)\mathbf{P}(B \,|\, A_n).$$

For an alternative view, consider an equivalent sequential model, as shown on the right. The probability of the leaf $A_i \cap B$ is the product $\mathbf{P}(A_i)\mathbf{P}(B \,|\, A_i)$ of the probabilities along the path leading to that leaf. The event B consists of the three highlighted leaves and $\mathbf{P}(B)$ is obtained by adding their probabilities.

Example 1.13. You enter a chess tournament where your probability of winning a game is 0.3 against half the players (call them type 1), 0.4 against a quarter of the players (call them type 2), and 0.5 against the remaining quarter of the players (call them type 3). You play a game against a randomly chosen opponent. What is the probability of winning?

Let A_i be the event of playing with an opponent of type i. We have

$$\mathbf{P}(A_1) = 0.5, \qquad \mathbf{P}(A_2) = 0.25, \qquad \mathbf{P}(A_3) = 0.25.$$

Also, let B be the event of winning. We have

$$\mathbf{P}(B \,|\, A_1) = 0.3, \qquad \mathbf{P}(B \,|\, A_2) = 0.4, \qquad \mathbf{P}(B \,|\, A_3) = 0.5.$$

Thus, by the total probability theorem, the probability of winning is

$$\begin{aligned}\mathbf{P}(B) &= \mathbf{P}(A_1)\mathbf{P}(B \mid A_1) + \mathbf{P}(A_2)\mathbf{P}(B \mid A_2) + \mathbf{P}(A_3)\mathbf{P}(B \mid A_3) \\ &= 0.5 \cdot 0.3 + 0.25 \cdot 0.4 + 0.25 \cdot 0.5 \\ &= 0.375.\end{aligned}$$

Example 1.14. You roll a fair four-sided die. If the result is 1 or 2, you roll once more but otherwise, you stop. What is the probability that the sum total of your rolls is at least 4?

Let A_i be the event that the result of first roll is i, and note that $\mathbf{P}(A_i) = 1/4$ for each i. Let B be the event that the sum total is at least 4. Given the event A_1, the sum total will be at least 4 if the second roll results in 3 or 4, which happens with probability 1/2. Similarly, given the event A_2, the sum total will be at least 4 if the second roll results in 2, 3, or 4, which happens with probability 3/4. Also, given the event A_3, you stop and the sum total remains below 4. Therefore,

$$\mathbf{P}(B \mid A_1) = \frac{1}{2}, \qquad \mathbf{P}(B \mid A_2) = \frac{3}{4}, \qquad \mathbf{P}(B \mid A_3) = 0, \qquad \mathbf{P}(B \mid A_4) = 1.$$

By the total probability theorem,

$$\mathbf{P}(B) = \frac{1}{4} \cdot \frac{1}{2} + \frac{1}{4} \cdot \frac{3}{4} + \frac{1}{4} \cdot 0 + \frac{1}{4} \cdot 1 = \frac{9}{16}.$$

The total probability theorem can be applied repeatedly to calculate probabilities in experiments that have a sequential character, as shown in the following example.

Example 1.15. Alice is taking a probability class and at the end of each week she can be either up-to-date or she may have fallen behind. If she is up-to-date in a given week, the probability that she will be up-to-date (or behind) in the next week is 0.8 (or 0.2, respectively). If she is behind in a given week, the probability that she will be up-to-date (or behind) in the next week is 0.4 (or 0.6, respectively). Alice is (by default) up-to-date when she starts the class. What is the probability that she is up-to-date after three weeks?

Let U_i and B_i be the events that Alice is up-to-date or behind, respectively, after i weeks. According to the total probability theorem, the desired probability $\mathbf{P}(U_3)$ is given by

$$\mathbf{P}(U_3) = \mathbf{P}(U_2)\mathbf{P}(U_3 \mid U_2) + \mathbf{P}(B_2)\mathbf{P}(U_3 \mid B_2) = \mathbf{P}(U_2) \cdot 0.8 + \mathbf{P}(B_2) \cdot 0.4.$$

The probabilities $\mathbf{P}(U_2)$ and $\mathbf{P}(B_2)$ can also be calculated using the total probability theorem:

$$\mathbf{P}(U_2) = \mathbf{P}(U_1)\mathbf{P}(U_2 \mid U_1) + \mathbf{P}(B_1)\mathbf{P}(U_2 \mid B_1) = \mathbf{P}(U_1) \cdot 0.8 + \mathbf{P}(B_1) \cdot 0.4,$$

$$\mathbf{P}(B_2) = \mathbf{P}(U_1)\mathbf{P}(B_2 \mid U_1) + \mathbf{P}(B_1)\mathbf{P}(B_2 \mid B_1) = \mathbf{P}(U_1) \cdot 0.2 + \mathbf{P}(B_1) \cdot 0.6.$$

Finally, since Alice starts her class up-to-date, we have

$$\mathbf{P}(U_1) = 0.8, \qquad \mathbf{P}(B_1) = 0.2.$$

We can now combine the preceding three equations to obtain

$$\mathbf{P}(U_2) = 0.8 \cdot 0.8 + 0.2 \cdot 0.4 = 0.72,$$

$$\mathbf{P}(B_2) = 0.8 \cdot 0.2 + 0.2 \cdot 0.6 = 0.28,$$

and by using the above probabilities in the formula for $\mathbf{P}(U_3)$:

$$\mathbf{P}(U_3) = 0.72 \cdot 0.8 + 0.28 \cdot 0.4 = 0.688.$$

Note that we could have calculated the desired probability $\mathbf{P}(U_3)$ by constructing a tree description of the experiment, then calculating the probability of every element of U_3 using the multiplication rule on the tree, and adding. However, there are cases where the calculation based on the total probability theorem is more convenient. For example, suppose we are interested in the probability $\mathbf{P}(U_{20})$ that Alice is up-to-date after 20 weeks. Calculating this probability using the multiplication rule is very cumbersome, because the tree representing the experiment is 20 stages deep and has 2^{20} leaves. On the other hand, with a computer, a sequential calculation using the total probability formulas

$$\mathbf{P}(U_{i+1}) = \mathbf{P}(U_i) \cdot 0.8 + \mathbf{P}(B_i) \cdot 0.4,$$

$$\mathbf{P}(B_{i+1}) = \mathbf{P}(U_i) \cdot 0.2 + \mathbf{P}(B_i) \cdot 0.6,$$

and the initial conditions $\mathbf{P}(U_1) = 0.8$, $\mathbf{P}(B_1) = 0.2$, is very simple.

Inference and Bayes' Rule

The total probability theorem is often used in conjunction with the following celebrated theorem, which relates conditional probabilities of the form $\mathbf{P}(A \mid B)$ with conditional probabilities of the form $\mathbf{P}(B \mid A)$, in which the order of the conditioning is reversed.

Bayes' Rule

Let $A_1, A_2, \ldots, A_n$ be disjoint events that form a partition of the sample space, and assume that $\mathbf{P}(A_i) > 0$, for all i. Then, for any event B such that $\mathbf{P}(B) > 0$, we have

$$\begin{aligned}\mathbf{P}(A_i \mid B) &= \frac{\mathbf{P}(A_i)\mathbf{P}(B \mid A_i)}{\mathbf{P}(B)} \\ &= \frac{\mathbf{P}(A_i)\mathbf{P}(B \mid A_i)}{\mathbf{P}(A_1)\mathbf{P}(B \mid A_1) + \cdots + \mathbf{P}(A_n)\mathbf{P}(B \mid A_n)}.\end{aligned}$$

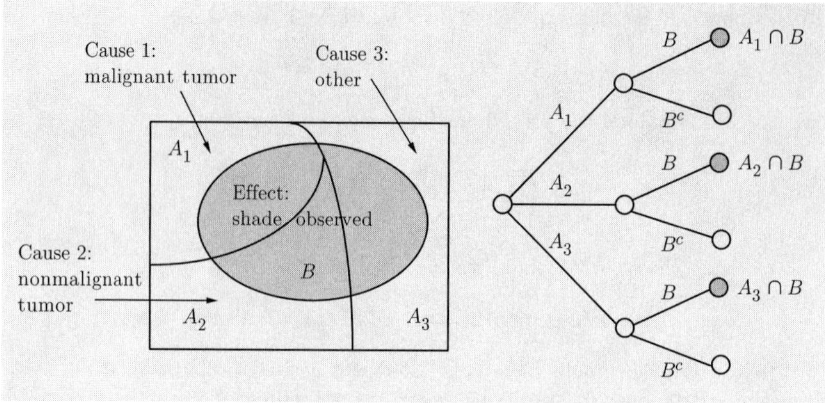

Figure 1.14: An example of the inference context that is implicit in Bayes' rule. We observe a shade in a person's X-ray (this is event B, the "effect") and we want to estimate the likelihood of three mutually exclusive and collectively exhaustive potential causes: cause 1 (event A_1) is that there is a malignant tumor, cause 2 (event A_2) is that there is a nonmalignant tumor, and cause 3 (event A_3) corresponds to reasons other than a tumor. We assume that we know the probabilities $\mathbf{P}(A_i)$ and $\mathbf{P}(B \mid A_i)$, $i = 1, 2, 3$. Given that we see a shade (event B occurs), Bayes' rule gives the posterior probabilities of the various causes as

$$\mathbf{P}(A_i \mid B) = \frac{\mathbf{P}(A_i)\mathbf{P}(B \mid A_i)}{\mathbf{P}(A_1)\mathbf{P}(B \mid A_1) + \mathbf{P}(A_2)\mathbf{P}(B \mid A_2) + \mathbf{P}(A_3)\mathbf{P}(B \mid A_3)}, \qquad i = 1, 2, 3.$$

For an alternative view, consider an equivalent sequential model, as shown on the right. The probability $\mathbf{P}(A_1 \mid B)$ of a malignant tumor is the probability of the first highlighted leaf, which is $\mathbf{P}(A_1 \cap B)$, divided by the total probability of the highlighted leaves, which is $\mathbf{P}(B)$.

To verify Bayes' rule, note that by the definition of conditional probability, we have

$$\mathbf{P}(A_i \cap B) = \mathbf{P}(A_i)\mathbf{P}(B \mid A_i) = \mathbf{P}(A_i \mid B)\mathbf{P}(B).$$

This yields the first equality. The second equality follows from the first by using the total probability theorem to rewrite $\mathbf{P}(B)$.

Bayes' rule is often used for **inference**. There are a number of "causes" that may result in a certain "effect." We observe the effect, and we wish to infer the cause. The events $A_1, \ldots, A_n$ are associated with the causes and the event B represents the effect. The probability $\mathbf{P}(B \mid A_i)$ that the effect will be observed when the cause A_i is present amounts to a probabilistic model of the cause-effect relation (cf. Fig. 1.14). Given that the effect B has been observed, we wish to evaluate the probability $\mathbf{P}(A_i \mid B)$ that the cause A_i is present. We refer to $\mathbf{P}(A_i \mid B)$ as the **posterior probability** of event A_i given the information, to be distinguished from $\mathbf{P}(A_i)$, which we call the **prior probability**.

Example 1.16. Let us return to the radar detection problem of Example 1.9 and Fig. 1.9. Let

$$A = \{\text{an aircraft is present}\},$$
$$B = \{\text{the radar generates an alarm}\}.$$

We are given that

$$\mathbf{P}(A) = 0.05, \qquad \mathbf{P}(B \,|\, A) = 0.99, \qquad \mathbf{P}(B \,|\, A^c) = 0.1.$$

Applying Bayes' rule, with $A_1 = A$ and $A_2 = A^c$, we obtain

$$\begin{aligned}\mathbf{P}(\text{aircraft present} \,|\, \text{alarm}) &= \mathbf{P}(A \,|\, B)\\ &= \frac{\mathbf{P}(A)\mathbf{P}(B \,|\, A)}{\mathbf{P}(B)}\\ &= \frac{\mathbf{P}(A)\mathbf{P}(B \,|\, A)}{\mathbf{P}(A)\mathbf{P}(B \,|\, A) + \mathbf{P}(A^c)\mathbf{P}(B \,|\, A^c)}\\ &= \frac{0.05 \cdot 0.99}{0.05 \cdot 0.99 + 0.95 \cdot 0.1}\\ &\approx 0.3426.\end{aligned}$$

Example 1.17. Let us return to the chess problem of Example 1.13. Here, A_i is the event of getting an opponent of type i, and

$$\mathbf{P}(A_1) = 0.5, \qquad \mathbf{P}(A_2) = 0.25, \qquad \mathbf{P}(A_3) = 0.25.$$

Also, B is the event of winning, and

$$\mathbf{P}(B \,|\, A_1) = 0.3, \qquad \mathbf{P}(B \,|\, A_2) = 0.4, \qquad \mathbf{P}(B \,|\, A_3) = 0.5.$$

Suppose that you win. What is the probability $\mathbf{P}(A_1 \,|\, B)$ that you had an opponent of type 1?

Using Bayes' rule, we have

$$\begin{aligned}\mathbf{P}(A_1 \,|\, B) &= \frac{\mathbf{P}(A_1)\mathbf{P}(B \,|\, A_1)}{\mathbf{P}(A_1)\mathbf{P}(B \,|\, A_1) + \mathbf{P}(A_2)\mathbf{P}(B \,|\, A_2) + \mathbf{P}(A_3)\mathbf{P}(B \,|\, A_3)}\\ &= \frac{0.5 \cdot 0.3}{0.5 \cdot 0.3 + 0.25 \cdot 0.4 + 0.25 \cdot 0.5}\\ &= 0.4.\end{aligned}$$

Example 1.18. The False-Positive Puzzle. A test for a certain rare disease is assumed to be correct 95% of the time: if a person has the disease, the test results are positive with probability 0.95, and if the person does not have the disease, the test results are negative with probability 0.95. A random person drawn from a certain population has probability 0.001 of having the disease. Given that the person just tested positive, what is the probability of having the disease?

If A is the event that the person has the disease, and B is the event that the test results are positive, the desired probability, $\mathbf{P}(A \mid B)$, is

$$\begin{aligned}\mathbf{P}(A \mid B) &= \frac{\mathbf{P}(A)\mathbf{P}(B \mid A)}{\mathbf{P}(A)\mathbf{P}(B \mid A) + \mathbf{P}(A^c)\mathbf{P}(B \mid A^c)} \\ &= \frac{0.001 \cdot 0.95}{0.001 \cdot 0.95 + 0.999 \cdot 0.05} \\ &= 0.0187.\end{aligned}$$

Note that even though the test was assumed to be fairly accurate, a person who has tested positive is still very unlikely (less than 2%) to have the disease. According to *The Economist* (February 20th, 1999), 80% of those questioned at a leading American hospital substantially missed the correct answer to a question of this type; most of them thought that the probability that the person has the disease is 0.95!

1.5 INDEPENDENCE

We have introduced the conditional probability $\mathbf{P}(A \mid B)$ to capture the partial information that event B provides about event A. An interesting and important special case arises when the occurrence of B provides no such information and does not alter the probability that A has occurred, i.e.,

$$\mathbf{P}(A \mid B) = \mathbf{P}(A).$$

When the above equality holds, we say that A is **independent** of B. Note that by the definition $\mathbf{P}(A \mid B) = \mathbf{P}(A \cap B)/\mathbf{P}(B)$, this is equivalent to

$$\mathbf{P}(A \cap B) = \mathbf{P}(A)\mathbf{P}(B).$$

We adopt this latter relation as the definition of independence because it can be used even when $\mathbf{P}(B) = 0$, in which case $\mathbf{P}(A \mid B)$ is undefined. The symmetry of this relation also implies that independence is a symmetric property; that is, if A is independent of B, then B is independent of A, and we can unambiguously say that A and B are **independent events**.

Independence is often easy to grasp intuitively. For example, if the occurrence of two events is governed by distinct and noninteracting physical processes, such events will turn out to be independent. On the other hand, independence is not easily visualized in terms of the sample space. A common first thought is that two events are independent if they are disjoint, but in fact the opposite is true: two disjoint events A and B with $\mathbf{P}(A) > 0$ and $\mathbf{P}(B) > 0$ are never independent, since their intersection $A \cap B$ is empty and has probability 0.

For example, an event A and its complement A^c are not independent [unless $\mathbf{P}(A) = 0$ or $\mathbf{P}(A) = 1$], since knowledge that A has occurred provides precise information about whether A^c has occurred.

Example 1.19. Consider an experiment involving two successive rolls of a 4-sided die in which all 16 possible outcomes are equally likely and have probability 1/16.

(a) Are the events

$$A_i = \{\text{1st roll results in } i\}, \qquad B_j = \{\text{2nd roll results in } j\},$$

independent? We have

$$\mathbf{P}(A_i \cap B_j) = \mathbf{P}\big(\text{the outcome of the two rolls is } (i,j)\big) = \frac{1}{16},$$

$$\mathbf{P}(A_i) = \frac{\text{number of elements of } A_i}{\text{total number of possible outcomes}} = \frac{4}{16},$$

$$\mathbf{P}(B_j) = \frac{\text{number of elements of } B_j}{\text{total number of possible outcomes}} = \frac{4}{16}.$$

We observe that $\mathbf{P}(A_i \cap B_j) = \mathbf{P}(A_i)\mathbf{P}(B_j)$, and the independence of A_i and B_j is verified. Thus, our choice of the discrete uniform probability law implies the independence of the two rolls.

(b) Are the events

$$A = \{\text{1st roll is a 1}\}, \qquad B = \{\text{sum of the two rolls is a 5}\},$$

independent? The answer here is not quite obvious. We have

$$\mathbf{P}(A \cap B) = \mathbf{P}\big(\text{the result of the two rolls is } (1,4)\big) = \frac{1}{16},$$

and also

$$\mathbf{P}(A) = \frac{\text{number of elements of } A}{\text{total number of possible outcomes}} = \frac{4}{16}.$$

The event B consists of the outcomes (1,4), (2,3), (3,2), and (4,1), and

$$\mathbf{P}(B) = \frac{\text{number of elements of } B}{\text{total number of possible outcomes}} = \frac{4}{16}.$$

Thus, we see that $\mathbf{P}(A \cap B) = \mathbf{P}(A)\mathbf{P}(B)$, and the events A and B are independent.

(c) Are the events

$$A = \{\text{maximum of the two rolls is 2}\}, \quad B = \{\text{minimum of the two rolls is 2}\},$$

independent? Intuitively, the answer is "no" because the minimum of the two rolls conveys some information about the maximum. For example, if the minimum is 2, the maximum cannot be 1. More precisely, to verify that A and B are not independent, we calculate

$$\mathbf{P}(A \cap B) = \mathbf{P}\big(\text{the result of the two rolls is } (2,2)\big) = \frac{1}{16},$$

and also

$$\mathbf{P}(A) = \frac{\text{number of elements of } A}{\text{total number of possible outcomes}} = \frac{3}{16},$$

$$\mathbf{P}(B) = \frac{\text{number of elements of } B}{\text{total number of possible outcomes}} = \frac{5}{16}.$$

We have $\mathbf{P}(A)\mathbf{P}(B) = 15/(16)^2$, so that $\mathbf{P}(A \cap B) \neq \mathbf{P}(A)\mathbf{P}(B)$, and A and B are not independent.

We finally note that, as mentioned earlier, if A and B are independent, the occurrence of B does not provide any new information on the probability of A occurring. It is then intuitive that the non-occurrence of B should also provide no information on the probability of A. Indeed, it can be verified that if A and B are independent, the same holds true for A and B^c (see the end-of-chapter problems).

Conditional Independence

We noted earlier that the conditional probabilities of events, conditioned on a particular event, form a legitimate probability law. We can thus talk about independence of various events with respect to this conditional law. In particular, given an event C, the events A and B are called **conditionally independent** if

$$\mathbf{P}(A \cap B \mid C) = \mathbf{P}(A \mid C)\mathbf{P}(B \mid C).$$

To derive an alternative characterization of conditional independence, we use the definition of the conditional probability and the multiplication rule, to write

$$\begin{aligned}\mathbf{P}(A \cap B \mid C) &= \frac{\mathbf{P}(A \cap B \cap C)}{\mathbf{P}(C)} \\ &= \frac{\mathbf{P}(C)\mathbf{P}(B \mid C)\mathbf{P}(A \mid B \cap C)}{\mathbf{P}(C)} \\ &= \mathbf{P}(B \mid C)\mathbf{P}(A \mid B \cap C).\end{aligned}$$

We now compare the preceding two expressions, and after eliminating the common factor $\mathbf{P}(B \mid C)$, assumed nonzero, we see that conditional independence is the same as the condition

$$\mathbf{P}(A \mid B \cap C) = \mathbf{P}(A \mid C).$$

In words, this relation states that if C is known to have occurred, the additional knowledge that B also occurred does not change the probability of A.

Interestingly, independence of two events A and B with respect to the unconditional probability law, does not imply conditional independence, and vice versa, as illustrated by the next two examples.

Example 1.20. Consider two independent fair coin tosses, in which all four possible outcomes are equally likely. Let

$$H_1 = \{\text{1st toss is a head}\},$$
$$H_2 = \{\text{2nd toss is a head}\},$$
$$D = \{\text{the two tosses have different results}\}.$$

The events H_1 and H_2 are (unconditionally) independent. But

$$\mathbf{P}(H_1 \,|\, D) = \frac{1}{2}, \qquad \mathbf{P}(H_2 \,|\, D) = \frac{1}{2}, \qquad \mathbf{P}(H_1 \cap H_2 \,|\, D) = 0,$$

so that $\mathbf{P}(H_1 \cap H_2 \,|\, D) \neq \mathbf{P}(H_1 \,|\, D)\mathbf{P}(H_2 \,|\, D)$, and H_1, H_2 are not conditionally independent.

This example can be generalized. For any probabilistic model, let A and B be independent events, and let C be an event such that $\mathbf{P}(C) > 0$, $\mathbf{P}(A \,|\, C) > 0$, and $\mathbf{P}(B \,|\, C) > 0$, while $A \cap B \cap C$ is empty. Then, A and B cannot be conditionally independent (given C) since $\mathbf{P}(A \cap B \,|\, C) = 0$ while $\mathbf{P}(A \,|\, C)\,\mathbf{P}(B \,|\, C) > 0$.

Example 1.21. There are two coins, a blue and a red one. We choose one of the two at random, each being chosen with probability 1/2, and proceed with two independent tosses. The coins are biased: with the blue coin, the probability of heads in any given toss is 0.99, whereas for the red coin it is 0.01.

Let B be the event that the blue coin was selected. Let also H_i be the event that the ith toss resulted in heads. Given the choice of a coin, the events H_1 and H_2 are independent, because of our assumption of independent tosses. Thus,

$$\mathbf{P}(H_1 \cap H_2 \,|\, B) = \mathbf{P}(H_1 \,|\, B)\mathbf{P}(H_2 \,|\, B) = 0.99 \cdot 0.99.$$

On the other hand, the events H_1 and H_2 are not independent. Intuitively, if we are told that the first toss resulted in heads, this leads us to suspect that the blue coin was selected, in which case, we expect the second toss to also result in heads. Mathematically, we use the total probability theorem to obtain

$$\mathbf{P}(H_1) = \mathbf{P}(B)\mathbf{P}(H_1 \,|\, B) + \mathbf{P}(B^c)\mathbf{P}(H_1 \,|\, B^c) = \frac{1}{2} \cdot 0.99 + \frac{1}{2} \cdot 0.01 = \frac{1}{2},$$

as should be expected from symmetry considerations. Similarly, we have $\mathbf{P}(H_2) = 1/2$. Now notice that

$$\begin{aligned}\mathbf{P}(H_1 \cap H_2) &= \mathbf{P}(B)\mathbf{P}(H_1 \cap H_2 \,|\, B) + \mathbf{P}(B^c)\mathbf{P}(H_1 \cap H_2 \,|\, B^c) \\ &= \frac{1}{2} \cdot 0.99 \cdot 0.99 + \frac{1}{2} \cdot 0.01 \cdot 0.01 \approx \frac{1}{2}.\end{aligned}$$

Thus, $\mathbf{P}(H_1 \cap H_2) \neq \mathbf{P}(H_1)\mathbf{P}(H_2)$, and the events H_1 and H_2 are dependent, even though they are conditionally independent given B.

We now summarize.

Independence

- Two events A and B are said to be **independent** if

$$\mathbf{P}(A \cap B) = \mathbf{P}(A)\mathbf{P}(B).$$

If in addition, $\mathbf{P}(B) > 0$, independence is equivalent to the condition

$$\mathbf{P}(A \mid B) = \mathbf{P}(A).$$

- If A and B are independent, so are A and B^c.
- Two events A and B are said to be **conditionally independent**, given another event C with $\mathbf{P}(C) > 0$, if

$$\mathbf{P}(A \cap B \mid C) = \mathbf{P}(A \mid C)\mathbf{P}(B \mid C).$$

If in addition, $\mathbf{P}(B \cap C) > 0$, conditional independence is equivalent to the condition

$$\mathbf{P}(A \mid B \cap C) = \mathbf{P}(A \mid C).$$

- Independence does not imply conditional independence, and vice versa.

Independence of a Collection of Events

The definition of independence can be extended to multiple events.

Definition of Independence of Several Events

We say that the events $A_1, A_2, \ldots, A_n$ are **independent** if

$$\mathbf{P}\left(\bigcap_{i \in S} A_i\right) = \prod_{i \in S} \mathbf{P}(A_i), \qquad \text{for every subset } S \text{ of } \{1, 2, \ldots, n\}.$$

For the case of three events, A_1, A_2, and A_3, independence amounts to satisfying the four conditions

$$\begin{aligned}\mathbf{P}(A_1 \cap A_2) &= \mathbf{P}(A_1)\,\mathbf{P}(A_2),\\ \mathbf{P}(A_1 \cap A_3) &= \mathbf{P}(A_1)\,\mathbf{P}(A_3),\\ \mathbf{P}(A_2 \cap A_3) &= \mathbf{P}(A_2)\,\mathbf{P}(A_3),\\ \mathbf{P}(A_1 \cap A_2 \cap A_3) &= \mathbf{P}(A_1)\,\mathbf{P}(A_2)\,\mathbf{P}(A_3).\end{aligned}$$

The first three conditions simply assert that any two events are independent, a property known as **pairwise independence**. But the fourth condition is also important and does not follow from the first three. Conversely, the fourth condition does not imply the first three; see the two examples that follow.

Example 1.22. Pairwise Independence does not Imply Independence. Consider two independent fair coin tosses, and the following events:

$$\begin{aligned}H_1 &= \{\text{1st toss is a head}\},\\ H_2 &= \{\text{2nd toss is a head}\},\\ D &= \{\text{the two tosses have different results}\}.\end{aligned}$$

The events H_1 and H_2 are independent, by definition. To see that H_1 and D are independent, we note that

$$\mathbf{P}(D \,|\, H_1) = \frac{\mathbf{P}(H_1 \cap D)}{\mathbf{P}(H_1)} = \frac{1/4}{1/2} = \frac{1}{2} = \mathbf{P}(D).$$

Similarly, H_2 and D are independent. On the other hand, we have

$$\mathbf{P}(H_1 \cap H_2 \cap D) = 0 \neq \frac{1}{2}\cdot\frac{1}{2}\cdot\frac{1}{2} = \mathbf{P}(H_1)\mathbf{P}(H_2)\mathbf{P}(D),$$

and these three events are not independent.

Example 1.23. The Equality $\mathbf{P}(A_1 \cap A_2 \cap A_3) = \mathbf{P}(A_1)\,\mathbf{P}(A_2)\,\mathbf{P}(A_3)$ is not Enough for Independence. Consider two independent rolls of a fair six-sided die, and the following events:

$$\begin{aligned}A &= \{\text{1st roll is 1, 2, or 3}\},\\ B &= \{\text{1st roll is 3, 4, or 5}\},\\ C &= \{\text{the sum of the two rolls is 9}\}.\end{aligned}$$

We have

$$\begin{aligned}\mathbf{P}(A \cap B) &= \frac{1}{6} \neq \frac{1}{2}\cdot\frac{1}{2} = \mathbf{P}(A)\mathbf{P}(B),\\ \mathbf{P}(A \cap C) &= \frac{1}{36} \neq \frac{1}{2}\cdot\frac{4}{36} = \mathbf{P}(A)\mathbf{P}(C),\\ \mathbf{P}(B \cap C) &= \frac{1}{12} \neq \frac{1}{2}\cdot\frac{4}{36} = \mathbf{P}(B)\mathbf{P}(C).\end{aligned}$$

Thus the three events A, B, and C are not independent, and indeed no two of these events are independent. On the other hand, we have

$$\mathbf{P}(A \cap B \cap C) = \frac{1}{36} = \frac{1}{2} \cdot \frac{1}{2} \cdot \frac{4}{36} = \mathbf{P}(A)\mathbf{P}(B)\mathbf{P}(C).$$

The intuition behind the independence of a collection of events is analogous to the case of two events. Independence means that the occurrence or non-occurrence of **any number** of the events from that collection carries no information on the remaining events or their complements. For example, if the events A_1, A_2, A_3, A_4 are independent, one obtains relations such as

$$\mathbf{P}(A_1 \cup A_2 \mid A_3 \cap A_4) = \mathbf{P}(A_1 \cup A_2)$$

or

$$\mathbf{P}(A_1 \cup A_2^c \mid A_3^c \cap A_4) = \mathbf{P}(A_1 \cup A_2^c);$$

see the end-of-chapter problems.

Reliability

In probabilistic models of complex systems involving several components, it is often convenient to assume that the behaviors of the components are uncoupled (independent). This typically simplifies the calculations and the analysis, as illustrated in the following example.

Example 1.24. Network Connectivity. A computer network connects two nodes A and B through intermediate nodes C, D, E, F, as shown in Fig. 1.15(a). For every pair of directly connected nodes, say i and j, there is a given probability p_{ij} that the link from i to j is up. We assume that link failures are independent of each other. What is the probability that there is a path connecting A and B in which all links are up?

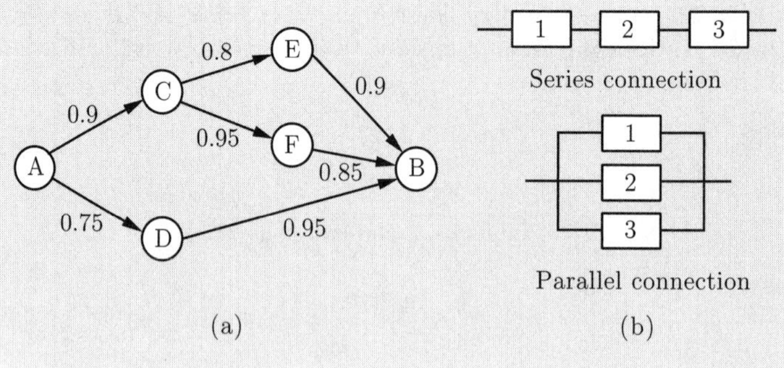

Figure 1.15: (a) Network for Example 1.24. The number next to each link indicates the probability that the link is up. (b) Series and parallel connections of three components in a reliability problem.

This is a typical problem of assessing the reliability of a system consisting of components that can fail independently. Such a system can often be divided into subsystems, where each subsystem consists in turn of several components that are connected either in **series** or in **parallel**; see Fig. 1.15(b).

Let a subsystem consist of components $1, 2, \ldots, m$, and let p_i be the probability that component i is up ("succeeds"). Then, a series subsystem succeeds if **all** of its components are up, so its probability of success is the product of the probabilities of success of the corresponding components, i.e.,

$$\mathbf{P}(\text{series subsystem succeeds}) = p_1 p_2 \cdots p_m.$$

A parallel subsystem succeeds if **any one** of its components succeeds, so its probability of failure is the product of the probabilities of failure of the corresponding components, i.e.,

$$\begin{aligned}\mathbf{P}(\text{parallel subsystem succeeds}) &= 1 - \mathbf{P}(\text{parallel subsystem fails})\\ &= 1 - (1 - p_1)(1 - p_2) \cdots (1 - p_m).\end{aligned}$$

Returning now to the network of Fig. 1.15(a), we can calculate the probability of success (a path from A to B is available) sequentially, using the preceding formulas, and starting from the end. Let us use the notation $X \to Y$ to denote the event that there is a (possibly indirect) connection from node X to node Y. Then,

$$\begin{aligned}\mathbf{P}(C \to B) &= 1 - \big(1 - \mathbf{P}(C \to E \text{ and } E \to B)\big)\big(1 - \mathbf{P}(C \to F \text{ and } F \to B)\big)\\ &= 1 - (1 - p_{CE}p_{EB})(1 - p_{CF}p_{FB})\\ &= 1 - (1 - 0.8 \cdot 0.9)(1 - 0.95 \cdot 0.85)\\ &= 0.946,\end{aligned}$$

$$\mathbf{P}(A \to C \text{ and } C \to B) = \mathbf{P}(A \to C)\mathbf{P}(C \to B) = 0.9 \cdot 0.946 = 0.851,$$

$$\mathbf{P}(A \to D \text{ and } D \to B) = \mathbf{P}(A \to D)\mathbf{P}(D \to B) = 0.75 \cdot 0.95 = 0.712,$$

and finally we obtain the desired probability

$$\begin{aligned}\mathbf{P}(A \to B) &= 1 - \big(1 - \mathbf{P}(A \to C \text{ and } C \to B)\big)\big(1 - \mathbf{P}(A \to D \text{ and } D \to B)\big)\\ &= 1 - (1 - 0.851)(1 - 0.712)\\ &= 0.957.\end{aligned}$$

Independent Trials and the Binomial Probabilities

If an experiment involves a sequence of independent but identical stages, we say that we have a sequence of **independent trials**. In the special case where there are only two possible results at each stage, we say that we have a sequence of independent **Bernoulli trials**. The two possible results can be anything, e.g., "it rains" or "it doesn't rain," but we will often think in terms of coin tosses and refer to the two results as "heads" (H) and "tails" (T).

Consider an experiment that consists of n independent tosses of a coin, in which the probability of heads is p, where p is some number between 0 and 1. In this context, independence means that the events $A_1, A_2, \ldots, A_n$ are independent, where $A_i = \{i\text{th toss is a head}\}$.

We can visualize independent Bernoulli trials by means of a sequential description, as shown in Fig. 1.16 for the case where $n = 3$. The conditional probability of any toss being a head, conditioned on the results of any preceding tosses, is p because of independence. Thus, by multiplying the conditional probabilities along the corresponding path of the tree, we see that any particular outcome (3-long sequence of heads and tails) that involves k heads and $3 - k$ tails has probability $p^k(1-p)^{3-k}$. This formula extends to the case of a general number n of tosses. We obtain that the probability of any particular n-long sequence that contains k heads and $n - k$ tails is $p^k(1-p)^{n-k}$, for all k from 0 to n.

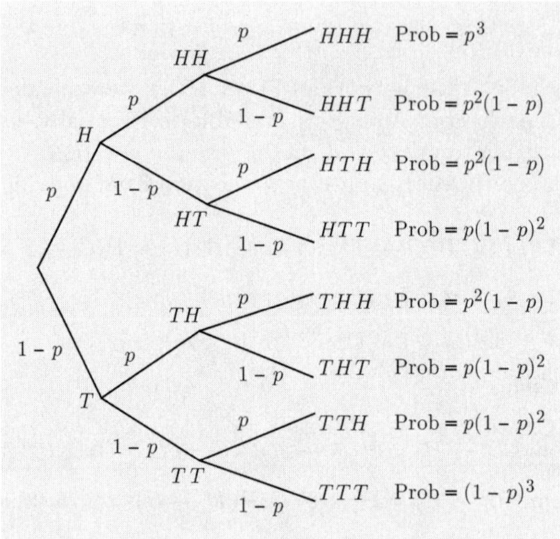

Figure 1.16: Sequential description of an experiment involving three independent tosses of a coin. Along the branches of the tree, we record the corresponding conditional probabilities, and by the multiplication rule, the probability of obtaining a particular 3-toss sequence is calculated by multiplying the probabilities recorded along the corresponding path of the tree.

Let us now consider the probability

$$p(k) = \mathbf{P}(k \text{ heads come up in an } n\text{-toss sequence}),$$

which will play an important role later. We showed above that the probability of any given sequence that contains k heads is $p^k(1-p)^{n-k}$, so we have

$$p(k) = \binom{n}{k} p^k(1-p)^{n-k},$$

where we use the notation

$$\binom{n}{k} = \text{number of distinct } n\text{-toss sequences that contain } k \text{ heads.}$$

The numbers $\binom{n}{k}$ (read as "n choose k") are known as the **binomial coefficients**, while the probabilities $p(k)$ are known as the **binomial probabilities**. Using a counting argument, to be given in Section 1.6, we can show that

$$\binom{n}{k} = \frac{n!}{k!\,(n-k)!}, \qquad k = 0, 1, \ldots, n,$$

where for any positive integer i we have

$$i! = 1 \cdot 2 \cdots (i-1) \cdot i,$$

and, by convention, $0! = 1$. An alternative verification is sketched in the end-of-chapter problems. Note that the binomial probabilities $p(k)$ must add to 1, thus showing the **binomial formula**

$$\sum_{k=0}^{n} \binom{n}{k} p^k (1-p)^{n-k} = 1.$$

Example 1.25. Grade of Service. An internet service provider has installed c modems to serve the needs of a population of n dialup customers. It is estimated that at a given time, each customer will need a connection with probability p, independent of the others. What is the probability that there are more customers needing a connection than there are modems?

Here we are interested in the probability that more than c customers simultaneously need a connection. It is equal to

$$\sum_{k=c+1}^{n} p(k),$$

where

$$p(k) = \binom{n}{k} p^k (1-p)^{n-k}$$

are the binomial probabilities. For instance, if $n = 100$, $p = 0.1$, and $c = 15$, the probability of interest turns out to be 0.0399.

This example is typical of problems of sizing a facility to serve the needs of a homogeneous population, consisting of independently acting customers. The problem is to select the facility size to guarantee a certain probability (sometimes called **grade of service**) that no user is left unserved.

1.6 COUNTING

The calculation of probabilities often involves counting the number of outcomes in various events. We have already seen two contexts where such counting arises.

(a) When the sample space Ω has a finite number of equally likely outcomes, so that the discrete uniform probability law applies. Then, the probability of any event A is given by

$$\mathbf{P}(A) = \frac{\text{number of elements of } A}{\text{number of elements of } \Omega},$$

and involves counting the elements of A and of Ω.

(b) When we want to calculate the probability of an event A with a finite number of equally likely outcomes, each of which has an already known probability p. Then the probability of A is given by

$$\mathbf{P}(A) = p \cdot (\text{number of elements of } A),$$

and involves counting the number of elements of A. An example of this type is the calculation of the probability of k heads in n coin tosses (the binomial probabilities). We saw in the preceding section that the probability of each distinct sequence involving k heads is easily obtained, but the calculation of the number of all such sequences, to be presented shortly, requires some thought.

While counting is in principle straightforward, it is frequently challenging; the art of counting constitutes a large portion of the field of **combinatorics**. In this section, we present the basic principle of counting and apply it to a number of situations that are often encountered in probabilistic models.

The Counting Principle

The counting principle is based on a divide-and-conquer approach, whereby the counting is broken down into stages through the use of a tree. For example, consider an experiment that consists of two consecutive stages. The possible results at the first stage are $a_1, a_2, \ldots, a_m$; the possible results at the second stage are $b_1, b_2, \ldots, b_n$. Then, the possible results of the two-stage experiment are all possible **ordered** pairs (a_i, b_j), $i = 1, \ldots, m$, $j = 1, \ldots, n$. Note that the number of such ordered pairs is equal to mn. This observation can be generalized as follows (see also Fig. 1.17).

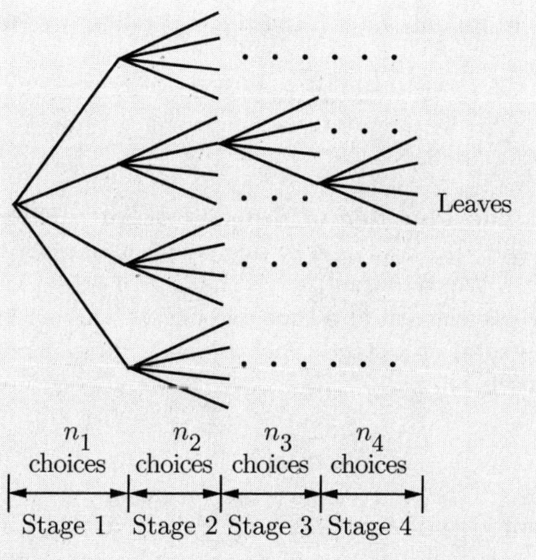

Figure 1.17: Illustration of the basic counting principle. The counting is carried out in r stages ($r = 4$ in the figure). The first stage has n_1 possible results. For every possible result at the first $i - 1$ stages, there are n_i possible results at the ith stage. The number of leaves is $n_1 n_2 \cdots n_r$. This is the desired count.

The Counting Principle

Consider a process that consists of r stages. Suppose that:

(a) There are n_1 possible results at the first stage.

(b) For every possible result at the first stage, there are n_2 possible results at the second stage.

(c) More generally, for any sequence of possible results at the first $i - 1$ stages, there are n_i possible results at the ith stage. Then, the total number of possible results of the r-stage process is

$$n_1 n_2 \cdots n_r.$$

Example 1.26. The Number of Telephone Numbers. A local telephone number is a 7-digit sequence, but the first digit has to be different from 0 or 1. How many distinct telephone numbers are there? We can visualize the choice of a sequence as a sequential process, where we select one digit at a time. We have a total of 7 stages, and a choice of one out of 10 elements at each stage, except for the first stage where we only have 8 choices. Therefore, the answer is

$$8 \cdot \underbrace{10 \cdot 10 \cdots 10}_{6 \text{ times}} = 8 \cdot 10^6.$$

Example 1.27. The Number of Subsets of an n-Element Set. Consider an n-element set $\{s_1, s_2, \ldots, s_n\}$. How many subsets does it have (including itself and the empty set)? We can visualize the choice of a subset as a sequential process where we examine one element at a time and decide whether to include it in the set or not. We have a total of n stages, and a binary choice at each stage. Therefore the number of subsets is

$$\underbrace{2 \cdot 2 \cdots 2}_{n \text{ times}} = 2^n.$$

It should be noted that the Counting Principle remains valid even if each first-stage result leads to a different set of potential second-stage results, etc. The only requirement is that the number of possible second-stage results is constant, regardless of the first-stage result.

In what follows, we will focus primarily on two types of counting arguments that involve the selection of k objects out of a collection of n objects. If the order of selection matters, the selection is called a **permutation**, and otherwise, it is called a **combination**. We will then discuss a more general type of counting, involving a **partition** of a collection of n objects into multiple subsets.

k-permutations

We start with n distinct objects, and let k be some positive integer, with $k \leq n$. We wish to count the number of different ways that we can pick k out of these n objects and arrange them in a sequence, i.e., the number of distinct k-object sequences. We can choose any of the n objects to be the first one. Having chosen the first, there are only $n-1$ possible choices for the second; given the choice of the first two, there only remain $n-2$ available objects for the third stage, etc. When we are ready to select the last (the kth) object, we have already chosen $k-1$ objects, which leaves us with $n-(k-1)$ choices for the last one. By the Counting Principle, the number of possible sequences, called **k-permutations**, is

$$\begin{aligned} n(n-1)\cdots(n-k+1) &= \frac{n(n-1)\cdots(n-k+1)(n-k)\cdots 2 \cdot 1}{(n-k)\cdots 2 \cdot 1} \\ &= \frac{n!}{(n-k)!}. \end{aligned}$$

In the special case where $k = n$, the number of possible sequences, simply called **permutations**, is

$$n(n-1)(n-2)\cdots 2 \cdot 1 = n!.$$

(Let $k = n$ in the formula for the number of k-permutations, and recall the convention $0! = 1$.)

Example 1.28. Let us count the number of words that consist of four distinct letters. This is the problem of counting the number of 4-permutations of the 26 letters in the alphabet. The desired number is

$$\frac{n!}{(n-k)!} = \frac{26!}{22!} = 26 \cdot 25 \cdot 24 \cdot 23 = 358{,}800.$$

The count for permutations can be combined with the Counting Principle to solve more complicated counting problems.

Example 1.29. You have n_1 classical music CDs, n_2 rock music CDs, and n_3 country music CDs. In how many different ways can you arrange them so that the CDs of the same type are contiguous?

We break down the problem in two stages, where we first select the order of the CD types, and then the order of the CDs of each type. There are 3! ordered sequences of the types of CDs (such as classical/rock/country, rock/country/classical, etc.), and there are $n_1!$ (or $n_2!$, or $n_3!$) permutations of the classical (or rock, or country, respectively) CDs. Thus for each of the 3! CD type sequences, there are $n_1!\,n_2!\,n_3!$ arrangements of CDs, and the desired total number is $3!\,n_1!\,n_2!\,n_3!$.

Suppose now that you offer to give k_i out of the n_i CDs of each type i to a friend, where $k_i < n_i$, $i = 1, 2, 3$. What is the number of all possible arrangements of the CDs that you are left with? The solution is similar, except that the number of $(n_i - k_i)$-permutations of CDs of type i replaces $n_i!$ in the estimate, so the number of possible arrangements is

$$3! \cdot \frac{n_1!}{k_1!} \cdot \frac{n_2!}{k_2!} \cdot \frac{n_3!}{k_3!}.$$

Combinations

There are n people and we are interested in forming a committee of k. How many different committees are possible? More abstractly, this is the same as the problem of counting the number of k-element subsets of a given n-element set. Notice that forming a combination is different than forming a k-permutation, because **in a combination there is no ordering of the selected elements.** For example, whereas the 2-permutations of the letters A, B, C, and D are

AB, BA, AC, CA, AD, DA, BC, CB, BD, DB, CD, DC,

the combinations of two out of these four letters are

AB, AC, AD, BC, BD, CD.

In the preceding example, the combinations are obtained from the permutations by grouping together "duplicates"; for example, AB and BA are not viewed as distinct, and are both associated with the combination AB. This reasoning can be generalized: each combination is associated with $k!$ "duplicate" k-permutations, so the number $n!/(n-k)!$ of k-permutations is equal to the number of combinations times $k!$. Hence, the number of possible combinations, is equal to

$$\frac{n!}{k!\,(n-k)!}.$$

Let us now relate the above expression to the binomial coefficient, which was denoted by $\binom{n}{k}$ and was defined in the preceding section as the number of n-toss sequences with k heads. We note that specifying an n-toss sequence with k heads is the same as selecting k elements (those that correspond to heads) out of the n-element set of tosses, i.e., a combination of k out of n objects. Hence, the binomial coefficient is also given by the same formula and we have

$$\binom{n}{k} = \frac{n!}{k!\,(n-k)!}.$$

Example 1.30. The number of combinations of two out of the four letters A, B, C, and D is found by letting $n = 4$ and $k = 2$. It is

$$\binom{4}{2} = \frac{4!}{2!\,2!} = 6,$$

consistent with the listing given earlier.

It is worth observing that counting arguments sometimes lead to formulas that are rather difficult to derive algebraically. One example is the **binomial formula**

$$\sum_{k=0}^{n} \binom{n}{k} p^k (1-p)^{n-k} = 1,$$

discussed in Section 1.5. In the special case where $p = 1/2$, this formula becomes

$$\sum_{k=0}^{n} \binom{n}{k} = 2^n,$$

and admits the following simple interpretation. Since $\binom{n}{k}$ is the number of k-element subsets of a given n-element subset, the sum over k of $\binom{n}{k}$ counts the number of subsets of all possible cardinalities. It is therefore equal to the number of all subsets of an n-element set, which is 2^n.

Example 1.31. We have a group of n persons. Consider clubs that consist of a special person from the group (the club leader) and a number (possibly zero) of additional club members. Let us count the number of possible clubs of this type in two different ways, thereby obtaining an algebraic identity.

There are n choices for club leader. Once the leader is chosen, we are left with a set of $n-1$ available persons, and we are free to choose any of the 2^{n-1} subsets. Thus the number of possible clubs is $n2^{n-1}$.

Alternatively, for fixed k, we can form a k-person club by first selecting k out of the n available persons [there are $\binom{n}{k}$ choices]. We can then select one of the members to be the leader (there are k choices). By adding over all possible club sizes k, we obtain the number of possible clubs as $\sum_{k=1}^{n} k\binom{n}{k}$, thereby showing the identity

$$\sum_{k=1}^{n} k\binom{n}{k} = n2^{n-1}.$$

Partitions

Recall that a combination is a choice of k elements out of an n-element set without regard to order. Thus, a combination can be viewed as a partition of the set in two: one part contains k elements and the other contains the remaining $n-k$. We now generalize by considering partitions into more than two subsets.

We are given an n-element set and nonnegative integers $n_1, n_2, \ldots, n_r$, whose sum is equal to n. We consider partitions of the set into r disjoint subsets, with the ith subset containing exactly n_i elements. Let us count in how many ways this can be done.

We form the subsets one at a time. We have $\binom{n}{n_1}$ ways of forming the first subset. Having formed the first subset, we are left with $n - n_1$ elements. We need to choose n_2 of them in order to form the second subset, and we have $\binom{n-n_1}{n_2}$ choices, etc. Using the Counting Principle for this r-stage process, the total number of choices is

$$\binom{n}{n_1}\binom{n-n_1}{n_2}\binom{n-n_1-n_2}{n_3}\cdots\binom{n-n_1-\cdots-n_{r-1}}{n_r},$$

which is equal to

$$\frac{n!}{n_1!\,(n-n_1)!}\cdot\frac{(n-n_1)!}{n_2!\,(n-n_1-n_2)!}\cdots\frac{(n-n_1-\cdots-n_{r-1})!}{(n-n_1-\cdots-n_{r-1}-n_r)!\,n_r!}.$$

We note that several terms cancel and we are left with

$$\frac{n!}{n_1!\,n_2!\cdots n_r!}.$$

This is called the **multinomial coefficient** and is usually denoted by

$$\binom{n}{n_1, n_2, \ldots, n_r}.$$

Example 1.32. Anagrams. How many different words (letter sequences) can be obtained by rearranging the letters in the word TATTOO? There are six positions to be filled by the available letters. Each rearrangement corresponds to a partition of the set of the six positions into a group of size 3 (the positions that get the letter T), a group of size 1 (the position that gets the letter A), and a group of size 2 (the positions that get the letter O). Thus, the desired number is

$$\frac{6!}{1!\,2!\,3!} = \frac{1 \cdot 2 \cdot 3 \cdot 4 \cdot 5 \cdot 6}{1 \cdot 1 \cdot 2 \cdot 1 \cdot 2 \cdot 3} = 60.$$

It is instructive to derive this answer using an alternative argument. (This argument can also be used to rederive the multinomial coefficient formula; see the end-of-chapter problems.) Let us write TATTOO in the form $\mathrm{T_1AT_2T_3O_1O_2}$ pretending for a moment that we are dealing with 6 distinguishable objects. These 6 objects can be rearranged in 6! different ways. However, any of the 3! possible permutations of $\mathrm{T_1}$, $\mathrm{T_2}$, and $\mathrm{T_3}$, as well as any of the 2! possible permutations of $\mathrm{O_1}$ and $\mathrm{O_2}$, lead to the same word. Thus, when the subscripts are removed, there are only $6!/(3!\,2!)$ different words.

Example 1.33. A class consisting of 4 graduate and 12 undergraduate students is randomly divided into four groups of 4. What is the probability that each group includes a graduate student? This is the same as Example 1.11 in Section 1.3, but we will now obtain the answer using a counting argument.

We first determine the nature of the sample space. A typical outcome is a particular way of partitioning the 16 students into four groups of 4. We take the term "randomly" to mean that every possible partition is equally likely, so that the probability question can be reduced to one of counting.

According to our earlier discussion, there are

$$\binom{16}{4,4,4,4} = \frac{16!}{4!\,4!\,4!\,4!}$$

different partitions, and this is the size of the sample space.

Let us now focus on the event that each group contains a graduate student. Generating an outcome with this property can be accomplished in two stages:

(a) Take the four graduate students and distribute them to the four groups; there are four choices for the group of the first graduate student, three choices for the second, two for the third. Thus, there is a total of 4! choices for this stage.

(b) Take the remaining 12 undergraduate students and distribute them to the four groups (3 students in each). This can be done in

$$\binom{12}{3,3,3,3} = \frac{12!}{3!\,3!\,3!\,3!}$$

different ways.

By the Counting Principle, the event of interest can occur in

$$\frac{4!\,12!}{3!\,3!\,3!\,3!}$$

different ways. The probability of this event is

$$\frac{\dfrac{4!\,12!}{3!\,3!\,3!\,3!}}{\dfrac{16!}{4!\,4!\,4!\,4!}}.$$

After some cancellations, we find that this is equal to

$$\frac{12\cdot 8\cdot 4}{15\cdot 14\cdot 13},$$

consistent with the answer obtained in Example 1.11.

Here is a summary of all the counting results we have developed.

Summary of Counting Results

- **Permutations** of n objects: $n!$.
- **k-permutations** of n objects: $n!/(n-k)!$.
- **Combinations** of k out of n objects: $\displaystyle\binom{n}{k} = \frac{n!}{k!\,(n-k)!}.$
- **Partitions** of n objects into r groups, with the ith group having n_i objects:
$$\binom{n}{n_1, n_2, \ldots, n_r} = \frac{n!}{n_1!\,n_2!\cdots n_r!}.$$

1.7 SUMMARY AND DISCUSSION

A probability problem can usually be broken down into a few basic steps:

(a) The description of the sample space, that is, the set of possible outcomes of a given experiment.

(b) The (possibly indirect) specification of the probability law (the probability of each event).

(c) The calculation of probabilities and conditional probabilities of various events of interest.

The probabilities of events must satisfy the nonnegativity, additivity, and normalization axioms. In the important special case where the set of possible outcomes is finite, one can just specify the probability of each outcome and obtain the probability of any event by adding the probabilities of the elements of the event.

Given a probability law, we are often interested in conditional probabilities, which allow us to reason based on partial information about the outcome of the experiment. We can view conditional probabilities as probability laws of a special type, under which only outcomes contained in the conditioning event can have positive conditional probability. Conditional probabilities can be derived from the (unconditional) probability law using the definition $\mathbf{P}(A \mid B) = \mathbf{P}(A \cap B)/\mathbf{P}(B)$. However, the reverse process is often convenient, that is, first specify some conditional probabilities that are natural for the real situation that we wish to model, and then use them to derive the (unconditional) probability law.

We have illustrated through examples three methods for calculating probabilities:

(a) The **counting method**. This method applies to the case where the number of possible outcomes is finite, and all outcomes are equally likely. To calculate the probability of an event, we count the number of elements of the event and divide by the number of elements of the sample space.

(b) The **sequential method**. This method applies when the experiment has a sequential character, and suitable conditional probabilities are specified or calculated along the branches of the corresponding tree (perhaps using the counting method). The probabilities of various events are then obtained by multiplying conditional probabilities along the corresponding paths of the tree, using the multiplication rule.

(c) The **divide-and-conquer method**. Here, the probabilities $\mathbf{P}(B)$ of various events B are obtained from conditional probabilities $\mathbf{P}(B \mid A_i)$, where the A_i are suitable events that form a partition of the sample space and have known probabilities $\mathbf{P}(A_i)$. The probabilities $\mathbf{P}(B)$ are then obtained by using the total probability theorem.

Finally, we have focused on a few side topics that reinforce our main themes. We have discussed the use of Bayes' rule in inference, which is an important application context. We have also discussed some basic principles of counting and combinatorics, which are helpful in applying the counting method.

PROBLEMS

SECTION 1.1. Sets

Problem 1. Consider rolling a six-sided die. Let A be the set of outcomes where the roll is an even number. Let B be the set of outcomes where the roll is greater than 3. Calculate and compare the sets on both sides of De Morgan's laws

$$(A \cup B)^c = A^c \cap B^c, \qquad \big(A \cap B\big)^c = A^c \cup B^c.$$

Problem 2. Let A and B be two sets.

(a) Show that

$$A^c = (A^c \cap B) \cup (A^c \cap B^c), \qquad B^c = (A \cap B^c) \cup (A^c \cap B^c).$$

(b) Show that

$$(A \cap B)^c = (A^c \cap B) \cup (A^c \cap B^c) \cup (A \cap B^c).$$

(c) Consider rolling a fair six-sided die. Let A be the set of outcomes where the roll is an odd number. Let B be the set of outcomes where the roll is less than 4. Calculate the sets on both sides of the equality in part (b), and verify that the equality holds.

Problem 3.* Prove the identity

$$A \cup \big(\cap_{n=1}^{\infty} B_n\big) = \cap_{n=1}^{\infty}(A \cup B_n).$$

Solution. If x belongs to the set on the left, there are two possibilities. Either $x \in A$, in which case x belongs to all of the sets $A \cup B_n$, and therefore belongs to the set on the right. Alternatively, x belongs to all of the sets B_n in which case, it belongs to all of the sets $A \cup B_n$, and therefore again belongs to the set on the right.

Conversely, if x belongs to the set on the right, then it belongs to $A \cup B_n$ for all n. If x belongs to A, then it belongs to the set on the left. Otherwise, x must belong to every set B_n and again belongs to the set on the left.

Problem 4.* **Cantor's diagonalization argument.** Show that the unit interval $[0, 1]$ is uncountable, i.e., its elements cannot be arranged in a sequence.

Solution. Any number x in $[0, 1]$ can be represented in terms of its decimal expansion, e.g., $1/3 = 0.3333\cdots$. Note that most numbers have a unique decimal expansion, but there are a few exceptions. For example, $1/2$ can be represented as $0.5000\cdots$ or as $0.49999\cdots$. It can be shown that this is the only kind of exception, i.e., decimal expansions that end with an infinite string of zeroes or an infinite string of nines.

Suppose, to obtain a contradiction, that the elements of $[0,1]$ can be arranged in a sequence $x_1, x_2, x_3, \ldots$, so that every element of $[0,1]$ appears in the sequence. Consider the decimal expansion of x_n:

$$x_n = 0.a_n^1 a_n^2 a_n^3 \cdots,$$

where each digit a_n^i belongs to $\{0, 1, \ldots, 9\}$. Consider now a number y constructed as follows. The nth digit of y can be 1 or 2, and is chosen so that it is different from the nth digit of x_n. Note that y has a unique decimal expansion since it does not end with an infinite sequence of zeroes or nines. The number y differs from each x_n, since it has a different nth digit. Therefore, the sequence $x_1, x_2, \ldots$ does not exhaust the elements of $[0,1]$, contrary to what was assumed. The contradiction establishes that the set $[0,1]$ is uncountable.

SECTION 1.2. Probabilistic Models

Problem 5. Out of the students in a class, 60% are geniuses, 70% love chocolate, and 40% fall into both categories. Determine the probability that a randomly selected student is neither a genius nor a chocolate lover.

Problem 6. A six-sided die is loaded in a way that each even face is twice as likely as each odd face. All even faces are equally likely, as are all odd faces. Construct a probabilistic model for a single roll of this die and find the probability that the outcome is less than 4.

Problem 7. A four-sided die is rolled repeatedly, until the first time (if ever) that an even number is obtained. What is the sample space for this experiment?

Problem 8. You enter a special kind of chess tournament, in which you play one game with each of three opponents, but you get to choose the order in which you play your opponents, knowing the probability of a win against each. You win the tournament if you win two games in a row, and you want to maximize the probability of winning. Show that it is optimal to play the weakest opponent second, and that the order of playing the other two opponents does not matter.

Problem 9. A partition of the sample space Ω is a collection of disjoint events $S_1, \ldots, S_n$ such that $\Omega = \cup_{i=1}^n S_i$.

(a) Show that for any event A, we have

$$\mathbf{P}(A) = \sum_{i=1}^{n} \mathbf{P}(A \cap S_i).$$

(b) Use part (a) to show that for any events A, B, and C, we have

$$\mathbf{P}(A) = \mathbf{P}(A \cap B) + \mathbf{P}(A \cap C) + \mathbf{P}\big(A \cap B^c \cap C^c\big) - \mathbf{P}(A \cap B \cap C).$$

Problem 10. Show the formula

$$\mathbf{P}\big((A \cap B^c) \cup (A^c \cap B)\big) = \mathbf{P}(A) + \mathbf{P}(B) - 2\mathbf{P}(A \cap B),$$

which gives the probability that exactly one of the events A and B will occur. [Compare with the formula $\mathbf{P}(A \cup B) = \mathbf{P}(A) + \mathbf{P}(B) - \mathbf{P}(A \cap B)$, which gives the probability that at least one of the events A and B will occur.]

Problem 11.* **Bonferroni's inequality.**

(a) Prove that for any two events A and B, we have

$$\mathbf{P}(A \cap B) \geq \mathbf{P}(A) + \mathbf{P}(B) - 1.$$

(b) Generalize to the case of n events $A_1, A_2, \ldots, A_n$, by showing that

$$\mathbf{P}(A_1 \cap A_2 \cap \cdots \cap A_n) \geq \mathbf{P}(A_1) + \mathbf{P}(A_2) + \cdots + \mathbf{P}(A_n) - (n-1).$$

Solution. We have $\mathbf{P}(A \cup B) = \mathbf{P}(A) + \mathbf{P}(B) - \mathbf{P}(A \cap B)$ and $\mathbf{P}(A \cup B) \leq 1$, which implies part (a). For part (b), we use De Morgan's law to obtain

$$\begin{aligned} 1 - \mathbf{P}(A_1 \cap \cdots \cap A_n) &= \mathbf{P}\big((A_1 \cap \cdots \cap A_n)^c\big) \\ &= \mathbf{P}(A_1^c \cup \cdots \cup A_n^c) \\ &\leq \mathbf{P}(A_1^c) + \cdots + \mathbf{P}(A_n^c) \\ &= \big(1 - \mathbf{P}(A_1)\big) + \cdots + \big(1 - \mathbf{P}(A_n)\big) \\ &= n - \mathbf{P}(A_1) - \cdots - \mathbf{P}(A_n). \end{aligned}$$

Problem 12.* **The inclusion-exclusion formula.** Show the following generalizations of the formula

$$\mathbf{P}(A \cup B) = \mathbf{P}(A) + \mathbf{P}(B) - \mathbf{P}(A \cap B).$$

(a) Let A, B, and C be events. Then,

$$\mathbf{P}(A \cup B \cup C) = \mathbf{P}(A) + \mathbf{P}(B) + \mathbf{P}(C) - \mathbf{P}(A \cap B) - \mathbf{P}(B \cap C) - \mathbf{P}(A \cap C) + \mathbf{P}(A \cap B \cap C).$$

(b) Let $A_1, A_2, \ldots, A_n$ be events. Let $S_1 = \{i \,|\, 1 \leq i \leq n\}$, $S_2 = \{(i_1, i_2) \,|\, 1 \leq i_1 < i_2 \leq n\}$, and more generally, let S_m be the set of all m-tuples $(i_1, \ldots, i_m)$ of indices that satisfy $1 \leq i_1 < i_2 < \cdots < i_m \leq n$. Then,

$$\begin{aligned} \mathbf{P}\left(\cup_{k=1}^n A_k\right) = \sum_{i \in S_1} \mathbf{P}(A_i) - \sum_{(i_1, i_2) \in S_2} \mathbf{P}(A_{i_1} \cap A_{i_2}) \\ + \sum_{(i_1, i_2, i_3) \in S_3} \mathbf{P}(A_{i_1} \cap A_{i_2} \cap A_{i_3}) - \cdots + (-1)^{n-1} \mathbf{P}\left(\cap_{k=1}^n A_k\right). \end{aligned}$$

Solution. (a) We use the formulas $\mathbf{P}(X \cup Y) = \mathbf{P}(X) + \mathbf{P}(Y) - \mathbf{P}(X \cap Y)$ and $(A \cup B) \cap C = (A \cap C) \cup (B \cap C)$. We have

$$\begin{aligned} \mathbf{P}(A \cup B \cup C) &= \mathbf{P}(A \cup B) + \mathbf{P}(C) - \mathbf{P}\big((A \cup B) \cap C\big) \\ &= \mathbf{P}(A \cup B) + \mathbf{P}(C) - \mathbf{P}\big((A \cap C) \cup (B \cap C)\big) \\ &= \mathbf{P}(A \cup B) + \mathbf{P}(C) - \mathbf{P}(A \cap C) - \mathbf{P}(B \cap C) + \mathbf{P}(A \cap B \cap C) \\ &= \mathbf{P}(A) + \mathbf{P}(B) - \mathbf{P}(A \cap B) + \mathbf{P}(C) - \mathbf{P}(A \cap C) - \mathbf{P}(B \cap C) \\ &\quad + \mathbf{P}(A \cap B \cap C) \\ &= \mathbf{P}(A) + \mathbf{P}(B) + \mathbf{P}(C) - \mathbf{P}(A \cap B) - \mathbf{P}(B \cap C) - \mathbf{P}(A \cap C) \\ &\quad + \mathbf{P}(A \cap B \cap C). \end{aligned}$$

(b) Use induction and verify the main induction step by emulating the derivation of part (a). For a different approach, see the problems at the end of Chapter 2.

Problem 13.* **Continuity property of probabilities.**

(a) Let $A_1, A_2, \ldots$ be an infinite sequence of events, which is "monotonically increasing," meaning that $A_n \subset A_{n+1}$ for every n. Let $A = \cup_{n=1}^{\infty} A_n$. Show that $\mathbf{P}(A) = \lim_{n\to\infty} \mathbf{P}(A_n)$. *Hint:* Express the event A as a union of countably many disjoint sets.

(b) Suppose now that the events are "monotonically decreasing," i.e., $A_{n+1} \subset A_n$ for every n. Let $A = \cap_{n=1}^{\infty} A_n$. Show that $\mathbf{P}(A) = \lim_{n\to\infty} \mathbf{P}(A_n)$. *Hint:* Apply the result of part (a) to the complements of the events.

(c) Consider a probabilistic model whose sample space is the real line. Show that

$$\mathbf{P}\big([0,\infty)\big) = \lim_{n\to\infty} \mathbf{P}\big([0,n]\big), \qquad \text{and} \qquad \lim_{n\to\infty} \mathbf{P}\big([n,\infty)\big) = 0.$$

Solution. (a) Let $B_1 = A_1$ and, for $n \geq 2$, $B_n = A_n \cap A_{n-1}^c$. The events B_n are disjoint, and we have $\cup_{k=1}^{n} B_k = A_n$, and $\cup_{k=1}^{\infty} B_k = A$. We apply the additivity axiom to obtain

$$\mathbf{P}(A) = \sum_{k=1}^{\infty} \mathbf{P}(B_k) = \lim_{n\to\infty} \sum_{k=1}^{n} \mathbf{P}(B_k) = \lim_{n\to\infty} \mathbf{P}(\cup_{k=1}^{n} B_k) = \lim_{n\to\infty} \mathbf{P}(A_n).$$

(b) Let $C_n = A_n^c$ and $C = A^c$. Since $A_{n+1} \subset A_n$, we obtain $C_n \subset C_{n+1}$, and the events C_n are increasing. Furthermore, $C = A^c = (\cap_{n=1}^{\infty} A_n)^c = \cup_{n=1}^{\infty} A_n^c = \cup_{n=1}^{\infty} C_n$. Using the result from part (a) for the sequence C_n, we obtain

$$1 - \mathbf{P}(A) = \mathbf{P}(A^c) = \mathbf{P}(C) = \lim_{n\to\infty} \mathbf{P}(C_n) = \lim_{n\to\infty} \big(1 - \mathbf{P}(A_n)\big),$$

from which we conclude that $\mathbf{P}(A) = \lim_{n\to\infty} \mathbf{P}(A_n)$.

(c) For the first equality, use the result from part (a) with $A_n = [0,n]$ and $A = [0,\infty)$. For the second, use the result from part (b) with $A_n = [n,\infty)$ and $A = \cap_{n=1}^{\infty} A_n = \emptyset$.

SECTION 1.3. Conditional Probability

Problem 14. We roll two fair 6-sided dice. Each one of the 36 possible outcomes is assumed to be equally likely.

(a) Find the probability that doubles are rolled.

(b) Given that the roll results in a sum of 4 or less, find the conditional probability that doubles are rolled.

(c) Find the probability that at least one die roll is a 6.

(d) Given that the two dice land on different numbers, find the conditional probability that at least one die roll is a 6.

Problem 15. A coin is tossed twice. Alice claims that the event of two heads is at least as likely if we know that the first toss is a head than if we know that at least one of the tosses is a head. Is she right? Does it make a difference if the coin is fair or unfair? How can we generalize Alice's reasoning?

Problem 16. We are given three coins: one has heads in both faces, the second has tails in both faces, and the third has a head in one face and a tail in the other. We choose a coin at random, toss it, and the result is heads. What is the probability that the opposite face is tails?

Problem 17. A batch of one hundred items is inspected by testing four randomly selected items. If one of the four is defective, the batch is rejected. What is the probability that the batch is accepted if it contains five defectives?

Problem 18. Let A and B be events. Show that $\mathbf{P}(A \cap B \mid B) = \mathbf{P}(A \mid B)$, assuming that $\mathbf{P}(B) > 0$.

SECTION 1.4. Total Probability Theorem and Bayes' Rule

Problem 19. Alice searches for her term paper in her filing cabinet, which has several drawers. She knows that she left her term paper in drawer j with probability $p_j > 0$. The drawers are so messy that even if she correctly guesses that the term paper is in drawer i, the probability that she finds it is only d_i. Alice searches in a particular drawer, say drawer i, but the search is unsuccessful. Conditioned on this event, show that the probability that her paper is in drawer j, is given by

$$\frac{p_j}{1 - p_i d_i}, \qquad \text{if } j \neq i, \qquad\qquad \frac{p_i(1 - d_i)}{1 - p_i d_i}, \qquad \text{if } j = i.$$

Problem 20. How an inferior player with a superior strategy can gain an advantage. Boris is about to play a two-game chess match with an opponent, and wants to find the strategy that maximizes his winning chances. Each game ends with either a win by one of the players, or a draw. If the score is tied at the end of the two games, the match goes into sudden-death mode, and the players continue to play until the first time one of them wins a game (and the match). Boris has two playing styles, *timid* and *bold*, and he can choose one of the two at will in each game, no matter what style he chose in previous games. With timid play, he draws with probability $p_d > 0$, and he loses with probability $1 - p_d$. With bold play, he wins with probability p_w, and he loses with probability $1 - p_w$. Boris will always play bold during sudden death, but may switch style between games 1 and 2.

(a) Find the probability that Boris wins the match for each of the following strategies:
   (i) Play bold in both games 1 and 2.
   (ii) Play timid in both games 1 and 2.
   (iii) Play timid whenever he is ahead in the score, and play bold otherwise.

(b) Assume that $p_w < 1/2$, so Boris is the worse player, regardless of the playing style he adopts. Show that with the strategy in (iii) above, and depending on the values of p_w and p_d, Boris may have a better than a 50-50 chance to win the match. How do you explain this advantage?

Problem 21. Two players take turns removing a ball from a jar that initially contains m white and n black balls. The first player to remove a white ball wins. Develop a recursive formula that allows the convenient computation of the probability that the starting player wins.

Problem 22. Each of k jars contains m white and n black balls. A ball is randomly chosen from jar 1 and transferred to jar 2, then a ball is randomly chosen from jar 2 and transferred to jar 3, etc. Finally, a ball is randomly chosen from jar k. Show that the probability that the last ball is white is the same as the probability that the first ball is white, i.e., it is $m/(m+n)$.

Problem 23. We have two jars, each initially containing an equal number of balls. We perform four successive ball exchanges. In each exchange, we pick simultaneously and at random a ball from each jar and move it to the other jar. What is the probability that at the end of the four exchanges all the balls will be in the jar where they started?

Problem 24. The prisoner's dilemma. The release of two out of three prisoners has been announced, but their identity is kept secret. One of the prisoners considers asking a friendly guard to tell him who is the prisoner other than himself that will be released, but hesitates based on the following rationale: at the prisoner's present state of knowledge, the probability of being released is 2/3, but after he knows the answer, the probability of being released will become 1/2, since there will be two prisoners (including himself) whose fate is unknown and exactly one of the two will be released. What is wrong with this line of reasoning?

Problem 25. A two-envelopes puzzle. You are handed two envelopes, and you know that each contains a positive integer dollar amount and that the two amounts are different. The values of these two amounts are modeled as constants that are unknown. Without knowing what the amounts are, you select at random one of the two envelopes, and after looking at the amount inside, you may switch envelopes if you wish. A friend claims that the following strategy will increase above 1/2 your probability of ending up with the envelope with the larger amount: toss a coin repeatedly, let X be equal to 1/2 plus the number of tosses required to obtain heads for the first time, and switch if the amount in the envelope you selected is less than the value of X. Is your friend correct?

Problem 26. The paradox of induction. Consider a statement whose truth is unknown. If we see many examples that are compatible with it, we are tempted to view the statement as more probable. Such reasoning is often referred to as *inductive inference* (in a philosophical, rather than mathematical sense). Consider now the statement that "all cows are white." An equivalent statement is that "everything that is not white is not a cow." We then observe several black crows. Our observations are clearly compatible with the statement, but do they make the hypothesis "all cows are white" more likely?

To analyze such a situation, we consider a probabilistic model. Let us assume that there are two possible states of the world, which we model as complementary events:

$$A: \text{all cows are white,}$$

$$A^c: 50\% \text{ of all cows are white.}$$

Let p be the prior probability $\mathbf{P}(A)$ that all cows are white. We make an observation of a cow or a crow, with probability q and $1-q$, respectively, independent of whether event A occurs or not. Assume that $0 < p < 1$, $0 < q < 1$, and that all crows are black.

(a) Given the event $B = \{\text{a black crow was observed}\}$, what is $\mathbf{P}(A \,|\, B)$?

(b) Given the event $C = \{\text{a white cow was observed}\}$, what is $\mathbf{P}(A \,|\, C)$?

Problem 27. Alice and Bob have $2n + 1$ coins, each coin with probability of heads equal to $1/2$. Bob tosses $n+1$ coins, while Alice tosses the remaining n coins. Assuming independent coin tosses, show that the probability that after all coins have been tossed, Bob will have gotten more heads than Alice is $1/2$.

Problem 28.* Conditional version of the total probability theorem. Let $C_1, \ldots, C_n$ be disjoint events that form a partition of the state space. Let also A and B be events such that $\mathbf{P}(B \cap C_i) > 0$ for all i. Show that

$$\mathbf{P}(A \,|\, B) = \sum_{i=1}^{n} \mathbf{P}(C_i \,|\, B)\mathbf{P}(A \,|\, B \cap C_i).$$

Solution. We have

$$\mathbf{P}(A \cap B) = \sum_{i=1}^{n} \mathbf{P}\big((A \cap B) \cap C_i\big),$$

and by using the multiplication rule,

$$\mathbf{P}\big((A \cap B) \cap C_i\big) = \mathbf{P}(B)\mathbf{P}(C_i \,|\, B)\mathbf{P}(A \,|\, B \cap C_i).$$

Combining these two equations, dividing by $\mathbf{P}(B)$, and using the formula $\mathbf{P}(A \,|\, B) = \mathbf{P}(A \cap B)/\mathbf{P}(B)$, we obtain the desired result.

Problem 29.* Let A and B be events with $\mathbf{P}(A) > 0$ and $\mathbf{P}(B) > 0$. We say that an event B *suggests* an event A if $\mathbf{P}(A \,|\, B) > \mathbf{P}(A)$, and *does not suggest* event A if $\mathbf{P}(A \,|\, B) < \mathbf{P}(A)$.

(a) Show that B suggests A if and only if A suggests B.

(b) Assume that $\mathbf{P}(B^c) > 0$. Show that B suggests A if and only if B^c does not suggest A.

(c) We know that a treasure is located in one of two places, with probabilities β and $1 - \beta$, respectively, where $0 < \beta < 1$. We search the first place and if the treasure is there, we find it with probability $p > 0$. Show that the event of not finding the treasure in the first place suggests that the treasure is in the second place.

Solution. (a) We have $\mathbf{P}(A \,|\, B) = \mathbf{P}(A \cap B)/\mathbf{P}(B)$, so B suggests A if and only if $\mathbf{P}(A \cap B) > \mathbf{P}(A)\mathbf{P}(B)$, which is equivalent to A suggesting B, by symmetry.

(b) Since $\mathbf{P}(B) + \mathbf{P}(B^c) = 1$, we have

$$\mathbf{P}(B)\mathbf{P}(A) + \mathbf{P}(B^c)\mathbf{P}(A) = \mathbf{P}(A) = \mathbf{P}(B)\mathbf{P}(A \,|\, B) + \mathbf{P}(B^c)\mathbf{P}(A \,|\, B^c),$$

which implies that

$$\mathbf{P}(B^c)\big(\mathbf{P}(A) - \mathbf{P}(A \,|\, B^c)\big) = \mathbf{P}(B)\big(\mathbf{P}(A \,|\, B) - \mathbf{P}(A)\big).$$

Thus, $\mathbf{P}(A \mid B) > \mathbf{P}(A)$ (B suggests A) if and only if $\mathbf{P}(A) > \mathbf{P}(A \mid B^c)$ (B^c does not suggest A).

(c) Let A and B be the events

$$A = \{\text{the treasure is in the second place}\},$$
$$B = \{\text{we don't find the treasure in the first place}\}.$$

Using the total probability theorem, we have

$$\mathbf{P}(B) = \mathbf{P}(A^c)\mathbf{P}(B \mid A^c) + \mathbf{P}(A)\mathbf{P}(B \mid A) = \beta(1-p) + (1-\beta),$$

so

$$\mathbf{P}(A \mid B) = \frac{\mathbf{P}(A \cap B)}{\mathbf{P}(B)} = \frac{1-\beta}{\beta(1-p)+(1-\beta)} = \frac{1-\beta}{1-\beta p} > 1-\beta = \mathbf{P}(A).$$

It follows that event B suggests event A.

SECTION 1.5. Independence

Problem 30. A hunter has two hunting dogs. One day, on the trail of some animal, the hunter comes to a place where the road diverges into two paths. He knows that each dog, independent of the other, will choose the correct path with probability p. The hunter decides to let each dog choose a path, and if they agree, take that one, and if they disagree, to randomly pick a path. Is his strategy better than just letting one of the two dogs decide on a path?

Problem 31. Communication through a noisy channel. A source transmits a message (a string of symbols) through a noisy communication channel. Each symbol is 0 or 1 with probability p and $1-p$, respectively, and is received incorrectly with probability ϵ_0 and ϵ_1, respectively (see Fig. 1.18). Errors in different symbol transmissions are independent.

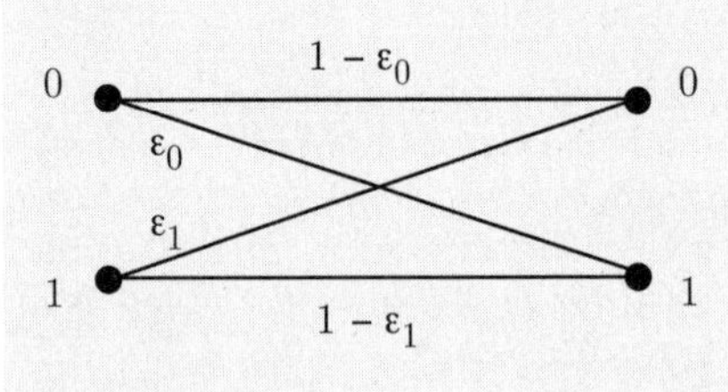

Figure 1.18: Error probabilities in a binary communication channel.

(a) What is the probability that the kth symbol is received correctly?

(b) What is the probability that the string of symbols 1011 is received correctly?

(c) In an effort to improve reliability, each symbol is transmitted three times and the received string is decoded by majority rule. In other words, a 0 (or 1) is transmitted as 000 (or 111, respectively), and it is decoded at the receiver as a 0 (or 1) if and only if the received three-symbol string contains at least two 0s (or 1s, respectively). What is the probability that a 0 is correctly decoded?

(d) For what values of ϵ_0 is there an improvement in the probability of correct decoding of a 0 when the scheme of part (c) is used?

(e) Suppose that the scheme of part (c) is used. What is the probability that a symbol was 0 given that the received string is 101?

Problem 32. The king's sibling. The king has only one sibling. What is the probability that the sibling is male? Assume that every birth results in a boy with probability 1/2, independent of other births. Be careful to state any additional assumptions you have to make in order to arrive at an answer.

Problem 33. Using a biased coin to make an unbiased decision. Alice and Bob want to choose between the opera and the movies by tossing a fair coin. Unfortunately, the only available coin is biased (though the bias is not known exactly). How can they use the biased coin to make a decision so that either option (opera or the movies) is equally likely to be chosen?

Problem 34. An electrical system consists of identical components, each of which is operational with probability p, independent of other components. The components are connected in three subsystems, as shown in Fig. 1.19. The system is operational if there is a path that starts at point A, ends at point B, and consists of operational components. What is the probability of this happening?

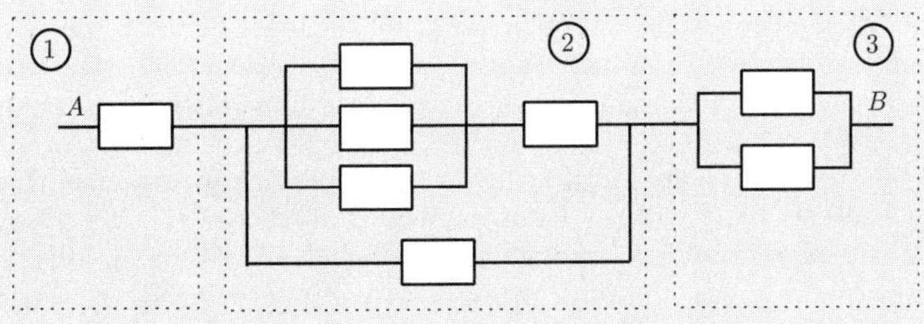

Figure 1.19: A system of identical components that consists of the three subsystems 1, 2, and 3. The system is operational if there is a path that starts at point A, ends at point B, and consists of operational components.

Problem 35. Reliability of a k-out-of-n system. A system consists of n identical components, each of which is operational with probability p, independent of other components. The system is operational if at least k out of the n components are operational. What is the probability that the system is operational?

Problem 36. A power utility can supply electricity to a city from n different power plants. Power plant i fails with probability p_i, independent of the others.

(a) Suppose that any one plant can produce enough electricity to supply the entire city. What is the probability that the city will experience a black-out?

(b) Suppose that two power plants are necessary to keep the city from a black-out. Find the probability that the city will experience a black-out.

Problem 37. A cellular phone system services a population of n_1 "voice users" (those who occasionally need a voice connection) and n_2 "data users" (those who occasionally need a data connection). We estimate that at a given time, each user will need to be connected to the system with probability p_1 (for voice users) or p_2 (for data users), independent of other users. The data rate for a voice user is r_1 bits/sec and for a data user is r_2 bits/sec. The cellular system has a total capacity of c bits/sec. What is the probability that more users want to use the system than the system can accommodate?

Problem 38. The problem of points. Telis and Wendy play a round of golf (18 holes) for a \$10 stake, and their probabilities of winning on any one hole are p and $1-p$, respectively, independent of their results in other holes. At the end of 10 holes, with the score 4 to 6 in favor of Wendy, Telis receives an urgent call and has to report back to work. They decide to split the stake in proportion to their probabilities of winning had they completed the round, as follows. If p_T and p_W are the conditional probabilities that Telis and Wendy, respectively, are ahead in the score after 18 holes given the 4-6 score after 10 holes, then Telis should get a fraction $p_T/(p_T+p_W)$ of the stake, and Wendy should get the remaining $p_W/(p_T+p_W)$. How much money should Telis get? *Note*: This is an example of the, so-called, problem of points, which played an important historical role in the development of probability theory. The problem was posed by Chevalier de Méré in the 17th century to Pascal, who introduced the idea that the stake of an interrupted game should be divided in proportion to the players' conditional probabilities of winning given the state of the game at the time of interruption. Pascal worked out some special cases and through a correspondence with Fermat, stimulated much thinking and several probability-related investigations.

Problem 39. A particular class has had a history of low attendance. The annoyed professor decides that she will not lecture unless at least k of the n students enrolled in the class are present. Each student will independently show up with probability p_g if the weather is good, and with probability p_b if the weather is bad. Given the probability of bad weather on a given day, obtain an expression for the probability that the professor will teach her class on that day.

Problem 40. Consider a coin that comes up heads with probability p and tails with probability $1-p$. Let q_n be the probability that after n independent tosses, there have been an even number of heads. Derive a recursion that relates q_n to q_{n-1}, and solve this recursion to establish the formula

$$q_n = \big(1 + (1-2p)^n\big)/2.$$

Problem 41. Consider a game show with an infinite pool of contestants, where at each round i, contestant i obtains a number by spinning a continuously calibrated wheel. The contestant with the smallest number thus far survives. Successive wheel spins are independent and we assume that there are no ties. Let N be the round at which contestant 1 is eliminated. For any positive integer n, find $\mathbf{P}(N = n)$.

Problem 42.* Gambler's ruin. A gambler makes a sequence of independent bets. In each bet, he wins \$1 with probability p, and loses \$1 with probability $1-p$. Initially, the gambler has \$$k$, and plays until he either accumulates \$$n$ or has no money left. What is the probability that the gambler will end up with \$$n$?

Solution. Let us denote by A the event that he ends up with \$$n$, and by F the event that he wins the first bet. Denote also by w_k the probability of event A, if he starts with \$$k$. We apply the total probability theorem to obtain

$$w_k = \mathbf{P}(A \mid F)\mathbf{P}(F) + \mathbf{P}(A \mid F^c)\mathbf{P}(F^c) = p\mathbf{P}(A \mid F) + q\mathbf{P}(A \mid F^c), \qquad 0 < k < n,$$

where $q = 1-p$. By the independence of past and future bets, having won the first bet is the same as if he were just starting now but with \$$(k+1)$, so that $\mathbf{P}(A \mid F) = w_{k+1}$ and similarly $\mathbf{P}(A \mid F^c) = w_{k-1}$. Thus, we have $w_k = pw_{k+1} + qw_{k-1}$, which can be written as

$$w_{k+1} - w_k = r(w_k - w_{k-1}), \qquad 0 < k < n,$$

where $r = q/p$. We will solve for w_k in terms of p and q using iteration, and the boundary values $w_0 = 0$ and $w_n = 1$.

We have $w_{k+1} - w_k = r^k(w_1 - w_0)$, and since $w_0 = 0$,

$$w_{k+1} = w_k + r^k w_1 = w_{k-1} + r^{k-1} w_1 + r^k w_1 = w_1 + rw_1 + \cdots + r^k w_1.$$

The sum in the right-hand side can be calculated separately for the two cases where $r = 1$ (or $p = q$) and $r \neq 1$ (or $p \neq q$). We have

$$w_k = \begin{cases} \dfrac{1-r^k}{1-r} w_1, & \text{if } p \neq q, \\ kw_1, & \text{if } p = q. \end{cases}$$

Since $w_n = 1$, we can solve for w_1 and therefore for w_k:

$$w_1 = \begin{cases} \dfrac{1-r}{1-r^n}, & \text{if } p \neq q, \\ \dfrac{1}{n}, & \text{if } p = q, \end{cases}$$

so that

$$w_k = \begin{cases} \dfrac{1-r^k}{1-r^n}, & \text{if } p \neq q, \\ \dfrac{k}{n}, & \text{if } p = q. \end{cases}$$

Problem 43.* Let A and B be independent events. Use the definition of independence to prove the following:

(a) The events A and B^c are independent.

(b) The events A^c and B^c are independent.

Solution. (a) The event A is the union of the disjoint events $A \cap B^c$ and $A \cap B$. Using the additivity axiom and the independence of A and B, we obtain

$$\mathbf{P}(A) = \mathbf{P}(A \cap B) + \mathbf{P}(A \cap B^c) = \mathbf{P}(A)\mathbf{P}(B) + \mathbf{P}(A \cap B^c).$$

It follows that

$$\mathbf{P}(A \cap B^c) = \mathbf{P}(A)\big(1 - \mathbf{P}(B)\big) = \mathbf{P}(A)\mathbf{P}(B^c),$$

so A and B^c are independent.

(b) Apply the result of part (a) twice: first on A and B, then on B^c and A.

Problem 44.* Let A, B, and C be independent events, with $\mathbf{P}(C) > 0$. Prove that A and B are conditionally independent given C.

Solution. We have

$$\begin{aligned}\mathbf{P}(A \cap B \,|\, C) &= \frac{\mathbf{P}(A \cap B \cap C)}{\mathbf{P}(C)} \\ &= \frac{\mathbf{P}(A)\mathbf{P}(B)\mathbf{P}(C)}{\mathbf{P}(C)} \\ &= \mathbf{P}(A)\mathbf{P}(B) \\ &= \mathbf{P}(A \,|\, C)\mathbf{P}(B \,|\, C),\end{aligned}$$

so A and B are conditionally independent given C. In the preceding calculation, the first equality uses the definition of conditional probabilities; the second uses the assumed independence; the fourth uses the independence of A from C, and of B from C.

Problem 45.* Assume that the events A_1, A_2, A_3, A_4 are independent and that $\mathbf{P}(A_3 \cap A_4) > 0$. Show that

$$\mathbf{P}(A_1 \cup A_2 \,|\, A_3 \cap A_4) = \mathbf{P}(A_1 \cup A_2).$$

Solution. We have

$$\mathbf{P}(A_1 \,|\, A_3 \cap A_4) = \frac{\mathbf{P}(A_1 \cap A_3 \cap A_4)}{\mathbf{P}(A_3 \cap A_4)} = \frac{\mathbf{P}(A_1)\mathbf{P}(A_3)\mathbf{P}(A_4)}{\mathbf{P}(A_3)\mathbf{P}(A_4)} = \mathbf{P}(A_1).$$

We similarly obtain $\mathbf{P}(A_2 \,|\, A_3 \cap A_4) = \mathbf{P}(A_2)$ and $\mathbf{P}(A_1 \cap A_2 \,|\, A_3 \cap A_4) = \mathbf{P}(A_1 \cap A_2)$, and finally,

$$\begin{aligned}\mathbf{P}(A_1 \cup A_2 \,|\, A_3 \cap A_4) &= \mathbf{P}(A_1 \,|\, A_3 \cap A_4) + \mathbf{P}(A_2 \,|\, A_3 \cap A_4) - \mathbf{P}(A_1 \cap A_2 \,|\, A_3 \cap A_4) \\ &= \mathbf{P}(A_1) + \mathbf{P}(A_2) - \mathbf{P}(A_1 \cap A_2) \\ &= \mathbf{P}(A_1 \cup A_2).\end{aligned}$$

Problem 46.* **Laplace's rule of succession.** Consider $m + 1$ boxes with the kth box containing k red balls and $m - k$ white balls, where k ranges from 0 to m. We choose a box at random (all boxes are equally likely) and then choose a ball at random from that box, n successive times (the ball drawn is replaced each time, and a new ball is selected independently). Suppose a red ball was drawn each of the n times. What is the probability that if we draw a ball one more time it will be red? Estimate this probability for large m.

Solution. We want to find the conditional probability $\mathbf{P}(E \,|\, R_n)$, where E is the event of a red ball drawn at time $n+1$, and R_n is the event of a red ball drawn each of the n preceding times. Intuitively, the consistent draw of a red ball indicates that a box with a high percentage of red balls was chosen, so we expect that $\mathbf{P}(E \mid R_n)$ is closer to 1 than to 0. In fact, Laplace used this example to calculate the probability that the sun will rise tomorrow given that it has risen for the preceding 5,000 years. (It is not clear how serious Laplace was about this calculation, but the story is part of the folklore of probability theory.)

We have

$$\mathbf{P}(E \mid R_n) = \frac{\mathbf{P}(E \cap R_n)}{\mathbf{P}(R_n)},$$

and by using the total probability theorem, we obtain

$$\mathbf{P}(R_n) = \sum_{k=0}^{m} \mathbf{P}(k\text{th box chosen}) \left(\frac{k}{m}\right)^n = \frac{1}{m+1} \sum_{k=0}^{m} \left(\frac{k}{m}\right)^n,$$

$$\mathbf{P}(E \cap R_n) = \mathbf{P}(R_{n+1}) = \frac{1}{m+1} \sum_{k=0}^{m} \left(\frac{k}{m}\right)^{n+1}.$$

For large m, we can view $\mathbf{P}(R_n)$ as a piecewise constant approximation to an integral:

$$\mathbf{P}(R_n) = \frac{1}{m+1} \sum_{k=0}^{m} \left(\frac{k}{m}\right)^n \approx \frac{1}{(m+1)m^n} \int_0^m x^n \, dx = \frac{1}{(m+1)m^n} \cdot \frac{m^{n+1}}{n+1} \approx \frac{1}{n+1}.$$

Similarly,

$$\mathbf{P}(E \cap R_n) = \mathbf{P}(R_{n+1}) \approx \frac{1}{n+2},$$

so that

$$\mathbf{P}(E \mid R_n) \approx \frac{n+1}{n+2}.$$

Thus, for large m, drawing a red ball one more time is almost certain when n is large.

Problem 47.* Binomial coefficient formula and the Pascal triangle.

(a) Use the definition of $\binom{n}{k}$ as the number of distinct n-toss sequences with k heads, to derive the recursion suggested by the so called Pascal triangle, given in Fig. 1.20.

(b) Use the recursion derived in part (a) and induction, to establish the formula

$$\binom{n}{k} = \frac{n!}{k!\,(n-k)!}.$$

Solution. (a) Note that n-toss sequences that contain k heads (for $0 < k < n$) can be obtained in two ways:

(1) By starting with an $(n-1)$-toss sequence that contains k heads and adding a tail at the end. There are $\binom{n-1}{k}$ different sequences of this type.

(2) By starting with an $(n-1)$-toss sequence that contains $k-1$ heads and adding a head at the end. There are $\binom{n-1}{k-1}$ different sequences of this type.

$$
\begin{array}{ccccccccc}
 & & & & \binom{0}{0} & & & & \\
 & & & \binom{1}{0} & & \binom{1}{1} & & & \\
 & & \binom{2}{0} & & \binom{2}{1} & & \binom{2}{2} & & \\
 & \binom{3}{0} & & \binom{3}{1} & & \binom{3}{2} & & \binom{3}{3} & \\
\binom{4}{0} & & \binom{4}{1} & & \binom{4}{2} & & \binom{4}{3} & & \binom{4}{4} \\
 & & & & \cdots\cdots\cdots & & & &
\end{array}
\qquad\qquad
\begin{array}{ccccccccc}
 & & & & 1 & & & & \\
 & & & 1 & & 1 & & & \\
 & & 1 & & 2 & & 1 & & \\
 & 1 & & 3 & & 3 & & 1 & \\
1 & & 4 & & 6 & & 4 & & 1 \\
 & & & & \cdots\cdots\cdots & & & &
\end{array}
$$

Figure 1.20: Sequential calculation method of the binomial coefficients using the Pascal triangle. Each term $\binom{n}{k}$ in the triangular array on the left is computed and placed in the triangular array on the right by adding its two neighbors in the row above it (except for the boundary terms with $k = 0$ or $k = n$, which are equal to 1).

Thus,

$$
\binom{n}{k} = \begin{cases} \binom{n-1}{k-1} + \binom{n-1}{k}, & \text{if } k = 1, 2, \ldots, n-1, \\ 1, & \text{if } k = 0, n. \end{cases}
$$

This is the formula corresponding to the Pascal triangle calculation, given in Fig. 1.20.

(b) We now use the recursion from part (a), to demonstrate the formula

$$
\binom{n}{k} = \frac{n!}{k!\,(n-k)!},
$$

by induction on n. Indeed, we have from the definition $\binom{1}{0} = \binom{1}{1} = 1$, so for $n = 1$ the above formula is seen to hold as long as we use the convention $0! = 1$. If the formula holds for each index up to $n - 1$, we have for $k = 1, 2, \ldots, n - 1$,

$$
\begin{aligned}
\binom{n}{k} &= \binom{n-1}{k-1} + \binom{n-1}{k} \\
&= \frac{(n-1)!}{(k-1)!\,(n-1-k+1)!} + \frac{(n-1)!}{k!\,(n-1-k)!} \\
&= \frac{k}{n} \cdot \frac{n!}{k!\,(n-k)!} + \frac{n-k}{n} \cdot \frac{n!}{k!\,(n-k)!} \\
&= \frac{n!}{k!\,(n-k)!},
\end{aligned}
$$

and the induction is complete.

Problem 48.* The Borel-Cantelli lemma. Consider an infinite sequence of trials. The probability of success at the ith trial is some positive number p_i. Let N be the event that there is no success, and let I be the event that there is an infinite number of successes.

(a) Assume that the trials are independent and that $\sum_{i=1}^{\infty} p_i = \infty$. Show that $\mathbf{P}(N) = 0$ and $\mathbf{P}(I) = 1$.

(b) Assume that $\sum_{i=1}^{\infty} p_i < \infty$. Show that $\mathbf{P}(I) = 0$.

Solution. (a) The event N is a subset of the event that there were no successes in the first n trials, so that

$$\mathbf{P}(N) \leq \prod_{i=1}^{n}(1 - p_i).$$

Taking logarithms,

$$\log \mathbf{P}(N) \leq \sum_{i=1}^{n} \log(1 - p_i) \leq \sum_{i=1}^{n}(-p_i).$$

Taking the limit as n tends to infinity, we obtain $\log \mathbf{P}(N) = -\infty$, or $\mathbf{P}(N) = 0$.

Let now L_n be the event that there is a finite number of successes and that the last success occurs at the nth trial. We use the already established result $\mathbf{P}(N) = 0$, and apply it to the sequence of trials after trial n, to obtain $\mathbf{P}(L_n) = 0$. The event I^c (finite number of successes) is the union of the disjoint events L_n, $n \geq 1$, and N, so that

$$\mathbf{P}(I^c) = \mathbf{P}(N) + \sum_{n=1}^{\infty} \mathbf{P}(L_n) = 0,$$

and $\mathbf{P}(I) = 1$.

(b) Let S_i be the event that the ith trial is a success. Fix some number n and for every $i > n$, let F_i be the event that the first success after time n occurs at time i. Note that $F_i \subset S_i$. Finally, let A_n be the event that there is at least one success after time n. Note that $I \subset A_n$, because an infinite number of successes implies that there are successes subsequent to time n. Furthermore, the event A_n is the union of the disjoint events F_i, $i > n$. Therefore,

$$\mathbf{P}(I) \leq \mathbf{P}(A_n) = \mathbf{P}\left(\bigcup_{i=n+1}^{\infty} F_i\right) = \sum_{i=n+1}^{\infty} \mathbf{P}(F_i) \leq \sum_{i=n+1}^{\infty} \mathbf{P}(S_i) = \sum_{i=n+1}^{\infty} p_i.$$

We take the limit of both sides as $n \to \infty$. Because of the assumption $\sum_{i=1}^{\infty} p_i < \infty$, the right-hand side converges to zero. This implies that $\mathbf{P}(I) = 0$.

SECTION 1.6. Counting

Problem 49. De Méré's puzzle. A six-sided die is rolled three times independently. Which is more likely: a sum of 11 or a sum of 12? (This question was posed by the French nobleman de Méré to his friend Pascal in the 17th century.)

Problem 50. The birthday problem. Consider n people who are attending a party. We assume that every person has an equal probability of being born on any day during the year, independent of everyone else, and ignore the additional complication presented by leap years (i.e., assume that nobody is born on February 29). What is the probability that each person has a distinct birthday?

Problem 51. An urn contains m red and n white balls.

(a) We draw two balls randomly and simultaneously. Describe the sample space and calculate the probability that the selected balls are of different color, by using two approaches: a counting approach based on the discrete uniform law, and a sequential approach based on the multiplication rule.

(b) We roll a fair 3-sided die whose faces are labeled 1,2,3, and if k comes up, we remove k balls from the urn at random and put them aside. Describe the sample space and calculate the probability that all of the balls drawn are red, using a divide-and-conquer approach and the total probability theorem.

Problem 52. We deal from a well-shuffled 52-card deck. Calculate the probability that the 13th card is the first king to be dealt.

Problem 53. Ninety students, including Joe and Jane, are to be split into three classes of equal size, and this is to be done at random. What is the probability that Joe and Jane end up in the same class?

Problem 54. Twenty distinct cars park in the same parking lot every day. Ten of these cars are US-made, while the other ten are foreign-made. The parking lot has exactly twenty spaces, all in a row, so the cars park side by side. However, the drivers have varying schedules, so the position any car might take on a certain day is random.

(a) In how many different ways can the cars line up?

(b) What is the probability that on a given day, the cars will park in such a way that they alternate (no two US-made are adjacent and no two foreign-made are adjacent)?

Problem 55. Eight rooks are placed in distinct squares of an 8×8 chessboard, with all possible placements being equally likely. Find the probability that all the rooks are safe from one another, i.e., that there is no row or column with more than one rook.

Problem 56. An academic department offers 8 lower level courses: $\{L_1, L_2, \ldots, L_8\}$ and 10 higher level courses: $\{H_1, H_2, \ldots, H_{10}\}$. A valid curriculum consists of 4 lower level courses, and 3 higher level courses.

(a) How many different curricula are possible?

(b) Suppose that $\{H_1, \ldots, H_5\}$ have L_1 as a prerequisite, and $\{H_6, \ldots H_{10}\}$ have L_2 and L_3 as prerequisites, i.e., any curricula which involve, say, one of $\{H_1, \ldots, H_5\}$ must also include L_1. How many different curricula are there?

Problem 57. How many 6-word sentences can be made using each of the 26 letters of the alphabet exactly once? A word is defined as a nonempty (possibly jibberish) sequence of letters.

Problem 58. We draw the top 7 cards from a well-shuffled standard 52-card deck. Find the probability that:

(a) The 7 cards include exactly 3 aces.

(b) The 7 cards include exactly 2 kings.

(c) The probability that the 7 cards include exactly 3 aces, or exactly 2 kings, or both.

Problem 59. A parking lot contains 100 cars, k of which happen to be lemons. We select m of these cars at random and take them for a test drive. Find the probability that n of the cars tested turn out to be lemons.

Problem 60. A well-shuffled 52-card deck is dealt to 4 players. Find the probability that each of the players gets an ace.

Problem 61.* Hypergeometric probabilities. An urn contains n balls, out of which m are red. We select k of the balls at random, without replacement (i.e., selected balls are not put back into the urn before the next selection). What is the probability that i of the selected balls are red?

Solution. The sample space consists of the $\binom{n}{k}$ different ways that we can select k out of the available balls. For the event of interest to occur, we have to select i out of the m red balls, which can be done in $\binom{m}{i}$ ways, and also select $k-i$ out of the $n-m$ balls that are not red, which can be done in $\binom{n-m}{k-i}$ ways. Therefore, the desired probability is

$$\frac{\binom{m}{i}\binom{n-m}{k-i}}{\binom{n}{k}},$$

for $i \geq 0$ satisfying $i \leq m$, $i \leq k$, and $k-i \leq n-m$. For all other i, the probability is zero.

Problem 62.* Correcting the number of permutations for indistinguishable objects. When permuting n objects, some of which are indistinguishable, different permutations may lead to indistinguishable object sequences, so the number of distinguishable object sequences is less than $n!$. For example, there are six permutations of the letters A, B, and C:

$$\text{ABC, ACB, BAC, BCA, CAB, CBA,}$$

but only three distinguishable sequences that can be formed using the letters A, D, and D:

$$\text{ADD, DAD, DDA.}$$

(a) Suppose that k out of the n objects are indistinguishable. Show that the number of distinguishable object sequences is $n!/k!$.

(b) Suppose that we have r types of indistinguishable objects, and for each i, k_i objects of type i. Show that the number of distinguishable object sequences is

$$\frac{n!}{k_1!\,k_2!\cdots k_r!}.$$

Solution. (a) Each one of the $n!$ permutations corresponds to $k!$ duplicates which are obtained by permuting the k indistinguishable objects. Thus, the $n!$ permutations can be grouped into $n!/k!$ groups of $k!$ indistinguishable permutations that result in the same object sequence. Therefore, the number of distinguishable object sequences is $n!/k!$. For example, the three letters A, D, and D give the $3! = 6$ permutations

$$\text{ADD, ADD, DAD, DDA, DAD, DDA,}$$

obtained by replacing B and C by D in the permutations of A, B, and C given earlier. However, these 6 permutations can be divided into the $n!/k! = 3!/2! = 3$ groups

$$\{\text{ADD}, \text{ADD}\}, \ \{\text{DAD}, \text{DAD}\}, \ \{\text{DDA}, \text{DDA}\},$$

each having $k! = 2! = 2$ indistinguishable permutations.

(b) One solution is to extend the argument in (a) above: for each object type i, there are $k_i!$ indistinguishable permutations of the k_i objects. Hence, each permutation belongs to a group of $k_1!\,k_2!\cdots k_r!$ indistinguishable permutations, all of which yield the same object sequence.

An alternative argument goes as follows. Choosing a distinguishable object sequence is the same as starting with n slots and for each i, choosing the k_i slots to be occupied by objects of type i. This is the same as partitioning the set $\{1, \ldots, n\}$ into groups of size $k_1, \ldots, k_r$, and the number of such partitions is given by the multinomial coefficient.

2

Discrete Random Variables

Contents

2.1 BASIC CONCEPTS

In many probabilistic models, the outcomes are numerical, e.g., when they correspond to instrument readings or stock prices. In other experiments, the outcomes are not numerical, but they may be associated with some numerical values of interest. For example, if the experiment is the selection of students from a given population, we may wish to consider their grade point average. When dealing with such numerical values, it is often useful to assign probabilities to them. This is done through the notion of a **random variable**, the focus of the present chapter.

Given an experiment and the corresponding set of possible outcomes (the sample space), a random variable associates a particular number with each outcome; see Fig. 2.1. We refer to this number as the **numerical value** or simply the **value** of the random variable. Mathematically, **a random variable is a real-valued function of the experimental outcome.**

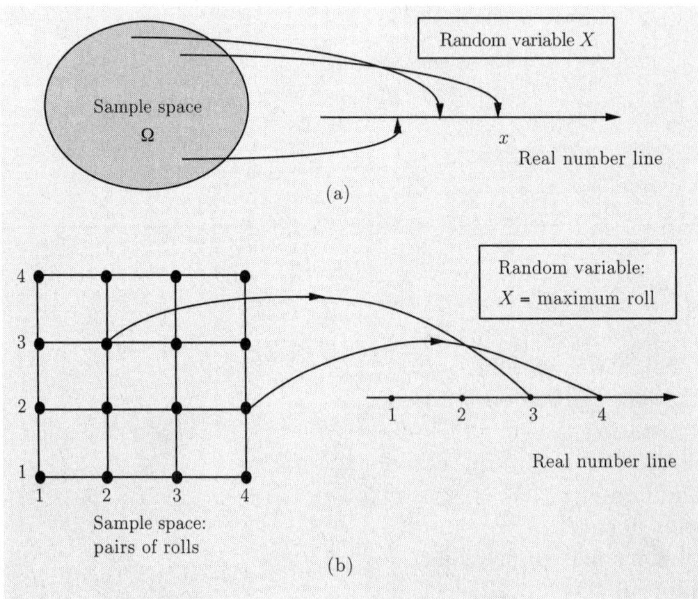

Figure 2.1: (a) Visualization of a random variable. It is a function that assigns a numerical value to each possible outcome of the experiment. (b) An example of a random variable. The experiment consists of two rolls of a 4-sided die, and the random variable is the maximum of the two rolls. If the outcome of the experiment is $(4, 2)$, the value of this random variable is 4.

Here are some examples of random variables:

(a) In an experiment involving a sequence of 5 tosses of a coin, the number of heads in the sequence is a random variable. However, the 5-long sequence of heads and tails is not considered a random variable because it does not have an explicit numerical value.

(b) In an experiment involving two rolls of a die, the following are examples of random variables:

(i) The sum of the two rolls.

(ii) The number of sixes in the two rolls.

(iii) The second roll raised to the fifth power.

(c) In an experiment involving the transmission of a message, the time needed to transmit the message, the number of symbols received in error, and the delay with which the message is received are all random variables.

There are several basic concepts associated with random variables, which are summarized below. These concepts will be discussed in detail in the present chapter.

Main Concepts Related to Random Variables

Starting with a probabilistic model of an experiment:

- A **random variable** is a real-valued function of the outcome of the experiment.
- A **function of a random variable** defines another random variable.
- We can associate with each random variable certain "averages" of interest, such as the **mean** and the **variance**.
- A random variable can be **conditioned** on an event or on another random variable.
- There is a notion of **independence** of a random variable from an event or from another random variable.

A random variable is called **discrete** if its range (the set of values that it can take) is either finite or countably infinite. For example, the random variables mentioned in (a) and (b) above can take at most a finite number of numerical values, and are therefore discrete.

A random variable that can take an uncountably infinite number of values is not discrete. For an example, consider the experiment of choosing a point a from the interval $[-1, 1]$. The random variable that associates the numerical value a^2 to the outcome a is not discrete. On the other hand, the random variable that associates with a the numerical value

$$\text{sgn}(a) = \begin{cases} 1, & \text{if } a > 0, \\ 0, & \text{if } a = 0, \\ -1, & \text{if } a < 0, \end{cases}$$

is discrete.

In this chapter, we focus exclusively on discrete random variables, even though we will typically omit the qualifier "discrete."

> **Concepts Related to Discrete Random Variables**
>
> Starting with a probabilistic model of an experiment:
>
> - A **discrete random variable** is a real-valued function of the outcome of the experiment that can take a finite or countably infinite number of values.
> - A discrete random variable has an associated **probability mass function (PMF)**, which gives the probability of each numerical value that the random variable can take.
> - A **function of a discrete random variable** defines another discrete random variable, whose PMF can be obtained from the PMF of the original random variable.

We will discuss each of the above concepts and the associated methodology in the following sections. In addition, we will provide examples of some important and frequently encountered random variables. In Chapter 3, we will discuss general (not necessarily discrete) random variables.

Even though this chapter may appear to be covering a lot of new ground, this is not really the case. The general line of development is to simply take the concepts from Chapter 1 (probabilities, conditioning, independence, etc.) and apply them to random variables rather than events, together with some convenient new notation. The only genuinely new concepts relate to means and variances.

2.2 PROBABILITY MASS FUNCTIONS

The most important way to characterize a random variable is through the probabilities of the values that it can take. For a discrete random variable X, these are captured by the **probability mass function** (PMF for short) of X, denoted p_X. In particular, if x is any possible value of X, the **probability mass** of x, denoted $p_X(x)$, is the probability of the event $\{X = x\}$ consisting of all outcomes that give rise to a value of X equal to x:

$$p_X(x) = \mathbf{P}\big(\{X = x\}\big).$$

For example, let the experiment consist of two independent tosses of a fair coin, and let X be the number of heads obtained. Then the PMF of X is

$$p_X(x) = \begin{cases} 1/4, & \text{if } x = 0 \text{ or } x = 2, \\ 1/2, & \text{if } x = 1, \\ 0, & \text{otherwise.} \end{cases}$$

In what follows, we will often omit the braces from the event/set notation when no ambiguity can arise. In particular, we will usually write $\mathbf{P}(X = x)$ in place of the more correct notation $\mathbf{P}(\{X = x\})$, and we will write $\mathbf{P}(X \in S)$ for the probability that X takes a value within a set S. We will also adhere to the following convention throughout: **we will use upper case characters to denote random variables, and lower case characters to denote real numbers such as the numerical values of a random variable.**

Note that

$$\sum_x p_X(x) = 1,$$

where in the summation above, x ranges over all the possible numerical values of X. This follows from the additivity and normalization axioms: as x ranges over all possible values of X, the events $\{X = x\}$ are disjoint and form a partition of the sample space. By a similar argument, for any set S of possible values of X, we have

$$\mathbf{P}(X \in S) = \sum_{x \in S} p_X(x).$$

For example, if X is the number of heads obtained in two independent tosses of a fair coin, as above, the probability of at least one head is

$$\mathbf{P}(X > 0) = \sum_{x=1}^{2} p_X(x) = \frac{1}{2} + \frac{1}{4} = \frac{3}{4}.$$

Calculating the PMF of X is conceptually straightforward, and is illustrated in Fig. 2.2.

Calculation of the PMF of a Random Variable X

For each possible value x of X:

1. Collect all the possible outcomes that give rise to the event $\{X = x\}$.
2. Add their probabilities to obtain $p_X(x)$.

The Bernoulli Random Variable

Consider the toss of a coin, which comes up a head with probability p, and a tail with probability $1 - p$. The **Bernoulli** random variable takes the two values 1

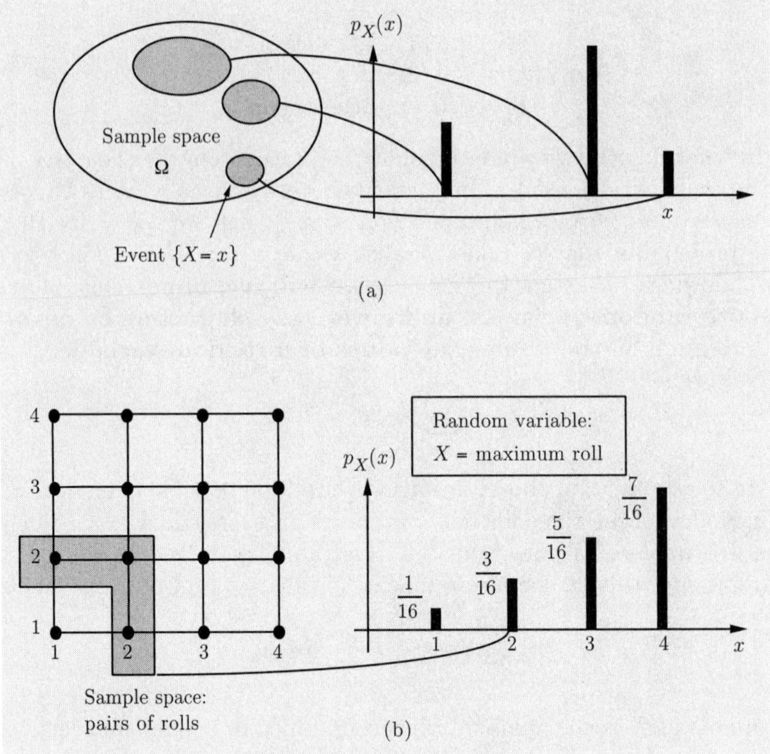

Figure 2.2: (a) Illustration of the method to calculate the PMF of a random variable X. For each possible value x, we collect all the outcomes that give rise to $X = x$ and add their probabilities to obtain $p_X(x)$. (b) Calculation of the PMF p_X of the random variable X = maximum roll in two independent rolls of a fair 4-sided die. There are four possible values x, namely, 1, 2, 3, 4. To calculate $p_X(x)$ for a given x, we add the probabilities of the outcomes that give rise to x. For example, there are three outcomes that give rise to $x = 2$, namely, $(1, 2), (2, 2), (2, 1)$. Each of these outcomes has probability 1/16, so $p_X(2) = 3/16$, as indicated in the figure.

and 0, depending on whether the outcome is a head or a tail:

$$X = \begin{cases} 1, & \text{if a head,} \\ 0, & \text{if a tail.} \end{cases}$$

Its PMF is

$$p_X(k) = \begin{cases} p, & \text{if } k = 1, \\ 1 - p, & \text{if } k = 0. \end{cases}$$

For all its simplicity, the Bernoulli random variable is very important. In practice, it is used to model generic probabilistic situations with just two outcomes, such as:

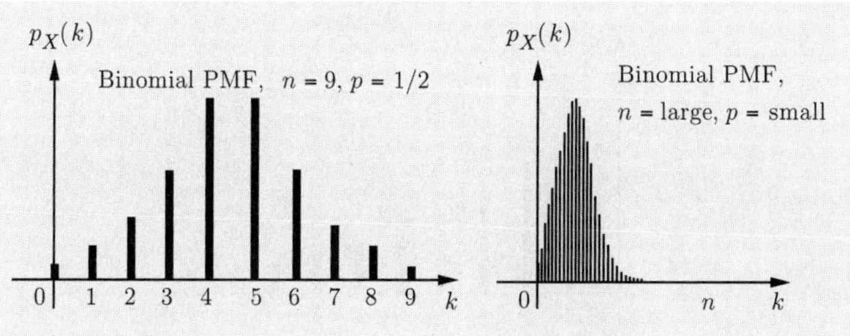

Figure 2.3: The PMF of a binomial random variable. If $p = 1/2$, the PMF is symmetric around $n/2$. Otherwise, the PMF is skewed towards 0 if $p < 1/2$, and towards n if $p > 1/2$.

(a) The state of a telephone at a given time that can be either free or busy.

(b) A person who can be either healthy or sick with a certain disease.

(c) The preference of a person who can be either for or against a certain political candidate.

Furthermore, by combining multiple Bernoulli random variables, one can construct more complicated random variables, such as the binomial random variable, which is discussed next.

The Binomial Random Variable

A coin is tossed n times. At each toss, the coin comes up a head with probability p, and a tail with probability $1 - p$, independent of prior tosses. Let X be the number of heads in the n-toss sequence. We refer to X as a **binomial** random variable **with parameters** n **and** p. The PMF of X consists of the binomial probabilities that were calculated in Section 1.5:

$$p_X(k) = \mathbf{P}(X = k) = \binom{n}{k} p^k (1-p)^{n-k}, \qquad k = 0, 1, \ldots, n.$$

(Note that here and elsewhere, we simplify notation and use k, instead of x, to denote the values of integer-valued random variables.) The normalization property, specialized to the binomial random variable, is written as

$$\sum_{k=0}^{n} \binom{n}{k} p^k (1-p)^{n-k} = 1.$$

Some special cases of the binomial PMF are sketched in Fig. 2.3.

The Geometric Random Variable

Suppose that we repeatedly and independently toss a coin with probability of a head equal to p, where $0 < p < 1$. The **geometric** random variable is the

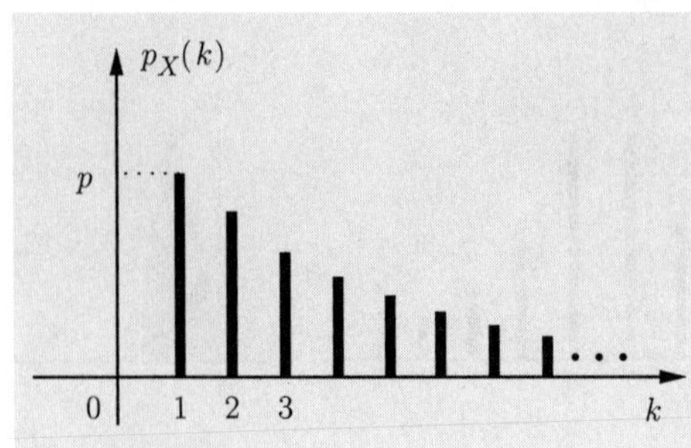

Figure 2.4: The PMF

$$p_X(k) = (1-p)^{k-1}p, \qquad k = 1, 2, \ldots,$$

of a geometric random variable. It decreases as a geometric progression with parameter $1 - p$.

number X of tosses needed for a head to come up for the first time. Its PMF is given by

$$p_X(k) = (1-p)^{k-1}p, \qquad k = 1, 2, \ldots,$$

since $(1-p)^{k-1}p$ is the probability of the sequence consisting of $k - 1$ successive tails followed by a head; see Fig. 2.4. This is a legitimate PMF because

$$\sum_{k=1}^{\infty} p_X(k) = \sum_{k=1}^{\infty}(1-p)^{k-1}p = p\sum_{k=0}^{\infty}(1-p)^k = p \cdot \frac{1}{1-(1-p)} = 1.$$

Naturally, the use of coin tosses here is just to provide insight. More generally, we can interpret the geometric random variable in terms of repeated independent trials until the first "success." Each trial has probability of success p and the number of trials until (and including) the first success is modeled by the geometric random variable. The meaning of "success" is context-dependent. For example, it could mean passing a test in a given try, finding a missing item in a given search, or finding the tax help information line free in a given attempt, etc.

The Poisson Random Variable

A Poisson random variable has a PMF given by

$$p_X(k) = e^{-\lambda}\frac{\lambda^k}{k!}, \qquad k = 0, 1, 2, \ldots,$$

where λ is a positive parameter characterizing the PMF, see Fig. 2.5. This is a legitimate PMF because

$$\sum_{k=0}^{\infty} e^{-\lambda}\frac{\lambda^k}{k!} = e^{-\lambda}\left(1 + \lambda + \frac{\lambda^2}{2!} + \frac{\lambda^3}{3!} + \cdots\right) = e^{-\lambda}e^{\lambda} = 1.$$

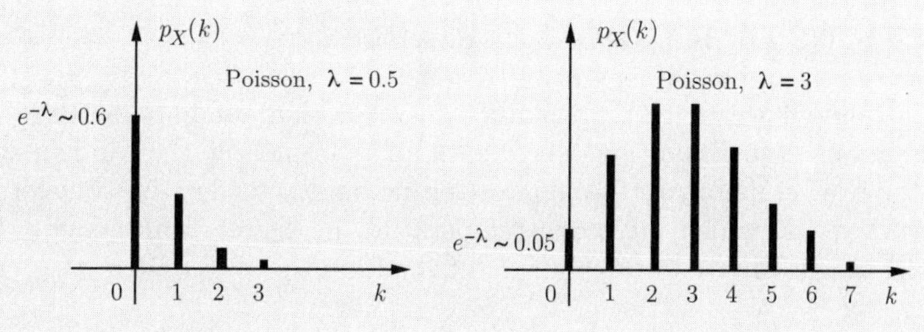

Figure 2.5: The PMF $e^{-\lambda}\lambda^k/k!$ of a Poisson random variable for different values of λ. Note that if $\lambda < 1$, then the PMF is monotonically decreasing with k, while if $\lambda > 1$, the PMF first increases and then decreases (this is shown in the end-of-chapter problems).

To get a feel for the Poisson random variable, think of a binomial random variable with very small p and very large n. For example, let X be the number of typos in a book with a total of n words. Then X is binomial, but since the probability p that any one word is misspelled is very small, X can also be well-modeled with a Poisson PMF (let p be the probability of heads in tossing a coin, and associate misspelled words with coin tosses that result in heads). There are many similar examples, such as the number of cars involved in accidents in a city on a given day.†

More precisely, the Poisson PMF with parameter λ is a good approximation for a binomial PMF with parameters n and p, i.e.,

$$e^{-\lambda}\frac{\lambda^k}{k!} \approx \frac{n!}{k!\,(n-k)!}p^k(1-p)^{n-k}, \qquad k = 0, 1, \ldots, n,$$

provided $\lambda = np$, n is very large, and p is very small. In this case, using the Poisson PMF may result in simpler models and calculations. For example, let $n = 100$ and $p = 0.01$. Then the probability of $k = 5$ successes in $n = 100$ trials is calculated using the binomial PMF as

$$\frac{100!}{95!\,5!} \cdot 0.01^5(1 - 0.01)^{95} = 0.00290.$$

Using the Poisson PMF with $\lambda = np = 100 \cdot 0.01 = 1$, this probability is approximated by

$$e^{-1}\frac{1}{5!} = 0.00306.$$

We provide a formal justification of the Poisson approximation property in the end-of-chapter problems and also in Chapter 6, where we will further interpret it, extend it, and use it in the context of the Poisson process.

† The first experimental verification of the connection between the binomial and the Poisson random variables reputedly occurred in the late 19th century, by matching the Poisson PMF to the number of horse kick accidents in the Polish cavalry over a period of several years.

2.3 FUNCTIONS OF RANDOM VARIABLES

Given a random variable X, one may generate other random variables by applying various transformations on X. As an example, let the random variable X be today's temperature in degrees Celsius, and consider the transformation $Y = 1.8X + 32$, which gives the temperature in degrees Fahrenheit. In this example, Y is a **linear** function of X, of the form

$$Y = g(X) = aX + b,$$

where a and b are scalars. We may also consider nonlinear functions of the general form

$$Y = g(X).$$

For example, if we wish to display temperatures on a logarithmic scale, we would want to use the function $g(X) = \log X$.

If $Y = g(X)$ is a function of a random variable X, then Y is also a random variable, since it provides a numerical value for each possible outcome. This is because every outcome in the sample space defines a numerical value x for X and hence also the numerical value $y = g(x)$ for Y. If X is discrete with PMF p_X, then Y is also discrete, and its PMF p_Y can be calculated using the PMF of X. In particular, to obtain $p_Y(y)$ for any y, we add the probabilities of all values of x such that $g(x) = y$:

$$p_Y(y) = \sum_{\{x \mid g(x)=y\}} p_X(x).$$

Example 2.1. Let $Y = |X|$ and let us apply the preceding formula for the PMF p_Y to the case where

$$p_X(x) = \begin{cases} 1/9, & \text{if } x \text{ is an integer in the range } [-4, 4], \\ 0, & \text{otherwise;} \end{cases}$$

see Fig. 2.6 for an illustration. The possible values of Y are $y = 0, 1, 2, 3, 4$. To compute $p_Y(y)$ for some given value y from this range, we must add $p_X(x)$ over all values x such that $|x| = y$. In particular, there is only one value of X that corresponds to $y = 0$, namely $x = 0$. Thus,

$$p_Y(0) = p_X(0) = \frac{1}{9}.$$

Also, there are two values of X that correspond to each $y = 1, 2, 3, 4$, so for example,

$$p_Y(1) = p_X(-1) + p_X(1) = \frac{2}{9}.$$

Thus, the PMF of Y is

$$p_Y(y) = \begin{cases} 2/9, & \text{if } y = 1, 2, 3, 4, \\ 1/9, & \text{if } y = 0, \\ 0, & \text{otherwise.} \end{cases}$$

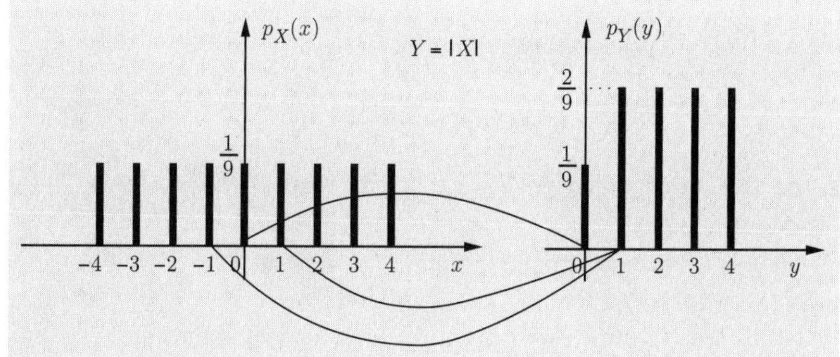

Figure 2.6: The PMFs of X and $Y = |X|$ in Example 2.1.

For another related example, let $Z = X^2$. To obtain the PMF of Z, we can view it either as the square of the random variable X or as the square of the random variable $Y = |X|$. By applying the formula $p_Z(z) = \sum_{\{x \mid x^2=z\}} p_X(x)$ or the formula $p_Z(z) = \sum_{\{y \mid y^2=z\}} p_Y(y)$, we obtain

$$p_Z(z) = \begin{cases} 2/9, & \text{if } z = 1, 4, 9, 16, \\ 1/9, & \text{if } z = 0, \\ 0, & \text{otherwise.} \end{cases}$$

2.4 EXPECTATION, MEAN, AND VARIANCE

The PMF of a random variable X provides us with several numbers, the probabilities of all the possible values of X. It is often desirable, however, to summarize this information in a single representative number. This is accomplished by the **expectation** of X, which is a weighted (in proportion to probabilities) average of the possible values of X.

As motivation, suppose you spin a wheel of fortune many times. At each spin, one of the numbers $m_1, m_2, \ldots, m_n$ comes up with corresponding probability $p_1, p_2, \ldots, p_n$, and this is your monetary reward from that spin. What is the amount of money that you "expect" to get "per spin"? The terms "expect" and "per spin" are a little ambiguous, but here is a reasonable interpretation.

Suppose that you spin the wheel k times, and that k_i is the number of times that the outcome is m_i. Then, the total amount received is $m_1k_1 + m_2k_2 + \cdots + m_nk_n$. The amount received per spin is

$$M = \frac{m_1k_1 + m_2k_2 + \cdots + m_nk_n}{k}.$$

If the number of spins k is very large, and if we are willing to interpret probabilities as relative frequencies, it is reasonable to anticipate that m_i comes up a fraction of times that is roughly equal to p_i:

$$\frac{k_i}{k} \approx p_i, \qquad i = 1, \ldots, n.$$

Thus, the amount of money per spin that you "expect" to receive is

$$M = \frac{m_1 k_1 + m_2 k_2 + \cdots + m_n k_n}{k} \approx m_1 p_1 + m_2 p_2 + \cdots + m_n p_n.$$

Motivated by this example, we introduce the following definition.†

Expectation

We define the **expected value** (also called the **expectation** or the **mean**) of a random variable X, with PMF p_X, by

$$\mathbf{E}[X] = \sum_x x p_X(x).$$

Example 2.2. Consider two independent coin tosses, each with a 3/4 probability of a head, and let X be the number of heads obtained. This is a binomial random variable with parameters $n = 2$ and $p = 3/4$. Its PMF is

$$p_X(k) = \begin{cases} (1/4)^2, & \text{if } k = 0, \\ 2 \cdot (1/4) \cdot (3/4), & \text{if } k = 1, \\ (3/4)^2, & \text{if } k = 2, \end{cases}$$

so the mean is

$$\mathbf{E}[X] = 0 \cdot \left(\frac{1}{4}\right)^2 + 1 \cdot \left(2 \cdot \frac{1}{4} \cdot \frac{3}{4}\right) + 2 \cdot \left(\frac{3}{4}\right)^2 = \frac{24}{16} = \frac{3}{2}.$$

† When dealing with random variables that take a countably infinite number of values, one has to deal with the possibility that the infinite sum $\sum_x x p_X(x)$ is not well-defined. More concretely, we will say that the expectation is well-defined if $\sum_x |x| p_X(x) < \infty$. In this case, it is known that the infinite sum $\sum_x x p_X(x)$ converges to a finite value that is independent of the order in which the various terms are summed.

For an example where the expectation is not well-defined, consider a random variable X that takes the value 2^k with probability 2^{-k}, for $k = 1, 2, \ldots$. For a more subtle example, consider a random variable X that takes the values 2^k and -2^k with probability 2^{-k}, for $k = 2, 3, \ldots$. The expectation is again undefined, even though the PMF is symmetric around zero and one might be tempted to say that $\mathbf{E}[X]$ is zero.

Throughout this book, in the absence of an indication to the contrary, we implicitly assume that the expected value of the random variables of interest is well-defined.

It is useful to view the mean of X as a "representative" value of X, which lies somewhere in the middle of its range. We can make this statement more precise, by viewing the mean as the **center of gravity** of the PMF, in the sense explained in Fig. 2.7. In particular, if the PMF is symmetric around a certain point, that point must be equal to the mean.

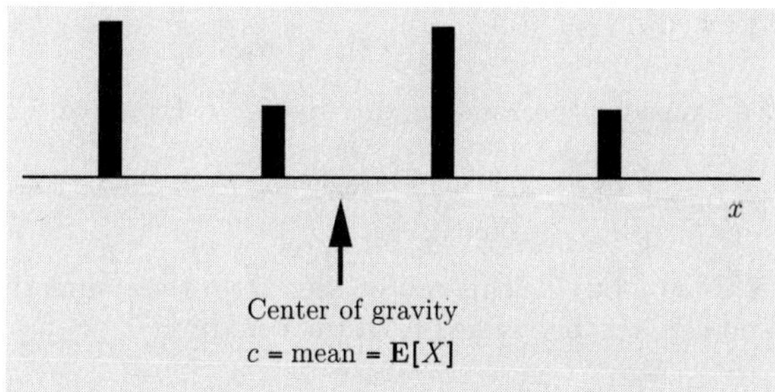

Figure 2.7: Interpretation of the mean as a center of gravity. Given a bar with a weight $p_X(x)$ placed at each point x with $p_X(x) > 0$, the center of gravity c is the point at which the sum of the torques from the weights to its left is equal to the sum of the torques from the weights to its right:

$$\sum_x (x - c)p_X(x) = 0.$$

Thus, $c = \sum_x x p_X(x)$, i.e., the center of gravity is equal to the mean $\mathbf{E}[X]$.

Variance, Moments, and the Expected Value Rule

Besides the mean, there are several other quantities that we can associate with a random variable and its PMF. For example, we define the **2nd moment** of the random variable X as the expected value of the random variable X^2. More generally, we define the **nth moment** as $\mathbf{E}[X^n]$, the expected value of the random variable X^n. With this terminology, the 1st moment of X is just the mean.

The most important quantity associated with a random variable X, other than the mean, is its **variance**, which is denoted by $\text{var}(X)$ and is defined as the expected value of the random variable $\left(X - \mathbf{E}[X]\right)^2$, i.e.,

$$\text{var}(X) = \mathbf{E}\left[\left(X - \mathbf{E}[X]\right)^2\right].$$

Since $\left(X - \mathbf{E}[X]\right)^2$ can only take nonnegative values, the variance is always nonnegative.

The variance provides a measure of dispersion of X around its mean. Another measure of dispersion is the **standard deviation** of X, which is defined as the square root of the variance and is denoted by σ_X:

$$\sigma_X = \sqrt{\text{var}(X)}.$$

The standard deviation is often easier to interpret because it has the same units as X. For example, if X measures length in meters, the units of variance are square meters, while the units of the standard deviation are meters.

One way to calculate $\text{var}(X)$, is to use the definition of expected value, after calculating the PMF of the random variable $\big(X - \mathbf{E}[X]\big)^2$. This latter random variable is a function of X, and its PMF can be obtained in the manner discussed in the preceding section.

Example 2.3. Consider the random variable X of Example 2.1, which has the PMF

$$p_X(x) = \begin{cases} 1/9, & \text{if } x \text{ is an integer in the range } [-4, 4], \\ 0, & \text{otherwise.} \end{cases}$$

The mean $\mathbf{E}[X]$ is equal to 0. This can be seen from the symmetry of the PMF of X around 0, and can also be verified from the definition:

$$\mathbf{E}[X] = \sum_x x p_X(x) = \frac{1}{9} \sum_{x=-4}^{4} x = 0.$$

Let $Z = \big(X - \mathbf{E}[X]\big)^2 = X^2$. As in Example 2.1, we have

$$p_Z(z) = \begin{cases} 2/9, & \text{if } z = 1, 4, 9, 16, \\ 1/9, & \text{if } z = 0, \\ 0, & \text{otherwise.} \end{cases}$$

The variance of X is then obtained by

$$\text{var}(X) = \mathbf{E}[Z] = \sum_z z p_Z(z) = 0 \cdot \frac{1}{9} + 1 \cdot \frac{2}{9} + 4 \cdot \frac{2}{9} + 9 \cdot \frac{2}{9} + 16 \cdot \frac{2}{9} = \frac{60}{9}.$$

It turns out that there is an easier method to calculate $\text{var}(X)$, which uses the PMF of X but **does not require the PMF of** $\big(X - \mathbf{E}[X]\big)^2$. This method is based on the following rule.

Expected Value Rule for Functions of Random Variables

Let X be a random variable with PMF p_X, and let $g(X)$ be a function of X. Then, the expected value of the random variable $g(X)$ is given by

$$\mathbf{E}\big[g(X)\big] = \sum_x g(x) p_X(x).$$

To verify this rule, we let $Y = g(X)$ and use the formula

$$p_Y(y) = \sum_{\{x \mid g(x) = y\}} p_X(x)$$

derived in the preceding section. We have

$$\begin{aligned}
\mathbf{E}\big[g(X)\big] &= \mathbf{E}[Y] \\
&= \sum_y y p_Y(y) \\
&= \sum_y y \sum_{\{x \,|\, g(x)=y\}} p_X(x) \\
&= \sum_y \sum_{\{x \,|\, g(x)=y\}} y p_X(x) \\
&= \sum_y \sum_{\{x \,|\, g(x)=y\}} g(x) p_X(x) \\
&= \sum_x g(x) p_X(x).
\end{aligned}$$

Using the expected value rule, we can write the variance of X as

$$\text{var}(X) = \mathbf{E}\Big[\big(X - \mathbf{E}[X]\big)^2\Big] = \sum_x \big(x - \mathbf{E}[X]\big)^2 p_X(x).$$

Similarly, the nth moment is given by

$$\mathbf{E}[X^n] = \sum_x x^n p_X(x),$$

and there is no need to calculate the PMF of X^n.

Example 2.3 (continued). For the random variable X with PMF

$$p_X(x) = \begin{cases} 1/9, & \text{if } x \text{ is an integer in the range } [-4, 4], \\ 0, & \text{otherwise,} \end{cases}$$

we have

$$\begin{aligned}
\text{var}(X) &= \mathbf{E}\Big[\big(X - \mathbf{E}[X]\big)^2\Big] \\
&= \sum_x \big(x - \mathbf{E}[X]\big)^2 p_X(x) \\
&= \frac{1}{9} \sum_{x=-4}^{4} x^2 \qquad\qquad (\text{since } \mathbf{E}[X] = 0) \\
&= \frac{1}{9}(16 + 9 + 4 + 1 + 0 + 1 + 4 + 9 + 16) \\
&= \frac{60}{9},
\end{aligned}$$

which is consistent with the result obtained earlier.

As we have already noted, the variance is always nonnegative, but could it be zero? Since every term in the formula $\sum_x \big(x - \mathbf{E}[X]\big)^2 p_X(x)$ for the variance is nonnegative, the sum is zero if and only if $\big(x - \mathbf{E}[X]\big)^2 p_X(x) = 0$ for every x. This condition implies that for any x with $p_X(x) > 0$, we must have $x = \mathbf{E}[X]$ and the random variable X is not really "random": its value is equal to the mean $\mathbf{E}[X]$, with probability 1.

Variance

The variance $\text{var}(X)$ of a random variable X is defined by

$$\text{var}(X) = \mathbf{E}\Big[\big(X - \mathbf{E}[X]\big)^2\Big],$$

and can be calculated as

$$\text{var}(X) = \sum_x \big(x - \mathbf{E}[X]\big)^2 p_X(x).$$

It is always nonnegative. Its square root is denoted by σ_X and is called the **standard deviation**.

Properties of Mean and Variance

We will now use the expected value rule in order to derive some important properties of the mean and the variance. We start with a random variable X and define a new random variable Y, of the form

$$Y = aX + b,$$

where a and b are given scalars. Let us derive the mean and the variance of the linear function Y. We have

$$\mathbf{E}[Y] = \sum_x (ax + b)p_X(x) = a\sum_x xp_X(x) + b\sum_x p_X(x) = a\mathbf{E}[X] + b.$$

Furthermore,

$$\begin{aligned}
\text{var}(Y) &= \sum_x \big(ax + b - \mathbf{E}[aX + b]\big)^2 p_X(x) \\
&= \sum_x \big(ax + b - a\mathbf{E}[X] - b\big)^2 p_X(x) \\
&= a^2 \sum_x \big(x - \mathbf{E}[X]\big)^2 p_X(x) \\
&= a^2\, \text{var}(X).
\end{aligned}$$

Mean and Variance of a Linear Function of a Random Variable

Let X be a random variable and let

$$Y = aX + b,$$

where a and b are given scalars. Then,

$$\mathbf{E}[Y] = a\mathbf{E}[X] + b, \qquad \text{var}(Y) = a^2 \, \text{var}(X).$$

Let us also give a convenient alternative formula for the variance of a random variable X.

Variance in Terms of Moments Expression

$$\text{var}(X) = \mathbf{E}[X^2] - \big(\mathbf{E}[X]\big)^2.$$

This expression is verified as follows:

$$\begin{aligned}
\text{var}(X) &= \sum_x \big(x - \mathbf{E}[X]\big)^2 p_X(x) \\
&= \sum_x \Big(x^2 - 2x\mathbf{E}[X] + \big(\mathbf{E}[X]\big)^2\Big) p_X(x) \\
&= \sum_x x^2 p_X(x) - 2\mathbf{E}[X] \sum_x x p_X(x) + \big(\mathbf{E}[X]\big)^2 \sum_x p_X(x) \\
&= \mathbf{E}[X^2] - 2\big(\mathbf{E}[X]\big)^2 + \big(\mathbf{E}[X]\big)^2 \\
&= \mathbf{E}[X^2] - \big(\mathbf{E}[X]\big)^2.
\end{aligned}$$

We finally illustrate by example a common pitfall: unless $g(X)$ is a linear function, it is not generally true that $\mathbf{E}\big[g(X)\big]$ is equal to $g\big(\mathbf{E}[X]\big)$.

Example 2.4. Average Speed Versus Average Time. If the weather is good (which happens with probability 0.6), Alice walks the 2 miles to class at a speed of $V = 5$ miles per hour, and otherwise rides her motorcycle at a speed of $V = 30$ miles per hour. What is the mean of the time T to get to class?

A correct way to solve the problem is to first derive the PMF of T,

$$p_T(t) = \begin{cases} 0.6, & \text{if } t = 2/5 \text{ hours}, \\ 0.4, & \text{if } t = 2/30 \text{ hours}, \end{cases}$$

and then calculate its mean by

$$\mathbf{E}[T] = 0.6 \cdot \frac{2}{5} + 0.4 \cdot \frac{2}{30} = \frac{4}{15} \text{ hours}.$$

However, it is wrong to calculate the mean of the speed V,

$$\mathbf{E}[V] = 0.6 \cdot 5 + 0.4 \cdot 30 = 15 \text{ miles per hour,}$$

and then claim that the mean of the time T is

$$\frac{2}{\mathbf{E}[V]} = \frac{2}{15} \text{ hours.}$$

To summarize, in this example we have

$$T = \frac{2}{V}, \qquad \text{and } \mathbf{E}[T] = \mathbf{E}\left[\frac{2}{V}\right] \neq \frac{2}{\mathbf{E}[V]}.$$

Mean and Variance of Some Common Random Variables

We will now derive formulas for the mean and the variance of a few important random variables. These formulas will be used repeatedly in a variety of contexts throughout the text.

Example 2.5. Mean and Variance of the Bernoulli. Consider the experiment of tossing a coin, which comes up a head with probability p and a tail with probability $1 - p$, and the Bernoulli random variable X with PMF

$$p_X(k) = \begin{cases} p, & \text{if } k = 1, \\ 1 - p, & \text{if } k = 0. \end{cases}$$

The mean, second moment, and variance of X are given by the following calculations:

$$\begin{aligned} \mathbf{E}[X] &= 1 \cdot p + 0 \cdot (1 - p) = p, \\ \mathbf{E}[X^2] &= 1^2 \cdot p + 0 \cdot (1 - p) = p, \\ \text{var}(X) &= \mathbf{E}[X^2] - \big(\mathbf{E}[X]\big)^2 = p - p^2 = p(1 - p). \end{aligned}$$

Example 2.6. Discrete Uniform Random Variable. What is the mean and variance associated with a roll of a fair six-sided die? If we view the result of the roll as a random variable X, its PMF is

$$p_X(k) = \begin{cases} 1/6, & \text{if } k = 1, 2, 3, 4, 5, 6, \\ 0, & \text{otherwise.} \end{cases}$$

Since the PMF is symmetric around 3.5, we conclude that $\mathbf{E}[X] = 3.5$. Regarding the variance, we have

$$\begin{aligned} \text{var}(X) &= \mathbf{E}[X^2] - \big(\mathbf{E}[X]\big)^2 \\ &= \frac{1}{6}(1^2 + 2^2 + 3^2 + 4^2 + 5^2 + 6^2) - (3.5)^2, \end{aligned}$$

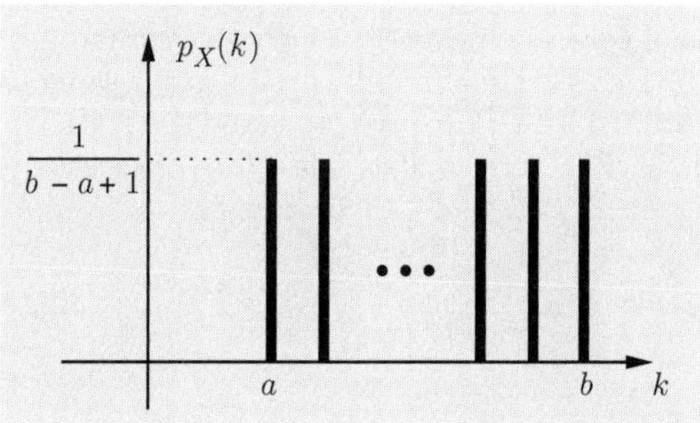

Figure 2.8: PMF of the discrete random variable that is uniformly distributed between two integers a and b. Its mean and variance are

$$\mathbf{E}[X] = \frac{a+b}{2}, \qquad \text{var}(X) = \frac{(b-a)(b-a+2)}{12}.$$

which yields $\text{var}(X) = 35/12$.

The above random variable is a special case of a **discrete uniformly distributed** random variable (or **discrete uniform** for short), which by definition, takes one out of a range of contiguous integer values, with equal probability. More precisely, a discrete uniform random variable has a PMF of the form

$$p_X(k) = \begin{cases} \dfrac{1}{b-a+1}, & \text{if } k = a, a+1, \ldots, b, \\ 0, & \text{otherwise,} \end{cases}$$

where a and b are two integers with $a < b$; see Fig. 2.8.

The mean is

$$\mathbf{E}[X] = \frac{a+b}{2},$$

as can be seen by inspection, since the PMF is symmetric around $(a+b)/2$. To calculate the variance of X, we first consider the simpler case where $a = 1$ and $b = n$. It can be verified by induction on n that

$$\mathbf{E}[X^2] = \frac{1}{n}\sum_{k=1}^{n} k^2 = \frac{1}{6}(n+1)(2n+1).$$

We leave the verification of this as an exercise for the reader. The variance can now be obtained in terms of the first and second moments

$$\begin{aligned} \text{var}(X) &= \mathbf{E}[X^2] - \big(\mathbf{E}[X]\big)^2 \\ &= \frac{1}{6}(n+1)(2n+1) - \frac{1}{4}(n+1)^2 \\ &= \frac{1}{12}(n+1)(4n+2-3n-3) \\ &= \frac{n^2-1}{12}. \end{aligned}$$

For the case of general integers a and b, we note that a random variable which is uniformly distributed over the interval $[a, b]$ has the same variance as one which is uniformly distributed over $[1, b-a+1]$, since the PMF of the second is just a shifted version of the PMF of the first. Therefore, the desired variance is given by the above formula with $n = b - a + 1$, which yields

$$\text{var}(X) = \frac{(b-a+1)^2 - 1}{12} = \frac{(b-a)(b-a+2)}{12}.$$

Example 2.7. The Mean of the Poisson. The mean of the Poisson PMF

$$p_X(k) = e^{-\lambda}\frac{\lambda^k}{k!}, \qquad k = 0, 1, 2, \ldots,$$

can be calculated is follows:

$$\begin{aligned}
\mathbf{E}[X] &= \sum_{k=0}^{\infty} k e^{-\lambda}\frac{\lambda^k}{k!} \\
&= \sum_{k=1}^{\infty} k e^{-\lambda}\frac{\lambda^k}{k!} && \text{(the } k = 0 \text{ term is zero)} \\
&= \lambda \sum_{k=1}^{\infty} e^{-\lambda}\frac{\lambda^{k-1}}{(k-1)!} \\
&= \lambda \sum_{m=0}^{\infty} e^{-\lambda}\frac{\lambda^m}{m!} && \text{(let } m = k - 1\text{)} \\
&= \lambda.
\end{aligned}$$

The last equality is obtained by noting that

$$\sum_{m=0}^{\infty} e^{-\lambda}\frac{\lambda^m}{m!} = \sum_{m=0}^{\infty} p_X(m) = 1$$

is the normalization property for the Poisson PMF.

A similar calculation shows that the variance of a Poisson random variable is also λ; see Example 2.20 in Section 2.7. We will derive this fact in a number of different ways in later chapters.

Decision Making Using Expected Values

Expected values often provide a convenient vehicle for optimizing the choice between several candidate decisions that result in random rewards. If we view the expected reward of a decision as its "average payoff over a large number of trials," it is reasonable to choose a decision with maximum expected reward. The following is an example.

Example 2.8. The Quiz Problem. This example, when generalized appropriately, is a prototypical formulation of the problem of optimal sequencing of a collection of tasks with uncertain rewards.

Consider a quiz game where a person is given two questions and must decide which one to answer first. Question 1 will be answered correctly with probability 0.8, and the person will then receive as prize \$100, while question 2 will be answered correctly with probability 0.5, and the person will then receive as prize \$200. If the first question attempted is answered incorrectly, the quiz terminates, i.e., the person is not allowed to attempt the second question. If the first question is answered correctly, the person is allowed to attempt the second question. Which question should be answered first to maximize the expected value of the total prize money received?

The answer is not obvious because there is a tradeoff: attempting first the more valuable but also more difficult question 2 carries the risk of never getting a chance to attempt the easier question 1. Let us view the total prize money received as a random variable X, and calculate the expected value $\mathbf{E}[X]$ under the two possible question orders (cf. Fig. 2.9):

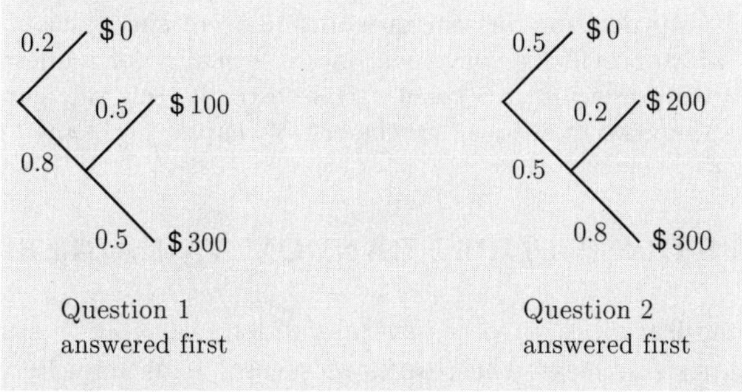

Figure 2.9: Sequential description of the sample space of the quiz problem for the two cases where we answer question 1 or question 2 first.

(a) *Answer question 1 first*: Then the PMF of X is (cf. the left side of Fig. 2.9)

$$p_X(0) = 0.2, \qquad p_X(100) = 0.8 \cdot 0.5, \qquad p_X(300) = 0.8 \cdot 0.5,$$

and we have

$$\mathbf{E}[X] = 0.8 \cdot 0.5 \cdot 100 + 0.8 \cdot 0.5 \cdot 300 = \$160.$$

(b) *Answer question 2 first*: Then the PMF of X is (cf. the right side of Fig. 2.9)

$$p_X(0) = 0.5, \qquad p_X(200) = 0.5 \cdot 0.2, \qquad p_X(300) = 0.5 \cdot 0.8,$$

and we have

$$\mathbf{E}[X] = 0.5 \cdot 0.2 \cdot 200 + 0.5 \cdot 0.8 \cdot 300 = \$140.$$

Thus, it is preferable to attempt the easier question 1 first.

Let us now generalize the analysis. Denote by p_1 and p_2 the probabilities of correctly answering questions 1 and 2, respectively, and by v_1 and v_2 the corresponding prizes. If question 1 is answered first, we have

$$\mathbf{E}[X] = p_1(1-p_2)v_1 + p_1p_2(v_1+v_2) = p_1v_1 + p_1p_2v_2,$$

while if question 2 is answered first, we have

$$\mathbf{E}[X] = p_2(1-p_1)v_2 + p_2p_1(v_2+v_1) = p_2v_2 + p_2p_1v_1.$$

It is thus optimal to answer question 1 first if and only if

$$p_1v_1 + p_1p_2v_2 \geq p_2v_2 + p_2p_1v_1,$$

or equivalently, if

$$\frac{p_1v_1}{1-p_1} \geq \frac{p_2v_2}{1-p_2}.$$

Therefore, it is optimal to order the questions in decreasing value of the expression $pv/(1-p)$, which provides a convenient index of quality for a question with probability of correct answer p and value v. Interestingly, this rule generalizes to the case of more than two questions (see the end-of-chapter problems).

2.5 JOINT PMFS OF MULTIPLE RANDOM VARIABLES

Probabilistic models often involve several random variables. For example, in a medical diagnosis context, the results of several tests may be significant, or in a networking context, the workloads of several routers may be of interest. All of these random variables are associated with the same experiment, sample space, and probability law, and their values may relate in interesting ways. This motivates us to consider probabilities of events involving simultaneously several random variables. In this section, we will extend the concepts of PMF and expectation developed so far to multiple random variables. Later on, we will also develop notions of conditioning and independence that closely parallel the ideas discussed in Chapter 1.

Consider two discrete random variables X and Y associated with the same experiment. The probabilities of the values that X and Y can take are captured by the **joint PMF** of X and Y, denoted $p_{X,Y}$. In particular, if (x, y) is a pair of possible values of X and Y, the probability mass of (x, y) is the probability of the event $\{X = x, Y = y\}$:

$$p_{X,Y}(x,y) = \mathbf{P}(X = x, Y = y).$$

Here and elsewhere, we use the abbreviated notation $\mathbf{P}(X = x, Y = y)$ instead of the more precise notations $\mathbf{P}(\{X = x\} \cap \{Y = y\})$ or $\mathbf{P}(X = x \text{ and } Y = y)$.

The joint PMF determines the probability of any event that can be specified in terms of the random variables X and Y. For example if A is the set of all pairs (x, y) that have a certain property, then

$$\mathbf{P}\big((X, Y) \in A\big) = \sum_{(x,y)\in A} p_{X,Y}(x, y).$$

In fact, we can calculate the PMFs of X and Y by using the formulas

$$p_X(x) = \sum_y p_{X,Y}(x, y), \qquad p_Y(y) = \sum_x p_{X,Y}(x, y).$$

The formula for $p_X(x)$ can be verified using the calculation

$$\begin{aligned} p_X(x) &= \mathbf{P}(X = x) \\ &= \sum_y \mathbf{P}(X = x, Y = y) \\ &= \sum_y p_{X,Y}(x, y), \end{aligned}$$

where the second equality follows by noting that the event $\{X = x\}$ is the union of the disjoint events $\{X = x, Y = y\}$ as y ranges over all the different values of Y. The formula for $p_Y(y)$ is verified similarly. We sometimes refer to p_X and p_Y as the **marginal** PMFs, to distinguish them from the joint PMF.

We can calculate the marginal PMFs from the joint PMF by using the **tabular method**. Here, the joint PMF of X and Y is arranged in a two-dimensional table, and *the marginal PMF of X or Y at a given value is obtained by adding the table entries along a corresponding column or row*, respectively. This method is illustrated by the following example and Fig. 2.10.

Example 2.9. Consider two random variables, X and Y, described by the joint PMF shown in Fig. 2.10. The marginal PMFs are calculated by adding the table entries along the columns (for the marginal PMF of X) and along the rows (for the marginal PMF of Y), as indicated.

Functions of Multiple Random Variables

When there are multiple random variables of interest, it is possible to generate new random variables by considering functions involving several of these random variables. In particular, a function $Z = g(X, Y)$ of the random variables X and Y defines another random variable. Its PMF can be calculated from the joint PMF $p_{X,Y}$ according to

$$p_Z(z) = \sum_{\{(x,y)\,|\,g(x,y)=z\}} p_{X,Y}(x, y).$$

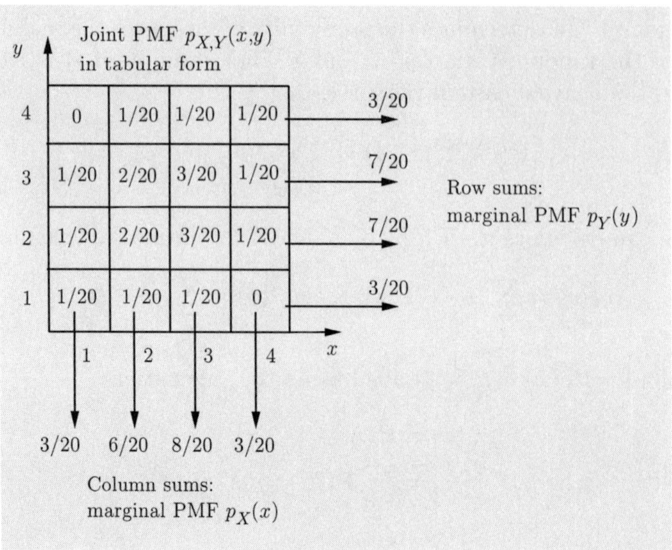

Figure 2.10: Illustration of the tabular method for calculating the marginal PMFs from the joint PMF in Example 2.9. The joint PMF is represented by the table, where the number in each square (x, y) gives the value of $p_{X,Y}(x, y)$. To calculate the marginal PMF $p_X(x)$ for a given value of x, we add the numbers in the column corresponding to x. For example $p_X(2) = 6/20$. Similarly, to calculate the marginal PMF $p_Y(y)$ for a given value of y, we add the numbers in the row corresponding to y. For example $p_Y(2) = 7/20$.

Furthermore, the expected value rule for functions naturally extends and takes the form

$$\mathbf{E}\big[g(X,Y)\big] = \sum_x \sum_y g(x,y)p_{X,Y}(x,y).$$

The verification of this is very similar to the earlier case of a function of a single random variable. In the special case where g is linear and of the form $aX+bY+c$, where a, b, and c are given scalars, we have

$$\mathbf{E}[aX + bY + c] = a\mathbf{E}[X] + b\mathbf{E}[Y] + c.$$

Example 2.9 (continued). Consider again the random variables X and Y whose joint PMF is given in Fig. 2.10, and a new random variable Z defined by

$$Z = X + 2Y.$$

The PMF of Z can be calculated using the formula

$$p_Z(z) = \sum_{\{(x,y)\,|\,x+2y=z\}} p_{X,Y}(x,y),$$

and we have, using the PMF given in Fig. 2.10,

$$p_Z(3) = \frac{1}{20}, \quad p_Z(4) = \frac{1}{20}, \quad p_Z(5) = \frac{2}{20}, \quad p_Z(6) = \frac{2}{20}, \quad p_Z(7) = \frac{4}{20},$$

$$p_Z(8) = \frac{3}{20}, \quad p_Z(9) = \frac{3}{20}, \quad p_Z(10) = \frac{2}{20}, \quad p_Z(11) = \frac{1}{20}, \quad p_Z(12) = \frac{1}{20}.$$

The expected value of Z can be obtained from its PMF:

$$\begin{aligned}\mathbf{E}[Z] &= \sum z p_Z(z) \\ &= 3 \cdot \frac{1}{20} + 4 \cdot \frac{1}{20} + 5 \cdot \frac{2}{20} + 6 \cdot \frac{2}{20} + 7 \cdot \frac{4}{20} \\ &\quad + 8 \cdot \frac{3}{20} + 9 \cdot \frac{3}{20} + 10 \cdot \frac{2}{20} + 11 \cdot \frac{1}{20} + 12 \cdot \frac{1}{20} \\ &= 7.55.\end{aligned}$$

Alternatively, we can obtain $\mathbf{E}[Z]$ using the formula

$$\mathbf{E}[Z] = \mathbf{E}[X] + 2\mathbf{E}[Y].$$

From the marginal PMFs, given in Fig. 2.10, we have

$$\mathbf{E}[X] = 1 \cdot \frac{3}{20} + 2 \cdot \frac{6}{20} + 3 \cdot \frac{8}{20} + 4 \cdot \frac{3}{20} = \frac{51}{20},$$

$$\mathbf{E}[Y] = 1 \cdot \frac{3}{20} + 2 \cdot \frac{7}{20} + 3 \cdot \frac{7}{20} + 4 \cdot \frac{3}{20} = \frac{50}{20},$$

so

$$\mathbf{E}[Z] = \frac{51}{20} + 2 \cdot \frac{50}{20} = 7.55.$$

More than Two Random Variables

The joint PMF of three random variables X, Y, and Z is defined in analogy with the above as

$$p_{X,Y,Z}(x, y, z) = \mathbf{P}(X = x, Y = y, Z = z),$$

for all possible triplets of numerical values (x, y, z). Corresponding marginal PMFs are analogously obtained by equations such as

$$p_{X,Y}(x, y) = \sum_z p_{X,Y,Z}(x, y, z),$$

and

$$p_X(x) = \sum_y \sum_z p_{X,Y,Z}(x, y, z).$$

The expected value rule for functions is given by

$$\mathbf{E}\big[g(X,Y,Z)\big] = \sum_x \sum_y \sum_z g(x,y,z)p_{X,Y,Z}(x,y,z),$$

and if g is linear and has the form $aX + bY + cZ + d$, then

$$\mathbf{E}[aX + bY + cZ + d] = a\mathbf{E}[X] + b\mathbf{E}[Y] + c\mathbf{E}[Z] + d.$$

Furthermore, there are obvious generalizations of the above to more than three random variables. For example, for any random variables $X_1, X_2, \ldots, X_n$ and any scalars $a_1, a_2, \ldots, a_n$, we have

$$\mathbf{E}[a_1X_1 + a_2X_2 + \cdots + a_nX_n] = a_1\mathbf{E}[X_1] + a_2\mathbf{E}[X_2] + \cdots + a_n\mathbf{E}[X_n].$$

Example 2.10. Mean of the Binomial. Your probability class has 300 students and each student has probability 1/3 of getting an A, independent of any other student. What is the mean of X, the number of students that get an A? Let

$$X_i = \begin{cases} 1, & \text{if the } i\text{th student gets an A,} \\ 0, & \text{otherwise.} \end{cases}$$

Thus $X_1, X_2, \ldots, X_n$ are Bernoulli random variables with common mean $p = 1/3$. Their sum

$$X = X_1 + X_2 + \cdots + X_n$$

is the number of students that get an A. Since X is the number of "successes" in n independent trials, it is a binomial random variable with parameters n and p.

Using the linearity of X as a function of the X_i, we have

$$\mathbf{E}[X] = \sum_{i=1}^{300} \mathbf{E}[X_i] = \sum_{i=1}^{300} \frac{1}{3} = 300 \cdot \frac{1}{3} = 100.$$

If we repeat this calculation for a general number of students n and probability of A equal to p, we obtain

$$\mathbf{E}[X] = \sum_{i=1}^{n} \mathbf{E}[X_i] = \sum_{i=1}^{n} p = np.$$

Example 2.11. The Hat Problem. Suppose that n people throw their hats in a box and then each picks one hat at random. (Each hat can be picked by only one person, and each assignment of hats to persons is equally likely.) What is the expected value of X, the number of people that get back their own hat?

For the ith person, we introduce a random variable X_i that takes the value 1 if the person selects his/her own hat, and takes the value 0 otherwise. Since $\mathbf{P}(X_i = 1) = 1/n$ and $\mathbf{P}(X_i = 0) = 1 - 1/n$, the mean of X_i is

$$\mathbf{E}[X_i] = 1 \cdot \frac{1}{n} + 0 \cdot \left(1 - \frac{1}{n}\right) = \frac{1}{n}.$$

We now have

$$X = X_1 + X_2 + \cdots + X_n,$$

so that

$$\mathbf{E}[X] = \mathbf{E}[X_1] + \mathbf{E}[X_2] + \cdots + \mathbf{E}[X_n] = n \cdot \frac{1}{n} = 1.$$

Summary of Facts About Joint PMFs

Let X and Y be random variables associated with the same experiment.

- The **joint PMF** $p_{X,Y}$ of X and Y is defined by

$$p_{X,Y}(x, y) = \mathbf{P}(X = x, Y = y).$$

- The **marginal PMFs** of X and Y can be obtained from the joint PMF, using the formulas

$$p_X(x) = \sum_y p_{X,Y}(x, y), \qquad p_Y(y) = \sum_x p_{X,Y}(x, y).$$

- A function $g(X, Y)$ of X and Y defines another random variable, and

$$\mathbf{E}\big[g(X, Y)\big] = \sum_x \sum_y g(x, y) p_{X,Y}(x, y).$$

  If g is linear, of the form $aX + bY + c$, we have

$$\mathbf{E}[aX + bY + c] = a\mathbf{E}[X] + b\mathbf{E}[Y] + c.$$

- The above have natural extensions to the case where more than two random variables are involved.

2.6 CONDITIONING

Similar to our discussion in Chapter 1, conditional probabilities can be used to capture the information conveyed by various events about the different possible values of a random variable. We are thus motivated to introduce conditional PMFs, given the occurrence of a certain event or given the value of another random variable. In this section, we develop this idea and we discuss the properties of conditional PMFs. In reality though, there is not much that is new, only an elaboration of concepts that are familiar from Chapter 1, together with some new notation.

Conditioning a Random Variable on an Event

The **conditional PMF** of a random variable X, conditioned on a particular event A with $\mathbf{P}(A) > 0$, is defined by

$$p_{X|A}(x) = \mathbf{P}(X = x \,|\, A) = \frac{\mathbf{P}(\{X = x\} \cap A)}{\mathbf{P}(A)}.$$

Note that the events $\{X = x\} \cap A$ are disjoint for different values of x, their union is A, and, therefore,

$$\mathbf{P}(A) = \sum_x \mathbf{P}(\{X = x\} \cap A).$$

Combining the above two formulas, we see that

$$\sum_x p_{X|A}(x) = 1,$$

so $p_{X|A}$ is a legitimate PMF.

The conditional PMF is calculated similar to its unconditional counterpart: to obtain $p_{X|A}(x)$, we add the probabilities of the outcomes that give rise to $X = x$ **and** belong to the conditioning event A, and then normalize by dividing with $\mathbf{P}(A)$.

Example 2.12. Let X be the roll of a fair six-sided die and let A be the event that the roll is an even number. Then, by applying the preceding formula, we obtain

$$\begin{aligned} p_{X|A}(k) &= \mathbf{P}(X = k \,|\, \text{roll is even}) \\ &= \frac{\mathbf{P}(X = k \text{ and } X \text{ is even})}{\mathbf{P}(\text{roll is even})} \\ &= \begin{cases} 1/3, & \text{if } k = 2, 4, 6, \\ 0, & \text{otherwise.} \end{cases} \end{aligned}$$

Example 2.13. A student will take a certain test repeatedly, up to a maximum of n times, each time with a probability p of passing, independent of the number of previous attempts. What is the PMF of the number of attempts, given that the student passes the test?

Let A be the event that the student passes the test (with at most n attempts). We introduce the random variable X, which is the number of attempts that would be needed if an unlimited number of attempts were allowed. Then, X is a geometric random variable with parameter p, and $A = \{X \leq n\}$. We have

$$\mathbf{P}(A) = \sum_{m=1}^{n} (1 - p)^{m-1} p,$$

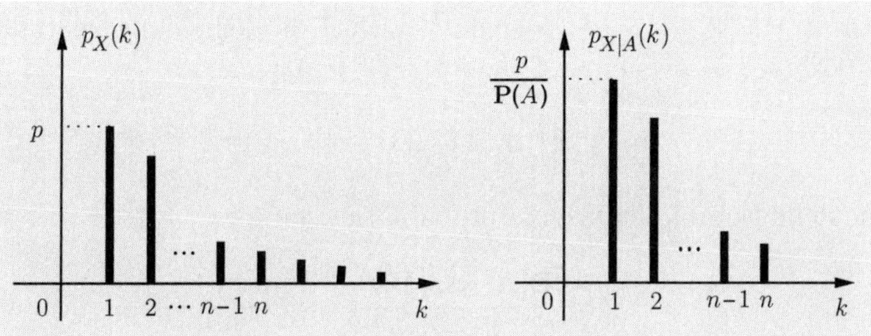

Figure 2.11: Visualization and calculation of the conditional PMF $p_{X|A}(k)$ in Example 2.13. We start with the PMF of X, we set to zero the PMF values for all k that do not belong to the conditioning event A, and we normalize the remaining values by dividing with $\mathbf{P}(A)$.

and

$$p_{X|A}(k) = \begin{cases} \dfrac{(1-p)^{k-1}p}{\displaystyle\sum_{m=1}^{n}(1-p)^{m-1}p}, & \text{if } k = 1, \ldots, n, \\ 0, & \text{otherwise,} \end{cases}$$

as illustrated in Fig. 2.11.

Figure 2.12 provides a more abstract visualization of the construction of the conditional PMF.

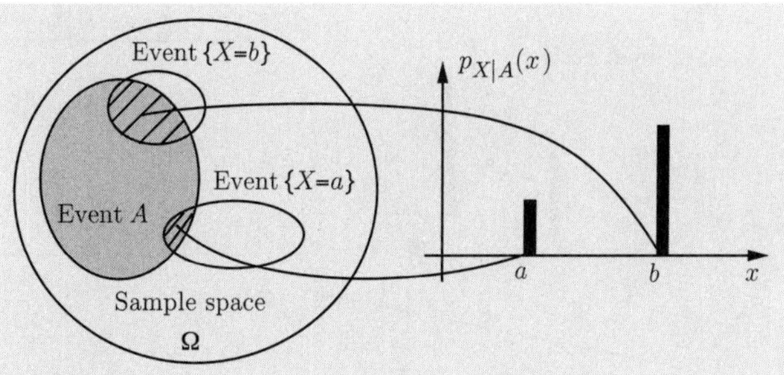

Figure 2.12: Visualization and calculation of the conditional PMF $p_{X|A}(x)$. For each x, we add the probabilities of the outcomes in the intersection $\{X = x\} \cap A$, and normalize by dividing with $\mathbf{P}(A)$.

Conditioning one Random Variable on Another

Let X and Y be two random variables associated with the same experiment. If we know that the value of Y is some particular y [with $p_Y(y) > 0$], this provides partial knowledge about the value of X. This knowledge is captured by the

conditional PMF $p_{X|Y}$ of X given Y, which is defined by specializing the definition of $p_{X|A}$ to events A of the form $\{Y = y\}$:

$$p_{X|Y}(x\,|\,y) = \mathbf{P}(X = x\,|\,Y = y).$$

Using the definition of conditional probabilities, we have

$$p_{X|Y}(x\,|\,y) = \frac{\mathbf{P}(X = x, Y = y)}{\mathbf{P}(Y = y)} = \frac{p_{X,Y}(x, y)}{p_Y(y)}.$$

Let us fix some y with $p_Y(y) > 0$, and consider $p_{X|Y}(x\,|\,y)$ as a function of x. This function is a valid PMF for X: it assigns nonnegative values to each possible x, and these values add to 1. Furthermore, this function of x has the same shape as $p_{X,Y}(x, y)$ except that it is divided by $p_Y(y)$, which enforces the normalization property

$$\sum_x p_{X|Y}(x\,|\,y) = 1.$$

Figure 2.13 provides a visualization of the conditional PMF.

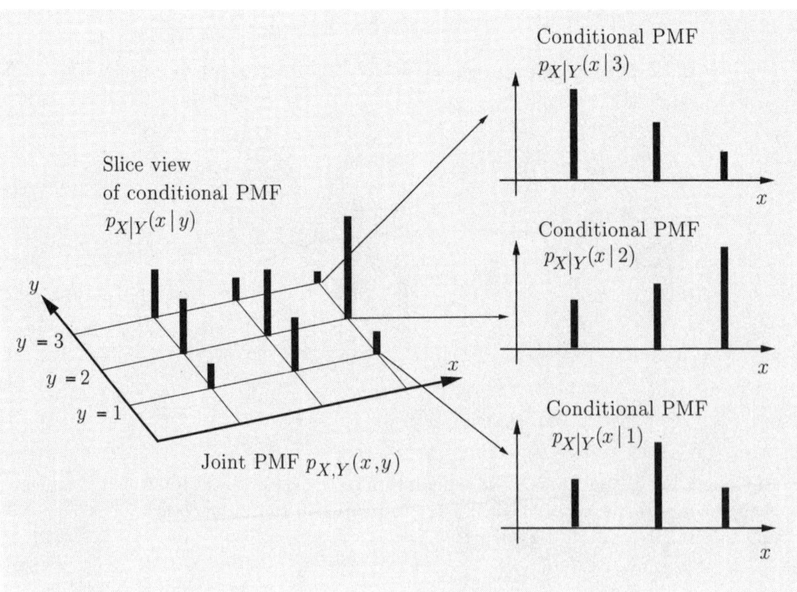

Figure 2.13: Visualization of the conditional PMF $p_{X|Y}(x\,|\,y)$. For each y, we view the joint PMF along the slice $Y = y$ and renormalize so that

$$\sum_x p_{X|Y}(x\,|\,y) = 1.$$

The conditional PMF is often convenient for the calculation of the joint PMF, using a sequential approach and the formula

$$p_{X,Y}(x,y) = p_Y(y)p_{X|Y}(x\,|\,y),$$

or its counterpart

$$p_{X,Y}(x,y) = p_X(x)p_{Y|X}(y\,|\,x).$$

This method is entirely similar to the use of the multiplication rule from Chapter 1. The following example provides an illustration.

Example 2.14. Professor May B. Right often has her facts wrong, and answers each of her students' questions incorrectly with probability 1/4, independent of other questions. In each lecture, May is asked 0, 1, or 2 questions with equal probability 1/3. Let X and Y be the number of questions May is asked and the number of questions she answers wrong in a given lecture, respectively. To construct the joint PMF $p_{X,Y}(x,y)$, we need to calculate the probability $\mathbf{P}(X = x, Y = y)$ for all combinations of values of x and y. This can be done by using a sequential description of the experiment and the multiplication rule, as shown in Fig. 2.14. For example, for the case where one question is asked and is answered wrong, we have

$$p_{X,Y}(1,1) = p_X(x)p_{Y|X}(y\,|\,x) = \frac{1}{3}\cdot\frac{1}{4} = \frac{1}{12}.$$

The joint PMF can be represented by a two-dimensional table, as shown in Fig. 2.14. It can be used to calculate the probability of any event of interest. For instance, we have

$$\begin{aligned}\mathbf{P}(\text{at least one wrong answer}) &= p_{X,Y}(1,1) + p_{X,Y}(2,1) + p_{X,Y}(2,2)\\ &= \frac{4}{48} + \frac{6}{48} + \frac{1}{48}.\end{aligned}$$

The conditional PMF can also be used to calculate the marginal PMFs. In particular, we have by using the definitions,

$$p_X(x) = \sum_y p_{X,Y}(x,y) = \sum_y p_Y(y)p_{X|Y}(x\,|\,y).$$

This formula provides a divide-and-conquer method for calculating marginal PMFs. It is in essence identical to the total probability theorem given in Chapter 1, but cast in different notation. The following example provides an illustration.

Example 2.15. Consider a transmitter that is sending messages over a computer network. Let us define the following two random variables:

X : the travel time of a given message, Y : the length of the given message.

We know the PMF of the travel time of a message that has a given length, and we know the PMF of the message length. We want to find the (unconditional) PMF of the travel time of a message.

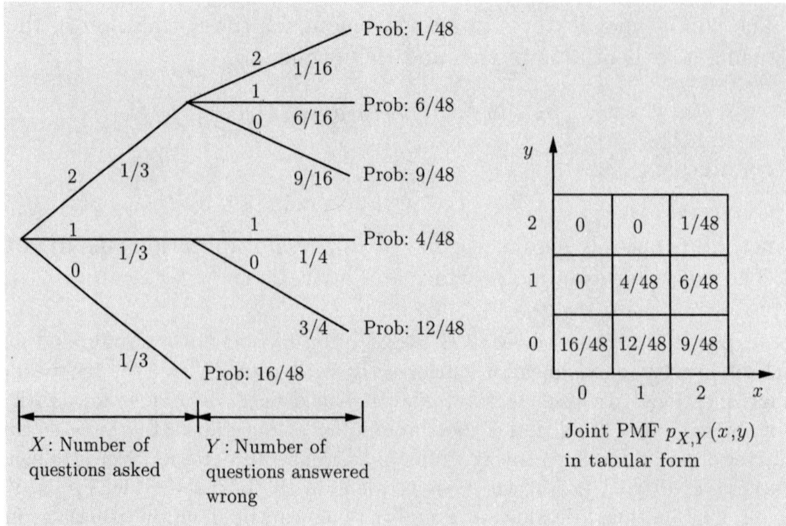

Figure 2.14: Calculation of the joint PMF $p_{X,Y}(x,y)$ in Example 2.14.

We assume that the length of a message can take two possible values: $y = 10^2$ bytes with probability 5/6, and $y = 10^4$ bytes with probability 1/6, so that

$$p_Y(y) = \begin{cases} 5/6, & \text{if } y = 10^2, \\ 1/6, & \text{if } y = 10^4. \end{cases}$$

We assume that the travel time X of the message depends on its length Y and the congestion in the network at the time of transmission. In particular, the travel time is $10^{-4}Y$ seconds with probability 1/2, $10^{-3}Y$ seconds with probability 1/3, and $10^{-2}Y$ seconds with probability 1/6. Thus, we have

$$p_{X|Y}(x\,|\,10^2) = \begin{cases} 1/2, & \text{if } x = 10^{-2}, \\ 1/3, & \text{if } x = 10^{-1}, \\ 1/6, & \text{if } x = 1, \end{cases} \qquad p_{X|Y}(x\,|\,10^4) = \begin{cases} 1/2, & \text{if } x = 1, \\ 1/3, & \text{if } x = 10, \\ 1/6, & \text{if } x = 100. \end{cases}$$

To find the PMF of X, we use the total probability formula

$$p_X(x) = \sum_y p_Y(y) p_{X|Y}(x\,|\,y).$$

We obtain

$$p_X(10^{-2}) = \frac{5}{6}\cdot\frac{1}{2}, \qquad p_X(10^{-1}) = \frac{5}{6}\cdot\frac{1}{3}, \qquad p_X(1) = \frac{5}{6}\cdot\frac{1}{6} + \frac{1}{6}\cdot\frac{1}{2},$$

$$p_X(10) = \frac{1}{6}\cdot\frac{1}{3}, \qquad p_X(100) = \frac{1}{6}\cdot\frac{1}{6}.$$

We finally note that one can define conditional PMFs involving more than two random variables, such as $p_{X,Y|Z}(x, y \,|\, z)$ or $p_{X|Y,Z}(x \,|\, y, z)$. The concepts and methods described above generalize easily.

Summary of Facts About Conditional PMFs

Let X and Y be random variables associated with the same experiment.

- Conditional PMFs are similar to ordinary PMFs, but pertain to a universe where the conditioning event is known to have occurred.
- The conditional PMF of X given an event A with $\mathbf{P}(A) > 0$, is defined by

$$p_{X|A}(x) = \mathbf{P}(X = x \,|\, A)$$

  and satisfies

$$\sum_x p_{X|A}(x) = 1.$$

- If $A_1, \ldots, A_n$ are disjoint events that form a partition of the sample space, with $\mathbf{P}(A_i) > 0$ for all i, then

$$p_X(x) = \sum_{i=1}^{n} \mathbf{P}(A_i) p_{X|A_i}(x).$$

  (This is a special case of the total probability theorem.) Furthermore, for any event B, with $\mathbf{P}(A_i \cap B) > 0$ for all i, we have

$$p_{X|B}(x) = \sum_{i=1}^{n} \mathbf{P}(A_i \,|\, B) p_{X|A_i \cap B}(x).$$

- The conditional PMF of X given $Y = y$ is related to the joint PMF by

$$p_{X,Y}(x, y) = p_Y(y) p_{X|Y}(x \,|\, y).$$

- The conditional PMF of X given Y can be used to calculate the marginal PMF of X through the formula

$$p_X(x) = \sum_y p_Y(y) p_{X|Y}(x \,|\, y).$$

- There are natural extensions of the above involving more than two random variables.

Conditional Expectation

A conditional PMF can be thought of as an ordinary PMF over a new universe determined by the conditioning event. In the same spirit, a conditional expectation is the same as an ordinary expectation, except that it refers to the new universe, and all probabilities and PMFs are replaced by their conditional counterparts. (Conditional variances can also be treated similarly.) We list the main definitions and relevant facts below.

Summary of Facts About Conditional Expectations

Let X and Y be random variables associated with the same experiment.

- The conditional expectation of X given an event A with $\mathbf{P}(A) > 0$, is defined by
$$\mathbf{E}[X \mid A] = \sum_x x p_{X|A}(x).$$
For a function $g(X)$, we have
$$\mathbf{E}\big[g(X) \mid A\big] = \sum_x g(x) p_{X|A}(x).$$
- The conditional expectation of X given a value y of Y is defined by
$$\mathbf{E}[X \mid Y = y] = \sum_x x p_{X|Y}(x \mid y).$$
- If $A_1, \ldots, A_n$ be disjoint events that form a partition of the sample space, with $\mathbf{P}(A_i) > 0$ for all i, then
$$\mathbf{E}[X] = \sum_{i=1}^n \mathbf{P}(A_i)\mathbf{E}[X \mid A_i].$$
Furthermore, for any event B with $\mathbf{P}(A_i \cap B) > 0$ for all i, we have
$$\mathbf{E}[X \mid B] = \sum_{i=1}^n \mathbf{P}(A_i \mid B)\mathbf{E}[X \mid A_i \cap B].$$
- We have
$$\mathbf{E}[X] = \sum_y p_Y(y)\mathbf{E}[X \mid Y = y].$$

The last three equalities above apply in different situations, but are essentially equivalent, and will be referred to collectively as the **total expectation**

theorem. They all follow from the total probability theorem, and express the fact that "the unconditional average can be obtained by averaging the conditional averages." They can be used to calculate the unconditional expectation $\mathbf{E}[X]$ from the conditional PMF or expectation, using a divide-and-conquer approach. To verify the first of the three equalities, we write

$$p_X(x) = \sum_{i=1}^{n} \mathbf{P}(A_i) p_{x|A_i}(x \mid A_i),$$

we multiply both sides by x, and we sum over x:

$$\begin{aligned}\mathbf{E}[X] &= \sum_x x\, p_X(x) \\ &= \sum_x x \sum_{i=1}^{n} \mathbf{P}(A_i) p_{x|A_i}(x \mid A_i) \\ &= \sum_{i=1}^{n} \mathbf{P}(A_i) \sum_x x\, p_{x|A_i}(x \mid A_i) \\ &= \sum_{i=1}^{n} \mathbf{P}(A_i) \mathbf{E}[X \mid A_i].\end{aligned}$$

The remaining two equalities are verified similarly.

Example 2.16. Messages transmitted by a computer in Boston through a data network are destined for New York with probability 0.5, for Chicago with probability 0.3, and for San Francisco with probability 0.2. The transit time X of a message is random. Its mean is 0.05 seconds if it is destined for New York, 0.1 seconds if it is destined for Chicago, and 0.3 seconds if it is destined for San Francisco. Then, $\mathbf{E}[X]$ is easily calculated using the total expectation theorem as

$$\mathbf{E}[X] = 0.5 \cdot 0.05 + 0.3 \cdot 0.1 + 0.2 \cdot 0.3 = 0.115 \text{ seconds}.$$

Example 2.17. Mean and Variance of the Geometric. You write a software program over and over, and each time there is probability p that it works correctly, independent of previous attempts. What is the mean and variance of X, the number of tries until the program works correctly?

We recognize X as a geometric random variable with PMF

$$p_X(k) = (1-p)^{k-1} p, \qquad k = 1, 2, \ldots.$$

The mean and variance of X are given by

$$\mathbf{E}[X] = \sum_{k=1}^{\infty} k(1-p)^{k-1} p, \qquad \text{var}(X) = \sum_{k=1}^{\infty} (k - \mathbf{E}[X])^2 (1-p)^{k-1} p,$$

but evaluating these infinite sums is somewhat tedious. As an alternative, we will apply the total expectation theorem, with $A_1 = \{X = 1\} = \{\text{first try is a success}\}$, $A_2 = \{X > 1\} = \{\text{first try is a failure}\}$, and end up with a much simpler calculation.

If the first try is successful, we have $X = 1$, and

$$\mathbf{E}[X \mid X = 1] = 1.$$

If the first try fails ($X > 1$), we have wasted one try, and we are back where we started. So, the expected number of remaining tries is $\mathbf{E}[X]$, and

$$\mathbf{E}[X \mid X > 1] = 1 + \mathbf{E}[X].$$

Thus,

$$\begin{aligned}\mathbf{E}[X] &= \mathbf{P}(X = 1)\mathbf{E}[X \mid X = 1] + \mathbf{P}(X > 1)\mathbf{E}[X \mid X > 1] \\ &= p + (1 - p)\big(1 + \mathbf{E}[X]\big),\end{aligned}$$

from which we obtain

$$\mathbf{E}[X] = \frac{1}{p}.$$

With similar reasoning, we also have

$$\mathbf{E}[X^2 \mid X = 1] = 1, \qquad \mathbf{E}[X^2 \mid X > 1] = \mathbf{E}\big[(1 + X)^2\big] = 1 + 2\mathbf{E}[X] + \mathbf{E}[X^2],$$

so that

$$\mathbf{E}[X^2] = p \cdot 1 + (1 - p)\big(1 + 2\mathbf{E}[X] + \mathbf{E}[X^2]\big),$$

from which we obtain

$$\mathbf{E}[X^2] = \frac{1 + 2(1 - p)\mathbf{E}[X]}{p},$$

and, using the formula $\mathbf{E}[X] = 1/p$ derived above,

$$\mathbf{E}[X^2] = \frac{2}{p^2} - \frac{1}{p}.$$

We conclude that

$$\text{var}(X) = \mathbf{E}[X^2] - \big(\mathbf{E}[X]\big)^2 = \frac{2}{p^2} - \frac{1}{p} - \frac{1}{p^2} = \frac{1 - p}{p^2}.$$

Example 2.18. The Two-Envelopes Paradox. This is a much discussed puzzle that involves a subtle mathematical point regarding conditional expectations.

You are handed two envelopes, and you are told that one of them contains m times as much money as the other, where m is an integer with $m > 1$. You open one of the envelopes and look at the amount inside. You may now keep this amount, or you may switch envelopes and keep the amount in the other envelope. What is the best strategy?

Here is a line of reasoning that argues in favor of switching. Let A be the envelope you open and B be the envelope that you may switch to. Let also x and y be the amounts in A and B, respectively. Then, as the argument goes, either $y = x/m$ or $y = mx$, with equal probability 1/2, so given x, the expected value of y is

$$\frac{1}{2} \cdot \frac{x}{m} + \frac{1}{2} \cdot mx = \frac{1}{2}\left(\frac{1}{m} + m\right)x = \frac{1+m^2}{2m}x > x,$$

since $1 + m^2 > 2m$ for $m > 1$. Therefore, you should always switch to envelope B! But then, since you should switch regardless of the amount found in A, you might as well open B to begin with; but once you do, you should switch again, etc.

There are two assumptions, both flawed to some extent, that underlie this paradoxical line of reasoning.

(a) You have no a priori knowledge about the amounts in the envelopes, so given x, the only thing you know about y is that it is either $1/m$ or m times x, and there is no reason to assume that one is more likely than the other.

(b) Given two random variables X and Y, representing monetary amounts, if

$$\mathbf{E}[Y \mid X = x] > x,$$

for all possible values x of X, then the strategy that always switches to Y yields a higher expected monetary gain.

Let us scrutinize these assumptions.

Assumption (a) is flawed because it relies on an incompletely specified probabilistic model. Indeed, in any correct model, all events, including the possible values of X and Y, must have well-defined probabilities. With such probabilistic knowledge about X and Y, the value of X may reveal a great deal of information about Y. For example, assume the following probabilistic model: someone chooses an integer dollar amount Z from a known range $[\underline{z}, \overline{z}]$ according to some distribution, places this amount in a randomly chosen envelope, and places m times this amount in the other envelope. You then choose to open one of the two envelopes (with equal probability), and look at the enclosed amount X. If X turns out to be larger than the upper range limit $\overline{z}$, you know that X is the larger of the two amounts, and hence you should not switch. On the other hand, for some other values of X, such as the lower range limit $\underline{z}$, you should switch envelopes. Thus, in this model, the choice to switch or not should depend on the value of X. Roughly speaking, if you have an idea about the range and likelihood of the values of X, you can judge whether the amount X found in A is relatively small or relatively large, and accordingly switch or not switch envelopes.

Mathematically, in a correct probabilistic model, we must have a joint PMF for the random variables X and Y, the amounts in envelopes A and B, respectively. This joint PMF is specified by introducing a PMF p_Z for the random variable Z, the minimum of the amounts in the two envelopes. Then, for all z,

$$p_{X,Y}(mz, z) = p_{X,Y}(z, mz) = \frac{1}{2}p_Z(z),$$

and

$$p_{X,Y}(x, y) = 0,$$

for every (x, y) that is not of the form (mz, z) or (z, mz). With this specification of $p_{X,Y}(x, y)$, and given that $X = x$, one can use the rule

$$\text{switch if and only if } \mathbf{E}[Y \mid X = x] > x.$$

According to this decision rule, one may or may not switch envelopes, depending on the value of X, as indicated earlier.

Is it true that, with the above described probabilistic model and decision rule, you should be switching for some values x but not for others? Ordinarily yes, as illustrated from the earlier example where Z takes values in a bounded range. However, here is a devilish example where because of a subtle mathematical quirk, you will always switch!

A fair coin is tossed until it comes up heads. Let N be the number of tosses. Then, m^N dollars are placed in one envelope and m^{N-1} dollars are placed in the other. Let X be the amount in the envelope you open (envelope A), and let Y be the amount in the other envelope (envelope B).

Now, if A contains \$1, clearly B contains \$$m$, so you should switch envelopes. If, on the other hand, A contains m^n dollars, where $n > 0$, then B contains either m^{n-1} or m^{n+1} dollars. Since N has a geometric PMF, we have

$$\frac{\mathbf{P}(Y = m^{n+1} \mid X = m^n)}{\mathbf{P}(Y = m^{n-1} \mid X = m^n)} = \frac{\mathbf{P}(Y = m^{n+1},\, X = m^n)}{\mathbf{P}(Y = m^{n-1},\, X = m^n)} = \frac{\mathbf{P}(N = n+1)}{\mathbf{P}(N = n)} = \frac{1}{2}.$$

Thus

$$\mathbf{P}(Y = m^{n-1} \mid X = m^n) = \frac{2}{3}, \qquad \mathbf{P}(Y = m^{n+1} \mid X = m^n) = \frac{1}{3},$$

and

$$\mathbf{E}[\text{amount in B} \mid X = m^n] = \frac{2}{3} \cdot m^{n-1} + \frac{1}{3} \cdot m^{n+1} = \frac{2 + m^2}{3m} \cdot m^n.$$

We have $(2 + m^2)/3m > 1$ if and only if $m^2 - 3m + 2 > 0$ or $(m - 1)(m - 2) > 0$. Thus if $m > 2$, then

$$\mathbf{E}[\text{amount in B} \mid X = m^n] > m^n,$$

and to maximize the expected monetary gain you should always switch to B!

What is happening in this example is that you switch for all values of x because

$$\mathbf{E}[Y \mid X = x] > x, \qquad \text{for all } x.$$

A naive application of the total expectation theorem might seem to indicate that $\mathbf{E}[Y] > \mathbf{E}[X]$. However, this cannot be true, since X and Y have identical PMFs. Instead, we have

$$\mathbf{E}[Y] = \mathbf{E}[X] = \infty,$$

which is not necessarily inconsistent with the relation $\mathbf{E}[Y \mid X = x] > x$ for all x.

The conclusion is that the decision rule that switches if and only if $\mathbf{E}[Y \mid X = x] > x$ does not improve the expected monetary gain in the case where $\mathbf{E}[Y] = \mathbf{E}[X] = \infty$, and the apparent paradox is resolved.

2.7 INDEPENDENCE

We now discuss concepts of independence related to random variables. These are analogous to the concepts of independence between events (cf. Chapter 1). They are developed by simply introducing suitable events involving the possible values of various random variables, and by considering the independence of these events.

Independence of a Random Variable from an Event

The independence of a random variable from an event is similar to the independence of two events. The idea is that knowing the occurrence of the conditioning event provides no new information on the value of the random variable. More formally, we say that the random variable X is **independent of the event** A if

$$\mathbf{P}(X = x \text{ and } A) = \mathbf{P}(X = x)\mathbf{P}(A) = p_X(x)\mathbf{P}(A), \qquad \text{for all } x,$$

which is the same as requiring that the two events $\{X = x\}$ and A be independent, for any choice x. From the definition of the conditional PMF, we have

$$\mathbf{P}(X = x \text{ and } A) = p_{X|A}(x)\mathbf{P}(A),$$

so that as long as $\mathbf{P}(A) > 0$, independence is the same as the condition

$$p_{X|A}(x) = p_X(x), \qquad \text{for all } x.$$

Example 2.19. Consider two independent tosses of a fair coin. Let X be the number of heads and let A be the event that the number of heads is even. The (unconditional) PMF of X is

$$p_X(x) = \begin{cases} 1/4, & \text{if } x = 0, \\ 1/2, & \text{if } x = 1, \\ 1/4, & \text{if } x = 2, \end{cases}$$

and $\mathbf{P}(A) = 1/2$. The conditional PMF is obtained from the definition $p_{X|A}(x) = \mathbf{P}\big(X = x \text{ and } A\big)/\mathbf{P}(A)$:

$$p_{X|A}(x) = \begin{cases} 1/2, & \text{if } x = 0, \\ 0, & \text{if } x = 1, \\ 1/2, & \text{if } x = 2. \end{cases}$$

Clearly, X and A are not independent, since the PMFs p_X and $p_{X|A}$ are different. For an example of a random variable that is independent of A, consider the random variable that takes the value 0 if the first toss is a head, and the value 1 if the first toss is a tail. This is intuitively clear and can also be verified by using the definition of independence.

Independence of Random Variables

The notion of independence of two random variables is similar to the independence of a random variable from an event. We say that two **random variables** X and Y are **independent** if

$$p_{X,Y}(x,y) = p_X(x)\,p_Y(y), \qquad \text{for all } x, y.$$

This is the same as requiring that the two events $\{X = x\}$ and $\{Y = y\}$ be independent for every x and y. Finally, the formula $p_{X,Y}(x,y) = p_{X|Y}(x\,|\,y)p_Y(y)$ shows that independence is equivalent to the condition

$$p_{X|Y}(x\,|\,y) = p_X(x), \qquad \text{for all } y \text{ with } p_Y(y) > 0 \text{ and all } x.$$

Intuitively, independence means that the value of Y provides no information on the value of X.

There is a similar notion of conditional independence of two random variables, given an event A with $\mathbf{P}(A) > 0$. The conditioning event A defines a new universe and all probabilities (or PMFs) have to be replaced by their conditional counterparts. For example, X and Y are said to be **conditionally independent**, given a positive probability event A, if

$$\mathbf{P}(X = x, Y = y\,|\,A) = \mathbf{P}(X = x\,|\,A)\mathbf{P}(Y = y\,|\,A), \qquad \text{for all } x \text{ and } y,$$

or, in this chapter's notation,

$$p_{X,Y|A}(x,y) = p_{X|A}(x)p_{Y|A}(y), \qquad \text{for all } x \text{ and } y.$$

Once more, this is equivalent to

$$p_{X|Y,A}(x\,|\,y) = p_{X|A}(x) \qquad \text{for all } x \text{ and } y \text{ such that } p_{Y|A}(y) > 0.$$

As in the case of events (Section 1.5), conditional independence may not imply unconditional independence and vice versa. This is illustrated by the example in Fig. 2.15.

If X and Y are independent random variables, then

$$\mathbf{E}[XY] = \mathbf{E}[X]\,\mathbf{E}[Y],$$

as shown by the following calculation:

$$\begin{aligned}
\mathbf{E}[XY] &= \sum_x \sum_y xy p_{X,Y}(x,y) \\
&= \sum_x \sum_y xy p_X(x) p_Y(y) \qquad \text{(by independence)} \\
&= \sum_x x p_X(x) \sum_y y p_Y(y) \\
&= \mathbf{E}[X]\,\mathbf{E}[Y].
\end{aligned}$$

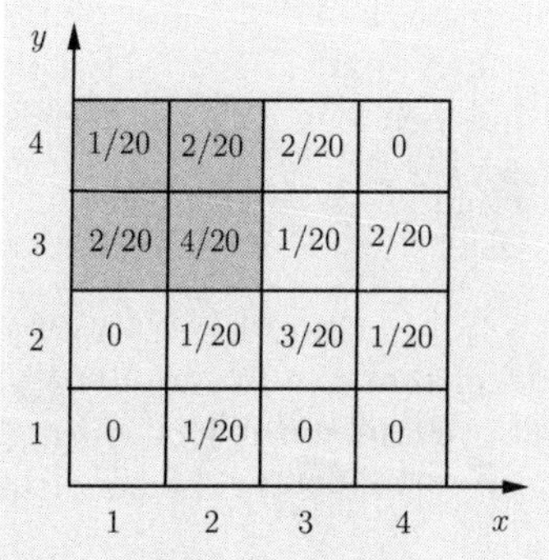

Figure 2.15: Example illustrating that conditional independence may not imply unconditional independence. For the PMF shown, the random variables X and Y are not independent. For example, we have

$$p_{X|Y}(1\,|\,1) = \mathbf{P}(X = 1\,|\,Y = 1) = 0 \neq \mathbf{P}(X = 1) = p_X(1).$$

On the other hand, conditional on the event $A = \{X \leq 2,\, Y \geq 3\}$ (the shaded set in the figure), the random variables X and Y can be seen to be independent. In particular, we have

$$p_{X|Y,A}(x\,|\,y) = \begin{cases} 1/3, & \text{if } x = 1, \\ 2/3, & \text{if } x = 2, \end{cases}$$

for both values $y = 3$ and $y = 4$.

A very similar calculation also shows that if X and Y are independent, then

$$\mathbf{E}\big[g(X)h(Y)\big] = \mathbf{E}\big[g(X)\big]\mathbf{E}\big[h(Y)\big],$$

for any functions g and h. In fact, this follows immediately once we realize that if X and Y are independent, then the same is true for $g(X)$ and $h(Y)$. This is intuitively clear and its formal verification is left as an end-of-chapter problem.

Consider now the sum $X + Y$ of two independent random variables X and Y, and let us calculate its variance. Since the variance of a random variable is unchanged when the random variable is shifted by a constant, it is convenient to work with the zero-mean random variables $\tilde{X} = X - \mathbf{E}[X]$ and $\tilde{Y} = Y - \mathbf{E}[Y]$. We have

$$\begin{aligned}
\text{var}(X + Y) &= \text{var}(\tilde{X} + \tilde{Y}) \\
&= \mathbf{E}\big[(\tilde{X} + \tilde{Y})^2\big] \\
&= \mathbf{E}[\tilde{X}^2 + 2\tilde{X}\tilde{Y} + \tilde{Y}^2] \\
&= \mathbf{E}[\tilde{X}^2] + 2\mathbf{E}[\tilde{X}\tilde{Y}] + \mathbf{E}[\tilde{Y}^2] \\
&= \mathbf{E}[\tilde{X}^2] + \mathbf{E}[\tilde{Y}^2] \\
&= \text{var}(\tilde{X}) + \text{var}(\tilde{Y}) \\
&= \text{var}(X) + \text{var}(Y).
\end{aligned}$$

We have used above the property $\mathbf{E}[\tilde{X}\,\tilde{Y}] = 0$, which is justified as follows. The random variables $\tilde{X} = X - \mathbf{E}[X]$ and $\tilde{Y} = Y - \mathbf{E}[Y]$ are independent (because they are functions of the independent random variables X and Y), and since they also have zero-mean, we obtain

$$\mathbf{E}[\tilde{X}\,\tilde{Y}] = \mathbf{E}[\tilde{X}]\,\mathbf{E}[\tilde{Y}] = 0.$$

In conclusion, the variance of the sum of two **independent** random variables is equal to the sum of their variances. For an interesting comparison, note that the mean of the sum of two random variables is always equal to the sum of their means, even if they are not independent.

Summary of Facts About Independent Random Variables

Let A be an event, with $\mathbf{P}(A) > 0$, and let X and Y be random variables associated with the same experiment.

- X is independent of the event A if

$$p_{X|A}(x) = p_X(x), \qquad \text{for all } x,$$

  that is, if for all x, the events $\{X = x\}$ and A are independent.

- X and Y are independent if for all pairs (x, y), the events $\{X = x\}$ and $\{Y = y\}$ are independent, or equivalently

$$p_{X,Y}(x, y) = p_X(x)p_Y(y), \qquad \text{for all } x, y.$$

- If X and Y are independent random variables, then

$$\mathbf{E}[XY] = \mathbf{E}[X]\,\mathbf{E}[Y].$$

  Furthermore, for any functions g and h, the random variables $g(X)$ and $h(Y)$ are independent, and we have

$$\mathbf{E}\big[g(X)h(Y)\big] = \mathbf{E}\big[g(X)\big]\,\mathbf{E}\big[h(Y)\big].$$

- If X and Y are independent, then

$$\text{var}(X + Y) = \text{var}(X) + \text{var}(Y).$$

Independence of Several Random Variables

The preceding discussion extends naturally to the case of more than two random variables. For example, three random variables X, Y, and Z are said to be independent if

$$p_{X,Y,Z}(x, y, z) = p_X(x)p_Y(y)p_Z(z), \qquad \text{for all } x, y, z.$$

If X, Y, and Z are independent random variables, then any three random variables of the form $f(X)$, $g(Y)$, and $h(Z)$, are also independent. Similarly, any two random variables of the form $g(X, Y)$ and $h(Z)$ are independent. On the other hand, two random variables of the form $g(X, Y)$ and $h(Y, Z)$ are usually not independent because they are both affected by Y. Properties such as the above are intuitively clear if we interpret independence in terms of noninteracting (sub)experiments. They can be formally verified but this is sometimes tedious. Fortunately, there is general agreement between intuition and what is mathematically correct. This is basically a testament that our definitions of independence adequately reflect the intended interpretation.

Variance of the Sum of Independent Random Variables

Sums of independent random variables are especially important in a variety of contexts. For example, they arise in statistical applications where we "average" a number of independent measurements, with the aim of minimizing the effects of measurement errors. They also arise when dealing with the cumulative effect of several independent sources of randomness. We provide some illustrations in the examples that follow and we will also return to this theme in later chapters.

In the examples below, we will make use of the following key property. If $X_1, X_2, \ldots, X_n$ are independent random variables, then

$$\text{var}(X_1 + X_2 + \cdots + X_n) = \text{var}(X_1) + \text{var}(X_2) + \cdots + \text{var}(X_n).$$

This can be verified by repeated use of the formula $\text{var}(X+Y) = \text{var}(X)+\text{var}(Y)$ for two independent random variables X and Y.

Example 2.20. Variance of the Binomial and the Poisson. We consider n independent coin tosses, with each toss having probability p of coming up a head. For each i, we let X_i be the Bernoulli random variable which is equal to 1 if the ith toss comes up a head, and is 0 otherwise. Then, $X = X_1 + X_2 + \cdots + X_n$ is a binomial random variable. Its mean is $\mathbf{E}[X] = np$, as derived in Example 2.10. By the independence of the coin tosses, the random variables $X_1, \ldots, X_n$ are independent, and

$$\text{var}(X) = \sum_{i=1}^{n} \text{var}(X_i) = np(1-p).$$

As we discussed in Section 2.2, a Poisson random variable Y with parameter λ can be viewed as the "limit" of the binomial as $n \to \infty$, $p \to 0$, while $np = \lambda$. Thus, taking the limit of the mean and the variance of the binomial, we informally obtain the mean and variance of the Poisson: $\mathbf{E}[Y] = \text{var}(Y) = \lambda$. We have indeed verified the formula $\mathbf{E}[Y] = \lambda$ in Example 2.7. To verify the formula $\text{var}(Y) = \lambda$, we write

$$\begin{aligned}
\mathbf{E}[Y^2] &= \sum_{k=1}^{\infty} k^2 e^{-\lambda} \frac{\lambda^k}{k!} \\
&= \lambda \sum_{k=1}^{\infty} k \frac{e^{-\lambda}\lambda^{k-1}}{(k-1)!} \\
&= \lambda \sum_{m=0}^{\infty} (m+1) \frac{e^{-\lambda}\lambda^m}{m!} \\
&= \lambda\big(\mathbf{E}[Y] + 1\big) \\
&= \lambda(\lambda + 1),
\end{aligned}$$

from which

$$\text{var}(Y) = \mathbf{E}[Y^2] - \big(\mathbf{E}[Y]\big)^2 = \lambda(\lambda+1) - \lambda^2 = \lambda.$$

The formulas for the mean and variance of a weighted sum of random variables form the basis for many statistical procedures that estimate the mean of a random variable by averaging many independent samples. A typical case is illustrated in the following example.

Example 2.21. Mean and Variance of the Sample Mean. We wish to estimate the approval rating of a president, to be called B. To this end, we ask n persons drawn at random from the voter population, and we let X_i be a random variable that encodes the response of the ith person:

$$X_i = \begin{cases} 1, & \text{if the } i\text{th person approves B's performance,} \\ 0, & \text{if the } i\text{th person disapproves B's performance.} \end{cases}$$

We model $X_1, X_2, \ldots, X_n$ as independent Bernoulli random variables with common mean p and variance $p(1-p)$. Naturally, we view p as the true approval rating of B. We "average" the responses and compute the **sample mean** S_n, defined as

$$S_n = \frac{X_1 + X_2 + \cdots + X_n}{n}.$$

Thus, the random variable S_n is the approval rating of B within our n-person sample.

We have, using the linearity of S_n as a function of the X_i,

$$\mathbf{E}[S_n] = \sum_{i=1}^{n} \frac{1}{n}\mathbf{E}[X_i] = \frac{1}{n}\sum_{i=1}^{n} p = p,$$

and making use of the independence of $X_1, \ldots, X_n$,

$$\text{var}(S_n) = \sum_{i=1}^{n} \frac{1}{n^2}\text{var}(X_i) = \frac{p(1-p)}{n}.$$

The sample mean S_n can be viewed as a "good" estimate of the approval rating. This is because it has the correct expected value, which is the approval rating p, and its accuracy, as reflected by its variance, improves as the sample size n increases.

Note that even if the random variables X_i are not Bernoulli, the same calculation yields

$$\text{var}(S_n) = \frac{\text{var}(X)}{n},$$

as long as the X_i are independent, with common mean $\mathbf{E}[X]$ and variance $\text{var}(X)$. Thus, again, the sample mean becomes a good estimate (in terms of variance) of the true mean $\mathbf{E}[X]$, as the sample size n increases. We will revisit the properties of the sample mean and discuss them in much greater detail in Chapter 5, in connection with the laws of large numbers.

Example 2.22. Estimating Probabilities by Simulation. In many practical situations, the analytical calculation of the probability of some event of interest is very difficult. However, if we have a physical or computer model that can generate outcomes of a given experiment in accordance with their true probabilities, we can use simulation to calculate with high accuracy the probability of any given event A. In particular, we independently generate with our model n outcomes, we record the number m of outcomes that belong to the event A of interest, and we approximate $\mathbf{P}(A)$ by m/n. For example, to calculate the probability $p = \mathbf{P}(\text{Heads})$ of a coin, we toss the coin n times, and we approximate p with the ratio (number of heads recorded)$/n$.

To see how accurate this process is, consider n independent Bernoulli random variables $X_1, \ldots, X_n$, each with PMF

$$p_{X_i}(k) = \begin{cases} \mathbf{P}(A), & \text{if } k = 1, \\ 1 - \mathbf{P}(A), & \text{if } k = 0. \end{cases}$$

In a simulation context, X_i corresponds to the ith outcome, and takes the value 1 if the ith outcome belongs to the event A. The value of the random variable

$$X = \frac{X_1 + X_2 + \cdots + X_n}{n}$$

is the estimate of $\mathbf{P}(A)$ provided by the simulation. According to Example 2.21, X has mean $\mathbf{P}(A)$ and variance $\mathbf{P}(A)\big(1 - \mathbf{P}(A)\big)/n$, so that for large n, it provides an accurate estimate of $\mathbf{P}(A)$.

2.8 SUMMARY AND DISCUSSION

Random variables provide the natural tools for dealing with probabilistic models in which the outcome determines certain numerical values of interest. In this chapter, we focused on discrete random variables, and developed a conceptual framework and some relevant tools.

In particular, we introduced concepts such as the PMF, the mean, and the variance, which describe in various degrees of detail the probabilistic character of a discrete random variable. We showed how to use the PMF of a random variable X to calculate the mean and the variance of a related random variable $Y = g(X)$ without calculating the PMF of Y. In the special case where g is a linear function, $Y = aX + b$, the means and the variances of X and Y are related by

$$\mathbf{E}[Y] = a\mathbf{E}[X] + b, \qquad \text{var}(Y) = a^2\text{var}(X).$$

We also discussed several special random variables, and derived their PMF, mean, and variance, as summarized in the table that follows.

Summary of Results for Special Random Variables

Discrete Uniform over $[a, b]$:

$$p_X(k) = \begin{cases} \dfrac{1}{b-a+1}, & \text{if } k = a, a+1, \ldots, b, \\ 0, & \text{otherwise,} \end{cases}$$

$$\mathbf{E}[X] = \frac{a+b}{2}, \qquad \text{var}(X) = \frac{(b-a)(b-a+2)}{12}.$$

Bernoulli with Parameter p: (Describes the success or failure in a single trial.)

$$p_X(k) = \begin{cases} p, & \text{if } k = 1, \\ 1-p, & \text{if } k = 0, \end{cases}$$

$$\mathbf{E}[X] = p, \qquad \text{var}(X) = p(1-p).$$

Binomial with Parameters p and n: (Describes the number of successes in n independent Bernoulli trials.)

$$p_X(k) = \binom{n}{k} p^k (1-p)^{n-k}, \qquad k = 0, 1, \ldots, n,$$

$$\mathbf{E}[X] = np, \qquad \text{var}(X) = np(1-p).$$

Geometric with Parameter p: (Describes the number of trials until the first success, in a sequence of independent Bernoulli trials.)

$$p_X(k) = (1-p)^{k-1}p, \qquad k = 1, 2, \ldots,$$

$$\mathbf{E}[X] = \frac{1}{p}, \qquad \text{var}(X) = \frac{1-p}{p^2}.$$

Poisson with Parameter λ: (Approximates the binomial PMF when n is large, p is small, and $\lambda = np$.)

$$p_X(k) = e^{-\lambda}\frac{\lambda^k}{k!}, \qquad k = 0, 1, \ldots,$$

$$\mathbf{E}[X] = \lambda, \qquad \text{var}(X) = \lambda.$$

We also considered multiple random variables, and introduced joint PMFs, conditional PMFs, and associated expected values. Conditional PMFs are often the starting point in probabilistic models and can be used to calculate other quantities of interest, such as marginal or joint PMFs and expectations, through a sequential or a divide-and-conquer approach. In particular, given the conditional PMF $p_{X|Y}(x\,|\,y)$:

(a) The joint PMF can be calculated by

$$p_{X,Y}(x,y) = p_Y(y)p_{X|Y}(x\,|\,y).$$

This can be extended to the case of three or more random variables, as in

$$p_{X,Y,Z}(x,y,z) = p_Z(z)p_{Y|Z}(y\,|\,z)p_{X|Y,Z}(x\,|\,y,z),$$

and is analogous to the sequential tree-based calculation method using the multiplication rule, discussed in Chapter 1.

(b) The marginal PMF can be calculated by

$$p_X(x) = \sum_y p_Y(y)p_{X|Y}(x\,|\,y),$$

which generalizes the divide-and-conquer calculation method we discussed in Chapter 1.

(c) The divide-and-conquer calculation method in (b) above can be extended to compute expected values using the total expectation theorem:

$$\mathbf{E}[X] = \sum_y p_Y(y)\mathbf{E}[X\,|\,Y = y].$$

We introduced the notion of independence of random variables, in analogy with the notion of independence of events. Among other topics, we focused on random variables X obtained by adding several independent random variables $X_1, \ldots, X_n$:

$$X = X_1 + \cdots + X_n.$$

We argued that the mean and the variance of the sum are equal to the sum of the means and the sum of the variances, respectively:

$$\mathbf{E}[X] = \mathbf{E}[X_1] + \cdots + \mathbf{E}[X_n], \qquad \text{var}(X) = \text{var}(X_1) + \cdots + \text{var}(X_n).$$

The formula for the mean does not require independence of the X_i, but the formula for the variance does.

The concepts and methods of this chapter extend appropriately to general random variables (see the next chapter), and are fundamental for our subject.

PROBLEMS

SECTION 2.2. Probability Mass Functions

Problem 1. The MIT soccer team has 2 games scheduled for one weekend. It has a 0.4 probability of not losing the first game, and a 0.7 probability of not losing the second game, independent of the first. If it does not lose a particular game, the team is equally likely to win or tie, independent of what happens in the other game. The MIT team will receive 2 points for a win, 1 for a tie, and 0 for a loss. Find the PMF of the number of points that the team earns over the weekend.

Problem 2. You go to a party with 500 guests. What is the probability that exactly one other guest has the same birthday as you? Calculate this exactly and also approximately by using the Poisson PMF. (For simplicity, exclude birthdays on February 29.)

Problem 3. Fischer and Spassky play a chess match in which the first player to win a game wins the match. After 10 successive draws, the match is declared drawn. Each game is won by Fischer with probability 0.4, is won by Spassky with probability 0.3, and is a draw with probability 0.3, independent of previous games.

(a) What is the probability that Fischer wins the match?

(b) What is the PMF of the duration of the match?

Problem 4. An internet service provider uses 50 modems to serve the needs of 1000 customers. It is estimated that at a given time, each customer will need a connection with probability 0.01, independent of the other customers.

(a) What is the PMF of the number of modems in use at the given time?

(b) Repeat part (a) by approximating the PMF of the number of customers that need a connection with a Poisson PMF.

(c) What is the probability that there are more customers needing a connection than there are modems? Provide an exact, as well as an approximate formula based on the Poisson approximation of part (b).

Problem 5. A packet communication system consists of a buffer that stores packets from some source, and a communication line that retrieves packets from the buffer and transmits them to a receiver. The system operates in time-slot pairs. In the first slot, the system stores a number of packets that are generated by the source according to a Poisson PMF with parameter λ; however, the maximum number of packets that can be stored is a given integer b, and packets arriving to a full buffer are discarded. In the second slot, the system transmits either all the stored packets or c packets (whichever is less). Here, c is a given integer with $0 < c < b$.

(a) Assuming that at the beginning of the first slot the buffer is empty, find the PMF of the number of packets stored at the end of the first slot and at the end of the second slot.

(b) What is the probability that some packets get discarded during the first slot?

Problem 6. The Celtics and the Lakers are set to play a playoff series of n basketball games, where n is odd. The Celtics have a probability p of winning any one game, independent of other games.

(a) Find the values of p for which $n = 5$ is better for the Celtics than $n = 3$.

(b) Generalize part (a), i.e., for any $k > 0$, find the values for p for which $n = 2k+1$ is better for the Celtics than $n = 2k - 1$.

Problem 7. You just rented a large house and the realtor gave you 5 keys, one for each of the 5 doors of the house. Unfortunately, all keys look identical, so to open the front door, you try them at random.

(a) Find the PMF of the number of trials you will need to open the door, under the following alternative assumptions: (1) after an unsuccessful trial, you mark the corresponding key, so that you never try it again, and (2) at each trial you are equally likely to choose any key.

(b) Repeat part (a) for the case where the realtor gave you an extra duplicate key for each of the 5 doors.

Problem 8. Recursive computation of the binomial PMF. Let X be a binomial random variable with parameters n and p. Show that its PMF can be computed by starting with $p_X(0) = (1-p)^n$, and then using the recursive formula

$$p_X(k+1) = \frac{p}{1-p} \cdot \frac{n-k}{k+1} \cdot p_X(k), \qquad k = 0, 1, \ldots, n-1.$$

Problem 9. Form of the binomial PMF. Consider a binomial random variable X with parameters n and p. Let k^* be the largest integer that is less than or equal to $(n+1)p$. Show that the PMF $p_X(k)$ is monotonically nondecreasing with k in the range from 0 to k^*, and is monotonically decreasing with k for $k \geq k^*$.

Problem 10. Form of the Poisson PMF. Let X be a Poisson random variable with parameter λ. Show that the PMF $p_X(k)$ increases monotonically with k up to the point where k reaches the largest integer not exceeding λ, and after that point decreases monotonically with k.

Problem 11.* The matchbox problem – inspired by Banach's smoking habits. A smoker mathematician carries one matchbox in his right pocket and one in his left pocket. Each time he wants to light a cigarette, he selects a matchbox from either pocket with probability $p = 1/2$, independent of earlier selections. The two matchboxes have initially n matches each. What is the PMF of the number of remaining matches at the moment when the mathematician reaches for a match and discovers that the corresponding matchbox is empty? How can we generalize to the case where the probabilities of a left and a right pocket selection are p and $1-p$, respectively?

Solution. Let X be the number of matches that remain when a matchbox is found empty. For $k = 0, 1, \ldots, n$, let L_k (or R_k) be the event that an empty box is first discovered in the left (respectively, right) pocket while the number of matches in the right (respectively, left) pocket is k at that time. The PMF of X is

$$p_X(k) = \mathbf{P}(L_k) + \mathbf{P}(R_k), \qquad k = 0, 1, \ldots, n.$$

Viewing a left and a right pocket selection as a "success" and a "failure," respectively, $\mathbf{P}(L_k)$ is the probability that there are n successes in the first $2n - k$ trials, and trial $2n - k + 1$ is a success, or

$$\mathbf{P}(L_k) = \frac{1}{2}\binom{2n-k}{n}\left(\frac{1}{2}\right)^{2n-k}, \qquad k = 0, 1, \ldots, n.$$

By symmetry, $\mathbf{P}(L_k) = \mathbf{P}(R_k)$, so

$$p_X(k) = \mathbf{P}(L_k) + \mathbf{P}(R_k) = \binom{2n-k}{n}\left(\frac{1}{2}\right)^{2n-k}, \qquad k = 0, 1, \ldots, n.$$

In the more general case, where the probabilities of a left and a right pocket selection are p and $1 - p$, using a similar reasoning, we obtain

$$\mathbf{P}(L_k) = p\binom{2n-k}{n}p^n(1-p)^{n-k}, \qquad k = 0, 1, \ldots, n,$$

and

$$\mathbf{P}(R_k) = (1-p)\binom{2n-k}{n}p^{n-k}(1-p)^n, \qquad k = 0, 1, \ldots, n,$$

which yields

$$\begin{aligned} p_X(k) &= \mathbf{P}(L_k) + \mathbf{P}(R_k) \\ &= \binom{2n-k}{n}\left(p^{n+1}(1-p)^{n-k} + p^{n-k}(1-p)^{n+1}\right), \qquad k = 0, 1, \ldots, n. \end{aligned}$$

Problem 12.* Justification of the Poisson approximation property. Consider the PMF of a binomial random variable with parameters n and p. Show that asymptotically, as

$$n \to \infty, \qquad p \to 0,$$

while np is fixed at a given value λ, this PMF approaches the PMF of a Poisson random variable with parameter λ.

Solution. Using the equation $\lambda = np$, write the binomial PMF as

$$\begin{aligned} p_X(k) &= \frac{n!}{(n-k)!\,k!}p^k(1-p)^{n-k} \\ &= \frac{n(n-1)\cdots(n-k+1)}{n^k} \cdot \frac{\lambda^k}{k!} \cdot \left(1 - \frac{\lambda}{n}\right)^{n-k}. \end{aligned}$$

Fix k and let $n \to \infty$. We have, for $j = 1, \ldots, k$,

$$\frac{n-k+j}{n} \to 1, \quad \left(1 - \frac{\lambda}{n}\right)^{-k} \to 1, \quad \left(1 - \frac{\lambda}{n}\right)^{n} \to e^{-\lambda}.$$

Thus, for each fixed k, as $n \to \infty$ we obtain

$$p_X(k) \to e^{-\lambda}\frac{\lambda^k}{k!}.$$

SECTION 2.3. Functions of Random Variables

Problem 13. A family has 5 natural children and has adopted 2 girls. Each natural child has equal probability of being a girl or a boy, independent of the other children. Find the PMF of the number of girls out of the 7 children.

Problem 14. Let X be a random variable that takes values from 0 to 9 with equal probability 1/10.

(a) Find the PMF of the random variable $Y = X \bmod(3)$.

(b) Find the PMF of the random variable $Y = 5 \bmod(\mathrm{X}+1)$.

Problem 15. Let K be a random variable that takes, with equal probability $1/(2n+1)$, the integer values in the interval $[-n, n]$. Find the PMF of the random variable $Y = \ln X$, where $X = a^{|K|}$, and a is a positive number.

SECTION 2.4. Expectation, Mean, and Variance

Problem 16. Let X be a random variable with PMF

$$p_X(x) = \begin{cases} x^2/a, & \text{if } x = -3, -2, -1, 0, 1, 2, 3, \\ 0, & \text{otherwise.} \end{cases}$$

(a) Find a and $\mathbf{E}[X]$.

(b) What is the PMF of the random variable $Z = \big(X - \mathbf{E}[X]\big)^2$?

(c) Using the result from part (b), find the variance of X.

(d) Find the variance of X using the formula $\text{var}(X) = \sum_x \big(x - \mathbf{E}[X]\big)^2 p_X(x)$.

Problem 17. A city's temperature is modeled as a random variable with mean and standard deviation both equal to 10 degrees Celsius. A day is described as "normal" if the temperature during that day ranges within one standard deviation from the mean. What would be the temperature range for a normal day if temperature were expressed in degrees Fahrenheit?

Problem 18. Let a and b be positive integers with $a \le b$, and let X be a random variable that takes as values, with equal probability, the powers of 2 in the interval $[2^a, 2^b]$. Find the expected value and the variance of X.

Problem 19. A prize is randomly placed in one of ten boxes, numbered from 1 to 10. You search for the prize by asking yes-no questions. Find the expected number of questions until you are sure about the location of the prize, under each of the following strategies.

(a) An enumeration strategy: you ask questions of the form "is it in box k?".

(b) A bisection strategy: you eliminate as close to half of the remaining boxes as possible by asking questions of the form "is it in a box numbered less than or equal to k?".

Solution. We will find the expected gain for each strategy, by computing the expected number of questions until we find the prize.

(a) With this strategy, the probability 1/10 of finding the location of the prize with i questions, where $i = 1, \ldots, 10$, is 1/10. Therefore, the expected number of questions is

$$\frac{1}{10} \sum_{i=1}^{10} i = \frac{1}{10} \cdot 55 = 5.5.$$

(b) It can be checked that for 4 of the 10 possible box numbers, exactly 4 questions will be needed, whereas for 6 of the 10 numbers, 3 questions will be needed. Therefore, with this strategy, the expected number of questions is

$$\frac{4}{10} \cdot 4 + \frac{6}{10} \cdot 3 = 3.4.$$

Problem 20. As an advertising campaign, a chocolate factory places golden tickets in some of its candy bars, with the promise that a golden ticket is worth a trip through the chocolate factory, and all the chocolate you can eat for life. If the probability of finding a golden ticket is p, find the mean and the variance of the number of candy bars you need to eat to find a ticket.

Problem 21. St. Petersburg paradox. You toss independently a fair coin and you count the number of tosses until the first tail appears. If this number is n, you receive 2^n dollars. What is the expected amount that you will receive? How much would you be willing to pay to play this game?

Problem 22. Two coins are simultaneously tossed until one of them comes up a head and the other a tail. The first coin comes up a head with probability p and the second with probability q. All tosses are assumed independent.

(a) Find the PMF, the expected value, and the variance of the number of tosses.

(b) What is the probability that the last toss of the first coin is a head?

Problem 23.

(a) A fair coin is tossed repeatedly and independently until two consecutive heads or two consecutive tails appear. Find the PMF, the expected value, and the variance of the number of tosses.

(b) Assume now that the coin is tossed until we obtain a tail that is immediately preceded by a head. Find the PMF and the expected value of the number of tosses.

SECTION 2.5. Joint PMFs of Multiple Random Variables

Problem 24. A stock market trader buys 100 shares of stock A and 200 shares of stock B. Let X and Y be the price changes of A and B, respectively, over a certain time period, and assume that the joint PMF of X and Y is uniform over the set of integers x and y satisfying

$$-2 \le x \le 4, \qquad -1 \le y - x \le 1.$$

(a) Find the marginal PMFs and the means of X and Y.

(b) Find the mean of the trader's profit.

Problem 25. A class of n students takes a test consisting of m questions. Suppose that student i submitted answers to the first m_i questions.

(a) The grader randomly picks one answer, call it (I, J), where I is the student ID number (taking values $1, \ldots, n$) and J is the question number (taking values $1, \ldots, m$). Assume that all answers are equally likely to be picked. Calculate the joint and the marginal PMFs of I and J.

(b) Assume that an answer to question j, if submitted by student i, is correct with probability p_{ij}. Each answer gets a points if it is correct and gets b points otherwise. Calculate the expected value of the score of student i.

Problem 26. PMF of the minimum of several random variables. On a given day, your golf score takes values from the range 101 to 110, with probability 0.1, independent of other days. Determined to improve your score, you decide to play on three different days and declare as your score the minimum X of the scores X_1, X_2, and X_3 on the different days.

(a) Calculate the PMF of X.

(b) By how much has your expected score improved as a result of playing on three days?

Problem 27.* The multinomial distribution. A die with r faces, numbered $1, \ldots, r$, is rolled a fixed number of times n. The probability that the ith face comes up on any one roll is denoted p_i, and the results of different rolls are assumed independent. Let X_i be the number of times that the ith face comes up.

(a) Find the joint PMF $p_{X_1,\ldots,X_r}(k_1, \ldots, k_r)$.

(b) Find the expected value and variance of X_i.

(c) Find $\mathbf{E}[X_i X_j]$ for $i \neq j$.

Solution. (a) The probability of a sequence of rolls where, for $i = 1, \ldots, r$, face i comes up k_i times is $p_1^{k_1} \cdots p_r^{k_r}$. Every such sequence determines a partition of the set of n rolls into r subsets with the ith subset having cardinality k_i (this is the set of rolls for which the ith face came up). The number of such partitions is the multinomial coefficient (cf. Section 1.6)

$$\binom{n}{k_1,\ldots,k_r} = \frac{n!}{k_1!\cdots k_r!}.$$

Thus, if $k_1 + \cdots + k_r = n$,

$$p_{X_1,\ldots,X_r}(k_1,\ldots,k_r) = \binom{n}{k_1,\ldots,k_r} p_1^{k_1}\cdots p_r^{k_r},$$

and otherwise, $p_{X_1,\ldots,X_r}(k_1,\ldots,k_r) = 0$.

(b) The random variable X_i is binomial with parameters n and p_i. Therefore, $\mathbf{E}[X_i] = np_i$, and $\text{var}(X_i) = np_i(1-p_i)$.

(c) Suppose that $i \neq j$, and let $Y_{i,k}$ (or $Y_{j,k}$) be the Bernoulli random variable that takes the value 1 if face i (respectively, j) comes up on the kth roll, and the value 0 otherwise. Note that $Y_{i,k}Y_{j,k} = 0$, and that for $l \neq k$, $Y_{i,k}$ and $Y_{j,l}$ are independent, so that $\mathbf{E}[Y_{i,k}Y_{j,l}] = p_ip_j$. Therefore,

$$\begin{aligned} \mathbf{E}[X_iX_j] &= \mathbf{E}\big[(Y_{i,1} + \cdots + Y_{i,n})(Y_{j,1} + \cdots + Y_{j,n})\big] \\ &= n(n-1)\mathbf{E}[Y_{i,1}Y_{j,2}] \\ &= n(n-1)p_ip_j. \end{aligned}$$

Problem 28.* The quiz problem. Consider a quiz contest where a person is given a list of n questions and can answer these questions in any order he or she chooses. Question i will be answered correctly with probability p_i, and the person will then receive a reward v_i. At the first incorrect answer, the quiz terminates and the person is allowed to keep his or her previous rewards. The problem is to choose the ordering of questions so as to maximize the expected value of the total reward obtained. Show that it is optimal to answer questions in a nonincreasing order of $p_iv_i/(1-p_i)$.

Solution. We will use a so-called interchange argument, which is often useful in sequencing problems. Let i and j be the kth and $(k+1)$st questions in an optimally ordered list

$$L = (i_1,\ldots,i_{k-1},i,j,i_{k+2},\ldots,i_n).$$

Consider the list

$$L' = (i_1,\ldots,i_{k-1},j,i,i_{k+2},\ldots,i_n)$$

obtained from L by interchanging the order of questions i and j. We compute the expected values of the rewards of L and L', and note that since L is optimally ordered, we have

$$\mathbf{E}[\text{reward of } L] \geq \mathbf{E}[\text{reward of } L'].$$

Define the *weight* of question i to be

$$w(i) = \frac{p_iv_i}{1-p_i}.$$

We will show that any permutation of the questions in a nonincreasing order of weights maximizes the expected reward.

If $L = (i_1, \ldots, i_n)$ is a permutation of the questions, define $L^{(k)}$ to be the permutation obtained from L by interchanging questions i_k and i_{k+1}. Let us first compute the difference between the expected reward of L and that of $L^{(k)}$. We have

$$\mathbf{E}[\text{reward of } L] = p_{i_1}v_{i_1} + p_{i_1}p_{i_2}v_{i_2} + \cdots + p_{i_1}\cdots p_{i_n}v_{i_n},$$

and

$$\begin{aligned}\mathbf{E}\big[\text{reward of } L^{(k)}\big] = {} & p_{i_1}v_{i_1} + p_{i_1}p_{i_2}v_{i_2} + \cdots + p_{i_1}\cdots p_{i_{k-1}}v_{i_{k-1}} \\ & + p_{i_1}\cdots p_{i_{k-1}}p_{i_{k+1}}v_{i_{k+1}} + p_{i_1}\cdots p_{i_{k-1}}p_{i_{k+1}}p_{i_k}v_{i_k} \\ & + p_{i_1}\cdots p_{i_{k+2}}v_{i_{k+2}} + \cdots + p_{i_1}\cdots p_{i_n}v_{i_n}.\end{aligned}$$

Therefore,

$$\begin{aligned}\mathbf{E}\big[\text{reward of } L^{(k)}\big] - \mathbf{E}[\text{reward of } L] &= p_{i_1}\cdots p_{i_{k-1}}(p_{i_{k+1}}v_{i_{k+1}} + p_{i_{k+1}}p_{i_k}v_{i_k} \\ &\qquad - p_{i_k}v_{i_k} - p_{i_k}p_{i_{k+1}}v_{i_{k+1}}) \\ &= p_{i_1}\cdots p_{i_{k-1}}(1-p_{i_k})(1-p_{i_{k+1}})\big(w(i_{k+1}) - w(i_k)\big).\end{aligned}$$

Now, let us go back to our problem. Consider any permutation L of the questions. If $w(i_k) < w(i_{k+1})$ for some k, it follows from the above equation that the permutation $L^{(k)}$ has an expected reward larger than that of L. So, an optimal permutation of the questions must be in a nonincreasing order of weights.

Let us finally show that any two such permutations have equal expected rewards. Assume that L is such a permutation and say that $w(i_k) = w(i_{k+1})$ for some k. We know that interchanging i_k and i_{k+1} preserves the expected reward. So, the expected reward of any permutation L' in a non-increasing order of weights is equal to that of L, because L' can be obtained from L by repeatedly interchanging adjacent questions having equal weights.

Problem 29.* The inclusion-exclusion formula. Let $A_1, A_2, \ldots, A_n$ be events. Let $S_1 = \{i \,|\, 1 \le i \le n\}$, $S_2 = \{(i_1, i_2) \,|\, 1 \le i_1 < i_2 \le n\}$, and more generally, let S_m be the set of all m-tuples $(i_1, \ldots, i_m)$ of indices that satisfy $1 \le i_1 < i_2 < \cdots < i_m \le n$. Show that

$$\begin{aligned}\mathbf{P}\left(\cup_{k=1}^n A_k\right) = {} & \sum_{i\in S_1}\mathbf{P}(A_i) - \sum_{(i_1,i_2)\in S_2}\mathbf{P}(A_{i_1}\cap A_{i_2}) \\ & + \sum_{(i_1,i_2,i_3)\in S_3}\mathbf{P}(A_{i_1}\cap A_{i_2}\cap A_{i_3}) - \cdots + (-1)^{n-1}\mathbf{P}\left(\cap_{k=1}^n A_k\right).\end{aligned}$$

Hint: Let X_i be a binary random variable which is equal to 1 when A_i occurs, and equal to 0 otherwise. Relate the event of interest to the random variable $(1-X_1)(1-X_2)\cdots(1-X_n)$.

Solution. Let us express the event $B = \cup_{k=1}^n A_k$ in terms of the random variables $X_1, \ldots, X_n$. The event B^c occurs when all of the random variables $X_1, \ldots, X_n$ are zero, which happens when the random variable $Y = (1-X_1)(1-X_2)\cdots(1-X_n)$ is equal to 1.

Note that Y can only take values in the set $\{0, 1\}$, so that $\mathbf{P}(B^c) = \mathbf{P}(Y = 1) = \mathbf{E}[Y]$. Therefore,

$$\mathbf{P}(B) = 1 - \mathbf{E}\big[(1 - X_1)(1 - X_2) \cdots (1 - X_n)\big]$$
$$= \mathbf{E}[X_1 + \cdots + X_n] - \mathbf{E}\left[\sum_{(i_1,i_2)\in S_2} X_{i_1}X_{i_2}\right] + \cdots + (-1)^{n-1}\mathbf{E}[X_1 \cdots X_n].$$

We note that

$$\mathbf{E}[X_i] = \mathbf{P}(A_i), \qquad \mathbf{E}[X_{i_1}X_{i_2}] = \mathbf{P}(A_{i_1} \cap A_{i_2}),$$
$$\mathbf{E}[X_{i_1}X_{i_2}X_{i_3}] = \mathbf{P}(A_{i_1} \cap A_{i_2} \cap A_{i_3}), \qquad \mathbf{E}\big[X_1X_2 \cdots X_n\big] = \mathbf{P}(\cap_{k=1}^n A_k),$$

etc., from which the desired formula follows.

Problem 30.* Alvin's database of friends contains n entries, but due to a software glitch, the addresses correspond to the names in a totally random fashion. Alvin writes a holiday card to each of his friends and sends it to the (software-corrupted) address. What is the probability that at least one of his friends will get the correct card? *Hint:* Use the inclusion-exclusion formula.

Solution. Let A_k be the event that the kth card is sent to the correct address. We have for any k, j, i,

$$\mathbf{P}(A_k) = \frac{1}{n} = \frac{(n-1)!}{n!},$$
$$\mathbf{P}(A_k \cap A_j) = \mathbf{P}(A_k)\mathbf{P}(A_j \mid A_k) = \frac{1}{n} \cdot \frac{1}{n-1} = \frac{(n-2)!}{n!},$$
$$\mathbf{P}(A_k \cap A_j \cap A_i) = \frac{1}{n} \cdot \frac{1}{n-1} \cdot \frac{1}{n-2} = \frac{(n-3)!}{n!},$$

etc., and

$$\mathbf{P}(\cap_{k=1}^n A_k) = \frac{1}{n!}.$$

Applying the inclusion-exclusion formula,

$$\mathbf{P}\left(\cup_{k=1}^n A_k\right) = \sum_{i\in S_1} \mathbf{P}(A_i) - \sum_{(i_1,i_2)\in S_2} \mathbf{P}(A_{i_1} \cap A_{i_2})$$
$$+ \sum_{(i_1,i_2,i_3)\in S_3} \mathbf{P}(A_{i_1} \cap A_{i_2} \cap A_{i_3}) - \cdots + (-1)^{n-1}\mathbf{P}\left(\cap_{k=1}^n A_k\right),$$

we obtain the desired probability

$$\mathbf{P}(\cup_{k=1}^n A_k) = \binom{n}{1}\frac{(n-1)!}{n!} - \binom{n}{2}\frac{(n-2)!}{n!} + \binom{n}{3}\frac{(n-3)!}{n!} - \cdots + (-1)^{n-1}\frac{1}{n!}$$
$$= 1 - \frac{1}{2!} + \frac{1}{3!} - \cdots + (-1)^{n-1}\frac{1}{n!}.$$

When n is large, this probability can be approximated by $1 - e^{-1}$.

SECTION 2.6. Conditioning

Problem 31. Consider four independent rolls of a 6-sided die. Let X be the number of 1s and let Y be the number of 2s obtained. What is the joint PMF of X and Y?

Problem 32. **D. Bernoulli's problem of joint lives.** Consider $2m$ persons forming m couples who live together at a given time. Suppose that at some later time, the probability of each person being alive is p, independent of other persons. At that later time, let A be the number of persons that are alive and let S be the number of couples in which both partners are alive. For any survivor number a, find $\mathbf{E}[S \mid A = a]$.

Problem 33.* A coin that has probability of heads equal to p is tossed successively and independently until a head comes twice in a row or a tail comes twice in a row. Find the expected value of the number of tosses.

Solution. One possibility here is to calculate the PMF of X, the number of tosses until the game is over, and use it to compute $\mathbf{E}[X]$. However, with an unfair coin, this turns out to be cumbersome, so we argue by using the total expectation theorem and a suitable partition of the sample space. Let H_k (or T_k) be the event that a head (or a tail, respectively) comes at the kth toss, and let p (respectively, q) be the probability of H_k (respectively, T_k). Since H_1 and T_1 form a partition of the sample space, and $\mathbf{P}(H_1) = p$ and $\mathbf{P}(T_1) = q$, we have

$$\mathbf{E}[X] = p\mathbf{E}[X \mid H_1] + q\mathbf{E}[X \mid T_1].$$

Using again the total expectation theorem, we have

$$\mathbf{E}[X \mid H_1] = p\mathbf{E}[X \mid H_1 \cap H_2] + q\mathbf{E}[X \mid H_1 \cap T_2] = 2p + q\big(1 + \mathbf{E}[X \mid T_1]\big),$$

where we have used the fact

$$\mathbf{E}[X \mid H_1 \cap H_2] = 2$$

(since the game ends after two successive heads), and

$$\mathbf{E}[X \mid H_1 \cap T_2] = 1 + \mathbf{E}[X \mid T_1]$$

(since if the game is not over, only the last toss matters in determining the number of additional tosses up to termination). Similarly, we obtain

$$\mathbf{E}[X \mid T_1] = 2q + p\big(1 + \mathbf{E}[X \mid H_1]\big).$$

Combining the above two relations, collecting terms, and using the fact $p + q = 1$, we obtain after some calculation

$$\mathbf{E}[X \mid T_1] = \frac{2 + p^2}{1 - pq},$$

and similarly

$$\mathbf{E}[X \mid H_1] = \frac{2 + q^2}{1 - pq}.$$

Thus,

$$\mathbf{E}[X] = p \cdot \frac{2+q^2}{1-pq} + q \cdot \frac{2+p^2}{1-pq},$$

and finally, using the fact $p + q = 1$,

$$\mathbf{E}[X] = \frac{2+pq}{1-pq}.$$

In the case of a fair coin ($p = q = 1/2$), we obtain $\mathbf{E}[X] = 3$. It can also be verified that $2 \leq \mathbf{E}[X] \leq 3$ for all values of p.

Problem 34.* A spider and a fly move along a straight line. At each second, the fly moves a unit step to the right or to the left with equal probability p, and stays where it is with probability $1 - 2p$. The spider always takes a unit step in the direction of the fly. The spider and the fly start D units apart, where D is a random variable taking positive integer values with a given PMF. If the spider lands on top of the fly, it's the end. What is the expected value of the time it takes for this to happen?

Solution. Let T be the time at which the spider lands on top of the fly. We define

A_d: the event that initially the spider and the fly are d units apart,

B_d: the event that after one second the spider and the fly are d units apart.

Our approach will be to first apply the (conditional version of the) total expectation theorem to compute $\mathbf{E}[T \mid A_1]$, then use the result to compute $\mathbf{E}[T \mid A_2]$, and similarly compute sequentially $\mathbf{E}[T \mid A_d]$ for all relevant values of d. We will then apply the (unconditional version of the) total expectation theorem to compute $\mathbf{E}[T]$.

We have

$$A_d = (A_d \cap B_d) \cup (A_d \cap B_{d-1}) \cup (A_d \cap B_{d-2}), \qquad \text{if } d > 1.$$

This is because if the spider and the fly are at a distance $d > 1$ apart, then one second later their distance will be d (if the fly moves away from the spider) or $d - 1$ (if the fly does not move) or $d - 2$ (if the fly moves towards the spider). We also have, for the case where the spider and the fly start one unit apart,

$$A_1 = (A_1 \cap B_1) \cup (A_1 \cap B_0).$$

Using the total expectation theorem, we obtain

$$\begin{aligned} \mathbf{E}[T \mid A_d] = \mathbf{P}(B_d \mid A_d)&\mathbf{E}[T \mid A_d \cap B_d] \\ &+ \mathbf{P}(B_{d-1} \mid A_d)\mathbf{E}[T \mid A_d \cap B_{d-1}] \\ &+ \mathbf{P}(B_{d-2} \mid A_d)\mathbf{E}[T \mid A_d \cap B_{d-2}], \qquad \text{if } d > 1, \end{aligned}$$

and

$$\mathbf{E}[T \mid A_1] = \mathbf{P}(B_1 \mid A_1)\mathbf{E}[T \mid A_1 \cap B_1] + \mathbf{P}(B_0 \mid A_1)\mathbf{E}[T \mid A_1 \cap B_0], \quad \text{if } d = 1.$$

It can be seen based on the problem data that

$$\mathbf{P}(B_1 \mid A_1) = 2p, \qquad \mathbf{P}(B_0 \mid A_1) = 1 - 2p,$$

$$\mathbf{E}[T \mid A_1 \cap B_1] = 1 + \mathbf{E}[T \mid A_1], \qquad \mathbf{E}[T \mid A_1 \cap B_0] = 1,$$

so by applying the formula for the case $d = 1$, we obtain

$$\mathbf{E}[T \mid A_1] = 2p\big(1 + \mathbf{E}[T \mid A_1]\big) + (1 - 2p),$$

or

$$\mathbf{E}[T \mid A_1] = \frac{1}{1 - 2p}.$$

By applying the formula with $d = 2$, we obtain

$$\mathbf{E}[T \mid A_2] = p\mathbf{E}[T \mid A_2 \cap B_2] + (1 - 2p)\mathbf{E}[T \mid A_2 \cap B_1] + p\mathbf{E}[T \mid A_2 \cap B_0].$$

We have

$$\begin{aligned} \mathbf{E}[T \mid A_2 \cap B_0] &= 1, \\ \mathbf{E}[T \mid A_2 \cap B_1] &= 1 + \mathbf{E}[T \mid A_1], \\ \mathbf{E}[T \mid A_2 \cap B_2] &= 1 + \mathbf{E}[T \mid A_2], \end{aligned}$$

so by substituting these relations in the expression for $\mathbf{E}[T \mid A_2]$, we obtain

$$\begin{aligned} \mathbf{E}[T \mid A_2] &= p\big(1 + \mathbf{E}[T \mid A_2]\big) + (1 - 2p)\big(1 + \mathbf{E}[T \mid A_1]\big) + p \\ &= p\big(1 + \mathbf{E}[T \mid A_2]\big) + (1 - 2p)\left(1 + \frac{1}{1 - 2p}\right) + p. \end{aligned}$$

This equation yields after some calculation

$$\mathbf{E}[T \mid A_2] = \frac{2}{1 - p}.$$

Generalizing, we obtain for $d > 2$,

$$\mathbf{E}[T \mid A_d] = p\big(1 + \mathbf{E}[T \mid A_d]\big) + (1 - 2p)\big(1 + \mathbf{E}[T \mid A_{d-1}]\big) + p\big(1 + \mathbf{E}[T \mid A_{d-2}]\big).$$

Thus, $\mathbf{E}[T \mid A_d]$ can be generated recursively for any initial distance d, using as initial conditions the values of $\mathbf{E}[T \mid A_1]$ and $\mathbf{E}[T \mid A_2]$ obtained earlier.

Finally, the expected value of T can be obtained using the given PMF for the initial distance D and the total expectation theorem:

$$\mathbf{E}[T] = \sum_d p_D(d)\mathbf{E}[T \mid A_d].$$

Problem 35.* Verify the expected value rule

$$\mathbf{E}\big[g(X, Y)\big] = \sum_x \sum_y g(x, y) p_{X,Y}(x, y),$$

using the expected value rule for a function of a single random variable. Then, use the rule for the special case of a linear function, to verify the formula

$$\mathbf{E}[aX + bY] = a\mathbf{E}[X] + b\mathbf{E}[Y],$$

where a and b are given scalars.

Solution. We use the total expectation theorem to reduce the problem to the case of a single random variable. In particular, we have

$$\begin{aligned}
\mathbf{E}\big[g(X,Y)\big] &= \sum_y p_Y(y)\mathbf{E}\big[g(X,Y) \,|\, Y=y\big] \\
&= \sum_y p_Y(y)\mathbf{E}\big[g(X,y) \,|\, Y=y\big] \\
&= \sum_y p_Y(y) \sum_x g(x,y) p_{X|Y}(x \,|\, y) \\
&= \sum_x \sum_y g(x,y) p_{X,Y}(x,y),
\end{aligned}$$

as desired. Note that the third equality above used the expected value rule for the function $g(X,y)$ of a single random variable X.

For the linear special case, the expected value rule gives

$$\begin{aligned}
\mathbf{E}[aX+bY] &= \sum_x \sum_y (ax+by) p_{X,Y}(x,y) \\
&= a\sum_x x \sum_y p_{X,Y}(x,y) + b\sum_y y \sum_x p_{X,Y}(x,y) \\
&= a\sum_x x p_X(x) + b\sum_y y p_Y(y) \\
&= a\mathbf{E}[X] + b\mathbf{E}[Y].
\end{aligned}$$

Problem 36.* The multiplication rule for conditional PMFs. Let X, Y, and Z be random variables.

(a) Show that

$$p_{X,Y,Z}(x,y,z) = p_X(x) p_{Y|X}(y \,|\, x) p_{Z|X,Y}(z \,|\, x,y).$$

(b) How can we interpret this formula as a special case of the multiplication rule given in Section 1.3?

(c) Generalize to the case of more than three random variables.

Solution. (a) We have

$$\begin{aligned}
p_{X,Y,Z}(x,y,z) &= \mathbf{P}(X=x, Y=y, Z=z) \\
&= \mathbf{P}(X=x)\mathbf{P}\big(Y=y, Z=z \,|\, X=x\big) \\
&= \mathbf{P}(X=x)\mathbf{P}(Y=y \,|\, X=x)\mathbf{P}(Z=z \,|\, X=x, Y=y) \\
&= p_X(x) p_{Y|X}(y \,|\, x) p_{Z|X,Y}(z \,|\, x,y).
\end{aligned}$$

(b) The formula can be written as

$$\mathbf{P}(X=x, Y=y, Z=z) = \mathbf{P}(X=x)\mathbf{P}(Y=y \,|\, X=x)\mathbf{P}(Z=z \,|\, X=x, Y=y),$$

which is a special case of the multiplication rule.

(c) The generalization is

$$p_{X_1,\ldots,X_n}(x_1,\ldots,x_n)$$
$$= p_{X_1}(x_1)p_{X_2|X_1}(x_2\,|\,x_1)\cdots p_{X_n|X_1,\ldots,X_{n-1}}(x_n\,|\,x_1,\ldots,x_{n-1}).$$

Problem 37.* **Splitting a Poisson random variable.** A transmitter sends out either a 1 with probability p, or a 0 with probability $1-p$, independent of earlier transmissions. If the number of transmissions within a given time interval has a Poisson PMF with parameter λ, show that the number of 1s transmitted in that same time interval has a Poisson PMF with parameter $p\lambda$.

Solution. Let X and Y be the numbers of 1s and 0s transmitted, respectively. Let $Z = X + Y$ be the total number of symbols transmitted. We have

$$\begin{aligned}
\mathbf{P}(X=n, Y=m) &= \mathbf{P}(X=n, Y=m\,|\,Z=n+m)\mathbf{P}(Z=n+m)\\
&= \binom{n+m}{n}p^n(1-p)^m\cdot\frac{e^{-\lambda}\lambda^{n+m}}{(n+m)!}\\
&= \frac{e^{-\lambda p}(\lambda p)^n}{n!}\cdot\frac{e^{-\lambda(1-p)}\big(\lambda(1-p)\big)^m}{m!}.
\end{aligned}$$

Thus,

$$\begin{aligned}
\mathbf{P}(X=n) &= \sum_{m=0}^{\infty}\mathbf{P}(X=n, Y=m)\\
&= \frac{e^{-\lambda p}(\lambda p)^n}{n!}e^{-\lambda(1-p)}\sum_{m=0}^{\infty}\frac{\big(\lambda(1-p)\big)^m}{m!}\\
&= \frac{e^{-\lambda p}(\lambda p)^n}{n!}e^{-\lambda(1-p)}e^{\lambda(1-p)}\\
&= \frac{e^{-\lambda p}(\lambda p)^n}{n!},
\end{aligned}$$

so that X is Poisson with parameter λp.

SECTION 2.7. Independence

Problem 38. Alice passes through four traffic lights on her way to work, and each light is equally likely to be green or red, independent of the others.

(a) What is the PMF, the mean, and the variance of the number of red lights that Alice encounters?

(b) Suppose that each red light delays Alice by exactly two minutes. What is the variance of Alice's commuting time?

Problem 39. Each morning, Hungry Harry eats some eggs. On any given morning, the number of eggs he eats is equally likely to be 1, 2, 3, 4, 5, or 6, independent of what he has done in the past. Let X be the number of eggs that Harry eats in 10 days. Find the mean and variance of X.

Problem 40. A particular professor is known for his arbitrary grading policies. Each paper receives a grade from the set $\{A, A-, B+, B, B-, C+\}$, with equal probability, independent of other papers. How many papers do you expect to hand in before you receive each possible grade at least once?

Problem 41. You drive to work 5 days a week for a full year (50 weeks), and with probability $p = 0.02$ you get a traffic ticket on any given day, independent of other days. Let X be the total number of tickets you get in the year.

(a) What is the probability that the number of tickets you get is exactly equal to the expected value of X?

(b) Calculate approximately the probability in (a) using a Poisson approximation.

(c) Any one of the tickets is \$10 or \$20 or \$50 with respective probabilities 0.5, 0.3, and 0.2, and independent of other tickets. Find the mean and the variance of the amount of money you pay in traffic tickets during the year.

(d) Suppose you don't know the probability p of getting a ticket, but you got 5 tickets during the year, and you estimate p by the sample mean

$$\hat{p} = \frac{5}{250} = 0.02.$$

What is the range of possible values of p assuming that the difference between p and the sample mean $\hat{p}$ is within 5 times the standard deviation of the sample mean?

Problem 42. **Computational problem.** Here is a probabilistic method for computing the area of a given subset S of the unit square. The method uses a sequence of independent random selections of points in the unit square $[0, 1] \times [0, 1]$, according to a uniform probability law. If the ith point belongs to the subset S the value of a random variable X_i is set to 1, and otherwise it is set to 0. Let $X_1, X_2, \ldots$ be the sequence of random variables thus defined, and for any n, let

$$S_n = \frac{X_1 + X_2 + \cdots + X_n}{n}.$$

(a) Show that $\mathbf{E}[S_n]$ is equal to the area of the subset S, and that $\text{var}(S_n)$ diminishes to 0 as n increases.

(b) Show that to calculate S_n, it is sufficient to know S_{n-1} and X_n, so the past values of X_k, $k = 1, \ldots, n-1$, do not need to be remembered. Give a formula.

(c) Write a computer program to generate S_n for $n = 1, 2, \ldots, 10000$, using the computer's random number generator, for the case where the subset S is the circle inscribed within the unit square. How can you use your program to measure experimentally the value of π?

(d) Use a similar computer program to calculate approximately the area of the set of all (x, y) that lie within the unit square and satisfy $0 \leq \cos \pi x + \sin \pi y \leq 1$.

Problem 43.* Suppose that X and Y are independent, identically distributed, geometric random variables with parameter p. Show that

$$\mathbf{P}(X = i \,|\, X + Y = n) = \frac{1}{n-1}, \qquad i = 1, \ldots, n-1.$$

Solution. Consider repeatedly and independently tossing a coin with probability of heads p. We can interpret $\mathbf{P}(X = i \,|\, X + Y = n)$ as the probability that we obtained a head for the first time on the ith toss given that we obtained a head for the second time on the nth toss. We can then argue, intuitively, that given that the second head occurred on the nth toss, the first head is equally likely to have come up at any toss between 1 and $n-1$. To establish this precisely, note that we have

$$\mathbf{P}(X = i \,|\, X + Y = n) = \frac{\mathbf{P}(X = i,\ X + Y = n)}{\mathbf{P}(X + Y = n)} = \frac{\mathbf{P}(X = i)\mathbf{P}(Y = n - i)}{\mathbf{P}(X + Y = n)}.$$

Also

$$\mathbf{P}(X = i) = p(1-p)^{i-1}, \qquad \text{for } i \geq 1,$$

and

$$\mathbf{P}(Y = n - i) = p(1-p)^{n-i-1}, \qquad \text{for } n - i \geq 1.$$

It follows that

$$\mathbf{P}(X = i)\mathbf{P}(Y = n - i) = \begin{cases} p^2(1-p)^{n-2}, & \text{if } i = 1, \ldots, n-1, \\ 0, & \text{otherwise.} \end{cases}$$

Therefore, for any i and j in the range $[1, n-1]$, we have

$$\mathbf{P}(X = i \,|\, X + Y = n) = \mathbf{P}(X = j \,|\, X + Y = n).$$

Hence

$$\mathbf{P}(X = i \,|\, X + Y = n) = \frac{1}{n-1}, \qquad i = 1, \ldots, n-1.$$

Problem 44.* Let X and Y be two random variables with given joint PMF, and let g and h be two functions of X and Y, respectively. Show that if X and Y are independent, then the same is true for the random variables $g(X)$ and $h(Y)$.

Solution. Let $U = g(X)$ and $V = h(Y)$. Then, we have

$$\begin{aligned} p_{U,V}(u, v) &= \sum_{\{(x,y) \,|\, g(x)=u,\, h(y)=v\}} p_{X,Y}(x, y) \\ &= \sum_{\{(x,y) \,|\, g(x)=u,\, h(y)=v\}} p_X(x) p_Y(y) \\ &= \sum_{\{x \,|\, g(x)=u\}} p_X(x) \sum_{\{x \,|\, h(y)=v\}} p_Y(y) \\ &= p_U(u) p_V(v), \end{aligned}$$

so U and V are independent.

Problem 45.* Variability extremes. Let $X_1, \ldots, X_n$ be independent random variables and let $X = X_1 + \cdots + X_n$ be their sum.

(a) Suppose that each X_i is Bernoulli with parameter p_i, and that $p_1, \ldots, p_n$ are chosen so that the mean of X is a given $\mu > 0$. Show that the variance of X is maximized if the p_i are chosen to be all equal to μ/n.

(b) Suppose that each X_i is geometric with parameter p_i, and that $p_1, \ldots, p_n$ are chosen so that the mean of X is a given $\mu > 0$. Show that the variance of X is minimized if the p_i are chosen to be all equal to n/μ. [Note the strikingly different character of the results of parts (a) and (b).]

Solution. (a) We have

$$\text{var}(X) = \sum_{i=1}^{n} \text{var}(X_i) = \sum_{i=1}^{n} p_i(1-p_i) = \mu - \sum_{i=1}^{n} p_i^2.$$

Thus maximizing the variance is equivalent to minimizing $\sum_{i=1}^n p_i^2$. It can be seen (using the constraint $\sum_{i=1}^n p_i = \mu$) that

$$\sum_{i=1}^{n} p_i^2 = \sum_{i=1}^{n} (\mu/n)^2 + \sum_{i=1}^{n} (p_i - \mu/n)^2,$$

so $\sum_{i=1}^n p_i^2$ is minimized when $p_i = \mu/n$ for all i.

(b) We have

$$\mu = \sum_{i=1}^{n} \mathbf{E}[X_i] = \sum_{i=1}^{n} \frac{1}{p_i},$$

and

$$\text{var}(X) = \sum_{i=1}^{n} \text{var}(X_i) = \sum_{i=1}^{n} \frac{1-p_i}{p_i^2}.$$

Introducing the change of variables $y_i = 1/p_i = \mathbf{E}[X_i]$, we see that the constraint becomes

$$\sum_{i=1}^{n} y_i = \mu,$$

and that we must minimize

$$\sum_{i=1}^{n} y_i(y_i - 1) = \sum_{i=1}^{n} y_i^2 - \mu,$$

subject to that constraint. This is the same problem as the one of part (a), so the method of proof given there applies.

Problem 46.* Entropy and uncertainty. Consider a random variable X that can take n values, $x_1, \ldots, x_n$, with corresponding probabilities $p_1, \ldots, p_n$. The **entropy** of X is defined to be

$$H(X) = -\sum_{i=1}^{n} p_i \log p_i.$$

(All logarithms in this problem are with respect to base two.) The entropy $H(X)$ provides a measure of the uncertainty about the value of X. To get a sense of this, note that $H(X) \geq 0$ and that $H(X)$ is very close to 0 when X is "nearly deterministic," i.e., takes one of its possible values with probability very close to 1 (since we have $p \log p \approx 0$ if either $p \approx 0$ or $p \approx 1$).

The notion of entropy is fundamental in information theory, which originated with C. Shannon's famous work and is described in many specialized textbooks. For example, it can be shown that $H(X)$ is a lower bound to the average number of yes-no questions (such as "is $X = x_1$?" or "is $X < x_5$?") that must be asked in order to determine the value of X. Furthermore, if k is the average number of questions required to determine the value of a string of independent identically distributed random variables $X_1, X_2, \ldots, X_n$, then, with a suitable strategy, k/n can be made as close to $H(X)$ as desired, when n is large.

(a) Show that if $q_1, \ldots, q_n$ are nonnegative numbers such that $\sum_{i=1}^n q_i = 1$, then

$$H(X) \leq -\sum_{i=1}^{n} p_i \log q_i,$$

with equality if and only if $p_i = q_i$ for all i. As a special case, show that $H(X) \leq \log n$, with equality if and only if $p_i = 1/n$ for all i. *Hint*: Use the inequality $\ln \alpha \leq \alpha - 1$, for $\alpha > 0$, which holds with equality if and only if $\alpha = 1$; here $\ln \alpha$ stands for the natural logarithm.

(b) Let X and Y be random variables taking a finite number of values, and having joint PMF $p_{X,Y}(x, y)$. Define

$$I(X, Y) = \sum_{x} \sum_{y} p_{X,Y}(x, y) \log \left(\frac{p_{X,Y}(x, y)}{p_X(x) p_Y(y)} \right).$$

Show that $I(X, Y) \geq 0$, and that $I(X, Y) = 0$ if and only if X and Y are independent.

(c) Show that

$$I(X, Y) = H(X) + H(Y) - H(X, Y),$$

where

$$H(X, Y) = -\sum_{x} \sum_{y} p_{X,Y}(x, y) \log p_{X,Y}(x, y),$$

$$H(X) = -\sum_{x} p_X(x) \log p_X(x), \qquad H(Y) = -\sum_{y} p_Y(y) \log p_Y(y).$$

(d) Show that

$$I(X, Y) = H(X) - H(X \,|\, Y),$$

where

$$H(X \,|\, Y) = -\sum_{y} p_Y(y) \sum_{x} p_{X|Y}(x \,|\, y) \log p_{X|Y}(x \,|\, y).$$

[Note that $H(X \,|\, Y)$ may be viewed as the conditional entropy of X given Y, that is, the entropy of the conditional distribution of X, given that $Y = y$, averaged over all possible values y. Thus, the quantity $I(X,Y) = H(X) - H(X \mid Y)$ is the reduction in the entropy (uncertainty) on X, when Y becomes known. It can be therefore interpreted as the information about X that is conveyed by Y, and is called the **mutual information** of X and Y.]

Solution. (a) We will use the inequality $\ln \alpha \le \alpha - 1$. (To see why this inequality is true, write $\ln \alpha = \int_1^\alpha \beta^{-1}\, d\beta < \int_1^\alpha d\beta = \alpha - 1$ for $\alpha > 1$, and write $\ln \alpha = -\int_\alpha^1 \beta^{-1}\, d\beta < -\int_\alpha^1 d\beta = \alpha - 1$ for $0 < \alpha < 1$.)

We have

$$-\sum_{i=1}^{n} p_i \ln p_i + \sum_{i=1}^{n} p_i \ln q_i = \sum_{i=1}^{n} p_i \ln\left(\frac{q_i}{p_i}\right) \le \sum_{i=1}^{n} p_i \left(\frac{q_i}{p_i} - 1\right) = 0,$$

with equality if and only if $p_i = q_i$ for all i. Since $\ln p = \log p \ln 2$, we obtain the desired relation $H(X) \le -\sum_{i=1}^{n} p_i \log q_i$. The inequality $H(X) \le \log n$ is obtained by setting $q_i = 1/n$ for all i.

(b) The numbers $p_X(x)p_Y(y)$ satisfy $\sum_x \sum_y p_X(x)p_Y(y) = 1$, so by part (a), we have

$$\sum_x \sum_y p_{X,Y}(x,y) \log\bigl(p_{X,Y}(x,y)\bigr) \ge \sum_x \sum_y p_{X,Y}(x,y) \log\bigl(p_X(x)p_Y(y)\bigr),$$

with equality if and only if

$$p_{X,Y}(x,y) = p_X(x)p_Y(y), \qquad \text{for all } x \text{ and } y,$$

which is equivalent to X and Y being independent.

(c) We have

$$I(X,Y) = \sum_x \sum_y p_{X,Y}(x,y) \log p_{X,Y}(x,y) - \sum_x \sum_y p_{X,Y}(x,y) \log\bigl(p_X(x)p_Y(y)\bigr),$$

and

$$\sum_x \sum_y p_{X,Y}(x,y) \log p_{X,Y}(x,y) = -H(X,Y),$$

$$\begin{aligned}
-\sum_x \sum_y p_{X,Y}(x,y) \log\bigl(p_X(x)p_Y(y)\bigr) &= -\sum_x \sum_y p_{X,Y}(x,y) \log p_X(x) \\
&\quad - \sum_x \sum_y p_{X,Y}(x,y) \log p_Y(y) \\
&= -\sum_x p_X(x) \log p_X(x) - \sum_y p_Y(y) \log p_Y(y) \\
&= H(X) + H(Y).
\end{aligned}$$

Combining the above three relations, we obtain $I(X,Y) = H(X) + H(Y) - H(X,Y)$.

(d) From the calculation in part (c), we have

$$\begin{aligned}
I(X,Y) &= \sum_x \sum_y p_{X,Y}(x,y) \log p_{X,Y}(x,y) - \sum_x p_X(x) \log p_X(x) \\
&\quad - \sum_x \sum_y p_{X,Y}(x,y) \log p_Y(y) \\
&= H(X) + \sum_x \sum_y p_{X,Y}(x,y) \log \left(\frac{p_{X,Y}(x,y)}{p_Y(y)} \right) \\
&= H(X) + \sum_x \sum_y p_Y(y) p_{X|Y}(x \,|\, y) \log p_{X|Y}(x \,|\, y) \\
&= H(X) - H(X \,|\, Y).
\end{aligned}$$

3

General Random Variables

Contents

Random variables with a continuous range of possible values are quite common; the velocity of a vehicle traveling along the highway could be one example. If the velocity is measured by a digital speedometer, we may view the speedometer's reading as a discrete random variable. But if we wish to model the exact velocity, a continuous random variable is called for. Models involving continuous random variables can be useful for several reasons. Besides being finer-grained and possibly more accurate, they allow the use of powerful tools from calculus and often admit an insightful analysis that would not be possible under a discrete model.

All of the concepts and methods introduced in Chapter 2, such as expectation, PMFs, and conditioning, have continuous counterparts. Developing and interpreting these counterparts is the subject of this chapter.

3.1 CONTINUOUS RANDOM VARIABLES AND PDFS

A random variable X is called **continuous** if there is a nonnegative function f_X, called the **probability density function of** X, or PDF for short, such that

$$\mathbf{P}(X \in B) = \int_B f_X(x)\,dx,$$

for every subset B of the real line.† In particular, the probability that the value of X falls within an interval is

$$\mathbf{P}(a \leq X \leq b) = \int_a^b f_X(x)\,dx,$$

and can be interpreted as the area under the graph of the PDF (see Fig. 3.1). For any single value a, we have $\mathbf{P}(X = a) = \int_a^a f_X(x)\,dx = 0$. For this reason, including or excluding the endpoints of an interval has no effect on its probability:

$$\mathbf{P}(a \leq X \leq b) = \mathbf{P}(a < X < b) = \mathbf{P}(a \leq X < b) = \mathbf{P}(a < X \leq b).$$

Note that to qualify as a PDF, a function f_X must be nonnegative, i.e., $f_X(x) \geq 0$ for every x, and must also have the normalization property

$$\int_{-\infty}^{\infty} f_X(x)\,dx = \mathbf{P}(-\infty < X < \infty) = 1.$$

---

† The integral $\int_B f_X(x)\,dx$ is to be interpreted in the usual calculus/Riemann sense and we implicitly assume that it is well-defined. For highly unusual functions and sets, this integral can be harder – or even impossible – to define, but such issues belong to a more advanced treatment of the subject. In any case, it is comforting to know that mathematical subtleties of this type do not arise if f_X is a piecewise continuous function with a finite or countable number of points of discontinuity, and B is the union of a finite or countable number of intervals.

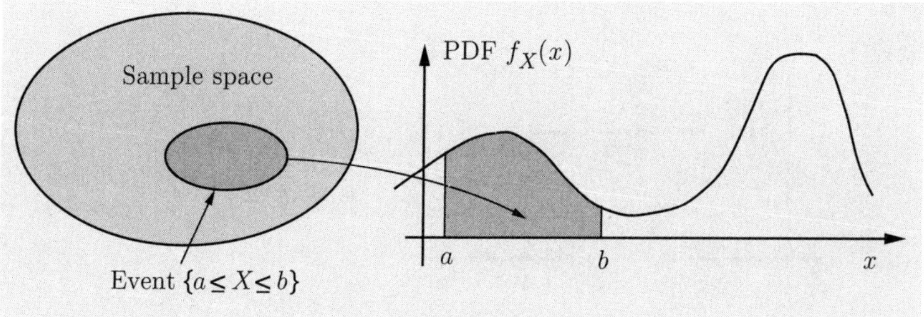

Figure 3.1: Illustration of a PDF. The probability that X takes a value in an interval $[a, b]$ is $\int_a^b f_X(x)\,dx$, which is the shaded area in the figure.

Graphically, this means that the entire area under the graph of the PDF must be equal to 1.

To interpret the PDF, note that for an interval $[x, x+\delta]$ with very small length δ, we have

$$\mathbf{P}\big([x, x+\delta]\big) = \int_x^{x+\delta} f_X(t)\,dt \approx f_X(x) \cdot \delta,$$

so we can view $f_X(x)$ as the "probability mass per unit length" near x (cf. Fig. 3.2). It is important to realize that even though a PDF is used to calculate event probabilities, $f_X(x)$ is not the probability of any particular event. In particular, it is not restricted to be less than or equal to one.

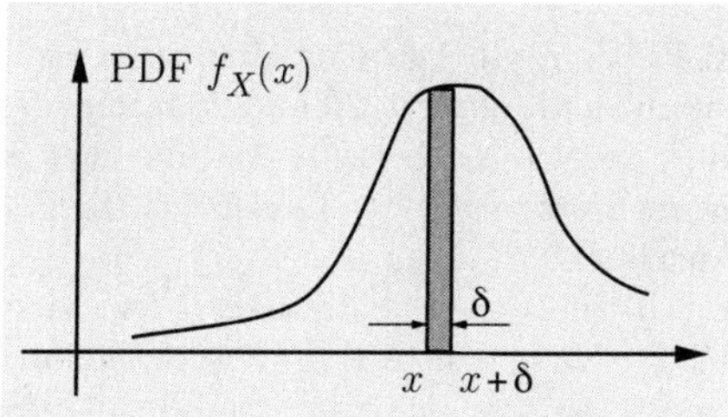

Figure 3.2: Interpretation of the PDF $f_X(x)$ as "probability mass per unit length" around x. If δ is very small, the probability that X takes a value in the interval $[x, x+\delta]$ is the shaded area in the figure, which is approximately equal to $f_X(x) \cdot \delta$.

Example 3.1. Continuous Uniform Random Variable. A gambler spins a wheel of fortune, continuously calibrated between 0 and 1, and observes the resulting number. Assuming that any two subintervals of [0,1] of the same length have the same probability, this experiment can be modeled in terms of a random variable X with PDF

$$f_X(x) = \begin{cases} c, & \text{if } 0 \leq x \leq 1, \\ 0, & \text{otherwise,} \end{cases}$$

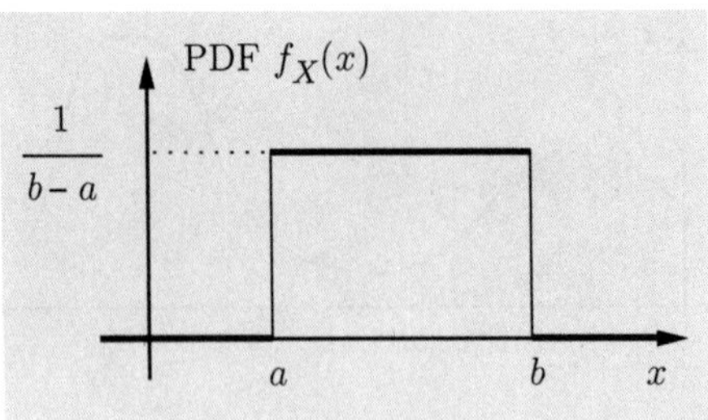

Figure 3.3: The PDF of a uniform random variable.

for some constant c. This constant can be determined by using the normalization property

$$1 = \int_{-\infty}^{\infty} f_X(x)\,dx = \int_0^1 c\,dx = c\int_0^1 dx = c,$$

so that $c = 1$.

More generally, we can consider a random variable X that takes values in an interval $[a, b]$, and again assume that any two subintervals of the same length have the same probability. We refer to this type of random variable as **uniform** or **uniformly distributed**. Its PDF has the form

$$f_X(x) = \begin{cases} \dfrac{1}{b-a}, & \text{if } a \le x \le b, \\ 0, & \text{otherwise,} \end{cases}$$

(cf. Fig. 3.3). The constant value of the PDF within $[a, b]$ is determined from the normalization property. Indeed, we have

$$1 = \int_{-\infty}^{\infty} f_X(x)\,dx = \int_a^b \frac{1}{b-a}\,dx.$$

Example 3.2. Piecewise Constant PDF. Alvin's driving time to work is between 15 and 20 minutes if the day is sunny, and between 20 and 25 minutes if the day is rainy, with all times being equally likely in each case. Assume that a day is sunny with probability 2/3 and rainy with probability 1/3. What is the PDF of the driving time, viewed as a random variable X?

We interpret the statement that "all times are equally likely" in the sunny and the rainy cases, to mean that the PDF of X is constant in each of the intervals [15, 20] and [20, 25]. Furthermore, since these two intervals contain all possible driving times, the PDF should be zero everywhere else:

$$f_X(x) = \begin{cases} c_1, & \text{if } 15 \le x < 20, \\ c_2, & \text{if } 20 \le x \le 25, \\ 0, & \text{otherwise,} \end{cases}$$

where c_1 and c_2 are some constants. We can determine these constants by using the given probabilities of a sunny and of a rainy day:

$$\frac{2}{3} = \mathbf{P}(\text{sunny day}) = \int_{15}^{20} f_X(x)\,dx = \int_{15}^{20} c_1\,dx = 5c_1,$$

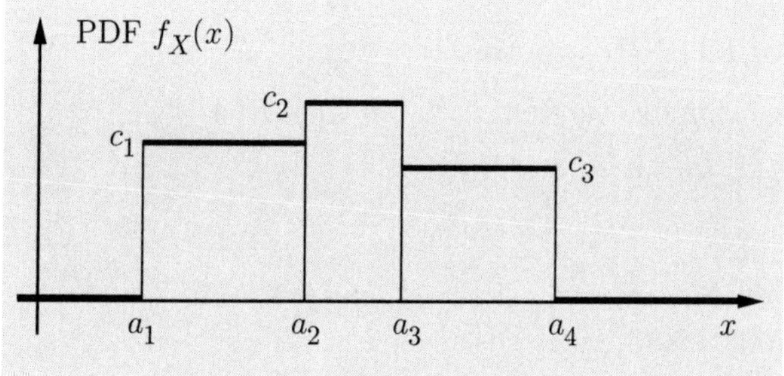

Figure 3.4: A piecewise constant PDF involving three intervals.

$$\frac{1}{3} = \mathbf{P}(\text{rainy day}) = \int_{20}^{25} f_X(x)\,dx = \int_{20}^{25} c_2\,dx = 5c_2,$$

so that

$$c_1 = \frac{2}{15}, \qquad c_2 = \frac{1}{15}.$$

Generalizing this example, consider a random variable X whose PDF has the piecewise constant form

$$f_X(x) = \begin{cases} c_i, & \text{if } a_i \le x < a_{i+1}, \quad i = 1, 2, \ldots, n-1, \\ 0, & \text{otherwise,} \end{cases}$$

where $a_1, a_2, \ldots, a_n$ are some scalars with $a_i < a_{i+1}$ for all i, and $c_1, c_2, \ldots, c_n$ are some nonnegative constants (cf. Fig. 3.4). The constants c_i may be determined by additional problem data, as in the preceding driving context. Generally, the c_i must be such that the normalization property holds:

$$1 = \int_{a_1}^{a_n} f_X(x)\,dx = \sum_{i=1}^{n-1} \int_{a_i}^{a_{i+1}} c_i\,dx = \sum_{i=1}^{n-1} c_i(a_{i+1} - a_i).$$

Example 3.3. A PDF Can Take Arbitrarily Large Values. Consider a random variable X with PDF

$$f_X(x) = \begin{cases} \dfrac{1}{2\sqrt{x}}, & \text{if } 0 < x \le 1, \\ 0, & \text{otherwise.} \end{cases}$$

Even though $f_X(x)$ becomes infinitely large as x approaches zero, this is still a valid PDF, because

$$\int_{-\infty}^{\infty} f_X(x)\,dx = \int_0^1 \frac{1}{2\sqrt{x}}\,dx = \sqrt{x}\,\Big|_0^1 = 1.$$

Summary of PDF Properties

Let X be a continuous random variable with PDF f_X.

- $f_X(x) \geq 0$ for all x.
- $\displaystyle\int_{-\infty}^{\infty} f_X(x)\,dx = 1.$
- If δ is very small, then $\mathbf{P}\big([x, x+\delta]\big) \approx f_X(x) \cdot \delta$.
- For any subset B of the real line,

$$\mathbf{P}(X \in B) = \int_B f_X(x)\,dx.$$

Expectation

The **expected value** or **expectation** or **mean** of a continuous random variable X is defined by†

$$\mathbf{E}[X] = \int_{-\infty}^{\infty} x f_X(x)\,dx.$$

This is similar to the discrete case except that the PMF is replaced by the PDF, and summation is replaced by integration. As in Chapter 2, $\mathbf{E}[X]$ can be interpreted as the "center of gravity" of the PDF and, also, as the anticipated average value of X in a large number of independent repetitions of the experiment. Its mathematical properties are similar to the discrete case – after all, an integral is just a limiting form of a sum.

If X is a continuous random variable with given PDF, any real-valued function $Y = g(X)$ of X is also a random variable. Note that Y can be a continuous random variable: for example, consider the trivial case where $Y = g(X) = X$. But Y can also turn out to be discrete. For example, suppose that

---

† One has to deal with the possibility that the integral $\int_{-\infty}^{\infty} x f_X(x)\,dx$ is infinite or undefined. More concretely, we will say that the expectation is well-defined if $\int_{-\infty}^{\infty} |x| f_X(x)\,dx < \infty$. In that case, it is known that the integral $\int_{-\infty}^{\infty} x f_X(x)\,dx$ takes a finite and unambiguous value.

For an example where the expectation is not well-defined, consider a random variable X with PDF $f_X(x) = c/(1+x^2)$, where c is a constant chosen to enforce the normalization condition. The expression $|x| f_X(x)$ can be approximated by $c/|x|$ when $|x|$ is large. Using the fact $\int_1^{\infty}(1/x)\,dx = \infty$, one can show that $\int_{-\infty}^{\infty} |x| f_X(x)\,dx = \infty$. Thus, $\mathbf{E}[X]$ is left undefined, despite the symmetry of the PDF around zero.

Throughout this book, in the absence of an indication to the contrary, we implicitly assume that the expected value of any random variable of interest is well-defined.

$g(x) = 1$ for $x > 0$, and $g(x) = 0$, otherwise. Then $Y = g(X)$ is a discrete random variable taking values in the finite set $\{0, 1\}$. In either case, the mean of $g(X)$ satisfies the **expected value rule**

$$\mathbf{E}\big[g(X)\big] = \int_{-\infty}^{\infty} g(x) f_X(x)\, dx,$$

in complete analogy with the discrete case; see the end-of-chapter problems.

The n**th moment** of a continuous random variable X is defined as $\mathbf{E}[X^n]$, the expected value of the random variable X^n. The **variance**, denoted by $\text{var}(X)$, is defined as the expected value of the random variable $\big(X - \mathbf{E}[X]\big)^2$.

We now summarize this discussion and list a number of additional facts that are practically identical to their discrete counterparts.

Expectation of a Continuous Random Variable and its Properties

Let X be a continuous random variable with PDF f_X.

- The expectation of X is defined by

$$\mathbf{E}[X] = \int_{-\infty}^{\infty} x f_X(x)\, dx.$$

- The expected value rule for a function $g(X)$ has the form

$$\mathbf{E}\big[g(X)\big] = \int_{-\infty}^{\infty} g(x) f_X(x)\, dx.$$

- The variance of X is defined by

$$\text{var}(X) = \mathbf{E}\Big[\big(X - \mathbf{E}[X]\big)^2\Big] = \int_{-\infty}^{\infty} \big(x - \mathbf{E}[X]\big)^2 f_X(x)\, dx.$$

- We have

$$0 \le \text{var}(X) = \mathbf{E}[X^2] - \big(\mathbf{E}[X]\big)^2.$$

- If $Y = aX + b$, where a and b are given scalars, then

$$\mathbf{E}[Y] = a\mathbf{E}[X] + b, \qquad \text{var}(Y) = a^2 \text{var}(X).$$

Example 3.4. Mean and Variance of the Uniform Random Variable. Consider a uniform PDF over an interval $[a, b]$, as in Example 3.1. We have

$$\mathbf{E}[X] = \int_{-\infty}^{\infty} x f_X(x)\, dx = \int_a^b x \cdot \frac{1}{b - a}\, dx$$

$$= \frac{1}{b-a} \cdot \frac{1}{2} x^2 \Big|_a^b$$

$$= \frac{1}{b-a} \cdot \frac{b^2 - a^2}{2}$$

$$= \frac{a+b}{2},$$

as one expects based on the symmetry of the PDF around $(a+b)/2$.

To obtain the variance, we first calculate the second moment. We have

$$\begin{aligned} \mathbf{E}[X^2] &= \int_a^b \frac{x^2}{b-a}\, dx = \frac{1}{b-a} \int_a^b x^2\, dx \\ &= \frac{1}{b-a} \cdot \frac{1}{3} x^3 \Big|_a^b = \frac{b^3 - a^3}{3(b-a)} \\ &= \frac{a^2 + ab + b^2}{3}. \end{aligned}$$

Thus, the variance is obtained as

$$\text{var}(X) = \mathbf{E}[X^2] - \big(\mathbf{E}[X]\big)^2 = \frac{a^2 + ab + b^2}{3} - \frac{(a+b)^2}{4} = \frac{(b-a)^2}{12},$$

after some calculation.

Exponential Random Variable

An **exponential** random variable has a PDF of the form

$$f_X(x) = \begin{cases} \lambda e^{-\lambda x}, & \text{if } x \geq 0, \\ 0, & \text{otherwise,} \end{cases}$$

where λ is a positive parameter characterizing the PDF (see Fig. 3.5). This is a legitimate PDF because

$$\int_{-\infty}^{\infty} f_X(x)\, dx = \int_0^{\infty} \lambda e^{-\lambda x}\, dx = -e^{-\lambda x} \Big|_0^{\infty} = 1.$$

Note that the probability that X exceeds a certain value decreases exponentially. Indeed, for any $a \geq 0$, we have

$$\mathbf{P}(X \geq a) = \int_a^{\infty} \lambda e^{-\lambda x}\, dx = -e^{-\lambda x} \Big|_a^{\infty} = e^{-\lambda a}.$$

An exponential random variable can, for example, be a good model for the amount of time until an incident of interest takes place, such as a message

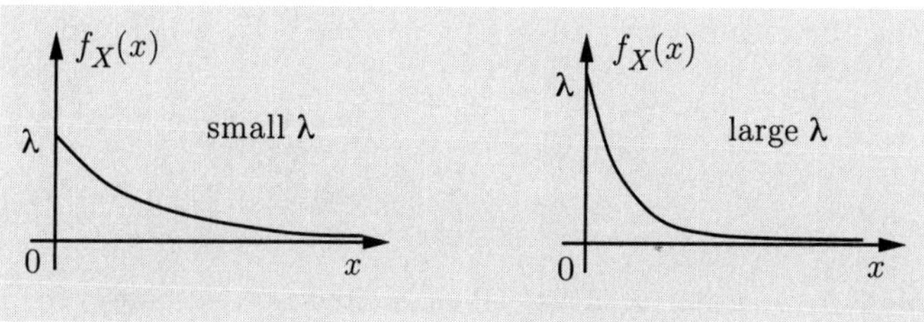

Figure 3.5: The PDF $\lambda e^{-\lambda x}$ of an exponential random variable.

arriving at a computer, some equipment breaking down, a light bulb burning out, an accident occurring, etc. We will see that it is closely connected to the geometric random variable, which also relates to the (discrete) time that will elapse until an incident of interest takes place. The exponential random variable will also play a major role in our study of stochastic processes in Chapter 6, but for the time being we will simply view it as a special random variable that is fairly tractable analytically.

The mean and the variance can be calculated to be

$$\mathbf{E}[X] = \frac{1}{\lambda}, \qquad \text{var}(X) = \frac{1}{\lambda^2}.$$

These formulas can be verified by straightforward calculation, as we now show. We have, using integration by parts,

$$\begin{aligned}\mathbf{E}[X] &= \int_0^\infty x\lambda e^{-\lambda x}\,dx \\ &= (-xe^{-\lambda x})\Big|_0^\infty + \int_0^\infty e^{-\lambda x}\,dx \\ &= 0 - \frac{e^{-\lambda x}}{\lambda}\Big|_0^\infty \\ &= \frac{1}{\lambda}.\end{aligned}$$

Using again integration by parts, the second moment is

$$\begin{aligned}\mathbf{E}[X^2] &= \int_0^\infty x^2\lambda e^{-\lambda x}\,dx \\ &= (-x^2e^{-\lambda x})\Big|_0^\infty + \int_0^\infty 2xe^{-\lambda x}\,dx \\ &= 0 + \frac{2}{\lambda}\mathbf{E}[X] \\ &= \frac{2}{\lambda^2}.\end{aligned}$$

Finally, using the formula $\text{var}(X) = \mathbf{E}[X^2] - \big(\mathbf{E}[X]\big)^2$, we obtain

$$\text{var}(X) = \frac{2}{\lambda^2} - \frac{1}{\lambda^2} = \frac{1}{\lambda^2}.$$

Example 3.5. The time until a small meteorite first lands anywhere in the Sahara desert is modeled as an exponential random variable with a mean of 10 days. The time is currently midnight. What is the probability that a meteorite first lands some time between 6 a.m. and 6 p.m. of the first day?

Let X be the time elapsed until the event of interest, measured in days. Then, X is exponential, with mean $1/\lambda = 10$, which yields $\lambda = 1/10$. The desired probability is

$$\mathbf{P}(1/4 \leq X \leq 3/4) = \mathbf{P}(X \geq 1/4) - \mathbf{P}(X > 3/4) = e^{-1/40} - e^{-3/40} = 0.0476,$$

where we have used the formula $\mathbf{P}(X \geq a) = \mathbf{P}(X > a) = e^{-\lambda a}$.

3.2 CUMULATIVE DISTRIBUTION FUNCTIONS

We have been dealing with discrete and continuous random variables in a somewhat different manner, using PMFs and PDFs, respectively. It would be desirable to describe all kinds of random variables with a single mathematical concept. This is accomplished with the **cumulative distribution function**, or CDF for short. The CDF of a random variable X is denoted by F_X and provides the probability $\mathbf{P}(X \leq x)$. In particular, for every x we have

$$F_X(x) = \mathbf{P}(X \leq x) = \begin{cases} \displaystyle\sum_{k \leq x} p_X(k), & \text{if } X \text{ is discrete,} \\ \displaystyle\int_{-\infty}^{x} f_X(t)\,dt, & \text{if } X \text{ is continuous.} \end{cases}$$

Loosely speaking, the CDF $F_X(x)$ "accumulates" probability "up to" the value x.

Any random variable associated with a given probability model has a CDF, regardless of whether it is discrete or continuous. This is because $\{X \leq x\}$ is always an event and therefore has a well-defined probability. In what follows, any unambiguous specification of the probabilities of all events of the form $\{X \leq x\}$, be it through a PMF, PDF, or CDF, will be referred to as the **probability law** of the random variable X.

Figures 3.6 and 3.7 illustrate the CDFs of various discrete and continuous random variables. From these figures, as well as from the definition, some general properties of the CDF can be observed.

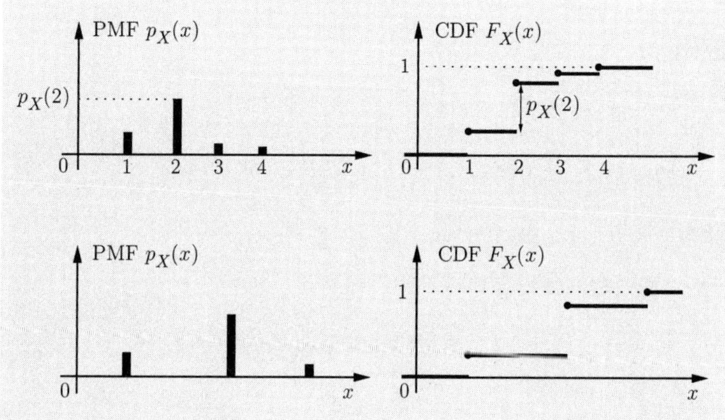

Figure 3.6: CDFs of some discrete random variables. The CDF is related to the PMF through the formula

$$F_X(x) = \mathbf{P}(X \leq x) = \sum_{k \leq x} p_X(k)$$

and has a staircase form, with jumps occurring at the values of positive probability mass. Note that at the points where a jump occurs, the value of F_X is the larger of the two corresponding values (i.e., F_X is continuous from the right).

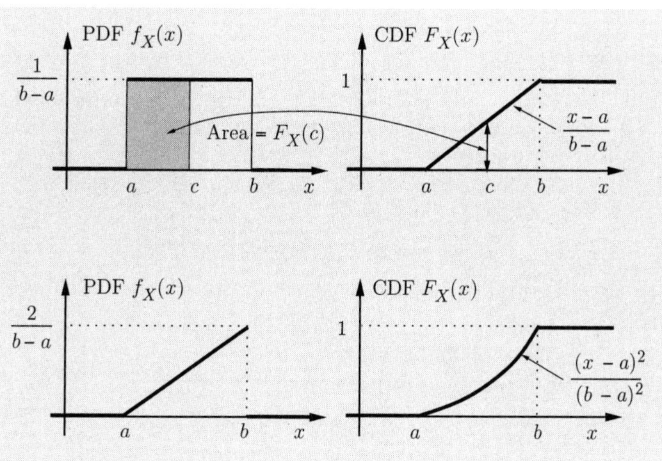

Figure 3.7: CDFs of some continuous random variables. The CDF is related to the PDF through the formula

$$F_X(x) = \mathbf{P}(X \leq x) = \int_{-\infty}^{x} f_X(t)\, dt.$$

Thus, the PDF f_X can be obtained from the CDF by differentiation:

$$f_X(x) = \frac{dF_X}{dx}(x).$$

For a continuous random variable, the CDF has no jumps, i.e., it is continuous.

Properties of a CDF

The CDF F_X of a random variable X is defined by

$$F_X(x) = \mathbf{P}(X \le x), \qquad \text{for all } x,$$

and has the following properties.

- F_X is monotonically nondecreasing:

$$\text{if } x \le y, \text{ then } F_X(x) \le F_X(y).$$

- $F_X(x)$ tends to 0 as $x \to -\infty$, and to 1 as $x \to \infty$.
- If X is discrete, then $F_X(x)$ is a piecewise constant function of x.
- If X is continuous, then $F_X(x)$ is a continuous function of x.
- If X is discrete and takes integer values, the PMF and the CDF can be obtained from each other by summing or differencing:

$$F_X(k) = \sum_{i=-\infty}^{k} p_X(i),$$

$$p_X(k) = \mathbf{P}(X \le k) - \mathbf{P}(X \le k-1) = F_X(k) - F_X(k-1),$$

for all integers k.
- If X is continuous, the PDF and the CDF can be obtained from each other by integration or differentiation:

$$F_X(x) = \int_{-\infty}^{x} f_X(t)\,dt, \qquad f_X(x) = \frac{dF_X}{dx}(x).$$

(The second equality is valid for those x at which the PDF is continuous.)

Sometimes, in order to calculate the PMF or PDF of a discrete or continuous random variable, respectively, it is more convenient to first calculate the CDF. The systematic use of this approach for functions of continuous random variables will be discussed in Section 4.1. The following is a discrete example.

Example 3.6. The Maximum of Several Random Variables. You are allowed to take a certain test three times, and your final score will be the maximum of the test scores. Thus,

$$X = \max\{X_1, X_2, X_3\},$$

where X_1, X_2, X_3 are the three test scores and X is the final score. Assume that your score in each test takes one of the values from 1 to 10 with equal probability 1/10, independently of the scores in other tests. What is the PMF p_X of the final score?

We calculate the PMF indirectly. We first compute the CDF F_X and then obtain the PMF as

$$p_X(k) = F_X(k) - F_X(k-1), \qquad k = 1, \ldots, 10.$$

We have

$$\begin{aligned} F_X(k) &= \mathbf{P}(X \le k) \\ &= \mathbf{P}(X_1 \le k,\ X_2 \le k,\ X_3 \le k) \\ &= \mathbf{P}(X_1 \le k)\,\mathbf{P}(X_2 \le k)\,\mathbf{P}(X_3 \le k) \\ &= \left(\frac{k}{10}\right)^3, \end{aligned}$$

where the third equality follows from the independence of the events $\{X_1 \le k\}$, $\{X_2 \le k\}$, $\{X_3 \le k\}$. Thus, the PMF is given by

$$p_X(k) = \left(\frac{k}{10}\right)^3 - \left(\frac{k-1}{10}\right)^3, \qquad k = 1, \ldots, 10.$$

The preceding line of argument can be generalized to any number of random variables $X_1, \ldots, X_n$. In particular, if the events $\{X_1 \le x\}, \ldots, \{X_n \le x\}$ are independent for every x, then the CDF of $X = \max\{X_1, \ldots, X_n\}$ is

$$F_X(x) = F_{X_1}(x) \cdots F_{X_n}(x).$$

From this formula, we can obtain $p_X(x)$ by differencing (if X is discrete), or $f_X(x)$ by differentiation (if X is continuous).

The Geometric and Exponential CDFs

Because the CDF is defined for any type of random variable, it provides a convenient means for exploring the relations between continuous and discrete random variables. A particularly interesting case in point is the relation between geometric and exponential random variables.

Let X be a geometric random variable with parameter p; that is, X is the number of trials until the first success in a sequence of independent Bernoulli trials, where the probability of success at each trial is p. Thus, for $k = 1, 2, \ldots$, we have $\mathbf{P}(X = k) = p(1-p)^{k-1}$ and the CDF is given by

$$F_{\text{geo}}(n) = \sum_{k=1}^{n} p(1-p)^{k-1} = p\frac{1-(1-p)^n}{1-(1-p)} = 1-(1-p)^n, \qquad \text{for } n = 1, 2, \ldots.$$

Suppose now that X is an exponential random variable with parameter $\lambda > 0$. Its CDF is given by

$$F_{\text{exp}}(x) = \mathbf{P}(X \leq x) = 0, \qquad \text{for } x \leq 0,$$

and

$$F_{\text{exp}}(x) = \int_0^x \lambda e^{-\lambda t} dt = -e^{-\lambda t}\Big|_0^x = 1 - e^{-\lambda x}, \qquad \text{for } x > 0.$$

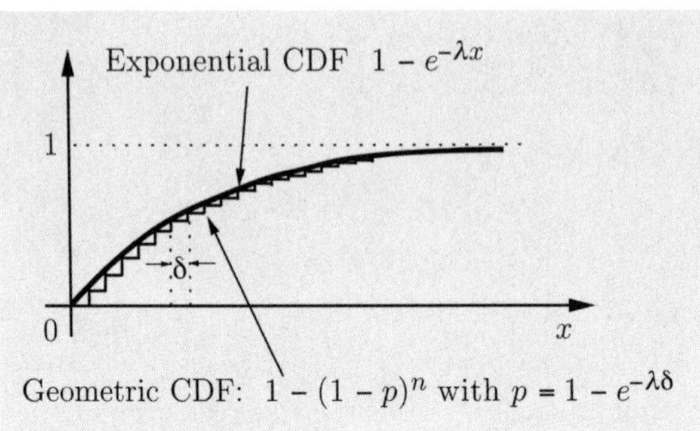

Figure 3.8: Relation of the geometric and the exponential CDFs. We have

$$F_{\text{exp}}(n\delta) = F_{\text{geo}}(n), \qquad n = 1, 2, \ldots,$$

where δ is chosen so that $e^{-\lambda\delta} = 1 - p$. As δ approaches 0, the exponential random variable can be interpreted as a "limit" of the geometric.

To compare the two CDFs above, let us define $\delta = -\ln(1-p)/\lambda$, so that

$$e^{-\lambda\delta} = 1 - p.$$

Then, we see that the values of the exponential and the geometric CDFs are equal whenever $x = n\delta$, with $n = 1, 2, \ldots$, i.e.,

$$F_{\text{exp}}(n\delta) = F_{\text{geo}}(n), \qquad n = 1, 2, \ldots,$$

and are close to each other for other values of x (see Fig. 3.8). Suppose now that we toss very quickly (every δ seconds, where $\delta \ll 1$) a biased coin with a very small probability of heads (equal to $p = 1 - e^{-\lambda\delta}$). Then, the first time to obtain a head (a geometric random variable with parameter p) is a close approximation to an exponential random variable with parameter λ, in the sense that the corresponding CDFs are very close to each other, as shown in Fig. 3.8. This relation between the geometric and the exponential random variables will play an important role when we study the Bernoulli and Poisson processes in Chapter 6.

3.3 NORMAL RANDOM VARIABLES

A continuous random variable X is said to be **normal** or **Gaussian** if it has a PDF of the form (see Fig. 3.9)

$$f_X(x) = \frac{1}{\sqrt{2\pi}\,\sigma} e^{-(x-\mu)^2/2\sigma^2},$$

where μ and σ are two scalar parameters characterizing the PDF, with σ assumed positive. It can be verified that the normalization property

$$\frac{1}{\sqrt{2\pi}\,\sigma} \int_{-\infty}^{\infty} e^{-(x-\mu)^2/2\sigma^2}\,dx = 1$$

holds (see the end-of-chapter problems).

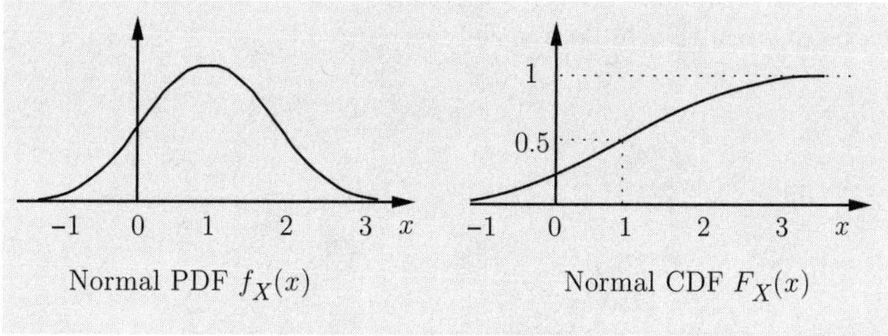

Figure 3.9: A normal PDF and CDF, with $\mu = 1$ and $\sigma^2 = 1$. We observe that the PDF is symmetric around its mean μ, and has a characteristic bell shape. As x gets further from μ, the term $e^{-(x-\mu)^2/2\sigma^2}$ decreases very rapidly. In this figure, the PDF is very close to zero outside the interval $[-1, 3]$.

The mean and the variance can be calculated to be

$$\mathbf{E}[X] = \mu, \qquad \text{var}(X) = \sigma^2.$$

To see this, note that the PDF is symmetric around μ, so the mean can only be μ. Furthermore, the variance is given by

$$\text{var}(X) = \frac{1}{\sqrt{2\pi}\,\sigma} \int_{-\infty}^{\infty} (x-\mu)^2 e^{-(x-\mu)^2/2\sigma^2}\,dx.$$

Using the change of variables $y = (x-\mu)/\sigma$ and integration by parts, we have

$$\begin{aligned}
\text{var}(X) &= \frac{\sigma^2}{\sqrt{2\pi}} \int_{-\infty}^{\infty} y^2 e^{-y^2/2}\,dy \\
&= \frac{\sigma^2}{\sqrt{2\pi}} \left(-ye^{-y^2/2}\right)\Big|_{-\infty}^{\infty} + \frac{\sigma^2}{\sqrt{2\pi}} \int_{-\infty}^{\infty} e^{-y^2/2}\,dy \\
&= \frac{\sigma^2}{\sqrt{2\pi}} \int_{-\infty}^{\infty} e^{-y^2/2}\,dy \\
&= \sigma^2.
\end{aligned}$$

The last equality above is obtained by using the fact

$$\frac{1}{\sqrt{2\pi}} \int_{-\infty}^{\infty} e^{-y^2/2}\, dy = 1,$$

which is just the normalization property of the normal PDF for the case where $\mu = 0$ and $\sigma = 1$.

A normal random variable has several special properties. The following one is particularly important and will be justified in Section 4.1.

Normality is Preserved by Linear Transformations

If X is a normal random variable with mean μ and variance σ^2, and if $a \neq 0$, b are scalars, then the random variable

$$Y = aX + b$$

is also normal, with mean and variance

$$\mathbf{E}[Y] = a\mu + b, \qquad \text{var}(Y) = a^2\sigma^2.$$

The Standard Normal Random Variable

A normal random variable Y with zero mean and unit variance is said to be a **standard normal**. Its CDF is denoted by Φ:

$$\Phi(y) = \mathbf{P}(Y \leq y) = \mathbf{P}(Y < y) = \frac{1}{\sqrt{2\pi}} \int_{-\infty}^{y} e^{-t^2/2}\, dt.$$

It is recorded in a table (given in the next page), and is a very useful tool for calculating various probabilities involving normal random variables; see also Fig. 3.10.

Note that the table only provides the values of $\Phi(y)$ for $y \geq 0$, because the omitted values can be found using the symmetry of the PDF. For example, if Y is a standard normal random variable, we have

$$\begin{aligned}\Phi(-0.5) &= \mathbf{P}(Y \leq -0.5) = \mathbf{P}(Y \geq 0.5) = 1 - \mathbf{P}(Y < 0.5)\\ &= 1 - \Phi(0.5) = 1 - .6915 = 0.3085.\end{aligned}$$

More generally, we have

$$\Phi(-y) = 1 - \Phi(y), \qquad \text{for all } y.$$

	.00	.01	.02	.03	.04	.05	.06	.07	.08	.09
0.0	.5000	.5040	.5080	.5120	.5160	.5199	.5239	.5279	.5319	.5359
0.1	.5398	.5438	.5478	.5517	.5557	.5596	.5636	.5675	.5714	.5753
0.2	.5793	.5832	.5871	.5910	.5948	.5987	.6026	.6064	.6103	.6141
0.3	.6179	.6217	.6255	.6293	.6331	.6368	.6406	.6443	.6480	.6517
0.4	.6554	.6591	.6628	.6664	.6700	.6736	.6772	.6808	.6844	.6879
0.5	.6915	.6950	.6985	.7019	.7054	.7088	.7123	.7157	.7190	.7224
0.6	.7257	.7291	.7324	.7357	.7389	.7422	.7454	.7486	.7517	.7549
0.7	.7580	.7611	.7642	.7673	.7704	.7734	.7764	.7794	.7823	.7852
0.8	.7881	.7910	.7939	.7967	.7995	.8023	.8051	.8078	.8106	.8133
0.9	.8159	.8186	.8212	.8238	.8264	.8289	.8315	.8340	.8365	.8389
1.0	.8413	.8438	.8461	.8485	.8508	.8531	.8554	.8577	.8599	.8621
1.1	.8643	.8665	.8686	.8708	.8729	.8749	.8770	.8790	.8810	.8830
1.2	.8849	.8869	.8888	.8907	.8925	.8944	.8962	.8980	.8997	.9015
1.3	.9032	.9049	.9066	.9082	.9099	.9115	.9131	.9147	.9162	.9177
1.4	.9192	.9207	.9222	.9236	.9251	.9265	.9279	.9292	.9306	.9319
1.5	.9332	.9345	.9357	.9370	.9382	.9394	.9406	.9418	.9429	.9441
1.6	.9452	.9463	.9474	.9484	.9495	.9505	.9515	.9525	.9535	.9545
1.7	.9554	.9564	.9573	.9582	.9591	.9599	.9608	.9616	.9625	.9633
1.8	.9641	.9649	.9656	.9664	.9671	.9678	.9686	.9693	.9699	.9706
1.9	.9713	.9719	.9726	.9732	.9738	.9744	.9750	.9756	.9761	.9767
2.0	.9772	.9778	.9783	.9788	.9793	.9798	.9803	.9808	.9812	.9817
2.1	.9821	.9826	.9830	.9834	.9838	.9842	.9846	.9850	.9854	.9857
2.2	.9861	.9864	.9868	.9871	.9875	.9878	.9881	.9884	.9887	.9890
2.3	.9893	.9896	.9898	.9901	.9904	.9906	.9909	.9911	.9913	.9916
2.4	.9918	.9920	.9922	.9925	.9927	.9929	.9931	.9932	.9934	.9936
2.5	.9938	.9940	.9941	.9943	.9945	.9946	.9948	.9949	.9951	.9952
2.6	.9953	.9955	.9956	.9957	.9959	.9960	.9961	.9962	.9963	.9964
2.7	.9965	.9966	.9967	.9968	.9969	.9970	.9971	.9972	.9973	.9974
2.8	.9974	.9975	.9976	.9977	.9977	.9978	.9979	.9979	.9980	.9981
2.9	.9981	.9982	.9982	.9983	.9984	.9984	.9985	.9985	.9986	.9986
3.0	.9987	.9987	.9987	.9988	.9988	.9989	.9989	.9989	.9990	.9990
3.1	.9990	.9991	.9991	.9991	.9992	.9992	.9992	.9992	.9993	.9993
3.2	.9993	.9993	.9994	.9994	.9994	.9994	.9994	.9995	.9995	.9995
3.3	.9995	.9995	.9995	.9996	.9996	.9996	.9996	.9996	.9996	.9997
3.4	.9997	.9997	.9997	.9997	.9997	.9997	.9997	.9997	.9997	.9998

The standard normal table. The entries in this table provide the numerical values of $\Phi(y) = \mathbf{P}(Y \leq y)$, where Y is a standard normal random variable, for y between 0 and 3.49. For example, to find $\Phi(1.71)$, we look at the row corresponding to 1.7 and the column corresponding to 0.01, so that $\Phi(1.71) = .9564$. When y is negative, the value of $\Phi(y)$ can be found using the formula $\Phi(y) = 1 - \Phi(-y)$.

Let X be a normal random variable with mean μ and variance σ^2. We "standardize" X by defining a new random variable Y given by

$$Y = \frac{X - \mu}{\sigma}.$$

Since Y is a linear function of X, it is normal. Furthermore,

$$\mathbf{E}[Y] = \frac{\mathbf{E}[X] - \mu}{\sigma} = 0, \qquad \text{var}(Y) = \frac{\text{var}(X)}{\sigma^2} = 1.$$

Thus, Y is a standard normal random variable. This fact allows us to calculate the probability of any event defined in terms of X: we redefine the event in terms of Y, and then use the standard normal table.

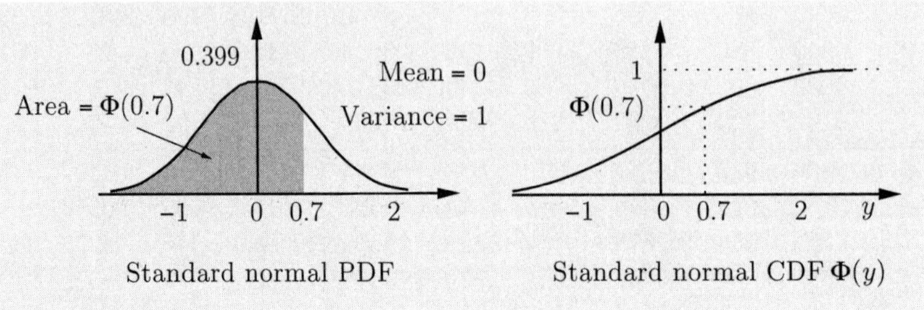

Figure 3.10: The PDF

$$f_Y(y) = \frac{1}{\sqrt{2\pi}} e^{-y^2/2}$$

of the standard normal random variable. The corresponding CDF, which is denoted by Φ, is recorded in a table.

Example 3.7. Using the Normal Table. The annual snowfall at a particular geographic location is modeled as a normal random variable with a mean of $\mu = 60$ inches and a standard deviation of $\sigma = 20$. What is the probability that this year's snowfall will be at least 80 inches?

Let X be the snow accumulation, viewed as a normal random variable, and let

$$Y = \frac{X - \mu}{\sigma} = \frac{X - 60}{20},$$

be the corresponding standard normal random variable. We have

$$\mathbf{P}(X \geq 80) = \mathbf{P}\left(\frac{X - 60}{20} \geq \frac{80 - 60}{20}\right) = \mathbf{P}\left(Y \geq \frac{80 - 60}{20}\right) = \mathbf{P}(Y \geq 1) = 1 - \Phi(1),$$

where Φ is the CDF of the standard normal. We read the value $\Phi(1)$ from the table:

$$\Phi(1) = 0.8413,$$

so that

$$\mathbf{P}(X \geq 80) = 1 - \Phi(1) = 0.1587.$$

Generalizing the approach in the preceding example, we have the following procedure.

CDF Calculation for a Normal Random Variable

For a normal random variable X with mean μ and variance σ^2, we use a two-step procedure.

(a) "Standardize" X, i.e., subtract μ and divide by σ to obtain a standard normal random variable Y.

(b) Read the CDF value from the standard normal table:

$$\mathbf{P}(X \leq x) = \mathbf{P}\left(\frac{X-\mu}{\sigma} \leq \frac{x-\mu}{\sigma}\right) = \mathbf{P}\left(Y \leq \frac{x-\mu}{\sigma}\right) = \Phi\left(\frac{x-\mu}{\sigma}\right).$$

Normal random variables are often used in signal processing and communications engineering to model noise and unpredictable distortions of signals. The following is a typical example.

Example 3.8. Signal Detection. A binary message is transmitted as a signal s, which is either -1 or $+1$. The communication channel corrupts the transmission with additive normal noise with mean $\mu = 0$ and variance σ^2. The receiver concludes that the signal -1 (or $+1$) was transmitted if the value received is < 0 (or ≥ 0, respectively); see Fig. 3.11. What is the probability of error?

An error occurs whenever -1 is transmitted and the noise N is at least 1 so that $s+N = -1+N \geq 0$, or whenever $+1$ is transmitted and the noise N is smaller than -1 so that $s+N = 1+N < 0$. In the former case, the probability of error is

$$\begin{aligned}\mathbf{P}(N \geq 1) = 1 - \mathbf{P}(N < 1) &= 1 - \mathbf{P}\left(\frac{N-\mu}{\sigma} < \frac{1-\mu}{\sigma}\right) \\ &= 1 - \Phi\left(\frac{1-\mu}{\sigma}\right) = 1 - \Phi\left(\frac{1}{\sigma}\right).\end{aligned}$$

In the latter case, the probability of error is the same, by symmetry. The value of $\Phi(1/\sigma)$ can be obtained from the normal table. For $\sigma = 1$, we have $\Phi(1/\sigma) = \Phi(1) = 0.8413$, and the probability of error is 0.1587.

Normal random variables play an important role in a broad range of probabilistic models. The main reason is that, generally speaking, they model well the additive effect of many independent factors in a variety of engineering, physical, and statistical contexts. Mathematically, the key fact is that *the sum of a large*

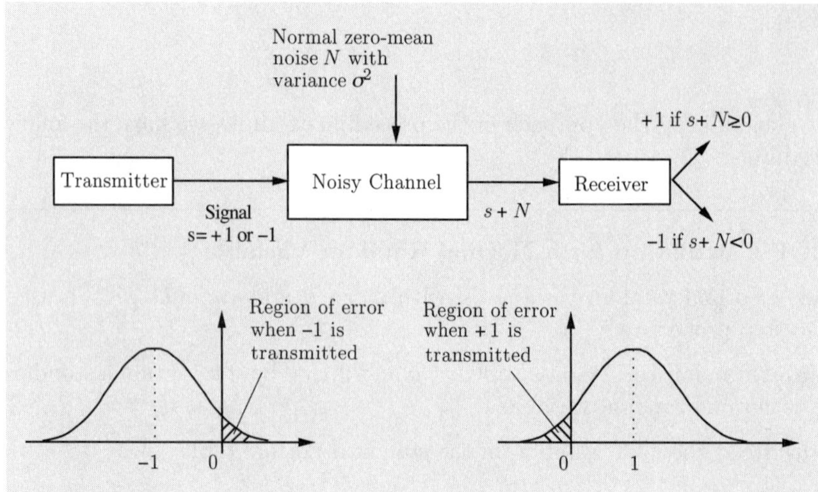

Figure 3.11: The signal detection scheme of Example 3.8. The area of the shaded region gives the probability of error in the two cases where -1 and $+1$ is transmitted.

number of independent and identically distributed (not necessarily normal) random variables has an approximately normal CDF, regardless of the CDF of the individual random variables. This property is captured in the celebrated *central limit theorem*, which will be discussed in Chapter 5.

3.4 JOINT PDFS OF MULTIPLE RANDOM VARIABLES

We will now extend the notion of a PDF to the case of multiple random variables. In complete analogy with discrete random variables, we introduce joint, marginal, and later on, conditional PDFs. Their intuitive interpretation as well as their main properties parallel the discrete case.

We say that two continuous random variables associated with the same experiment are **jointly continuous** and can be described in terms of a **joint PDF** $f_{X,Y}$ if $f_{X,Y}$ is a nonnegative function that satisfies

$$\mathbf{P}\big((X,Y)\in B\big) = \int\int_{(x,y)\in B} f_{X,Y}(x,y)\,dx\,dy,$$

for every subset B of the two-dimensional plane. The notation above means that the integration is carried over the set B. In the particular case where B is a rectangle of the form $B = \big\{(x,y)\,|\,a\leq x\leq b,\ c\leq y\leq d\big\}$, we have

$$\mathbf{P}(a\leq X\leq b,\, c\leq Y\leq d) = \int_c^d\int_a^b f_{X,Y}(x,y)\,dx\,dy.$$

Furthermore, by letting B be the entire two-dimensional plane, we obtain the normalization property

$$\int_{-\infty}^{\infty}\int_{-\infty}^{\infty} f_{X,Y}(x,y)\,dx\,dy = 1.$$

To interpret the joint PDF, we let δ be a small positive number and consider the probability of a small rectangle. We have

$$\mathbf{P}(a \le X \le a+\delta,\ c \le Y \le c+\delta) = \int_{c}^{c+\delta}\int_{a}^{a+\delta} f_{X,Y}(x,y)\,dx\,dy \approx f_{X,Y}(a,c)\cdot\delta^2,$$

so we can view $f_{X,Y}(a,c)$ as the "probability per unit area" in the vicinity of (a,c).

The joint PDF contains all relevant probabilistic information on the random variables X, Y, and their dependencies. It allows us to calculate the probability of any event that can be defined in terms of these two random variables. As a special case, it can be used to calculate the probability of an event involving only one of them. For example, let A be a subset of the real line and consider the event $\{X \in A\}$. We have

$$\mathbf{P}(X \in A) = \mathbf{P}\big(X \in A \text{ and } Y \in (-\infty,\infty)\big) = \int_{A}\int_{-\infty}^{\infty} f_{X,Y}(x,y)\,dy\,dx.$$

Comparing with the formula

$$\mathbf{P}(X \in A) = \int_{A} f_X(x)\,dx,$$

we see that the **marginal** PDF f_X of X is given by

$$f_X(x) = \int_{-\infty}^{\infty} f_{X,Y}(x,y)\,dy.$$

Similarly,

$$f_Y(y) = \int_{-\infty}^{\infty} f_{X,Y}(x,y)\,dx.$$

Example 3.9. Two-Dimensional Uniform PDF. Romeo and Juliet have a date at a given time, and each will arrive at the meeting place with a delay between 0 and 1 hour (recall the example given in Section 1.2). Let X and Y denote the delays of Romeo and Juliet, respectively. Assuming that no pairs (x,y) in the unit square are more likely than others, a natural model involves a joint PDF of the form

$$f_{X,Y}(x,y) = \begin{cases} c, & \text{if } 0 \le x \le 1 \text{ and } 0 \le y \le 1,\\ 0, & \text{otherwise,}\end{cases}$$

where c is a constant. For this PDF to satisfy the normalization property

$$\int_{-\infty}^{\infty}\int_{-\infty}^{\infty} f_{X,Y}(x,y)\,dx\,dy = \int_{0}^{1}\int_{0}^{1} c\,dx\,dy = 1,$$

we must have

$$c = 1.$$

This is an example of a uniform joint PDF. More generally, let us fix some subset S of the two-dimensional plane. The corresponding uniform joint PDF on S is defined to be

$$f_{X,Y}(x,y) = \begin{cases} \dfrac{1}{\text{area of } S}, & \text{if } (x,y) \in S, \\ 0, & \text{otherwise.} \end{cases}$$

For any set $A \subset S$, the probability that (X,Y) lies in A is

$$\mathbf{P}\big((X,Y) \in A\big) = \int\int_{(x,y)\in A} f_{X,Y}(x,y)\,dx\,dy = \frac{1}{\text{area of } S}\int\int_{(x,y)\in A} dx\,dy = \frac{\text{area of } A}{\text{area of } S}.$$

Example 3.10. We are told that the joint PDF of the random variables X and Y is a constant c on the set S shown in Fig. 3.12 and is zero outside. We wish to determine the value of c and the marginal PDFs of X and Y.

The area of the set S is equal to 4 and, therefore, $f_{X,Y}(x,y) = c = 1/4$, for $(x,y) \in S$. To find the marginal PDF $f_X(x)$ for some particular x, we integrate (with respect to y) the joint PDF over the vertical line corresponding to that x. The resulting PDF is shown in the figure. We can compute f_Y similarly.

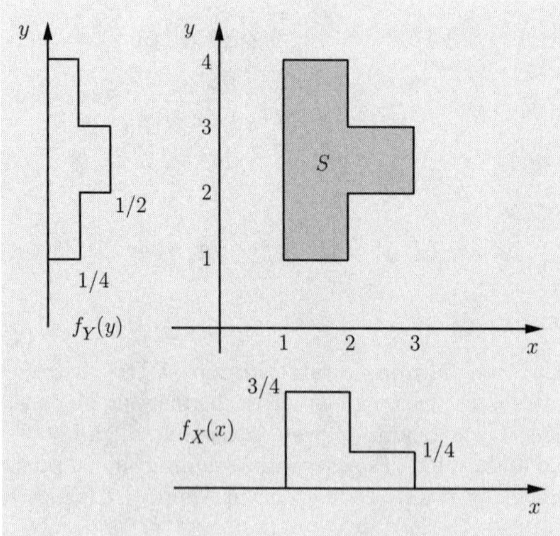

Figure 3.12: The joint PDF in Example 3.10 and the resulting marginal PDFs.

Example 3.11. Buffon's Needle.† This is a famous example, which marks the origin of the subject of geometric probability, that is, the analysis of the geometric configuration of randomly placed objects.

A surface is ruled with parallel lines, which are at distance d from each other (see Fig. 3.13). Suppose that we throw a needle of length l on the surface at random. What is the probability that the needle will intersect one of the lines?

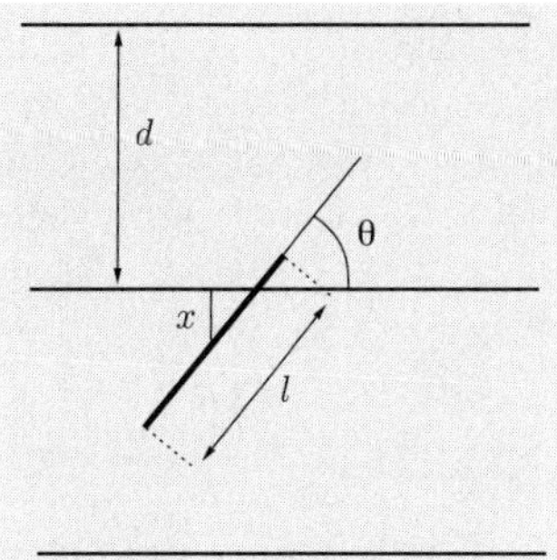

Figure 3.13: Buffon's needle. The length of the line segment between the midpoint of the needle and the point of intersection of the axis of the needle with the closest parallel line is $x/\sin\theta$. The needle will intersect the closest parallel line if and only if this length is less than $l/2$.

We assume here that $l < d$ so that the needle cannot intersect two lines simultaneously. Let X be the vertical distance from the midpoint of the needle to the nearest of the parallel lines, and let Θ be the acute angle formed by the axis of the needle and the parallel lines (see Fig. 3.13). We model the pair of random variables (X, Θ) with a uniform joint PDF over the rectangular set $\{(x,\theta) \mid 0 \le x \le d/2,\ 0 \le \theta \le \pi/2\}$, so that

$$f_{X,\Theta}(x,\theta) = \begin{cases} 4/(\pi d), & \text{if } x \in [0, d/2] \text{ and } \theta \in [0, \pi/2], \\ 0, & \text{otherwise.} \end{cases}$$

As can be seen from Fig. 3.13, the needle will intersect one of the lines if and only if

$$X \le \frac{l}{2} \sin\Theta,$$

so the probability of intersection is

$$\mathbf{P}\big(X \le (l/2)\sin\Theta\big) = \int\!\!\int_{x \le (l/2)\sin\theta} f_{X,\Theta}(x,\theta)\,dx\,d\theta$$

† This problem was posed and solved in 1777 by the French naturalist Buffon. A number of variants of the problem have been investigated, including the case where the surface is ruled with two sets of perpendicular lines (Laplace, 1812); see the end-of-chapter problems. The problem has long fascinated scientists, and has been used as a basis for experimental evaluations of π (among others, it has been reported that a captain named Fox measured π experimentally using needles, while recovering from wounds suffered in the American Civil War). The internet contains several graphical simulation programs for computing π using Buffon's ideas.

$$
\begin{aligned}
&= \frac{4}{\pi d} \int_0^{\pi/2} \int_0^{(l/2)\sin\theta} dx\, d\theta \\
&= \frac{4}{\pi d} \int_0^{\pi/2} \frac{l}{2} \sin\theta\, d\theta \\
&= \frac{2l}{\pi d} (-\cos\theta)\Big|_0^{\pi/2} \\
&= \frac{2l}{\pi d}.
\end{aligned}
$$

The probability of intersection can be empirically estimated, by repeating the experiment a large number of times. Since it is equal to $2l/\pi d$, this provides us with a method for the experimental evaluation of π.

Joint CDFs

If X and Y are two random variables associated with the same experiment, we define their joint CDF by

$$F_{X,Y}(x,y) = \mathbf{P}(X \leq x,\, Y \leq y).$$

As in the case of a single random variable, the advantage of working with the CDF is that it applies equally well to discrete and continuous random variables. In particular, if X and Y are described by a joint PDF $f_{X,Y}$, then

$$F_{X,Y}(x,y) = \mathbf{P}(X \leq x,\, Y \leq y) = \int_{-\infty}^{x} \int_{-\infty}^{y} f_{X,Y}(s,t)\, dt\, ds.$$

Conversely, the PDF can be recovered from the CDF by differentiating:

$$f_{X,Y}(x,y) = \frac{\partial^2 F_{X,Y}}{\partial x \partial y}(x,y).$$

Example 3.12. Let X and Y be described by a uniform PDF on the unit square. The joint CDF is given by

$$F_{X,Y}(x,y) = \mathbf{P}(X \leq x,\, Y \leq y) = xy, \qquad \text{for } 0 \leq x, y \leq 1.$$

We then verify that

$$\frac{\partial^2 F_{X,Y}}{\partial x \partial y}(x,y) = \frac{\partial^2 (xy)}{\partial x \partial y}(x,y) = 1 = f_{X,Y}(x,y),$$

for all (x,y) in the unit square.

Expectation

If X and Y are jointly continuous random variables and g is some function, then $Z = g(X, Y)$ is also a random variable. We will see in Section 4.1 methods for computing the PDF of Z, if it has one. For now, let us note that the expected value rule is still applicable and

$$\mathbf{E}\big[g(X,Y)\big] = \int_{-\infty}^{\infty}\int_{-\infty}^{\infty} g(x,y) f_{X,Y}(x,y)\,dx\,dy.$$

As an important special case, for any scalars a, b, and c, we have

$$\mathbf{E}[aX + bY + c] = a\mathbf{E}[X] + b\mathbf{E}[Y] + c.$$

More than Two Random Variables

The joint PDF of three random variables X, Y, and Z is defined in analogy with the case of two random variables. For example, we have

$$\mathbf{P}\big((X,Y,Z) \in B\big) = \iiint_{(x,y,z)\in B} f_{X,Y,Z}(x,y,z)\,dx\,dy\,dz,$$

for any set B. We also have relations such as

$$f_{X,Y}(x,y) = \int_{-\infty}^{\infty} f_{X,Y,Z}(x,y,z)\,dz,$$

and

$$f_X(x) = \int_{-\infty}^{\infty}\int_{-\infty}^{\infty} f_{X,Y,Z}(x,y,z)\,dy\,dz.$$

The expected value rule takes the form

$$\mathbf{E}\big[g(X,Y,Z)\big] = \int_{-\infty}^{\infty}\int_{-\infty}^{\infty}\int_{-\infty}^{\infty} g(x,y,z) f_{X,Y,Z}(x,y,z)\,dx\,dy\,dz,$$

and if g is linear, of the form $aX + bY + cZ$, then

$$\mathbf{E}[aX + bY + cZ] = a\mathbf{E}[X] + b\mathbf{E}[Y] + c\mathbf{E}[Z].$$

Furthermore, there are obvious generalizations of the above to the case of more than three random variables. For example, for any random variables $X_1, X_2, \ldots,$ X_n and any scalars $a_1, a_2, \ldots, a_n$, we have

$$\mathbf{E}[a_1X_1 + a_2X_2 + \cdots + a_nX_n] = a_1\mathbf{E}[X_1] + a_2\mathbf{E}[X_2] + \cdots + a_n\mathbf{E}[X_n].$$

Summary of Facts about Joint PDFs

Let X and Y be jointly continuous random variables with joint PDF $f_{X,Y}$.

- The **joint PDF** is used to calculate probabilities:

$$\mathbf{P}\big((X,Y)\in B\big) = \int\int_{(x,y)\in B} f_{X,Y}(x,y)\,dx\,dy.$$

- The **marginal PDF**s of X and Y can be obtained from the joint PDF, using the formulas

$$f_X(x) = \int_{-\infty}^{\infty} f_{X,Y}(x,y)\,dy, \qquad f_Y(y) = \int_{-\infty}^{\infty} f_{X,Y}(x,y)\,dx.$$

- The **joint CDF** is defined by $F_{X,Y}(x,y) = \mathbf{P}(X \le x,\, Y \le y)$, and determines the joint PDF through the formula

$$f_{X,Y}(x,y) = \frac{\partial^2 F_{X,Y}}{\partial x \partial y}(x,y),$$

for every (x,y) at which the joint PDF is continuous.

- A function $g(X,Y)$ of X and Y defines a new random variable, and

$$\mathbf{E}\big[g(X,Y)\big] = \int_{-\infty}^{\infty}\int_{-\infty}^{\infty} g(x,y) f_{X,Y}(x,y)\,dx\,dy.$$

If g is linear, of the form $aX+bY+c$, we have

$$\mathbf{E}[aX+bY+c] = a\mathbf{E}[X] + b\mathbf{E}[Y] + c.$$

- The above have natural extensions to the case where more than two random variables are involved.

3.5 CONDITIONING

Similar to the case of discrete random variables, we can condition a random variable on an event or on another random variable, and define the concepts of conditional PDF and conditional expectation. The various definitions and formulas parallel the ones for the discrete case, and their interpretation is similar, except for some subtleties that arise when we condition on an event of the form $\{Y = y\}$, which has zero probability.

Conditioning a Random Variable on an Event

The **conditional PDF** of a continuous random variable X, given an event A with $\mathbf{P}(A) > 0$, is defined as a nonnegative function $f_{X|A}$ that satisfies

$$\mathbf{P}(X \in B \,|\, A) = \int_B f_{X|A}(x)\,dx,$$

for any subset B of the real line. In particular, by letting B be the entire real line, we obtain the normalization property

$$\int_{-\infty}^{\infty} f_{X|A}(x)\,dx = 1,$$

so that $f_{X|A}$ is a legitimate PDF.

In the important special case where we condition on an event of the form $\{X \in A\}$, with $\mathbf{P}(X \in A) > 0$, the definition of conditional probabilities yields

$$\mathbf{P}(X \in B \,|\, X \in A) = \frac{\mathbf{P}(X \in B,\ X \in A)}{\mathbf{P}(X \in A)} = \frac{\displaystyle\int_{A \cap B} f_X(x)\,dx}{\mathbf{P}(X \in A)}.$$

By comparing with the earlier formula, we conclude that

$$f_{X|\{X \in A\}}(x) = \begin{cases} \dfrac{f_X(x)}{\mathbf{P}(X \in A)}, & \text{if } x \in A, \\ 0, & \text{otherwise.} \end{cases}$$

As in the discrete case, the conditional PDF is zero outside the conditioning set. Within the conditioning set, the conditional PDF has exactly the same shape as the unconditional one, except that it is scaled by the constant factor $1/\mathbf{P}(X \in A)$, so that $f_{X|\{X \in A\}}$ integrates to 1; see Fig. 3.14. Thus, the conditional PDF is similar to an ordinary PDF, except that it refers to a new universe in which the event $\{X \in A\}$ is known to have occurred.

Example 3.13. The Exponential Random Variable is Memoryless. The time T until a new light bulb burns out is an exponential random variable with parameter λ. Ariadne turns the light on, leaves the room, and when she returns, t time units later, finds that the light bulb is still on, which corresponds to the event $A = \{T > t\}$. Let X be the additional time until the light bulb burns out. What is the conditional CDF of X, given the event A?

We have, for $x \geq 0$,

$$\mathbf{P}(X > x \,|\, A) = \mathbf{P}(T > t + x \,|\, T > t) = \frac{\mathbf{P}(T > t + x \text{ and } T > t)}{\mathbf{P}(T > t)}$$

$$= \frac{\mathbf{P}(T > t + x)}{\mathbf{P}(T > t)} = \frac{e^{-\lambda(t+x)}}{e^{-\lambda t}} = e^{-\lambda x},$$

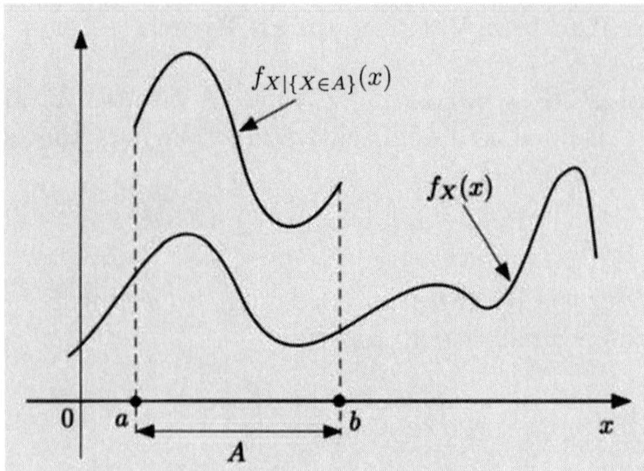

Figure 3.14: The unconditional PDF f_X and the conditional PDF $f_{X|\{X\in A\}}$, where A is the interval $[a, b]$. Note that within the conditioning event A, $f_{X|\{X\in A\}}$ retains the same shape as f_X, except that it is scaled along the vertical axis.

where we have used the expression for the CDF of an exponential random variable derived in Section 3.2.

Thus, the conditional CDF of X is exponential with parameter λ, regardless of the time t that elapsed between the lighting of the bulb and Ariadne's arrival. This is known as the *memorylessness property* of the exponential. Generally, if we model the time to complete a certain operation by an exponential random variable X, this property implies that as long as the operation has not been completed, the remaining time up to completion has the same exponential CDF, no matter when the operation started.

When multiple random variables are involved, there is a similar notion of a joint conditional PDF. Suppose, for example, that X and Y are jointly continuous random variables, with joint PDF $f_{X,Y}$. If we condition on a positive probability event of the form $C = \{(X, Y) \in A\}$, we have

$$f_{X,Y|C}(x, y) = \begin{cases} \dfrac{f_{X,Y}(x, y)}{\mathbf{P}(C)}, & \text{if } (x, y) \in A, \\ 0, & \text{otherwise.} \end{cases}$$

In this case, the conditional PDF of X, given this event, can be obtained from the formula

$$f_{X|C}(x) = \int_{-\infty}^{\infty} f_{X,Y|C}(x, y)\, dy.$$

These two formulas provide one possible method for obtaining the conditional PDF of a random variable X when the conditioning event is not of the form $\{X \in A\}$, but is instead defined in terms of multiple random variables.

We finally note that there is a version of the total probability theorem, which involves conditional PDFs: if the events $A_1, \ldots, A_n$ form a partition of the sample space, then

$$f_X(x) = \sum_{i=1}^{n} \mathbf{P}(A_i) f_{X|A_i}(x).$$

To justify this statement, we use the total probability theorem from Chapter 1, and obtain

$$\mathbf{P}(X \leq x) = \sum_{i=1}^{n} \mathbf{P}(A_i)\mathbf{P}(X \leq x \,|\, A_i).$$

This formula can be rewritten as

$$\int_{-\infty}^{x} f_X(t)\,dt = \sum_{i=1}^{n} \mathbf{P}(A_i) \int_{-\infty}^{x} f_{X|A_i}(t)\,dt.$$

We then take the derivative of both sides, with respect to x, and obtain the desired result.

Conditional PDF Given an Event

- The conditional PDF $f_{X|A}$ of a continuous random variable X, given an event A with $\mathbf{P}(A) > 0$, satisfies

$$\mathbf{P}(X \in B \,|\, A) = \int_B f_{X|A}(x)\,dx.$$

- If A is a subset of the real line with $\mathbf{P}(X \in A) > 0$, then

$$f_{X|\{X \in A\}}(x) = \begin{cases} \dfrac{f_X(x)}{\mathbf{P}(X \in A)}, & \text{if } x \in A, \\ 0, & \text{otherwise.} \end{cases}$$

- Let $A_1, A_2, \ldots, A_n$ be disjoint events that form a partition of the sample space, and assume that $\mathbf{P}(A_i) > 0$ for all i. Then,

$$f_X(x) = \sum_{i=1}^{n} \mathbf{P}(A_i) f_{X|A_i}(x)$$

(a version of the total probability theorem).

The following example illustrates a divide-and-conquer approach that uses the total probability theorem to calculate a PDF.

Example 3.14. The metro train arrives at the station near your home every quarter hour starting at 6:00 a.m. You walk into the station every morning between 7:10 and 7:30 a.m., and your arrival time is a uniform random variable over this interval. What is the PDF of the time you have to wait for the first train to arrive?

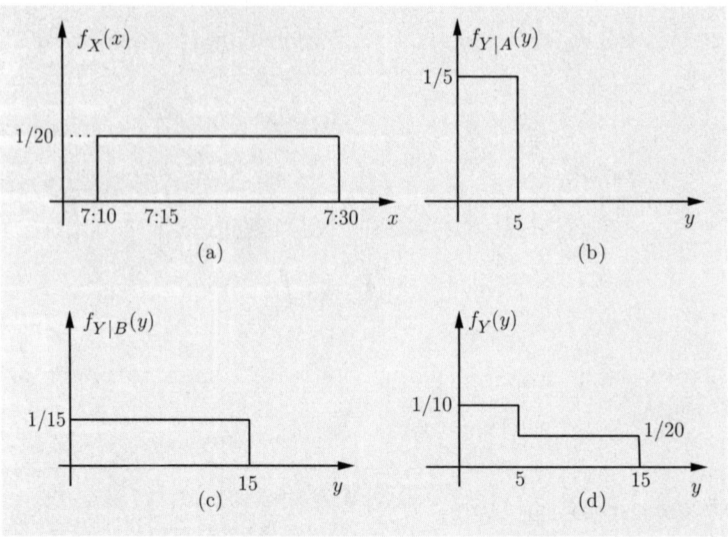

Figure 3.15: The PDFs f_X, $f_{Y|A}$, $f_{Y|B}$, and f_Y in Example 3.14.

The time of your arrival, denoted by X, is a uniform random variable over the interval from 7:10 to 7:30; see Fig. 3.15(a). Let Y be the waiting time. We calculate the PDF f_Y using a divide-and-conquer strategy. Let A and B be the events

$$A = \{7{:}10 \le X \le 7{:}15\} = \{\text{you board the 7:15 train}\},$$

$$B = \{7{:}15 < X \le 7{:}30\} = \{\text{you board the 7:30 train}\}.$$

Conditioned on the event A, your arrival time is uniform over the interval from 7:10 to 7:15. In this case, the waiting time Y is also uniform and takes values between 0 and 5 minutes; see Fig. 3.15(b). Similarly, conditioned on B, Y is uniform and takes values between 0 and 15 minutes; see Fig. 3.15(c). The PDF of Y is obtained using the total probability theorem,

$$f_Y(y) = \mathbf{P}(A)f_{Y|A}(y) + \mathbf{P}(B)f_{Y|B}(y),$$

and is shown in Fig. 3.15(d). We have

$$f_Y(y) = \frac{1}{4}\cdot\frac{1}{5} + \frac{3}{4}\cdot\frac{1}{15} = \frac{1}{10}, \qquad \text{for } 0 \le y \le 5,$$

$$f_Y(y) = \frac{1}{4}\cdot 0 + \frac{3}{4}\cdot\frac{1}{15} = \frac{1}{20}, \qquad \text{for } 5 < y \le 15.$$

Conditioning one Random Variable on Another

Let X and Y be continuous random variables with joint PDF $f_{X,Y}$. For any y with $f_Y(y) > 0$, the **conditional PDF** of X given that $Y = y$, is defined by

$$f_{X|Y}(x\,|\,y) = \frac{f_{X,Y}(x, y)}{f_Y(y)}.$$

This definition is analogous to the formula $p_{X|Y}(x\,|\,y) = p_{X,Y}(x, y)/p_Y(y)$ for the discrete case.

When thinking about the conditional PDF, it is best to view y as a fixed number and consider $f_{X|Y}(x\,|\,y)$ as a function of the single variable x. Viewed as a function of x, $f_{X|Y}(x\,|\,y)$ has the same shape as the joint PDF $f_{X,Y}(x, y)$, because the denominator $f_Y(y)$ does not depend on x; see Fig. 3.16. Furthermore, the formula

$$f_Y(y) = \int_{-\infty}^{\infty} f_{X,Y}(x, y)\, dx$$

implies the normalization property

$$\int_{-\infty}^{\infty} f_{X|Y}(x\,|\,y)\, dx = 1,$$

so *for any fixed* y, $f_{X|Y}(x\,|\,y)$ is a legitimate PDF.

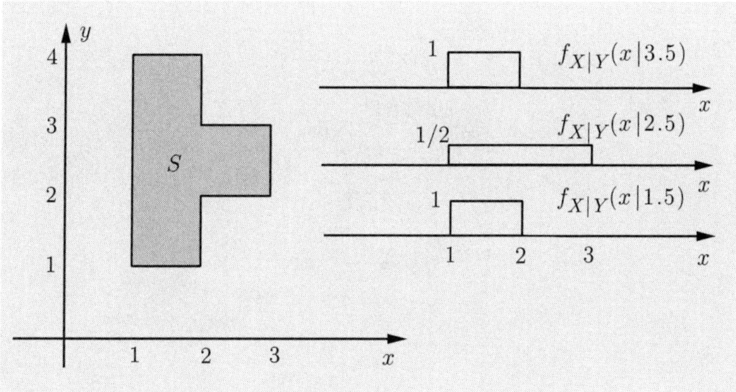

Figure 3.16: Visualization of the conditional PDF $f_{X|Y}(x\,|\,y)$. Let X and Y have a joint PDF which is uniform on the set S. For each fixed y, we consider the joint PDF along the slice $Y = y$ and normalize it so that it integrates to 1.

Example 3.15. Circular Uniform PDF. Ben throws a dart at a circular target of radius r (see Fig. 3.17). We assume that he always hits the target, and that all points of impact (x, y) are equally likely, so that the joint PDF of the random

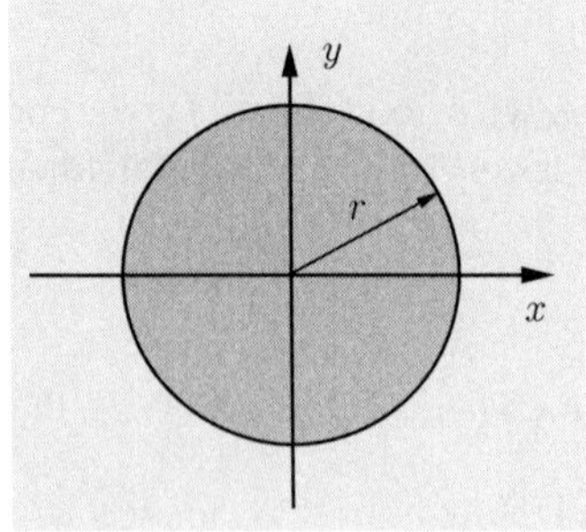

Figure 3.17: Circular target for Example 3.15.

variables X and Y is uniform. Following Example 3.9, and since the area of the circle is πr^2, we have

$$f_{X,Y}(x,y) = \begin{cases} \dfrac{1}{\text{area of the circle}}, & \text{if } (x,y) \text{ is in the circle,} \\ 0, & \text{otherwise,} \end{cases}$$
$$= \begin{cases} \dfrac{1}{\pi r^2}, & \text{if } x^2+y^2 \le r^2, \\ 0, & \text{otherwise.} \end{cases}$$

To calculate the conditional PDF $f_{X|Y}(x\,|\,y)$, let us first find the marginal PDF $f_Y(y)$. For $|y| > r$, it is zero. For $|y| \le r$, it is given by

$$\begin{aligned} f_Y(y) &= \int_{-\infty}^{\infty} f_{X,Y}(x,y)\,dx \\ &= \frac{1}{\pi r^2}\int_{x^2+y^2\le r^2} dx \\ &= \frac{1}{\pi r^2}\int_{-\sqrt{r^2-y^2}}^{\sqrt{r^2-y^2}} dx \\ &= \frac{2}{\pi r^2}\sqrt{r^2-y^2}, \qquad \text{if } |y| \le r. \end{aligned}$$

Note that the marginal PDF f_Y is not uniform.

The conditional PDF is

$$f_{X|Y}(x\,|\,y) = \frac{f_{X,Y}(x,y)}{f_Y(y)} = \frac{\dfrac{1}{\pi r^2}}{\dfrac{2}{\pi r^2}\sqrt{r^2-y^2}} = \frac{1}{2\sqrt{r^2-y^2}}, \qquad \text{if } x^2+y^2 \le r^2.$$

Thus, for a fixed value of y, the conditional PDF $f_{X|Y}$ is uniform.

To interpret the conditional PDF, let us fix some small positive numbers δ_1 and δ_2, and condition on the event $B = \{y \leq Y \leq y + \delta_2\}$. We have

$$\begin{aligned}\mathbf{P}(x \leq X \leq x+\delta_1 \,|\, y \leq Y \leq y+\delta_2) &= \frac{\mathbf{P}(x \leq X \leq x+\delta_1 \text{ and } y \leq Y \leq y+\delta_2)}{\mathbf{P}(y \leq Y \leq y+\delta_2)} \\ &\approx \frac{f_{X,Y}(x,y)\delta_1\delta_2}{f_Y(y)\delta_2} \\ &= f_{X|Y}(x\,|\,y)\delta_1.\end{aligned}$$

In words, $f_{X|Y}(x\,|\,y)\delta_1$ provides us with the probability that X belongs to a small interval $[x, x+\delta_1]$, given that Y belongs to a small interval $[y, y+\delta_2]$. Since $f_{X|Y}(x\,|\,y)\delta_1$ does not depend on δ_2, we can think of the limiting case where δ_2 decreases to zero and write

$$\mathbf{P}(x \leq X \leq x+\delta_1 \,|\, Y = y) \approx f_{X|Y}(x\,|\,y)\delta_1, \qquad (\delta_1 \text{ small}),$$

and, more generally,

$$\mathbf{P}(X \in A \,|\, Y = y) = \int_A f_{X|Y}(x\,|\,y)\,dx.$$

Conditional probabilities, given the zero probability event $\{Y = y\}$, were left undefined in Chapter 1. But the above formula provides a natural way of defining such conditional probabilities in the present context. In addition, it allows us to view the conditional PDF $f_{X|Y}(x\,|\,y)$ (as a function of x) as a description of the probability law of X, given that the event $\{Y = y\}$ has occurred.

As in the discrete case, the conditional PDF $f_{X|Y}$, together with the marginal PDF f_Y are sometimes used to calculate the joint PDF. Furthermore, this approach can also be used for modeling: instead of directly specifying $f_{X,Y}$, it is often natural to provide a probability law for Y, in terms of a PDF f_Y, and then provide a conditional PDF $f_{X|Y}(x\,|\,y)$ for X, given any possible value y of Y.

Example 3.16. The speed of a typical vehicle that drives past a police radar is modeled as an exponentially distributed random variable X with mean 50 miles per hour. The police radar's measurement Y of the vehicle's speed has an error which is modeled as a normal random variable with zero mean and standard deviation equal to one tenth of the vehicle's speed. What is the joint PDF of X and Y?

We have $f_X(x) = (1/50)e^{-x/50}$, for $x \geq 0$. Also, conditioned on $X = x$, the measurement Y has a normal PDF with mean x and variance $x^2/100$. Therefore,

$$f_{Y|X}(y\,|\,x) = \frac{1}{\sqrt{2\pi}\,(x/10)} e^{-(y-x)^2/(2x^2/100)}.$$

Thus, for all $x \geq 0$ and all y,

$$f_{X,Y}(x,y) = f_X(x)f_{Y|X}(y\,|\,x) = \frac{1}{50}e^{-x/50}\frac{10}{\sqrt{2\pi}\,x}e^{-50(y-x)^2/x^2}.$$

Conditional PDF Given a Random Variable

Let X and Y be jointly continuous random variables with joint PDF $f_{X,Y}$.

- The joint, marginal, and conditional PDFs are related to each other by the formulas

$$f_{X,Y}(x,y) = f_Y(y) f_{X|Y}(x\,|\,y),$$

$$f_X(x) = \int_{-\infty}^{\infty} f_Y(y) f_{X|Y}(x\,|\,y)\,dy.$$

  The conditional PDF $f_{X|Y}(x\,|\,y)$ is defined only for those y for which $f_Y(y) > 0$.

- We have

$$\mathbf{P}(X \in A\,|\,Y = y) = \int_A f_{X|Y}(x\,|\,y)\,dx.$$

For the case of more than two random variables, there are natural extensions to the above. For example, we can define conditional PDFs by formulas such as

$$f_{X,Y|Z}(x,y\,|\,z) = \frac{f_{X,Y,Z}(x,y,z)}{f_Z(z)}, \qquad \text{if } f_Z(z) > 0,$$

$$f_{X|Y,Z}(x\,|\,y,z) = \frac{f_{X,Y,Z}(x,y,z)}{f_{Y,Z}(y,z)}, \qquad \text{if } f_{Y,Z}(y,z) > 0.$$

There is also an analog of the multiplication rule,

$$f_{X,Y,Z}(x,y,z) = f_{X|Y,Z}(x\,|\,y,z) f_{Y|Z}(y\,|\,z) f_Z(z),$$

and of other formulas developed in this section.

Conditional Expectation

For a continuous random variable X, we define its **conditional expectation** $\mathbf{E}[X\,|\,A]$ given an event A, similar to the unconditional case, except that we now need to use the conditional PDF $f_{X|A}$. The conditional expectation $\mathbf{E}[X\,|\,Y = y]$ is defined similarly, in terms of the conditional PDF $f_{X|Y}$. Various familiar properties of expectations carry over to the present context and are summarized below. We note that all formulas are analogous to corresponding formulas for the case of discrete random variables, except that sums are replaced by integrals, and PMFs are replaced by PDFs.

Summary of Facts About Conditional Expectations

Let X snd Y be jointly continuous random variables, and let A be an event with $\mathbf{P}(A) > 0$.

- **Definitions:** The conditional expectation of X given the event A is defined by

$$\mathbf{E}[X \mid A] = \int_{-\infty}^{\infty} x f_{X|A}(x)\, dx.$$

The conditional expectation of X given that $Y = y$ is defined by

$$\mathbf{E}[X \mid Y = y] = \int_{-\infty}^{\infty} x f_{X|Y}(x \mid y)\, dx.$$

- **The expected value rule:** For a function $g(X)$, we have

$$\mathbf{E}\big[g(X) \mid A\big] = \int_{-\infty}^{\infty} g(x) f_{X|A}(x)\, dx,$$

and

$$\mathbf{E}\big[g(X) \mid Y = y\big] = \int_{-\infty}^{\infty} g(x) f_{X|Y}(x \mid y)\, dx.$$

- **Total expectation theorem:** Let $A_1, A_2, \ldots, A_n$ be disjoint events that form a partition of the sample space, and assume that $\mathbf{P}(A_i) > 0$ for all i. Then,

$$\mathbf{E}[X] = \sum_{i=1}^{n} \mathbf{P}(A_i)\mathbf{E}[X \mid A_i].$$

Similarly,

$$\mathbf{E}[X] = \int_{-\infty}^{\infty} \mathbf{E}[X \mid Y = y] f_Y(y)\, dy.$$

- There are natural analogs for the case of functions of several random variables. For example,

$$\mathbf{E}\big[g(X,Y) \mid Y = y\big] = \int g(x,y) f_{X|Y}(x \mid y)\, dx,$$

and

$$\mathbf{E}\big[g(X,Y)\big] = \int \mathbf{E}\big[g(X,Y) \mid Y = y\big] f_Y(y)\, dy.$$

The expected value rule is established in the same manner as for the case of unconditional expectations. To justify the first version of the total expectation theorem, we start with the total probability theorem

$$f_X(x) = \sum_{i=1}^{n} \mathbf{P}(A_i) f_{X|A_i}(x),$$

multiply both sides by x, and then integrate from $-\infty$ to ∞.

To justify the second version of the total expectation theorem, we observe that

$$\begin{aligned}
\int_{-\infty}^{\infty} \mathbf{E}[X \,|\, Y = y] f_Y(y)\, dy &= \int_{-\infty}^{\infty} \left[\int_{-\infty}^{\infty} x f_{X|Y}(x \,|\, y)\, dx \right] f_Y(y)\, dy \\
&= \int_{-\infty}^{\infty} \int_{-\infty}^{\infty} x f_{X|Y}(x \,|\, y) f_Y(y)\, dx\, dy \\
&= \int_{-\infty}^{\infty} \int_{-\infty}^{\infty} x f_{X,Y}(x, y)\, dx\, dy \\
&= \int_{-\infty}^{\infty} x \left[\int_{-\infty}^{\infty} f_{X,Y}(x, y)\, dy \right] dx \\
&= \int_{-\infty}^{\infty} x f_X(x)\, dx \\
&= \mathbf{E}[X].
\end{aligned}$$

The total expectation theorem can often facilitate the calculation of the mean, variance, and other moments of a random variable, using a divide-and-conquer approach.

Example 3.17. Mean and Variance of a Piecewise Constant PDF. Suppose that the random variable X has the piecewise constant PDF

$$f_X(x) = \begin{cases} 1/3, & \text{if } 0 \le x \le 1, \\ 2/3, & \text{if } 1 < x \le 2, \\ 0, & \text{otherwise,} \end{cases}$$

(see Fig. 3.18). Consider the events

$$\begin{aligned}
A_1 &= \big\{X \text{ lies in the first interval } [0, 1]\big\}, \\
A_2 &= \big\{X \text{ lies in the second interval } (1, 2]\big\}.
\end{aligned}$$

We have from the given PDF,

$$\mathbf{P}(A_1) = \int_0^1 f_X(x)\, dx = \frac{1}{3}, \qquad \mathbf{P}(A_2) = \int_1^2 f_X(x)\, dx = \frac{2}{3}.$$

Furthermore, the conditional mean and second moment of X, conditioned on A_1 and A_2, are easily calculated since the corresponding conditional PDFs $f_{X|A_1}$ and

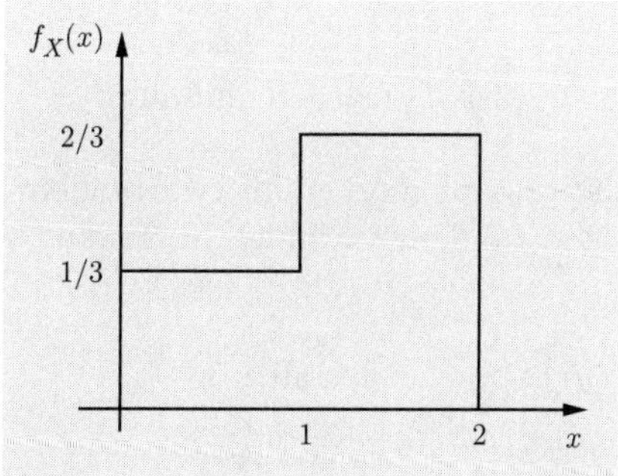

Figure 3.18: Piecewise constant PDF for Example 3.17.

$f_{X|A_2}$ are uniform. We recall from Example 3.4 that the mean of a uniform random variable over an interval $[a, b]$ is $(a+b)/2$ and its second moment is $(a^2+ab+b^2)/3$. Thus,

$$\mathbf{E}[X \,|\, A_1] = \frac{1}{2}, \qquad \mathbf{E}[X \,|\, A_2] = \frac{3}{2},$$

$$\mathbf{E}\big[X^2 \,|\, A_1\big] = \frac{1}{3}, \qquad \mathbf{E}\big[X^2 \,|\, A_2\big] = \frac{7}{3}.$$

We now use the total expectation theorem to obtain

$$\mathbf{E}[X] = \mathbf{P}(A_1)\mathbf{E}[X \,|\, A_1] + \mathbf{P}(A_2)\mathbf{E}[X \,|\, A_2] = \frac{1}{3} \cdot \frac{1}{2} + \frac{2}{3} \cdot \frac{3}{2} = \frac{7}{6},$$

$$\mathbf{E}[X^2] = \mathbf{P}(A_1)\mathbf{E}[X^2 \,|\, A_1] + \mathbf{P}(A_2)\mathbf{E}[X^2 \,|\, A_2] = \frac{1}{3} \cdot \frac{1}{3} + \frac{2}{3} \cdot \frac{7}{3} = \frac{15}{9}.$$

The variance is given by

$$\text{var}(X) = \mathbf{E}[X^2] - \big(\mathbf{E}[X]\big)^2 = \frac{15}{9} - \frac{49}{36} = \frac{11}{36}.$$

Note that this approach to the mean and variance calculation is easily generalized to piecewise constant PDFs with more than two pieces.

Independence

In full analogy with the discrete case, we say that two continuous random variables X and Y are **independent** if their joint PDF is the product of the marginal PDFs:

$$f_{X,Y}(x, y) = f_X(x) f_Y(y), \qquad \text{for all } x, y.$$

Comparing with the formula $f_{X,Y}(x, y) = f_{X|Y}(x \,|\, y) f_Y(y)$, we see that independence is the same as the condition

$$f_{X|Y}(x \,|\, y) = f_X(x), \qquad \text{for all } y \text{ with } f_Y(y) > 0 \text{ and all } x,$$

or, symmetrically,

$$f_{Y|X}(y\,|\,x) = f_Y(y), \qquad \text{for all } x \text{ with } f_X(x) > 0 \text{ and all } y.$$

There is a natural generalization to the case of more than two random variables. For example, we say that the three random variables X, Y, and Z are independent if

$$f_{X,Y,Z}(x, y, z) = f_X(x)f_Y(y)f_Z(z), \qquad \text{for all } x, y, z.$$

Example 3.18. Independent Normal Random Variables. Let X and Y be independent normal random variables with means μ_x, μ_y, and variances σ_x^2, σ_y^2, respectively. Their joint PDF is of the form

$$f_{X,Y}(x, y) = f_X(x)f_Y(y) = \frac{1}{2\pi\sigma_x\sigma_y} \exp\left\{ -\frac{(x-\mu_x)^2}{2\sigma_x^2} - \frac{(y-\mu_y)^2}{2\sigma_y^2} \right\}.$$

This joint PDF has the shape of a bell centered at (μ_x, μ_y), and whose width in the x and y directions is proportional to σ_x and σ_y, respectively. We can get some additional insight into the form of this PDF by considering its contours, i.e., sets of points at which the PDF takes a constant value. These contours are described by an equation of the form

$$\frac{(x-\mu_x)^2}{\sigma_X^2} + \frac{(y-\mu_y)^2}{\sigma_Y^2} = \text{constant},$$

and are ellipses whose two axes are horizontal and vertical (see Fig. 3.19). In the special case where $\sigma_x = \sigma_y$, the contours are circles.

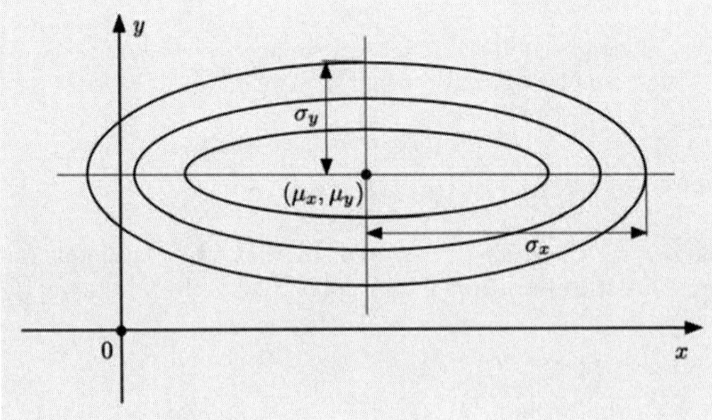

Figure 3.19: Contours of the joint PDF of two independent normal random variables X and Y with means μ_x, μ_y, and variances σ_x^2, σ_y^2, respectively.

If X and Y are independent, then any two events of the form $\{X \in A\}$ and $\{Y \in B\}$ are independent. Indeed,

$$\begin{aligned}
\mathbf{P}(X \in A \text{ and } Y \in B) &= \int_{x \in A} \int_{y \in B} f_{X,Y}(x, y)\, dy\, dx \\
&= \int_{x \in A} \int_{y \in B} f_X(x) f_Y(y)\, dy\, dx \\
&= \int_{x \in A} f_X(x)\, dx \int_{y \in D} f_Y(y)\, dy \\
&= \mathbf{P}(X \in A)\, \mathbf{P}(Y \in B).
\end{aligned}$$

In particular, independence implies that

$$F_{X,Y}(x, y) = \mathbf{P}(X \le x,\ Y \le y) = \mathbf{P}(X \le x)\, \mathbf{P}(Y \le y) = F_X(x) F_Y(y).$$

The converse of these statements is also true; see the end-of-chapter problems. The property

$$F_{X,Y}(x, y) = F_X(x) F_Y(y), \qquad \text{for all } x,\ y,$$

can be used to provide a general definition of independence between two random variables, e.g., if X is discrete and Y is continuous.

An argument similar to the discrete case shows that if X and Y are independent, then

$$\mathbf{E}\big[g(X)h(Y)\big] = \mathbf{E}\big[g(X)\big]\, \mathbf{E}\big[h(Y)\big],$$

for any two functions g and h. Finally, the variance of the sum of *independent* random variables is equal to the sum of their variances.

Independence of Continuous Random Variables

Let X and Y be jointly continuous random variables.

- X and Y are **independent** if

$$f_{X,Y}(x, y) = f_X(x) f_Y(y), \qquad \text{for all } x, y.$$

- If X and Y are independent, then

$$\mathbf{E}[XY] = \mathbf{E}[X]\, \mathbf{E}[Y].$$

Furthermore, for any functions g and h, the random variables $g(X)$ and $h(Y)$ are independent, and we have

$$\mathbf{E}\big[g(X)h(Y)\big] = \mathbf{E}\big[g(X)\big]\,\mathbf{E}\big[h(Y)\big].$$

- If X and Y are independent, then

$$\text{var}(X+Y) = \text{var}(X) + \text{var}(Y).$$

3.6 THE CONTINUOUS BAYES' RULE

In many situations, we represent an unobserved phenomenon by a random variable X with PDF f_X and we make a noisy measurement Y, which is modeled in terms of a conditional PDF $f_{Y|X}$. Once the value of Y is measured, what information does it provide on the unknown value of X? This setting is similar to the one of Section 1.4, where we introduced Bayes' rule and used it to solve inference problems; see Fig. 3.20. The only difference is that we are now dealing with continuous random variables.

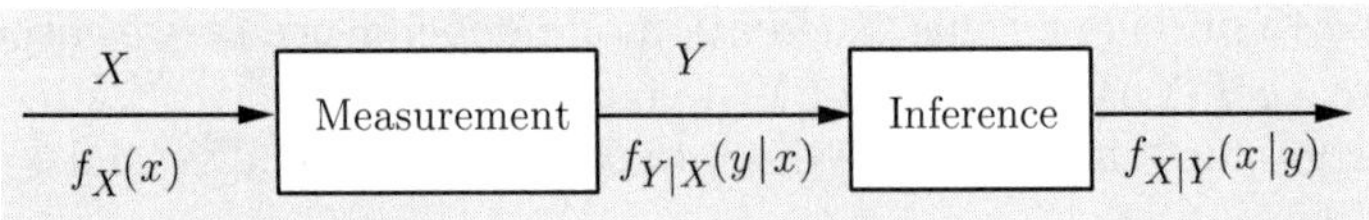

Figure 3.20: Schematic description of the inference problem. We have an unobserved random variable X with known PDF, and we obtain a measurement Y according to a conditional PDF $f_{Y|X}$. Given an observed value y of Y, the inference problem is to evaluate the conditional PDF $f_{X|Y}(x\,|\,y)$.

Note that whatever information is provided by the event $\{Y = y\}$ is captured by the conditional PDF $f_{X|Y}(x\,|\,y)$. It thus suffices to evaluate this PDF. From the formulas $f_X f_{Y|X} = f_{X,Y} = f_Y f_{X|Y}$, it follows that

$$f_{X|Y}(x\,|\,y) = \frac{f_X(x) f_{Y|X}(y\,|\,x)}{f_Y(y)}.$$

Based on the normalization property $\int_{-\infty}^{\infty} f_{X|Y}(x\,|\,y)\,dx = 1$, an equivalent expression is

$$f_{X|Y}(x\,|\,y) = \frac{f_X(x) f_{Y|X}(y\,|\,x)}{\displaystyle\int_{-\infty}^{\infty} f_X(t) f_{Y|X}(y\,|\,t)\,dt}.$$

Example 3.19. A light bulb produced by the General Illumination Company is known to have an exponentially distributed lifetime Y. However, the company has been experiencing quality control problems. On any given day, the parameter λ of the PDF of Y is actually a random variable, uniformly distributed in the interval $[1, 3/2]$. We test a light bulb and record its lifetime. What can we say about the underlying parameter λ?

We model the parameter λ in terms of a uniform random variable Λ with PDF

$$f_\Lambda(\lambda) = 2, \qquad \text{for } 1 \le \lambda \le \frac{3}{2}.$$

The available information about Λ is captured by the conditional PDF $f_{\Lambda|Y}(\lambda\,|\,y)$, which using the continuous Bayes' rule, is given by

$$f_{\Lambda|Y}(\lambda\,|\,y) = \frac{f_\Lambda(\lambda)f_{Y|\Lambda}(y\,|\,\lambda)}{\displaystyle\int_{-\infty}^{\infty} f_\Lambda(t)f_{Y|\Lambda}(y\,|\,t)\,dt} = \frac{2\lambda e^{-\lambda y}}{\displaystyle\int_1^{3/2} 2te^{-ty}\,dt}, \qquad \text{for } 1 \le \lambda \le \frac{3}{2}.$$

Inference about a Discrete Random Variable

In some cases, the unobserved phenomenon is inherently discrete. For some examples, consider a binary signal which is observed in the presence of normally distributed noise, or a medical diagnosis that is made on the basis of continuous measurements such as temperature and blood counts. In such cases, a somewhat different version of Bayes' rule applies.

We first consider the case where the unobserved phenomenon is described in terms of an event A whose occurrence is unknown. Let $\mathbf{P}(A)$ be the probability of event A. Let Y be a continuous random variable, and assume that the conditional PDFs $f_{Y|A}(y)$ and $f_{Y|A^c}(y)$ are known. We are interested in the conditional probability $\mathbf{P}(A\,|\,Y = y)$ of the event A, given the value y of Y.

Instead of working with the conditioning event $\{Y = y\}$, which has zero probability, let us instead condition on the event $\{y \le Y \le y + \delta\}$, where δ is a small positive number, and then take the limit as δ tends to zero. We have, using Bayes' rule, and assuming that $f_Y(y) > 0$,

$$\begin{aligned}\mathbf{P}(A\,|\,Y = y) &\approx \mathbf{P}(A\,|\,y \le Y \le y + \delta)\\ &= \frac{\mathbf{P}(A)\mathbf{P}(y \le Y \le y + \delta\,|\,A)}{\mathbf{P}(y \le Y \le y + \delta)}\\ &\approx \frac{\mathbf{P}(A)f_{Y|A}(y)\delta}{f_Y(y)\delta}\\ &= \frac{\mathbf{P}(A)f_{Y|A}(y)}{f_Y(y)}.\end{aligned}$$

The denominator can be evaluated using the following version of the total probability theorem:

$$f_Y(y) = \mathbf{P}(A)f_{Y|A}(y) + \mathbf{P}(A^c)f_{Y|A^c}(y),$$

so that

$$\mathbf{P}(A \,|\, Y = y) = \frac{\mathbf{P}(A) f_{Y|A}(y)}{\mathbf{P}(A) f_{Y|A}(y) + \mathbf{P}(A^c) f_{Y|A^c}(y)}.$$

In a variant of this formula, we consider an event A of the form $\{N = n\}$, where N is a discrete random variable that represents the different discrete possibilities for the unobserved phenomenon of interest. Let p_N be the PMF of N. Let also Y be a continuous random variable which, for any given value n of N, is described by a conditional PDF $f_{Y|N}(y \,|\, n)$. The above formula becomes

$$\mathbf{P}(N = n \,|\, Y = y) = \frac{p_N(n) f_{Y|N}(y \,|\, n)}{f_Y(y)}.$$

The denominator can be evaluated using the following version of the total probability theorem:

$$f_Y(y) = \sum_i p_N(i) f_{Y|N}(y \,|\, i),$$

so that

$$\mathbf{P}(N = n \,|\, Y = y) = \frac{p_N(n) f_{Y|N}(y \,|\, n)}{\sum_i p_N(i) f_{Y|N}(y \,|\, i)}.$$

Example 3.20. Signal Detection. A binary signal S is transmitted, and we are given that $\mathbf{P}(S = 1) = p$ and $\mathbf{P}(S = -1) = 1 - p$. The received signal is $Y = N + S$, where N is normal noise, with zero mean and unit variance, independent of S. What is the probability that $S = 1$, as a function of the observed value y of Y?

Conditioned on $S = s$, the random variable Y has a normal distribution with mean s and unit variance. Applying the formulas given above, we obtain

$$\mathbf{P}(S = 1 \,|\, Y = y) = \frac{p_S(1) f_{Y|S}(y \,|\, 1)}{f_Y(y)} = \frac{\dfrac{p}{\sqrt{2\pi}} e^{-(y-1)^2/2}}{\dfrac{p}{\sqrt{2\pi}} e^{-(y-1)^2/2} + \dfrac{1-p}{\sqrt{2\pi}} e^{-(y+1)^2/2}},$$

which simplifies to

$$\mathbf{P}(S = 1 \,|\, Y = y) = \frac{pe^y}{pe^y + (1 - p)e^{-y}}.$$

Note that the probability $\mathbf{P}(S = 1 \,|\, Y = y)$ goes to zero as y decreases to $-\infty$, goes to 1 as y increases to ∞, and is monotonically increasing in between, which is consistent with intuition.

Inference Based on Discrete Observations

We finally note that our earlier formula expressing $\mathbf{P}(A \mid Y = y)$ in terms of $f_{Y|A}(y)$ can be turned around to yield

$$f_{Y|A}(y) = \frac{f_Y(y)\,\mathbf{P}(A \mid Y = y)}{\mathbf{P}(A)}.$$

Based on the normalization property $\int_{-\infty}^{\infty} f_{Y|A}(y)\,dy = 1$, an equivalent expression is

$$f_{Y|A}(y) = \frac{f_Y(y)\,\mathbf{P}(A \mid Y = y)}{\displaystyle\int_{-\infty}^{\infty} f_Y(t)\,\mathbf{P}(A \mid Y = t)\,dt}.$$

This formula can be used to make an inference about a random variable Y when an event A is observed. There is a similar formula for the case where the event A is of the form $\{N = n\}$, where N is an observed discrete random variable that depends on Y in a manner described by a conditional PMF $p_{N|Y}(n \mid y)$.

Bayes' Rule for Continuous Random Variables

Let Y be a continuous random variable.

- If X is a continuous random variable, we have

$$f_Y(y) f_{X|Y}(x \mid y) = f_X(x) f_{Y|X}(y \mid x),$$

and

$$f_{X|Y}(x \mid y) = \frac{f_X(x) f_{Y|X}(y \mid x)}{f_Y(y)} = \frac{f_X(x) f_{Y|X}(y \mid x)}{\displaystyle\int_{-\infty}^{\infty} f_X(t) f_{Y|X}(y \mid t)\,dt}.$$

- If N is a discrete random variable, we have

$$f_Y(y)\,\mathbf{P}(N = n \mid Y = y) = p_N(n) f_{Y|N}(y \mid n),$$

resulting in the formulas

$$\mathbf{P}(N = n \mid Y = y) = \frac{p_N(n) f_{Y|N}(y \mid n)}{f_Y(y)} = \frac{p_N(n) f_{Y|N}(y \mid n)}{\displaystyle\sum_i p_N(i) f_{Y|N}(y \mid i)},$$

and

$$f_{Y|N}(y\,|\,n) = \frac{f_Y(y)\,\mathbf{P}(N=n\,|\,Y=y)}{p_N(n)} = \frac{f_Y(y)\,\mathbf{P}(N=n\,|\,Y=y)}{\displaystyle\int_{-\infty}^{\infty} f_Y(t)\,\mathbf{P}(N=n\,|\,Y=t)\,dt}.$$

- There are similar formulas for $\mathbf{P}(A\,|\,Y=y)$ and $f_{Y|A}(y)$.

3.7 SUMMARY AND DISCUSSION

Continuous random variables are characterized by PDFs, which are used to calculate event probabilities. This is similar to the use of PMFs for the discrete case, except that now we need to integrate instead of summing. Joint PDFs are similar to joint PMFs and are used to determine the probability of events that are defined in terms of multiple random variables. Furthermore, conditional PDFs are similar to conditional PMFs and are used to calculate conditional probabilities, given the value of the conditioning random variable. An important application is in problems of inference, using various forms of Bayes' rule that were developed in this chapter.

There are several special continuous random variables which frequently arise in probabilistic models. We introduced some of them, and derived their mean and variance. A summary is provided in the table that follows.

Summary of Results for Special Random Variables

Continuous Uniform Over $[a, b]$:

$$f_X(x) = \begin{cases} \dfrac{1}{b-a}, & \text{if } a \le x \le b, \\ 0, & \text{otherwise,} \end{cases}$$

$$\mathbf{E}[X] = \frac{a+b}{2}, \qquad \text{var}(X) = \frac{(b-a)^2}{12}.$$

Exponential with Parameter λ:

$$f_X(x) = \begin{cases} \lambda e^{-\lambda x}, & \text{if } x \ge 0, \\ 0, & \text{otherwise,} \end{cases} \qquad F_X(x) = \begin{cases} 1 - e^{-\lambda x}, & \text{if } x \ge 0, \\ 0, & \text{otherwise,} \end{cases}$$

$$\mathbf{E}[X] = \frac{1}{\lambda}, \qquad \text{var}(X) = \frac{1}{\lambda^2}.$$

Normal with Parameters μ and $\sigma^2 > 0$:

$$f_X(x) = \frac{1}{\sqrt{2\pi}\,\sigma} e^{-(x-\mu)^2/2\sigma^2},$$

$$\mathbf{E}[X] = \mu, \qquad \text{var}(X) = \sigma^2.$$

We have also introduced CDFs, which can be used to characterize general random variables that are neither discrete nor continuous. CDFs are related to PMFs and PDFs, but are more general. For a discrete random variable, we can obtain the PMF by differencing the CDF; for a continuous random variable, we can obtain the PDF by differentiating the CDF.

PROBLEMS

SECTION 3.1. Continuous Random Variables and PDFs

Problem 1. Let X be uniformly distributed in the unit interval $[0, 1]$. Consider the random variable $Y = g(X)$, where

$$g(x) = \begin{cases} 1, & \text{if } x \le 1/3, \\ 2, & \text{if } x > 1/3. \end{cases}$$

Find the expected value of Y by first deriving its PMF. Verify the result using the expected value rule.

Problem 2. Laplace random variable. Let X have the PDF

$$f_X(x) = \frac{\lambda}{2} e^{-\lambda|x|},$$

where λ is a positive scalar. Verify that f_X satisfies the normalization condition, and evaluate the mean and variance of X.

Problem 3.* Show that the expected value of a discrete or continuous random variable X satisfies

$$\mathbf{E}[X] = \int_0^\infty \mathbf{P}(X > x)\, dx - \int_0^\infty \mathbf{P}(X < -x)\, dx.$$

Solution. Suppose that X is continuous. We then have

$$\begin{aligned} \int_0^\infty \mathbf{P}(X > x)\, dx &= \int_0^\infty \left(\int_x^\infty f_X(y)\, dy \right) dx \\ &= \int_0^\infty \left(\int_0^y f_X(y)\, dx \right) dy \\ &= \int_0^\infty f_X(y) \left(\int_0^y dx \right) dy \\ &= \int_0^\infty y f_X(y)\, dy, \end{aligned}$$

where for the second equality we have reversed the order of integration by writing the set $\{(x, y) \,|\, 0 \le x < \infty,\ x \le y < \infty\}$ as $\{(x, y) \,|\, 0 \le x \le y,\ 0 \le y < \infty\}$. Similarly, we can show that

$$\int_0^\infty \mathbf{P}(X < -x)\, dx = -\int_{-\infty}^0 y f_X(y)\, dy.$$

Combining the two relations above, we obtain the desired result.

If X is discrete, we have

$$\begin{aligned}\mathbf{P}(X > x) &= \int_0^\infty \sum_{y>x} p_X(y) \\ &= \sum_{y>0} \left(\int_0^y p_X(y)\,dx \right) \\ &= \sum_{y>0} p_X(y) \left(\int_0^y dx \right) \\ &= \sum_{y>0} p_X(y) y,\end{aligned}$$

and the rest of the argument is similar to the continuous case.

Problem 4.* Establish the validity of the expected value rule

$$\mathbf{E}\big[g(X)\big] = \int_{-\infty}^{\infty} g(x) f_X(x)\,dx,$$

where X is a continuous random variable with PDF f_X.

Solution. Let us express the function g as the difference of two nonnegative functions,

$$g(x) = g^+(x) - g^-(x),$$

where $g^+(x) = \max\{g(x), 0\}$, and $g^-(x) = \max\{-g(x), 0\}$. In particular, for any $t \geq 0$, we have $g(x) > t$ if and only if $g^+(x) > t$.

We will use the result

$$\mathbf{E}\big[g(X)\big] = \int_0^\infty \mathbf{P}\big(g(X) > t\big)\,dt - \int_0^\infty \mathbf{P}\big(g(X) < -t\big)\,dt$$

from the preceding problem. The first term in the right-hand side is equal to

$$\int_0^\infty \int_{\{x \,|\, g(x) > t\}} f_X(x)\,dx\,dt = \int_{-\infty}^\infty \int_{\{t \,|\, 0 \leq t < g(x)\}} f_X(x)\,dt\,dx = \int_{-\infty}^\infty g^+(x) f_X(x)\,dx.$$

By a symmetrical argument, the second term in the right-hand side is given by

$$\int_0^\infty \mathbf{P}\big(g(X) < -t\big)\,dt = \int_{-\infty}^\infty g^-(x) f_X(x)\,dx.$$

Combining the above equalities, we obtain

$$\mathbf{E}\big[g(X)\big] = \int_{-\infty}^\infty g^+(x) f_X(x)\,dx - \int_{-\infty}^\infty g^-(x) f_X(x)\,dx = \int_{-\infty}^\infty g(x) f_X(x)\,dx.$$

SECTION 3.2. Cumulative Distribution Functions

Problem 5. Consider a triangle and a point chosen within the triangle according to the uniform probability law. Let X be the distance from the point to the base of the triangle. Given the height of the triangle, find the CDF and the PDF of X.

Problem 6. Calamity Jane goes to the bank to make a withdrawal, and is equally likely to find 0 or 1 customers ahead of her. The service time of the customer ahead, if present, is exponentially distributed with parameter λ. What is the CDF of Jane's waiting time?

Problem 7. Alvin throws darts at a circular target of radius r and is equally likely to hit any point in the target. Let X be the distance of Alvin's hit from the center.

(a) Find the PDF, the mean, and the variance of X.

(b) The target has an inner circle of radius t. If $X \leq t$, Alvin gets a score of $S = 1/X$. Otherwise his score is $S = 0$. Find the CDF of S. Is S a continuous random variable?

Problem 8. Consider two continuous random variables Y and Z, and a random variable X that is equal to Y with probability p and to Z with probability $1 - p$.

(a) Show that the PDF of X is given by

$$f_X(x) = pf_Y(x) + (1-p)f_Z(x).$$

(b) Calculate the CDF of the two-sided exponential random variable that has PDF given by

$$f_X(x) = \begin{cases} p\lambda e^{\lambda x}, & \text{if } x < 0, \\ (1-p)\lambda e^{-\lambda x}, & \text{if } x \geq 0, \end{cases}$$

where $\lambda > 0$ and $0 < p < 1$.

Problem 9.* Mixed random variables. Probabilistic models sometimes involve random variables that can be viewed as a mixture of a discrete random variable Y and a continuous random variable Z. By this we mean that the value of X is obtained according to the probability law of Y with a given probability p, and according to the probability law of Z with the complementary probability $1 - p$. Then, X is called a *mixed random variable* and its CDF is given, using the total probability theorem, by

$$\begin{aligned} F_X(x) &= \mathbf{P}(X \leq x) \\ &= p\mathbf{P}(Y \leq x) + (1-p)\mathbf{P}(Z \leq x) \\ &= pF_Y(x) + (1-p)F_Z(x). \end{aligned}$$

Its expected value is defined in a way that conforms to the total expectation theorem:

$$\mathbf{E}[X] = p\mathbf{E}[Y] + (1-p)\mathbf{E}[Z].$$

The taxi stand and the bus stop near Al's home are in the same location. Al goes there at a given time and if a taxi is waiting (this happens with probability 2/3) he boards it. Otherwise he waits for a taxi or a bus to come, whichever comes first. The next taxi will arrive in a time that is uniformly distributed between 0 and 10 minutes, while the next bus will arrive in exactly 5 minutes. Find the CDF and the expected value of Al's waiting time.

Solution. Let A be the event that Al will find a taxi waiting or will be picked up by the bus after 5 minutes. Note that the probability of boarding the next bus, given that Al has to wait, is

$$\mathbf{P}(\text{a taxi will take more than 5 minutes to arrive}) = \frac{1}{2}.$$

Al's waiting time, call it X, is a mixed random variable. With probability

$$\mathbf{P}(A) = \frac{2}{3} + \frac{1}{3} \cdot \frac{1}{2} = \frac{5}{6},$$

it is equal to its discrete component Y (corresponding to either finding a taxi waiting, or boarding the bus), which has PMF

$$p_Y(y) = \begin{cases} \dfrac{2}{3\mathbf{P}(A)}, & \text{if } y = 0, \\ \dfrac{1}{6\mathbf{P}(A)}, & \text{if } y = 5, \end{cases}$$

$$= \begin{cases} \dfrac{12}{15}, & \text{if } y = 0, \\ \dfrac{3}{15}, & \text{if } y = 5. \end{cases}$$

[This equation follows from the calculation

$$p_Y(0) = \mathbf{P}(Y = 0 \mid A) = \frac{\mathbf{P}(Y = 0, A)}{\mathbf{P}(A)} = \frac{2}{3\mathbf{P}(A)}.$$

The calculation for $p_Y(5)$ is similar.] With the complementary probability $1 - \mathbf{P}(A)$, the waiting time is equal to its continuous component Z (corresponding to boarding a taxi after having to wait for some time less than 5 minutes), which has PDF

$$f_Z(z) = \begin{cases} 1/5, & \text{if } 0 \le z \le 5, \\ 0, & \text{otherwise.} \end{cases}$$

The CDF is given by $F_X(x) = \mathbf{P}(A)F_Y(x) + \big(1 - \mathbf{P}(A)\big)F_Z(x)$, from which

$$F_X(x) = \begin{cases} 0, & \text{if } x < 0, \\ \dfrac{5}{6} \cdot \dfrac{12}{15} + \dfrac{1}{6} \cdot \dfrac{x}{5}, & \text{if } 0 \le x < 5, \\ 1, & \text{if } 5 \le x. \end{cases}$$

The expected value of the waiting time is

$$\mathbf{E}[X] = \mathbf{P}(A)\mathbf{E}[Y] + \big(1 - \mathbf{P}(A)\big)\mathbf{E}[Z] = \frac{5}{6} \cdot \frac{3}{15} \cdot 5 + \frac{1}{6} \cdot \frac{5}{2} = \frac{15}{12}.$$

Problem 10.* **Simulating a continuous random variable.** A computer has a subroutine that can generate values of a random variable U that is uniformly distributed in the interval $[0, 1]$. Such a subroutine can be used to generate values of a continuous random variable with given CDF $F(x)$ as follows. If U takes a value u, we let the value of X be a number x that satisfies $F(x) = u$. For simplicity, we assume that the given CDF is strictly increasing over the range S of values of interest, where $S = \{x \,|\, 0 < F(x) < 1\}$. This condition guarantees that for any $u \in (0, 1)$, there is a unique x that satisfies $F(x) = u$.

(a) Show that the CDF of the random variable X thus generated is indeed equal to the given CDF.

(b) Describe how this procedure can be used to simulate an exponential random variable with parameter λ.

(c) How can this procedure be generalized to simulate a discrete integer-valued random variable?

Solution. (a) By definition, the random variables X and U satisfy the relation $F(X) = U$. Since F is strictly increasing, we have for every x,

$$X \leq x \qquad \text{if and only if} \qquad F(X) \leq F(x).$$

Therefore,

$$\mathbf{P}(X \leq x) = \mathbf{P}\big(F(X) \leq F(x)\big) = \mathbf{P}\big(U \leq F(x)\big) = F(x),$$

where the last equality follows because U is uniform. Thus, X has the desired CDF.

(b) The exponential CDF has the form $F(x) = 1 - e^{-\lambda x}$ for $x \geq 0$. Thus, to generate values of X, we should generate values $u \in (0, 1)$ of a uniformly distributed random variable U, and set X to the value for which $1 - e^{-\lambda x} = u$, or $x = -\ln(1 - u)/\lambda$.

(c) Let again F be the desired CDF. To any $u \in (0, 1)$, there corresponds a unique integer x_u such that $F(x_u - 1) < u \leq F(x_u)$. This correspondence defines a random variable X as a function of the random variable U. We then have, for every integer k,

$$\mathbf{P}(X = k) = \mathbf{P}\big(F(k - 1) < U \leq F(k)\big) = F(k) - F(k - 1).$$

Therefore, the CDF of X is equal to F, as desired.

SECTION 3.3. Normal Random Variables

Problem 11. Let X and Y be normal random variables with means 0 and 1, respectively, and variances 1 and 4, respectively.

(a) Find $\mathbf{P}(X \leq 1.5)$ and $\mathbf{P}(X \leq -1)$.

(b) Find the PDF of $(Y - 1)/2$.

(c) Find $\mathbf{P}(-1 \leq Y \leq 1)$.

Problem 12. Let X be a normal random variable with zero mean and standard deviation σ. Use the normal tables to compute the probabilities of the events $\{X \geq k\sigma\}$ and $\{|X| \leq k\sigma\}$ for $k = 1, 2, 3$.

Problem 13. A city's temperature is modeled as a normal random variable with mean and standard deviation both equal to 10 degrees Celsius. What is the probability that the temperature at a randomly chosen time will be less than or equal to 59 degrees Fahrenheit?

Problem 14.* Show that the normal PDF satisfies the normalization property. *Hint:* The integral $\int_{-\infty}^{\infty} e^{-x^2/2}\,dx$ is equal to the square root of

$$\int_{-\infty}^{\infty}\int_{-\infty}^{\infty} e^{-x^2/2}e^{-y^2/2}\,dx\,dy,$$

and the latter integral can be evaluated by transforming to polar coordinates.

Solution. We note that

$$\begin{aligned}
\left(\int_{-\infty}^{\infty}\frac{1}{\sqrt{2\pi}}e^{-x^2/2}\,dx\right)^2 &= \frac{1}{2\pi}\int_{-\infty}^{\infty}e^{-x^2/2}\,dx\int_{-\infty}^{\infty}e^{-y^2/2}\,dy\\
&= \frac{1}{2\pi}\int_{-\infty}^{\infty}\int_{-\infty}^{\infty}e^{-(x^2+y^2)/2}\,dx\,dy\\
&= \frac{1}{2\pi}\int_{0}^{2\pi}\int_{0}^{\infty}e^{-r^2/2}r\,dr\,d\theta\\
&= \int_{0}^{\infty}e^{-r^2/2}\,r\,dr\\
&= \int_{0}^{\infty}e^{-u}\,du\\
&= -e^{-u}\Big|_0^{\infty}\\
&= 1,
\end{aligned}$$

where for the third equality, we use a transformation into polar coordinates, and for the fifth equality, we use the change of variables $u = r^2/2$. Thus, we have

$$\int_{-\infty}^{\infty}\frac{1}{\sqrt{2\pi}}e^{-x^2/2}\,dx = 1,$$

because the integral is positive. Using the change of variables $u = (x-\mu)/\sigma$, it follows that

$$\int_{-\infty}^{\infty} f_X(x)\,dx = \int_{-\infty}^{\infty}\frac{1}{\sqrt{2\pi}\,\sigma}e^{-(x-\mu)^2/2\sigma^2}\,dx = \int_{-\infty}^{\infty}\frac{1}{\sqrt{2\pi}}e^{-u^2/2}du = 1.$$

SECTION 3.4. Joint PDFs of Multiple Random Variables

Problem 15. A point is chosen at random (according to a uniform PDF) within a semicircle of the form $\big\{(x,y)\,|\,x^2+y^2\le r^2,\ y\ge 0\big\}$, for some given $r>0$.

(a) Find the joint PDF of the coordinates X and Y of the chosen point.

(b) Find the marginal PDF of Y and use it to find $\mathbf{E}[Y]$.

(c) Check your answer in (b) by computing $\mathbf{E}[Y]$ directly without using the marginal PDF of Y.

Problem 16. Consider the following variant of Buffon's needle problem (Example 3.11), which was investigated by Laplace. A needle of length l is dropped on a plane surface that is partitioned in rectangles by horizontal lines that are a apart and vertical lines that are b apart. Suppose that the needle's length l satisfies $l < a$ and $l < b$. What is the expected number of rectangle sides crossed by the needle? What is the probability that the needle will cross at least one side of some rectangle?

Problem 17.* Estimating an expected value by simulation using samples of another random variable. Let $Y_1, \ldots, Y_n$ be independent random variables drawn from a common and known PDF f_Y. Let S be the set of all possible values of Y_i, $S = \{y \mid f_Y(y) > 0\}$. Let X be a random variable with known PDF f_X, such that $f_X(y) = 0$, for all $y \notin S$. Consider the random variable

$$Z = \frac{1}{n} \sum_{i=1}^{n} Y_i \frac{f_X(Y_i)}{f_Y(Y_i)}.$$

Show that

$$\mathbf{E}[Z] = \mathbf{E}[X].$$

Solution. We have

$$\mathbf{E}\left[Y_i \frac{f_X(Y_i)}{f_Y(Y_i)}\right] = \int_S y \frac{f_X(y)}{f_Y(y)} f_Y(y)\, dy = \int_S y f_X(y)\, dy = \mathbf{E}[X].$$

Thus,

$$\mathbf{E}[Z] = \frac{1}{n} \sum_{i=1}^{n} \mathbf{E}\left[Y_i \frac{f_X(Y_i)}{f_Y(Y_i)}\right] = \frac{1}{n} \sum_{i=1}^{n} \mathbf{E}[X] = \mathbf{E}[X].$$

SECTION 3.5. Conditioning

Problem 18. Let X be a random variable with PDF

$$f_X(x) = \begin{cases} x/4, & \text{if } 1 < x \leq 3, \\ 0, & \text{otherwise,} \end{cases}$$

and let A be the event $\{X \geq 2\}$.

(a) Find $\mathbf{E}[X]$, $\mathbf{P}(A)$, $f_{X|A}(x)$, and $\mathbf{E}[X \mid A]$.

(b) Let $Y = X^2$. Find $\mathbf{E}[Y]$ and $\text{var}(Y)$.

Problem 19. The random variable X has the PDF

$$f_X(x) = \begin{cases} cx^{-2}, & \text{if } 1 \leq x \leq 2, \\ 0, & \text{otherwise.} \end{cases}$$

(a) Determine the value of c.

(b) Let A be the event $\{X > 1.5\}$. Calculate $\mathbf{P}(A)$ and the conditional PDF of X given that A has occurred.

(c) Let $Y = X^2$. Calculate the conditional expectation and the conditional variance of Y given A.

Problem 20. An absent-minded professor schedules two student appointments for the same time. The appointment durations are independent and exponentially distributed with mean thirty minutes. The first student arrives on time, but the second student arrives five minutes late. What is the expected value of the time between the arrival of the first student and the departure of the second student?

Problem 21. We start with a stick of length ℓ. We break it at a point which is chosen according to a uniform distribution and keep the piece, of length Y, that contains the left end of the stick. We then repeat the same process on the piece that we were left with, and let X be the length of the remaining piece after breaking for the second time.

(a) Find the joint PDF of Y and X.

(b) Find the marginal PDF of X.

(c) Use the PDF of X to evaluate $\mathbf{E}[X]$.

(d) Evaluate $\mathbf{E}[X]$, by exploiting the relation $X = Y \cdot (X/Y)$.

Problem 22. We have a stick of unit length, and we consider breaking it in three pieces using one of the following three methods.

(i) We choose randomly and independently two points on the stick using a uniform PDF, and we break the stick at these two points.

(ii) We break the stick at a random point chosen by using a uniform PDF, and then we break the piece that contains the right end of the stick, at a random point chosen by using a uniform PDF.

(iii) We break the stick at a random point chosen by using a uniform PDF, and then we break the larger of the two pieces at a random point chosen by using a uniform PDF.

For each of the methods (i), (ii), and (iii), what is the probability that the three pieces we are left with can form a triangle?

Problem 23. Let the random variables X and Y have a joint PDF which is uniform over the triangle with vertices at $(0,0)$, $(0,1)$, and $(1,0)$.

(a) Find the joint PDF of X and Y.

(b) Find the marginal PDF of Y.

(c) Find the conditional PDF of X given Y.

(d) Find $\mathbf{E}[X \mid Y = y]$, and use the total expectation theorem to find $\mathbf{E}[X]$ in terms of $\mathbf{E}[Y]$.

(e) Use the symmetry of the problem to find the value of $\mathbf{E}[X]$.

Problem 24. Let X and Y be two random variables that are uniformly distributed over the triangle formed by the points $(0,0)$, $(1,0)$, and $(0,2)$ (this is an asymmetric version of the PDF in the previous problem). Calculate $\mathbf{E}[X]$ and $\mathbf{E}[Y]$ by following the same steps as in the previous problem.

Problem 25. The coordinates X and Y of a point are independent zero mean normal random variables with common variance σ^2. Given that the point is at a distance of at least c from the origin, find the conditional joint PDF of X and Y.

Problem 26.* Let $X_1, \ldots, X_n$ be independent random variables. Show that

$$\frac{\text{var}\left(\prod_{i=1}^n X_i\right)}{\prod_{i=1}^n \mathbf{E}[X_i^2]} = \prod_{i=1}^n \left(\frac{\text{var}(X_i)}{\mathbf{E}[X_i^2]} + 1\right) - 1.$$

Solution. We have

$$\begin{aligned}\text{var}\left(\prod_{i=1}^n X_i\right) &= \mathbf{E}\left[\prod_{i=1}^n X_i^2\right] - \prod_{i=1}^n \big(\mathbf{E}[X_i]\big)^2 \\ &= \prod_{i=1}^n \mathbf{E}\big[X_i^2\big] - \prod_{i=1}^n \big(\mathbf{E}[X_i]\big)^2 \\ &= \prod_{i=1}^n \left(\text{var}(X_i) + \big(\mathbf{E}[X_i]\big)^2\right) - \prod_{i=1}^n \big(\mathbf{E}[X_i]\big)^2.\end{aligned}$$

The desired result follows by dividing both sides by

$$\prod_{i=1}^n \big(\mathbf{E}[X_i]\big)^2.$$

Problem 27.* **Conditioning multiple random variables on events.** Let X and Y be continuous random variables with joint PDF $f_{X,Y}$, let A be a subset of the two-dimensional plane, and let $C = \{(X,Y) \in A\}$. Assume that $\mathbf{P}(C) > 0$, and define

$$f_{X,Y|C}(x,y) = \begin{cases} \dfrac{f_{X,Y}(x,y)}{\mathbf{P}(C)}, & \text{if } (x,y) \in A, \\ 0, & \text{otherwise.} \end{cases}$$

(a) Show that $f_{X,Y|C}$ is a legitimate joint PDF.

(b) Consider a partition of the two-dimensional plane into disjoint subsets A_i, $i = 1, \ldots, n$, let $C_i = \{(X,Y) \in A_i\}$, and assume that $\mathbf{P}(C_i) > 0$ for all i. Derive the following version of the total probability theorem

$$f_{X,Y}(x,y) = \sum_{i=1}^n \mathbf{P}(C_i) f_{X,Y|C_i}(x,y).$$

Problem 28.* Consider the following two-sided exponential PDF

$$f_X(x) = \begin{cases} p\lambda e^{-\lambda x}, & \text{if } x \geq 0, \\ (1-p)\lambda e^{\lambda x}, & \text{if } x < 0, \end{cases}$$

where λ and p are scalars with $\lambda > 0$ and $p \in [0, 1]$. Find the mean and the variance of X in two ways:

(a) By straightforward calculation of the associated expected values.

(b) By using a divide-and-conquer strategy, and the mean and variance of the (one-sided) exponential random variable.

Solution. (a)

$$\begin{aligned} \mathbf{E}[X] &= \int_{-\infty}^{\infty} x f_X(x)\, dx \\ &= \int_{-\infty}^{0} x(1-p)\lambda e^{\lambda x}\, dx + \int_{0}^{\infty} xp\lambda e^{-\lambda x}\, dx \\ &= -\frac{1-p}{\lambda} + \frac{p}{\lambda} \\ &= \frac{2p-1}{\lambda}, \end{aligned}$$

$$\begin{aligned} \mathbf{E}[X^2] &= \int_{-\infty}^{\infty} x^2 f_X(x)\, dx \\ &= \int_{-\infty}^{0} x^2(1-p)\lambda e^{\lambda x}\, dx + \int_{0}^{\infty} x^2 p\lambda e^{-\lambda x}\, dx \\ &= \frac{2(1-p)}{\lambda^2} + \frac{2p}{\lambda^2} \\ &= \frac{2}{\lambda^2}, \end{aligned}$$

and

$$\mathrm{var}(X) = \frac{2}{\lambda^2} - \left(\frac{2p-1}{\lambda}\right)^2.$$

(b) Let A be the event $\{X \geq 0\}$, and note that $\mathbf{P}(A) = p$. Conditioned on A, the random variable X has a (one-sided) exponential distribution with parameter λ. Also, conditioned on A^c, the random variable $-X$ has the same one-sided exponential distribution. Thus,

$$\mathbf{E}[X \mid A] = \frac{1}{\lambda}, \qquad \mathbf{E}[X \mid A^c] = -\frac{1}{\lambda},$$

and

$$\mathbf{E}[X^2 \mid A] = \mathbf{E}[X^2 \mid A^c] = \frac{2}{\lambda^2}.$$

It follows that

$$\begin{aligned} \mathbf{E}[X] &= \mathbf{P}(A)\mathbf{E}[X \mid A] + \mathbf{P}(A^c)\mathbf{E}[X \mid A^c] \\ &= \frac{p}{\lambda} - \frac{1-p}{\lambda} \\ &= \frac{2p-1}{\lambda}, \end{aligned}$$

$$\begin{aligned}\mathbf{E}[X^2] &= \mathbf{P}(A)\mathbf{E}[X^2 \mid A] + \mathbf{P}(A^c)\mathbf{E}[X^2 \mid A^c] \\ &= \frac{2p}{\lambda^2} + \frac{2(1-p)}{\lambda^2} \\ &= \frac{2}{\lambda^2},\end{aligned}$$

and

$$\text{var}(X) = \frac{2}{\lambda^2} - \left(\frac{2p-1}{\lambda}\right)^2.$$

Problem 29.* Let X, Y, and Z be three random variables with joint PDF $f_{X,Y,Z}$. Show the multiplication rule:

$$f_{X,Y,Z}(x,y,z) = f_{X|Y,Z}(x \mid y,z) f_{Y|Z}(y \mid z) f_Z(z).$$

Solution. We have, using the definition of conditional density,

$$f_{X|Y,Z}(x \mid y,z) = \frac{f_{X,Y,Z}(x,y,z)}{f_{Y,Z}(y,z)},$$

and

$$f_{Y,Z}(y,z) = f_{Y|Z}(y \mid z) f_Z(z).$$

Combining these two relations, we obtain the multiplication rule.

Problem 30.* **The Beta PDF.** The beta PDF with parameters $\alpha > 0$ and $\beta > 0$ has the form

$$f_X(x) = \begin{cases} \dfrac{1}{\mathrm{B}(\alpha,\beta)} x^{\alpha-1}(1-x)^{\beta-1}, & \text{if } 0 \le x \le 1, \\ 0, & \text{otherwise.} \end{cases}$$

The normalizing constant is

$$\mathrm{B}(\alpha,\beta) = \int_0^1 x^{\alpha-1}(1-x)^{\beta-1}\,dx,$$

and is known as the Beta function.

(a) Show that for any $m > 0$, the mth moment of X is given by

$$\mathbf{E}[X^m] = \frac{\mathrm{B}(\alpha+m,\beta)}{\mathrm{B}(\alpha,\beta)}.$$

(b) Assume that α and β are integer. Show that

$$\mathrm{B}(\alpha,\beta) = \frac{(\alpha-1)!\,(\beta-1)!}{(\alpha+\beta-1)!},$$

so that

$$\mathbf{E}[X^m] = \frac{\alpha(\alpha+1)\cdots(\alpha+m-1)}{(\alpha+\beta)(\alpha+\beta+1)\cdots(\alpha+\beta+m-1)}.$$

(Recall here the convention that $0! = 1$.)

Solution. (a) We have

$$\mathbf{E}[X^m] = \frac{1}{\mathrm{B}(\alpha,\beta)} \int_0^1 x^m x^{\alpha-1}(1-x)^{\beta-1}\,dx = \frac{\mathrm{B}(\alpha+m,\beta)}{\mathrm{B}(\alpha,\beta)}.$$

(b) In the special case where $\alpha = 1$ or $\beta = 1$, we can carry out the straightforward integration in the definition of $\mathrm{B}(\alpha,\beta)$, and verify the result. We will now deal with the general case. Let $Y, Y_1, \ldots, Y_{\alpha+\beta}$ be independent random variables, uniformly distributed over the interval $[0,1]$, and let A be the event

$$A = \{Y_1 \leq \cdots \leq Y_\alpha \leq Y \leq Y_{\alpha+1} \leq \cdots \leq Y_{\alpha+\beta}\}.$$

Then,

$$\mathbf{P}(A) = \frac{1}{(\alpha+\beta+1)!},$$

because all ways of ordering these $\alpha+\beta+1$ random variables are equally likely.

Consider the following two events:

$$B = \Big\{\max\{Y_1,\ldots,Y_\alpha\} \leq Y\Big\}, \qquad C = \Big\{Y \leq \min\{Y_{\alpha+1},\ldots,Y_{\alpha+\beta}\}\Big\}.$$

We have, using the total probability theorem,

$$\begin{aligned}
\mathbf{P}(B\cap C) &= \int_0^1 \mathbf{P}(B\cap C \,|\, Y=y) f_Y(y)\,dy \\
&= \int_0^1 \mathbf{P}\Big(\max\{Y_1,\ldots,Y_\alpha\} \leq y \leq \min\{Y_{\alpha+1},\ldots,Y_{\alpha+\beta}\}\Big)\,dy \\
&= \int_0^1 \mathbf{P}\Big(\max\{Y_1,\ldots,Y_\alpha\} \leq y\Big)\,\mathbf{P}\Big(y \leq \min\{Y_{\alpha+1},\ldots,Y_{\alpha+\beta}\}\Big)\,dy \\
&= \int_0^1 y^\alpha (1-y)^\beta\,dy.
\end{aligned}$$

We also have

$$\mathbf{P}(A \,|\, B\cap C) = \frac{1}{\alpha!\,\beta!},$$

because given the events B and C, all $\alpha!$ possible orderings of $Y_1,\ldots,Y_\alpha$ are equally likely, and all $\beta!$ possible orderings of $Y_{\alpha+1},\ldots,Y_{\alpha+\beta}$ are equally likely.

By writing the equation

$$\mathbf{P}(A) = \mathbf{P}(B\cap C)\,\mathbf{P}(A \,|\, B\cap C)$$

in terms of the preceding relations, we finally obtain

$$\frac{1}{(\alpha+\beta+1)!} = \frac{1}{\alpha!\,\beta!}\int_0^1 y^\alpha(1-y)^\beta\,dy,$$

or

$$\int_0^1 y^\alpha (1-y)^\beta \, dy = \frac{\alpha! \, \beta!}{(\alpha+\beta+1)!}.$$

This equation can be written as

$$\mathrm{B}(\alpha+1, \beta+1) = \frac{\alpha! \, \beta!}{(\alpha+\beta+1)!}, \qquad \text{for all integer } \alpha > 0, \ \beta > 0.$$

Problem 31.* Estimating an expected value by simulation. Let $f_X(x)$ be a PDF such that for some nonnegative scalars a, b, and c, we have $f_X(x) = 0$ for all x outside the interval $[a, b]$, and $x f_X(x) \leq c$ for all x. Let Y_i, $i = 1, \ldots, n$, be independent random variables with values generated as follows: a point (V_i, W_i) is chosen at random (according to a uniform PDF) within the rectangle whose corners are $(a, 0)$, $(b, 0)$, (a, c), and (b, c), and if $W_i \leq V_i f_X(V_i)$, the value of Y_i is set to 1, and otherwise it is set to 0. Consider the random variable

$$Z = \frac{Y_1 + \cdots + Y_n}{n}.$$

Show that

$$\mathbf{E}[Z] = \frac{\mathbf{E}[X]}{c(b-a)}$$

and

$$\mathrm{var}(Z) \leq \frac{1}{4n}.$$

In particular, we have $\mathrm{var}(Z) \to 0$ as $n \to \infty$.

Solution. We have

$$\begin{aligned}
\mathbf{P}(Y_i = 1) &= \mathbf{P}\big(W_i \leq V_i f_X(V_i)\big) \\
&= \int_a^b \int_0^{v f_X(v)} \frac{1}{c(b-a)} \, dw \, dv \\
&= \frac{\displaystyle\int_a^b v f_X(v) \, dv}{c(b-a)} \\
&= \frac{\mathbf{E}[X]}{c(b-a)}.
\end{aligned}$$

The random variable Z has mean $\mathbf{P}(Y_i = 1)$ and variance

$$\mathrm{var}(Z) = \frac{\mathbf{P}(Y_i = 1)\big(1 - \mathbf{P}(Y_i = 1)\big)}{n}.$$

Since $0 \leq (1-2p)^2 = 1 - 4p(1-p)$, we have $p(1-p) \leq 1/4$ for any p in $[0, 1]$, so it follows that $\mathrm{var}(Z) \leq 1/(4n)$.

Problem 32.* Let X and Y be continuous random variables with joint PDF $f_{X,Y}$. Suppose that for any subsets A and B of the real line, the events $\{X \in A\}$ and $\{Y \in B\}$ are independent. Show that the random variables X and Y are independent.

Solution. For any two real numbers x and y, using the independence of the events $\{X \le x\}$ and $\{Y \le y\}$, we have

$$F_{X,Y}(x,y) = \mathbf{P}(X \le x,\, Y \le y) = \mathbf{P}(X \le x)\,\mathbf{P}(Y \le y) = F_X(x)F_Y(y).$$

Taking derivatives of both sides, we obtain

$$f_{X,Y}(x,y) = \frac{\partial^2 F_{X,Y}}{\partial x \partial y}(x,y) = \frac{\partial F_X}{\partial x}(x)\frac{\partial F_Y}{\partial y}(y) = f_X(x)f_Y(y),$$

which establishes that X and Y are independent.

Problem 33.* The sum of a random number of random variables. You visit a random number N of stores and in the ith store, you spend a random amount of money X_i. Let

$$T = X_1 + X_2 + \cdots + X_N$$

be the total amount of money that you spend. We assume that N is a positive integer random variable with a given PMF, and that the X_i are random variables with the same mean $\mathbf{E}[X]$ and variance $\text{var}(X)$. Furthermore, we assume that N and all the X_i are independent. Show that

$$\mathbf{E}[T] = \mathbf{E}[X]\,\mathbf{E}[N], \qquad \text{and} \qquad \text{var}(T) = \text{var}(X)\,\mathbf{E}[N] + \big(\mathbf{E}[X]\big)^2 \text{var}(N).$$

Solution. We have for all i,

$$\mathbf{E}[T \,|\, N = i] = i\mathbf{E}[X],$$

since conditional on $N = i$, you will visit exactly i stores, and you will spend an expected amount of money $\mathbf{E}[X]$ in each.

We now apply the total expectation theorem. We have

$$\begin{aligned}
\mathbf{E}[T] &= \sum_{i=1}^{\infty} \mathbf{P}(N = i)\,\mathbf{E}[T \,|\, N = i] \\
&= \sum_{i=1}^{\infty} \mathbf{P}(N = i) i\mathbf{E}[X] \\
&= \mathbf{E}[X] \sum_{i=1}^{\infty} i\mathbf{P}(N = i) \\
&= \mathbf{E}[X]\,\mathbf{E}[N].
\end{aligned}$$

Similarly, using also the independence of the X_i, which implies that $\mathbf{E}[X_iX_j] = \big(\mathbf{E}[X]\big)^2$ if $i \neq j$, the second moment of T is calculated as

$$\begin{aligned}
\mathbf{E}[T^2] &= \sum_{i=1}^{\infty} \mathbf{P}(N=i)\,\mathbf{E}[T^2 \,|\, N=i] \\
&= \sum_{i=1}^{\infty} \mathbf{P}(N=i)\,\mathbf{E}\big[(X_1+\cdots+X_N)^2 \,|\, N=i\big] \\
&= \sum_{i=1}^{\infty} \mathbf{P}(N=i)\big(i\mathbf{E}[X^2] + i(i-1)\big(\mathbf{E}[X]\big)^2\big) \\
&= \mathbf{E}[X^2]\sum_{i=1}^{\infty} i\mathbf{P}(N=i) + \big(\mathbf{E}[X]\big)^2 \sum_{i=1}^{\infty} i(i-1)\mathbf{P}(N=i) \\
&= \mathbf{E}[X^2]\,\mathbf{E}[N] + \big(\mathbf{E}[X]\big)^2\big(\mathbf{E}[N^2] - \mathbf{E}[N]\big) \\
&= \text{var}(X)\,\mathbf{E}[N] + \big(\mathbf{E}[X]\big)^2\mathbf{E}[N^2].
\end{aligned}$$

The variance is then obtained by

$$\begin{aligned}
\text{var}(T) &= \mathbf{E}[T^2] - \big(\mathbf{E}[T]\big)^2 \\
&= \text{var}(X)\,\mathbf{E}[N] + \big(\mathbf{E}[X]\big)^2\mathbf{E}[N^2] - \big(\mathbf{E}[X]\big)^2\big(\mathbf{E}[N]\big)^2 \\
&= \text{var}(X)\,\mathbf{E}[N] + \big(\mathbf{E}[X]\big)^2\big(\mathbf{E}[N^2] - \big(\mathbf{E}[N]\big)^2\big),
\end{aligned}$$

so finally

$$\text{var}(T) = \text{var}(X)\,\mathbf{E}[N] + \big(\mathbf{E}[X]\big)^2\text{var}(N).$$

Note: The formulas for $\mathbf{E}[T]$ and $\text{var}(T)$ will also be obtained in Chapter 4, using a more abstract approach.

SECTION 3.6. The Continuous Bayes' Rule

Problem 34. A defective coin minting machine produces coins whose probability of heads is a random variable P with PDF

$$f_P(p) = \begin{cases} pe^p, & p \in [0,1], \\ 0, & \text{otherwise}. \end{cases}$$

A coin produced by this machine is selected and tossed repeatedly, with successive tosses assumed independent.

(a) Find the probability that a coin toss results in heads.

(b) Given that a coin toss resulted in heads, find the conditional PDF of P.

(c) Given that the first coin toss resulted in heads, find the conditional probability of heads on the next toss.

Problem 35.* Let X and Y be independent continuous random variables with PDFs f_X and f_Y, respectively, and let $Z = X + Y$.

(a) Show that $f_{Z|X}(z\,|\,x) = f_Y(z-x)$. *Hint:* Write an expression for the conditional CDF of Z given X, and differentiate.

(b) Assume that X and Y are exponentially distributed with parameter λ. Find the conditional PDF of X, given that $Z = z$.

(c) Assume that X and Y are normal random variables with mean zero and variances σ_x^2 and σ_y^2, respectively. Find the conditional PDF of X, given that $Z = z$.

Solution. (a) We have

$$\begin{aligned}\mathbf{P}(Z \le z\,|\,X = x) &= \mathbf{P}(X + Y \le z\,|\,X = x)\\ &= \mathbf{P}(x + Y \le z\,|\,X = x)\\ &= \mathbf{P}(x + Y \le z)\\ &= \mathbf{P}(Y \le z - x),\end{aligned}$$

where the third equality follows from the independence of X and Y. By differentiating both sides with respect to z, the result follows.

(b) We have, for $0 \le x \le z$,

$$f_{X|Z}(x\,|\,z) = \frac{f_{Z|X}(z\,|\,x)f_X(x)}{f_Z(z)} = \frac{f_Y(z-x)f_X(x)}{f_Z(z)} = \frac{\lambda e^{-\lambda(z-x)}\lambda e^{-\lambda x}}{f_Z(z)} = \frac{\lambda^2 e^{-\lambda z}}{f_Z(z)}.$$

Since this is the same for all x, it follows that the conditional distribution of X is uniform on the interval $[0, z]$, with PDF $f_{X|Z}(x\,|\,z) = 1/z$.

(c) We have

$$f_{X|Z}(x\,|\,z) = \frac{f_Y(z-x)f_X(x)}{f_Z(z)} = \frac{1}{f_Z(z)} \cdot \frac{1}{\sqrt{2\pi}\,\sigma_y} e^{-(z-x)^2/2\sigma_y^2} \frac{1}{\sqrt{2\pi}\,\sigma_x} e^{-x^2/2\sigma_x^2}.$$

We focus on the terms in the exponent. By completing the square, we find that the negative of the exponent is of the form

$$\frac{(z-x)^2}{2\sigma_y^2} + \frac{x^2}{2\sigma_x^2} = \frac{\sigma_x^2+\sigma_y^2}{2\sigma_x^2\sigma_y^2}\left(x - \frac{z\sigma_x^2}{\sigma_x^2+\sigma_y^2}\right)^2 + \frac{z^2}{2\sigma_y^2}\left(1 - \frac{\sigma_x^2}{\sigma_x^2+\sigma_y^2}\right).$$

Thus, the conditional PDF of X is of the form

$$f_{X|Z}(x\,|\,z) = c(z)\cdot \exp\left\{-\frac{\sigma_x^2+\sigma_y^2}{2\sigma_x^2\sigma_y^2}\left(x - \frac{z\sigma_x^2}{\sigma_x^2+\sigma_y^2}\right)^2\right\},$$

where $c(z)$ does not depend on x and plays the role of a normalizing constant. We recognize this as a normal distribution with mean

$$\mathbf{E}[X\,|\,Z = z] = \frac{\sigma_x^2}{\sigma_x^2+\sigma_y^2}\,z,$$

and variance

$$\text{var}(X\,|\,Z = z) = \frac{\sigma_x^2\sigma_y^2}{\sigma_x^2+\sigma_y^2}.$$

4

Further Topics on Random Variables

Contents

In this chapter, we develop a number of more advanced topics. We introduce methods that are useful in:

(a) deriving the distribution of a function of one or multiple random variables;

(b) dealing with the sum of independent random variables, including the case where the number of random variables is itself random;

(c) quantifying the degree of dependence between two random variables.

With these goals in mind, we introduce a number of tools, including transforms and convolutions, and we refine our understanding of the concept of conditional expectation.

The material in this chapter is not needed for Chapters 5-7, with the exception of the solutions of a few problems, and may be viewed as optional in a first reading of the book. On the other hand, the concepts and methods discussed here constitute essential background for a more advanced treatment of probability and stochastic processes, and provide powerful tools in several disciplines that rely on probabilistic models. Furthermore, the concepts introduced in Sections 4.2 and 4.3 will be required in our study of inference and statistics, in Chapters 8 and 9.

4.1 DERIVED DISTRIBUTIONS

In this section, we consider functions $Y = g(X)$ of a continuous random variable X. We discuss techniques whereby, given the PDF of X, we calculate the PDF of Y (also called a *derived distribution*). The principal method for doing so is the following two-step approach.

Calculation of the PDF of a Function $Y = g(X)$ of a Continuous Random Variable X

1. Calculate the CDF F_Y of Y using the formula

$$F_Y(y) = \mathbf{P}\big(g(X) \leq y\big) = \int_{\{x \mid g(x) \leq y\}} f_X(x)\, dx.$$

2. Differentiate to obtain the PDF of Y:

$$f_Y(y) = \frac{dF_Y}{dy}(y).$$

Example 4.1. Let X be uniform on $[0, 1]$, and let $Y = \sqrt{X}$. We note that for every $y \in [0, 1]$, we have

$$F_Y(y) = \mathbf{P}(Y \leq y) = \mathbf{P}(\sqrt{X} \leq y) = \mathbf{P}(X \leq y^2) = y^2.$$

We then differentiate and obtain

$$f_Y(y) = \frac{dF_Y}{dy}(y) = \frac{d(y^2)}{dy} = 2y, \qquad 0 \le y \le 1.$$

Outside the range $[0, 1]$, the CDF $F_Y(y)$ is constant, with $F_Y(y) = 0$ for $y \le 0$, and $F_Y(y) = 1$ for $y \ge 1$. By differentiating, we see that $f_Y(y) = 0$ for y outside $[0, 1]$.

Example 4.2. John Slow is driving from Boston to the New York area, a distance of 180 miles at a constant speed, whose value is uniformly distributed between 30 and 60 miles per hour. What is the PDF of the duration of the trip?

Let X be the speed and let $Y = g(X)$ be the trip duration:

$$g(X) = \frac{180}{X}.$$

To find the CDF of Y, we must calculate

$$\mathbf{P}(Y \le y) = \mathbf{P}\left(\frac{180}{X} \le y\right) = \mathbf{P}\left(\frac{180}{y} \le X\right).$$

We use the given uniform PDF of X, which is

$$f_X(x) = \begin{cases} 1/30, & \text{if } 30 \le x \le 60, \\ 0, & \text{otherwise,} \end{cases}$$

and the corresponding CDF, which is

$$F_X(x) = \begin{cases} 0, & \text{if } x \le 30, \\ (x-30)/30, & \text{if } 30 \le x \le 60, \\ 1, & \text{if } 60 \le x. \end{cases}$$

Thus,

$$\begin{aligned} F_Y(y) &= \mathbf{P}\left(\frac{180}{y} \le X\right) \\ &= 1 - F_X\left(\frac{180}{y}\right) \\ &= \begin{cases} 0, & \text{if } y \le 180/60, \\ 1 - \dfrac{\dfrac{180}{y} - 30}{30}, & \text{if } 180/60 \le y \le 180/30, \\ 1, & \text{if } 180/30 \le y, \end{cases} \\ &= \begin{cases} 0, & \text{if } y \le 3, \\ 2 - (6/y), & \text{if } 3 \le y \le 6, \\ 1, & \text{if } 6 \le y, \end{cases} \end{aligned}$$

(see Fig. 4.1). Differentiating this expression, we obtain the PDF of Y:

$$f_Y(y) = \begin{cases} 0, & \text{if } y < 3, \\ 6/y^2, & \text{if } 3 < y < 6, \\ 0, & \text{if } 6 < y. \end{cases}$$

Example 4.3. Let $Y = g(X) = X^2$, where X is a random variable with known PDF. For any $y \geq 0$, we have

$$\begin{aligned} F_Y(y) &= \mathbf{P}(Y \leq y) \\ &= \mathbf{P}(X^2 \leq y) \\ &= \mathbf{P}(-\sqrt{y} \leq X \leq \sqrt{y}) \\ &= F_X(\sqrt{y}) - F_X(-\sqrt{y}), \end{aligned}$$

and therefore, by differentiating and using the chain rule,

$$f_Y(y) = \frac{1}{2\sqrt{y}} f_X(\sqrt{y}) + \frac{1}{2\sqrt{y}} f_X(-\sqrt{y}), \qquad y \geq 0.$$

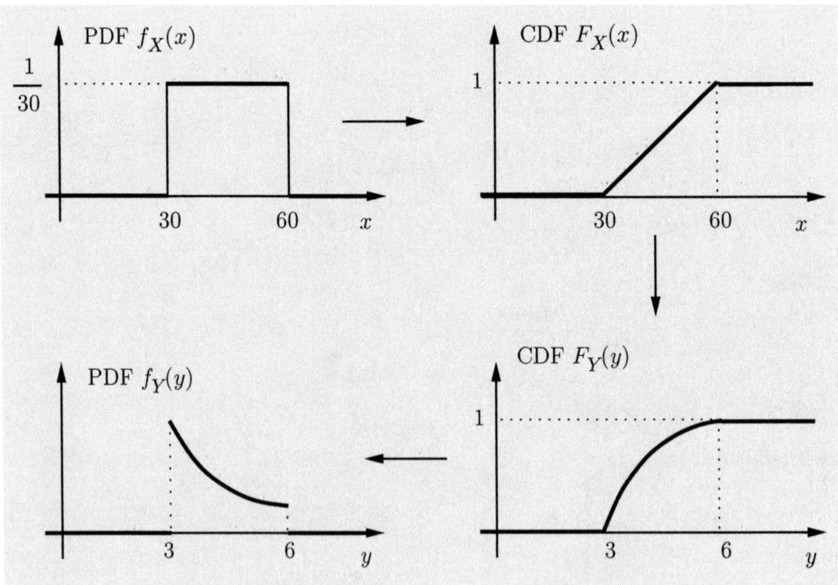

Figure 4.1: The calculation of the PDF of $Y = 180/X$ in Example 4.20. The arrows indicate the flow of the calculation.

The Linear Case

We now focus on the important special case where Y is a linear function of X; see Fig. 4.2 for a graphical interpretation.

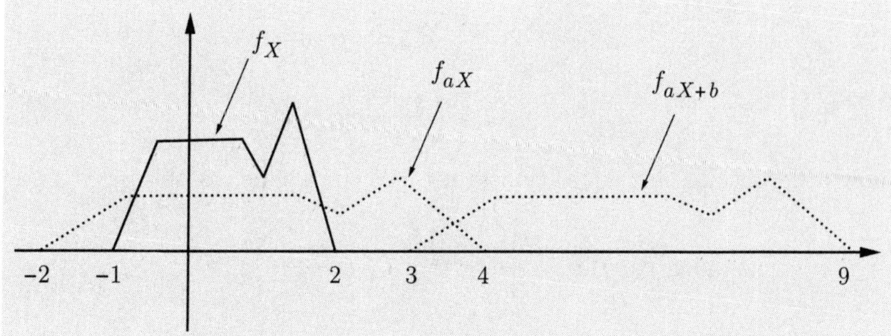

Figure 4.2: The PDF of $aX + b$ in terms of the PDF of X. In this figure, $a = 2$ and $b = 5$. As a first step, we obtain the PDF of aX. The range of Y is wider than the range of X, by a factor of a. Thus, the PDF f_X must be stretched (scaled horizontally) by this factor. But in order to keep the total area under the PDF equal to 1, we need to scale down the PDF (vertically) by the same factor a. The random variable $aX + b$ is the same as aX except that its values are shifted by b. Accordingly, we take the PDF of aX and shift it (horizontally) by b. The end result of these operations is the PDF of $Y = aX + b$ and is given mathematically by

$$f_Y(y) = \frac{1}{|a|} f_X\left(\frac{y-b}{a}\right).$$

If a were negative, the procedure would be the same except that the PDF of X would first need to be reflected around the vertical axis ("flipped") yielding f_{-X}. Then a horizontal and vertical scaling (by a factor of $|a|$ and $1/|a|$, respectively) yields the PDF of $-|a|X = aX$. Finally, a horizontal shift of b would again yield the PDF of $aX + b$.

The PDF of a Linear Function of a Random Variable

Let X be a continuous random variable with PDF f_X, and let

$$Y = aX + b,$$

where a and b are scalars, with $a \neq 0$. Then,

$$f_Y(y) = \frac{1}{|a|} f_X\left(\frac{y-b}{a}\right).$$

To verify this formula, we first calculate the CDF of Y and then differentiate. We only show the steps for the case where $a > 0$; the case $a < 0$ is similar. We have

$$\begin{aligned} F_Y(y) &= \mathbf{P}(Y \le y) \\ &= \mathbf{P}(aX + b \le y) \\ &= \mathbf{P}\left(X \le \frac{y-b}{a}\right) \\ &= F_X\left(\frac{y-b}{a}\right). \end{aligned}$$

We now differentiate this equality and use the chain rule, to obtain

$$f_Y(y) = \frac{dF_Y}{dy}(y) = \frac{1}{a} f_X\left(\frac{y-b}{a}\right).$$

Example 4.4. A Linear Function of an Exponential Random Variable. Suppose that X is an exponential random variable with PDF

$$f_X(x) = \begin{cases} \lambda e^{-\lambda x}, & \text{if } x \ge 0, \\ 0, & \text{otherwise}, \end{cases}$$

where λ is a positive parameter. Let $Y = aX + b$. Then,

$$f_Y(y) = \begin{cases} \dfrac{\lambda}{|a|} e^{-\lambda(y-b)/a}, & \text{if } (y-b)/a \ge 0, \\ 0, & \text{otherwise}. \end{cases}$$

Note that if $b = 0$ and $a > 0$, then Y is an exponential random variable with parameter λ/a. In general, however, Y need not be exponential. For example, if $a < 0$ and $b = 0$, then the range of Y is the negative real axis.

Example 4.5. A Linear Function of a Normal Random Variable is Normal. Suppose that X is a normal random variable with mean μ and variance σ^2, and let $Y = aX + b$, where a and b are scalars, with $a \neq 0$. We have

$$f_X(x) = \frac{1}{\sqrt{2\pi}\,\sigma} e^{-(x-\mu)^2/2\sigma^2}.$$

Therefore,

$$\begin{aligned} f_Y(y) &= \frac{1}{|a|} f_X\left(\frac{y-b}{a}\right) \\ &= \frac{1}{|a|}\frac{1}{\sqrt{2\pi}\,\sigma} \exp\left\{-\left(\frac{y-b}{a} - \mu\right)^2 \Big/ 2\sigma^2\right\} \\ &= \frac{1}{\sqrt{2\pi}\,|a|\sigma} \exp\left\{-\frac{(y-b-a\mu)^2}{2a^2\sigma^2}\right\}. \end{aligned}$$

We recognize this as a normal PDF with mean $a\mu + b$ and variance $a^2\sigma^2$. In particular, Y is a normal random variable.

The Monotonic Case

The calculation and the formula for the linear case can be generalized to the case where g is a monotonic function. Let X be a continuous random variable and suppose that its range is contained in a certain interval I, in the sense that $f_X(x) = 0$ for $x \notin I$. We consider the random variable $Y = g(X)$, and assume that g is **strictly monotonic** over the interval I, so that either

(a) $g(x) < g(x')$ for all $x, x' \in I$ satisfying $x < x'$ (monotonically increasing case), or

(b) $g(x) > g(x')$ for all $x, x' \in I$ satisfying $x < x'$ (monotonically decreasing case).

Furthermore, we assume that the function g is differentiable. Its derivative will necessarily be nonnegative in the increasing case and nonpositive in the decreasing case.

An important fact is that a strictly monotonic function can be "inverted" in the sense that there is some function h, called the inverse of g, such that for all $x \in I$, we have

$$y = g(x) \qquad \text{if and only if} \qquad x = h(y).$$

For example, the inverse of the function $g(x) = 180/x$ considered in Example 4.2 is $h(y) = 180/y$, because we have $y = 180/x$ if and only if $x = 180/y$. Other such examples of pairs of inverse functions include

$$g(x) = ax + b, \qquad h(y) = \frac{y - b}{a},$$

where a and b are scalars with $a \neq 0$, and

$$g(x) = e^{ax}, \qquad h(y) = \frac{\ln y}{a},$$

where a is a nonzero scalar.

For strictly monotonic functions g, the following is a convenient analytical formula for the PDF of the function $Y = g(X)$.

PDF Formula for a Strictly Monotonic Function of a Continuous Random Variable

Suppose that g is strictly monotonic and that for some function h and all x in the range of X we have

$$y = g(x) \qquad \text{if and only if} \qquad x = h(y).$$

Assume that h is differentiable. Then, the PDF of Y in the region where $f_Y(y) > 0$ is given by

$$f_Y(y) = f_X\big(h(y)\big) \left| \frac{dh}{dy}(y) \right|.$$

For a verification of the above formula, assume first that g is monotonically increasing. Then, we have

$$F_Y(y) = \mathbf{P}\big(g(X) \leq y\big) = \mathbf{P}\big(X \leq h(y)\big) = F_X\big(h(y)\big),$$

where the second equality can be justified using the monotonically increasing property of g (see Fig. 4.3). By differentiating this relation, using also the chain rule, we obtain

$$f_Y(y) = \frac{dF_Y}{dy}(y) = f_X\big(h(y)\big)\frac{dh}{dy}(y).$$

Because g is monotonically increasing, h is also monotonically increasing, so its derivative is nonnegative:

$$\frac{dh}{dy}(y) = \left| \frac{dh}{dy}(y) \right|.$$

This justifies the PDF formula for a monotonically increasing function g. The justification for the case of monotonically decreasing function is similar: we differentiate instead the relation

$$F_Y(y) = \mathbf{P}\big(g(X) \leq y\big) = \mathbf{P}\big(X \geq h(y)\big) = 1 - F_X\big(h(y)\big),$$

and use the chain rule.

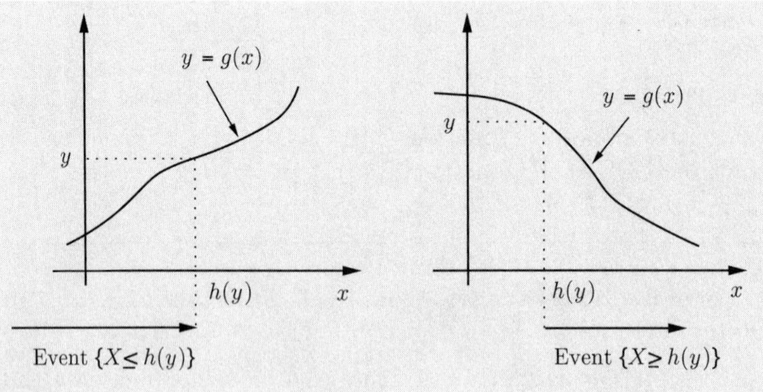

Figure 4.3: Calculating the probability $\mathbf{P}\big(g(X) \leq y\big)$. When g is monotonically increasing (left figure), the event $\{g(X) \leq y\}$ is the same as the event $\{X \leq h(y)\}$. When g is monotonically decreasing (right figure), the event $\{g(X) \leq y\}$ is the same as the event $\{X \geq h(y)\}$.

Example 4.2 (continued). To check the PDF formula, let us apply it to the problem of Example 4.2. In the region of interest, $x \in [30, 60]$, we have $h(y) = 180/y$, and

$$f_X\big(h(y)\big) = \frac{1}{30}, \qquad \left|\frac{dh}{dy}(y)\right| = \frac{180}{y^2}.$$

Thus, in the region of interest $y \in [3, 6]$, the PDF formula yields

$$f_Y(y) = f_X\big(h(y)\big)\left|\frac{dh}{dy}(y)\right| = \frac{1}{30}\cdot\frac{180}{y^2} = \frac{6}{y^2},$$

consistent with the expression obtained earlier.

Example 4.6. Let $Y = g(X) = X^2$, where X is a continuous uniform random variable on the interval $(0, 1]$. Within this interval, g is strictly monotonic, and its inverse is $h(y) = \sqrt{y}$. Thus, for any $y \in (0, 1]$, we have

$$f_X(\sqrt{y}) = 1, \qquad \left|\frac{dh}{dy}(y)\right| = \frac{1}{2\sqrt{y}},$$

and

$$f_Y(y) = \begin{cases} \dfrac{1}{2\sqrt{y}}, & \text{if } y \in (0, 1], \\ 0, & \text{otherwise.} \end{cases}$$

We finally note that if we interpret PDFs in terms of probabilities of small intervals, the content of our formulas becomes pretty intuitive; see Fig. 4.4.

Functions of Two Random Variables

The two-step procedure that first calculates the CDF and then differentiates to obtain the PDF also applies to functions of more than one random variable.

Example 4.7. Two archers shoot at a target. The distance of each shot from the center of the target is uniformly distributed from 0 to 1, independent of the other shot. What is the PDF of the distance of the losing shot from the center?

Let X and Y be the distances from the center of the first and second shots, respectively. Let also Z be the distance of the losing shot:

$$Z = \max\{X, Y\}.$$

We know that X and Y are uniformly distributed over $[0, 1]$, so that for all $z \in [0, 1]$, we have

$$\mathbf{P}(X \le z) = \mathbf{P}(Y \le z) = z.$$

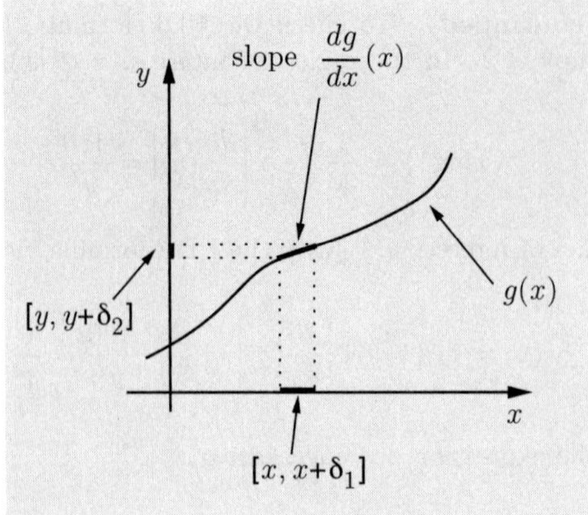

Figure 4.4: Illustration of the PDF formula for a monotonically increasing function g. Consider an interval $[x, x + \delta_1]$, where δ_1 is a small number. Under the mapping g, the image of this interval is another interval $[y, y + \delta_2]$. Since $(dg/dx)(x)$ is the slope of g, we have

$$\frac{\delta_2}{\delta_1} \approx \frac{dg}{dx}(x),$$

or in terms of the inverse function,

$$\frac{\delta_1}{\delta_2} \approx \frac{dh}{dy}(y),$$

We now note that the event $\{x \leq X \leq x + \delta_1\}$ is the same as the event $\{y \leq Y \leq y + \delta_2\}$. Thus,

$$\begin{aligned} f_Y(y)\delta_2 &\approx \mathbf{P}(y \leq Y \leq y + \delta_2) \\ &= \mathbf{P}(x \leq X \leq x + \delta_1) \\ &\approx f_X(x)\delta_1. \end{aligned}$$

We move δ_1 to the left-hand side and use our earlier formula for the ratio δ_2/δ_1, to obtain

$$f_Y(y)\frac{dg}{dx}(x) = f_X(x).$$

Alternatively, if we move δ_2 to the right-hand side and use the formula for δ_1/δ_2, we obtain

$$f_Y(y) = f_X\big(h(y)\big)\frac{dh}{dy}(y).$$

Thus, using the independence of X and Y, we have for all $z \in [0, 1]$,

$$\begin{aligned} F_Z(z) &= \mathbf{P}\big(\max\{X, Y\} \leq z\big) \\ &= \mathbf{P}(X \leq z,\, Y \leq z) \\ &= \mathbf{P}(X \leq z)\,\mathbf{P}(Y \leq z) \\ &= z^2. \end{aligned}$$

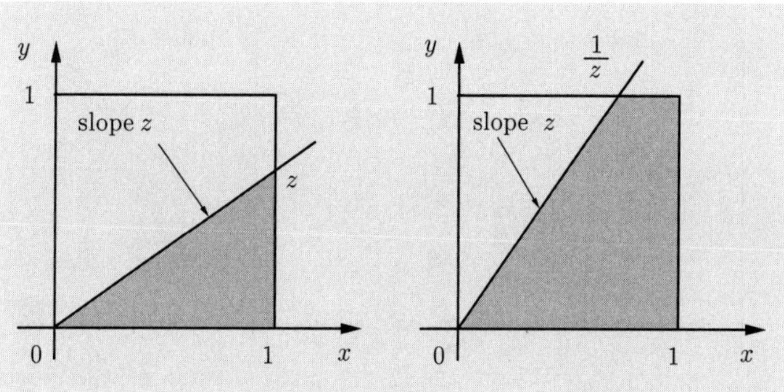

Figure 4.5: The calculation of the CDF of $Z = Y/X$ in Example 4.8. The value $\mathbf{P}(Y/X \le z)$ is equal to the shaded subarea of the unit square. The figure on the left deals with the case where $0 \le z \le 1$ and the figure on the right refers to the case where $z > 1$.

Differentiating, we obtain

$$f_Z(z) = \begin{cases} 2z, & \text{if } 0 \le z \le 1, \\ 0, & \text{otherwise.} \end{cases}$$

Example 4.8. Let X and Y be independent random variables that are uniformly distributed on the interval $[0, 1]$. What is the PDF of the random variable $Z = Y/X$?

We will find the PDF of Z by first finding its CDF and then differentiating. We consider separately the cases $0 \le z \le 1$ and $z > 1$. As shown in Fig. 4.5, we have

$$F_Z(z) = \mathbf{P}\left(\frac{Y}{X} \le z\right) = \begin{cases} z/2, & \text{if } 0 \le z \le 1, \\ 1 - 1/(2z), & \text{if } z > 1, \\ 0, & \text{otherwise.} \end{cases}$$

By differentiating, we obtain

$$f_Z(z) = \begin{cases} 1/2, & \text{if } 0 \le z \le 1, \\ 1/(2z^2), & \text{if } z > 1, \\ 0, & \text{otherwise.} \end{cases}$$

Example 4.9. Romeo and Juliet have a date at a given time, and each, independently, will be late by an amount of time that is exponentially distributed with parameter λ. What is the PDF of the difference between their times of arrival?

Let us denote by X and Y the amounts by which Romeo and Juliet are late, respectively. We want to find the PDF of $Z = X - Y$, assuming that X and Y are independent and exponentially distributed with parameter λ. We will first calculate the CDF $F_Z(z)$ by considering separately the cases $z \ge 0$ and $z < 0$ (see Fig. 4.6).

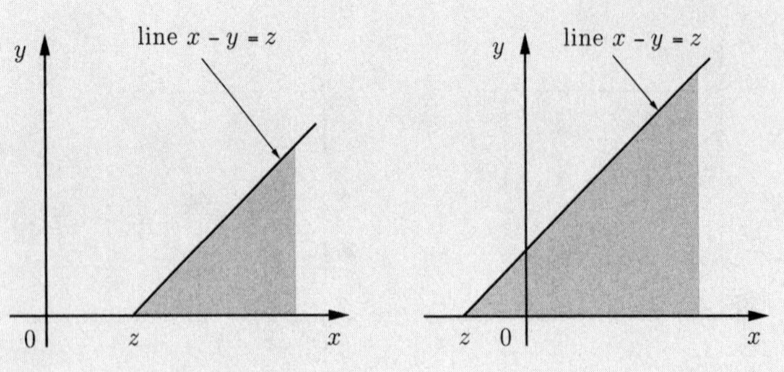

Figure 4.6: The calculation of the CDF of $Z = X - Y$ in Example 4.9. To obtain the value $\mathbf{P}(X - Y > z)$ we must integrate the joint PDF $f_{X,Y}(x, y)$ over the shaded area in the above figures, which correspond to $z \geq 0$ (left side) and $z < 0$ (right side).

For $z \geq 0$, we have (see the left side of Fig. 4.6)

$$\begin{aligned}
F_Z(z) &= \mathbf{P}(X - Y \leq z) \\
&= 1 - \mathbf{P}(X - Y > z) \\
&= 1 - \int_0^\infty \left(\int_{z+y}^\infty f_{X,Y}(x, y)\, dx \right) dy \\
&= 1 - \int_0^\infty \lambda e^{-\lambda y} \left(\int_{z+y}^\infty \lambda e^{-\lambda x}\, dx \right) dy \\
&= 1 - \int_0^\infty \lambda e^{-\lambda y} e^{-\lambda(z+y)}\, dy \\
&= 1 - e^{-\lambda z} \int_0^\infty \lambda e^{-2\lambda y}\, dy \\
&= 1 - \frac{1}{2} e^{-\lambda z}.
\end{aligned}$$

For the case $z < 0$, we can use a similar calculation, but we can also argue using symmetry. Indeed, the symmetry of the situation implies that the random variables $Z = X - Y$ and $-Z = Y - X$ have the same distribution. We have

$$F_Z(z) = \mathbf{P}(Z \leq z) = \mathbf{P}(-Z \geq -z) = \mathbf{P}(Z \geq -z) = 1 - F_Z(-z).$$

With $z < 0$, we have $-z > 0$ and using the formula derived earlier,

$$F_Z(z) = 1 - F_Z(-z) = 1 - \left(1 - \frac{1}{2} e^{-\lambda(-z)}\right) = \frac{1}{2} e^{\lambda z}.$$

Combining the two cases $z \geq 0$ and $z < 0$, we obtain

$$F_Z(z) = \begin{cases} 1 - \dfrac{1}{2} e^{-\lambda z}, & \text{if } z \geq 0, \\ \dfrac{1}{2} e^{\lambda z}, & \text{if } z < 0, \end{cases}$$

We now calculate the PDF of Z by differentiating its CDF. We have

$$f_Z(z) = \begin{cases} \dfrac{\lambda}{2}e^{-\lambda z}, & \text{if } z \geq 0, \\ \dfrac{\lambda}{2}e^{\lambda z}, & \text{if } z < 0, \end{cases}$$

or

$$f_Z(z) = \frac{\lambda}{2}e^{-\lambda |z|}.$$

This is known as a **two-sided exponential** PDF, also called the **Laplace** PDF.

Sums of Independent Random Variables — Convolution

We now consider an important example of a function Z of two random variables, namely, the case where $Z = X + Y$, for independent X and Y. For some initial insight, we start by deriving a PMF formula for the case where X and Y are discrete.

Let $Z = X + Y$, where X and Y are independent integer-valued random variables with PMFs p_X and p_Y, respectively. Then, for any integer z,

$$\begin{aligned} p_Z(z) &= \mathbf{P}(X + Y = z) \\ &= \sum_{\{(x,y)\,|\,x+y=z\}} \mathbf{P}(X = x, Y = y) \\ &= \sum_x \mathbf{P}(X = x, Y = z - x) \\ &= \sum_x p_X(x)p_Y(z - x). \end{aligned}$$

The resulting PMF p_Z is called the **convolution** of the PMFs of X and Y. See Fig. 4.7 for an illustration.

Suppose now that X and Y are independent continuous random variables with PDFs f_X and f_Y, respectively. We wish to find the PDF of $Z = X + Y$. Towards this goal, we will first find the joint PDF of X and Z, and then integrate to find the PDF of Z.

We first note that

$$\begin{aligned} \mathbf{P}(Z \leq z \,|\, X = x) &= \mathbf{P}(X + Y \leq z \,|\, X = x) \\ &= \mathbf{P}(x + Y \leq z \,|\, X = x) \\ &= \mathbf{P}(x + Y \leq z) \\ &= \mathbf{P}(Y \leq z - x), \end{aligned}$$

where the third equality follows from the independence of X and Y. By differentiating both sides with respect to z, we see that $f_{Z|X}(z \,|\, x) = f_Y(z - x)$. Using the multiplication rule, we have

$$f_{X,Z}(x, z) = f_X(x)f_{Z|X}(z \,|\, x) = f_X(x)f_Y(z - x),$$

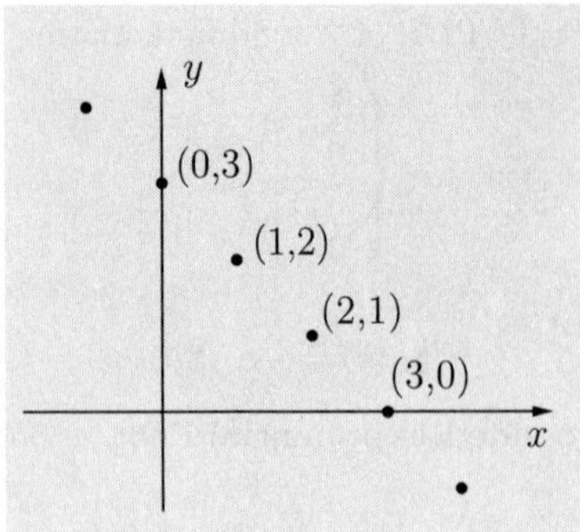

Figure 4.7: The probability $p_Z(3)$ that $X+Y=3$ is the sum of the probabilities of all pairs (x,y) such that $x+y=3$, which are the points indicated in the figure. The probability of a generic such point is of the form

$$p_{X,Y}(x, 3-x) = p_X(x)p_Y(3-x).$$

from which we finally obtain

$$f_Z(z) = \int_{-\infty}^{\infty} f_{X,Z}(x,z)\,dx = \int_{-\infty}^{\infty} f_X(x)f_Y(z-x)\,dx.$$

This formula is entirely analogous to the one for the discrete case, except that the summation is replaced by an integral and the PMFs are replaced by PDFs. For an intuitive interpretation, see Fig. 4.8.

Example 4.10. The random variables X and Y are independent and uniformly distributed in the interval $[0,1]$. The PDF of $Z = X+Y$ is

$$f_Z(z) = \int_{-\infty}^{\infty} f_X(x)f_Y(z-x)\,dx.$$

The integrand $f_X(x)f_Y(z-x)$ is nonzero (and equal to 1) for $0 \le x \le 1$ and $0 \le z-x \le 1$. Combining these two inequalities, the integrand is nonzero for $\max\{0, z-1\} \le x \le \min\{1, z\}$. Thus,

$$f_Z(z) = \begin{cases} \min\{1,z\} - \max\{0, z-1\}, & 0 \le z \le 2, \\ 0, & \text{otherwise,} \end{cases}$$

which has the triangular shape shown in Fig. 4.9.

We next describe an important application of the convolution formula.

Example 4.11. The Sum of Two Independent Normal Random Variables is Normal. Let X and Y be independent normal random variables with means

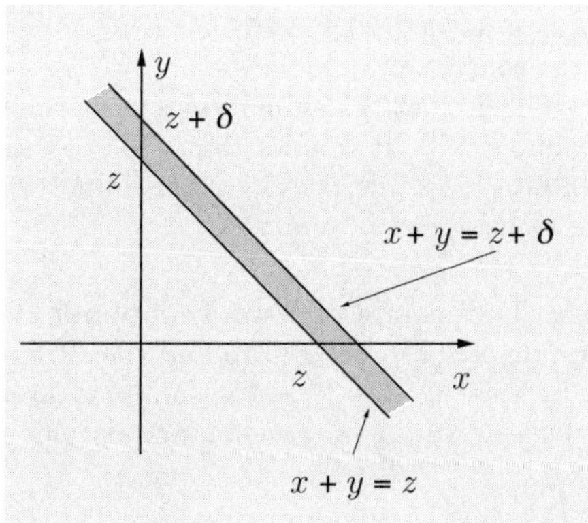

Figure 4.8: Illustration of the convolution formula for the case of continuous random variables (compare with Fig. 4.7). For small δ, the probability of the strip indicated in the figure is $\mathbf{P}(z \le X+Y \le z+\delta) \approx f_Z(z)\delta$. Thus,

$$\begin{aligned} f_Z(z)\delta &= \mathbf{P}(z \le X+Y \le z+\delta) \\ &= \int_{-\infty}^{\infty} \int_{z-x}^{z-x+\delta} f_X(x) f_Y(y)\, dy\, dx \\ &\approx \int_{-\infty}^{\infty} f_X(x) f_Y(z-x)\delta\, dx. \end{aligned}$$

The desired formula follows by canceling δ from both sides.

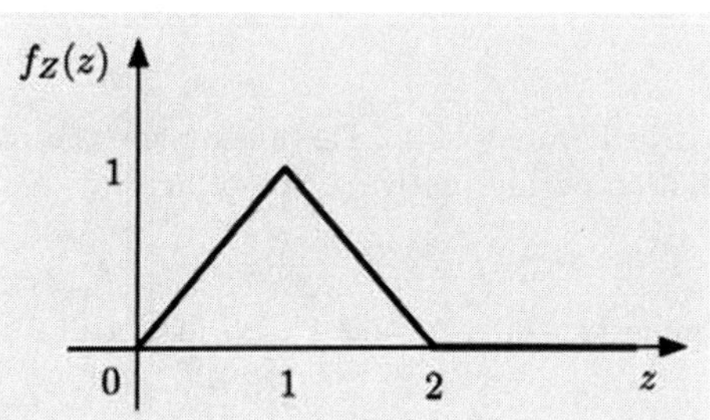

Figure 4.9: The PDF of the sum of two independent uniform random variables in $[0,1]$.

μ_x, μ_y, and variances σ_x^2, σ_y^2, respectively, and let $Z = X+Y$. We have

$$f_Z(z) = \int_{-\infty}^{\infty} \frac{1}{\sqrt{2\pi}\,\sigma_x} \exp\left\{ -\frac{(x-\mu_x)^2}{2\sigma_x^2} \right\} \frac{1}{\sqrt{2\pi}\,\sigma_y} \exp\left\{ -\frac{(z-x-\mu_y)^2}{2\sigma_y^2} \right\} dx.$$

This integral can be evaluated in closed form, but the details are tedious and are omitted. The answer turns out to be of the form

$$f_Z(z) = \frac{1}{\sqrt{2\pi(\sigma_x^2+\sigma_y^2)}} \exp\left\{ -\frac{(z-\mu_x-\mu_y)^2}{2(\sigma_x^2+\sigma_y^2)} \right\},$$

which we recognize as a normal PDF with mean $\mu_x + \mu_y$ and variance $\sigma_x^2 + \sigma_y^2$. We therefore reach the conclusion that the sum of two independent normal random variables is normal. Given that scalar multiples of normal random variables are also normal (cf. Example 4.5), it follows that $aX + bY$ is also normal, for any nonzero a and b. An alternative derivation of this important fact will be provided in Section 4.4, using transforms.

Example 4.12. The Difference of Two Independent Random Variables. The convolution formula can also be used to find the PDF of $X - Y$, when X and Y are independent, by viewing $X - Y$ as the sum of X and $-Y$. We observe that the PDF of $-Y$ is given by $f_{-Y}(y) = f_Y(-y)$, and obtain

$$f_{X-Y}(z) = \int_{-\infty}^{\infty} f_X(x) f_{-Y}(z-x)\,dx = \int_{-\infty}^{\infty} f_X(x) f_Y(x-z)\,dx.$$

As an illustration, consider the case where X and Y are independent exponential random variables with parameter λ, as in Example 4.9. Fix some $z \geq 0$ and note that $f_Y(x-z)$ is nonzero only when $x \geq z$. Thus,

$$\begin{aligned} f_{X-Y}(z) &= \int_{-\infty}^{\infty} f_X(x) f_Y(x-z)\,dx \\ &= \int_z^{\infty} \lambda e^{-\lambda x} \lambda e^{-\lambda(x-z)}\,dx \\ &= \lambda^2 e^{\lambda z} \int_z^{\infty} e^{-2\lambda x}\,dx \\ &= \lambda^2 e^{\lambda z} \frac{1}{2\lambda} e^{-2\lambda z} \\ &= \frac{\lambda}{2} e^{-\lambda z}, \end{aligned}$$

in agreement with the result obtained in Example 4.9. The answer for the case $z < 0$ is obtained with a similar calculation or, alternatively, by noting that

$$f_{X-Y}(z) = f_{Y-X}(z) = f_{-(X-Y)}(z) = f_{X-Y}(-z),$$

where the first equality holds by symmetry, since X and Y are identically distributed.

When applying the convolution formula, often the most delicate step was to determine the correct limits for the integration. This is often tedious and error prone, but can be bypassed using a graphical method described next.

Graphical Calculation of Convolutions

We use a dummy variable t as the argument of the different functions involved in this discussion; see also Fig. 4.10. Consider two PDFs $f_X(t)$ and $f_Y(t)$. For a fixed value of z, the graphical evaluation of the convolution

$$f_Z(z) = \int_{-\infty}^{\infty} f_X(t) f_Y(z-t)\,dt$$

consists of the following steps:

(a) We plot $f_Y(z-t)$ as a function of t. This plot has the same shape as the plot of $f_Y(t)$ except that it is first "flipped" and then shifted by an amount z. If $z > 0$, this is a shift to the right, if $z < 0$, this is a shift to the left.

(b) We place the plots of $f_X(t)$ and $f_Y(z-t)$ on top of each other, and form their product.

(c) We calculate the value of $f_Z(z)$ by calculating the integral of the product of these two plots.

By varying the amount z by which we are shifting, we obtain $f_Z(z)$ for any z.

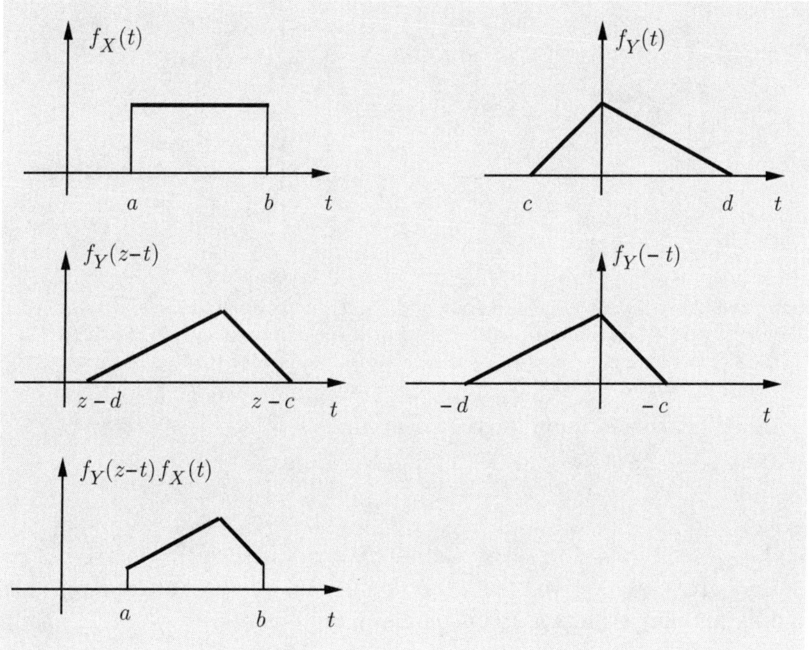

Figure 4.10: Illustration of the convolution calculation. For the value of z under consideration, $f_Z(z)$ is equal to the integral of the function shown in the last plot.

4.2 COVARIANCE AND CORRELATION

In this section, we introduce a quantitative measure of the strength and direction of the relationship between two random variables. It plays an important role in many contexts, and it will be used in the estimation methodology to be developed in Chapters 8 and 9.

The **covariance** of two random variables X and Y, denoted by $\text{cov}(X, Y)$, is defined by

$$\text{cov}(X, Y) = \mathbf{E}\Big[\big(X - \mathbf{E}[X]\big)\big(Y - \mathbf{E}[Y]\big)\Big].$$

When $\text{cov}(X, Y) = 0$, we say that X and Y are **uncorrelated**. Roughly speaking, a positive or negative covariance indicates that the values of $X - \mathbf{E}[X]$ and $Y - \mathbf{E}[Y]$ obtained in a single experiment "tend" to have the same or the opposite sign, respectively (see Fig. 4.11). Thus the sign of the covariance provides an important qualitative indicator of the relationship between X and Y.

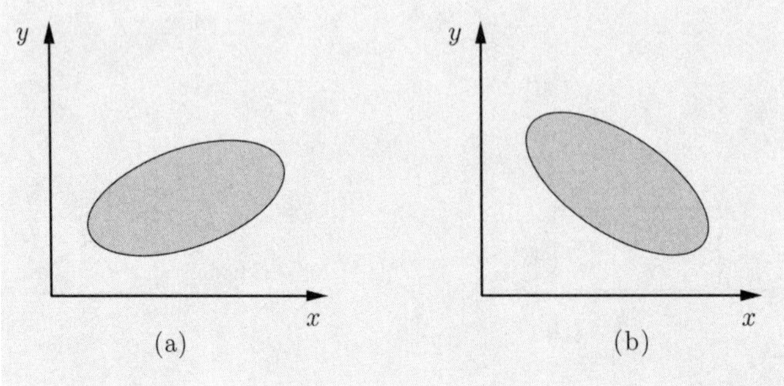

Figure 4.11: Examples of positively and negatively correlated random variables. Here, X and Y are uniformly distributed over the ellipses shown in the figure. In case (a) the covariance $\text{cov}(X, Y)$ is positive, while in case (b) it is negative.

An alternative formula for the covariance is

$$\text{cov}(X, Y) = \mathbf{E}[XY] - \mathbf{E}[X]\,\mathbf{E}[Y],$$

as can be verified by a simple calculation. We record a few properties of covariances that are easily derived from the definition: for any random variables X, Y, and Z, and any scalars a and b, we have

$$\begin{aligned}\text{cov}(X, X) &= \text{var}(X),\\ \text{cov}(X, aY + b) &= a \cdot \text{cov}(X, Y),\\ \text{cov}(X, Y + Z) &= \text{cov}(X, Y) + \text{cov}(X, Z).\end{aligned}$$

Note that if X and Y are independent, we have $\mathbf{E}[XY] = \mathbf{E}[X]\,\mathbf{E}[Y]$, which implies that $\text{cov}(X, Y) = 0$. Thus, if X and Y are independent, they are also uncorrelated. However, the converse is generally not true, as illustrated by the following example.

Example 4.13. The pair of random variables (X, Y) takes the values $(1, 0)$, $(0, 1)$, $(-1, 0)$, and $(0, -1)$, each with probability $1/4$ (see Fig. 4.12). Thus, the marginal

PMFs of X and Y are symmetric around 0, and $\mathbf{E}[X] = \mathbf{E}[Y] = 0$. Furthermore, for all possible value pairs (x, y), either x or y is equal to 0, which implies that $XY = 0$ and $\mathbf{E}[XY] = 0$. Therefore,

$$\text{cov}(X, Y) = \mathbf{E}[XY] - \mathbf{E}[X]\,\mathbf{E}[Y] = 0,$$

and X and Y are uncorrelated. However, X and Y are not independent since, for example, a nonzero value of X fixes the value of Y to zero.

This example can be generalized. In particular, assume that X and Y satisfy

$$\mathbf{E}[X \mid Y = y] = \mathbf{E}[X], \qquad \text{for all } y.$$

Then, assuming X and Y are discrete, the total expectation theorem implies that

$$\mathbf{E}[XY] = \sum_y y p_Y(y) \mathbf{E}[X \mid Y = y] = \mathbf{E}[X] \sum_y y p_Y(y) = \mathbf{E}[X]\,\mathbf{E}[Y],$$

so X and Y are uncorrelated. The argument for the continuous case is similar.

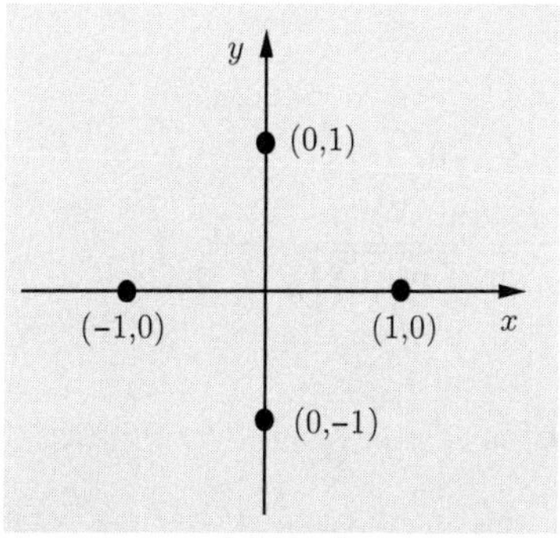

Figure 4.12: Joint PMF of X and Y for Example 4.13. Each of the four points shown has probability 1/4. Here X and Y are uncorrelated but not independent.

The **correlation coefficient** $\rho(X, Y)$ of two random variables X and Y that have nonzero variances is defined as

$$\rho(X, Y) = \frac{\text{cov}(X, Y)}{\sqrt{\text{var}(X)\text{var}(Y)}}.$$

(The simpler notation ρ will also be used when X and Y are clear from the context.) It may be viewed as a normalized version of the covariance $\text{cov}(X, Y)$, and in fact, it can be shown that ρ ranges from -1 to 1 (see the end-of-chapter problems).

If $\rho > 0$ (or $\rho < 0$), then the values of $X - \mathbf{E}[X]$ and $Y - \mathbf{E}[Y]$ "tend" to have the same (or opposite, respectively) sign. The size of $|\rho|$ provides a normalized measure of the extent to which this is true. In fact, always assuming that X and Y have positive variances, it can be shown that $\rho = 1$ (or $\rho = -1$) if and only if there exists a positive (or negative, respectively) constant c such that

$$Y - \mathbf{E}[Y] = c\big(X - \mathbf{E}[X]\big)$$

(see the end-of-chapter problems). The following example illustrates in part this property.

Example 4.14. Consider n independent tosses of a coin with probability of a head equal to p. Let X and Y be the numbers of heads and of tails, respectively, and let us look at the correlation coefficient of X and Y. Here, we have $X + Y = n$, and also $\mathbf{E}[X] + \mathbf{E}[Y] = n$. Thus,

$$X - \mathbf{E}[X] = -\big(Y - \mathbf{E}[Y]\big).$$

We will calculate the correlation coefficient of X and Y, and verify that it is indeed equal to -1.

We have

$$\begin{aligned} \text{cov}(X, Y) &= \mathbf{E}\Big[\big(X - \mathbf{E}[X]\big)\big(Y - \mathbf{E}[Y]\big)\Big] \\ &= -\mathbf{E}\left[\big(X - \mathbf{E}[X]\big)^2\right] \\ &= -\text{var}(X). \end{aligned}$$

Hence, the correlation coefficient is

$$\rho(X, Y) = \frac{\text{cov}(X, Y)}{\sqrt{\text{var}(X)\text{var}(Y)}} = \frac{-\text{var}(X)}{\sqrt{\text{var}(X)\text{var}(X)}} = -1.$$

Variance of the Sum of Random Variables

The covariance can be used to obtain a formula for the variance of the sum of several (not necessarily independent) random variables. In particular, if $X_1, X_2, \ldots, X_n$ are random variables with finite variance, we have

$$\text{var}(X_1 + X_2) = \text{var}(X_1) + \text{var}(X_2) + 2\text{cov}(X_1, X_2),$$

and, more generally,

$$\text{var}\left(\sum_{i=1}^n X_i\right) = \sum_{i=1}^n \text{var}(X_i) + \sum_{\{(i,j) \,|\, i \neq j\}} \text{cov}(X_i, X_j).$$

This can be seen from the following calculation, where for brevity, we denote $\tilde{X}_i = X_i - \mathbf{E}[X_i]$:

$$\text{var}\left(\sum_{i=1}^n X_i\right) = \mathbf{E}\left[\left(\sum_{i=1}^n \tilde{X}_i\right)^2\right]$$

$$= \mathbf{E}\left[\sum_{i=1}^{n}\sum_{j=1}^{n}\tilde{X}_i\tilde{X}_j\right]$$
$$= \sum_{i=1}^{n}\sum_{j=1}^{n}\mathbf{E}[\tilde{X}_i\tilde{X}_j]$$
$$= \sum_{i=1}^{n}\mathbf{E}\left[\tilde{X}_i^{\,2}\right] + \sum_{\{(i,j)\,|\,i\neq j\}}\mathbf{E}[\tilde{X}_i\tilde{X}_j]$$
$$= \sum_{i=1}^{n}\mathrm{var}(X_i) + \sum_{\{(i,j)\,|\,i\neq j\}}\mathrm{cov}(X_i, X_j).$$

The following example illustrates the use of this formula.

Example 4.15. Consider the hat problem discussed in Section 2.5, where n people throw their hats in a box and then pick a hat at random. Let us find the variance of X, the number of people who pick their own hat. We have

$$X = X_1 + \cdots + X_n,$$

where X_i is the random variable that takes the value 1 if the ith person selects his/her own hat, and takes the value 0 otherwise. Noting that X_i is Bernoulli with parameter $p = \mathbf{P}(X_i = 1) = 1/n$, we obtain

$$\mathbf{E}[X_i] = \frac{1}{n}, \qquad \mathrm{var}(X_i) = \frac{1}{n}\left(1 - \frac{1}{n}\right).$$

For $i \neq j$, we have

$$\begin{aligned}
\mathrm{cov}(X_i, X_j) &= \mathbf{E}[X_iX_j] - \mathbf{E}[X_i]\,\mathbf{E}[X_j] \\
&= \mathbf{P}(X_i = 1 \text{ and } X_j = 1) - \frac{1}{n}\cdot\frac{1}{n} \\
&= \mathbf{P}(X_i = 1)\mathbf{P}(X_j = 1\,|\,X_i = 1) - \frac{1}{n^2} \\
&= \frac{1}{n}\cdot\frac{1}{n-1} - \frac{1}{n^2} \\
&= \frac{1}{n^2(n-1)}.
\end{aligned}$$

Therefore,

$$\begin{aligned}
\mathrm{var}(X) &= \mathrm{var}\left(\sum_{i=1}^{n}X_i\right) \\
&= \sum_{i=1}^{n}\mathrm{var}(X_i) + \sum_{\{(i,j)\,|\,i\neq j\}}\mathrm{cov}(X_i, X_j) \\
&= n\cdot\frac{1}{n}\left(1 - \frac{1}{n}\right) + n(n-1)\cdot\frac{1}{n^2(n-1)} \\
&= 1.
\end{aligned}$$

Covariance and Correlation

- The **covariance** of X and Y is given by

$$\text{cov}(X,Y) = \mathbf{E}\Big[\big(X - \mathbf{E}[X]\big)\big(Y - \mathbf{E}[Y]\big)\Big] = \mathbf{E}[XY] - \mathbf{E}[X]\,\mathbf{E}[Y].$$

- If $\text{cov}(X,Y) = 0$, we say that X and Y are **uncorrelated**.
- If X and Y are independent, they are uncorrelated. The converse is not always true.
- We have

$$\text{var}(X + Y) = \text{var}(X) + \text{var}(Y) + 2\text{cov}(X,Y).$$

- The **correlation coefficient** $\rho(X,Y)$ of two random variables X and Y with positive variances is defined by

$$\rho(X,Y) = \frac{\text{cov}(X,Y)}{\sqrt{\text{var}(X)\text{var}(Y)}},$$

  and satisfies

$$-1 \le \rho(X,Y) \le 1.$$

4.3 CONDITIONAL EXPECTATION AND VARIANCE REVISITED

In this section, we revisit the conditional expectation of a random variable X given another random variable Y, and view it as a random variable determined by Y. We derive a reformulation of the total expectation theorem, called the **law of iterated expectations**. We also obtain a new formula, the **law of total variance**, that relates conditional and unconditional variances.

We introduce a random variable, denoted by $\mathbf{E}[X \,|\, Y]$, that takes the value $\mathbf{E}[X \,|\, Y = y]$ when Y takes the value y. Since $\mathbf{E}[X \,|\, Y = y]$ is a function of y, $\mathbf{E}[X \,|\, Y]$ is a function of Y, and its distribution is determined by the distribution of Y. The properties of $\mathbf{E}[X \,|\, Y]$ will be important in this section but also later, particularly in the context of estimation and statistical inference, in Chapters 8 and 9.

Example 4.16. We are given a biased coin and we are told that because of manufacturing defects, the probability of heads, denoted by Y, is itself random, with a known distribution over the interval $[0, 1]$. We toss the coin a fixed number n of times, and we let X be the number of heads obtained. Then, for any $y \in [0, 1]$, we have $\mathbf{E}[X \,|\, Y = y] = ny$, so $\mathbf{E}[X \,|\, Y]$ is the random variable nY.

Since $\mathbf{E}[X \mid Y]$ is a random variable, it has an expectation $\mathbf{E}\big[\mathbf{E}[X \mid Y]\big]$ of its own, which can be calculated using the expected value rule:

$$\mathbf{E}\big[\mathbf{E}[X \mid Y]\big] = \begin{cases} \displaystyle\sum_y \mathbf{E}[X \mid Y = y] p_Y(y), & Y \text{ discrete,} \\ \displaystyle\int_{-\infty}^{\infty} \mathbf{E}[X \mid Y = y] f_Y(y)\, dy, & Y \text{ continuous.} \end{cases}$$

Both expressions in the right-hand side are familiar from Chapters 2 and 3, respectively. By the corresponding versions of the total expectation theorem, they are equal to $\mathbf{E}[X]$. This brings us to the following conclusion, which is actually valid for every type of random variable Y (discrete, continuous, or mixed), as long as X has a well-defined and finite expectation $\mathbf{E}[X]$.

Law of Iterated Expectations: $\mathbf{E}\big[\mathbf{E}[X \mid Y]\big] = \mathbf{E}[X]$.

The following examples illustrate how the law of iterated expectations facilitates the calculation of expected values when the problem data include conditional probabilities.

Example 4.16 (continued). Suppose that Y, the probability of heads for our coin is uniformly distributed over the interval $[0, 1]$. Since $\mathbf{E}[X \mid Y] = nY$ and $\mathbf{E}[Y] = 1/2$, by the law of iterated expectations, we have

$$\mathbf{E}[X] = \mathbf{E}\big[\mathbf{E}[X \mid Y]\big] = \mathbf{E}[nY] = n\mathbf{E}[Y] = \frac{n}{2}.$$

Example 4.17. We start with a stick of length ℓ. We break it at a point which is chosen randomly and uniformly over its length, and keep the piece that contains the left end of the stick. We then repeat the same process on the piece that we were left with. What is the expected length of the piece that we are left with after breaking twice?

Let Y be the length of the piece after we break for the first time. Let X be the length after we break for the second time. We have $\mathbf{E}[X \mid Y] = Y/2$, since the breakpoint is chosen uniformly over a piece of length Y. For a similar reason, we also have $\mathbf{E}[Y] = \ell/2$. Thus,

$$\mathbf{E}[X] = \mathbf{E}\big[\mathbf{E}[X \mid Y]\big] = \mathbf{E}\left[\frac{Y}{2}\right] = \frac{\mathbf{E}[Y]}{2} = \frac{\ell}{4}.$$

Example 4.18. Averaging Quiz Scores by Section. A class has n students and the quiz score of student i is x_i. The average quiz score is

$$m = \frac{1}{n} \sum_{i=1}^{n} x_i.$$

The students are divided into k disjoint subsets $A_1, \ldots, A_k$, and are accordingly assigned to different sections. We use n_s to denote the number of students in section s. The average score in section s is

$$m_s = \frac{1}{n_s} \sum_{i \in A_s} x_i.$$

The average score over the whole class can be computed by taking the average score m_s of each section, and then forming a *weighted average*; the weight given to section s is proportional to the number of students in that section, and is n_s/n. We verify that this gives the correct result:

$$\begin{aligned}
\sum_{s=1}^{k} \frac{n_s}{n} m_s &= \sum_{s=1}^{k} \frac{n_s}{n} \cdot \frac{1}{n_s} \sum_{i \in A_s} x_i \\
&= \frac{1}{n} \sum_{s=1}^{k} \sum_{i \in A_s} x_i \\
&= \frac{1}{n} \sum_{i=1}^{n} x_i. \\
&= m.
\end{aligned}$$

How is this related to conditional expectations? Consider an experiment in which a student is selected at random, each student having probability $1/n$ of being selected. Consider the following two random variables:

$$\begin{aligned}
X &= \text{quiz score of a student}, \\
Y &= \text{section of a student}, \quad \big(Y \in \{1, \ldots, k\}\big).
\end{aligned}$$

We then have

$$\mathbf{E}[X] = m.$$

Conditioning on $Y = s$ is the same as assuming that the selected student is in section s. Conditional on that event, every student in that section has the same probability $1/n_s$ of being chosen. Therefore,

$$\mathbf{E}[X \mid Y = s] = \frac{1}{n_s} \sum_{i \in A_s} x_i = m_s.$$

A randomly selected student belongs to section s with probability n_s/n, i.e., $\mathbf{P}(Y = s) = n_s/n$. Hence,

$$m = \mathbf{E}[X] = \mathbf{E}\big[\mathbf{E}[X \mid Y]\big] = \sum_{s=1}^{k} \mathbf{E}[X \mid Y = s]\mathbf{P}(Y = s) = \sum_{s=1}^{k} \frac{n_s}{n} m_s.$$

Thus, averaging by section can be viewed as a special case of the law of iterated expectations.

Example 4.19. Forecast Revisions. Let Y be the sales of a company in the first semester of the coming year, and let X be the sales over the entire year. The company has constructed a statistical model of sales, and so the joint distribution of X and Y is assumed to be known. In the beginning of the year, the expected value $\mathbf{E}[X]$ serves as a forecast of the actual sales X. In the middle of the year, the first semester sales have been realized and the value of the random variable Y is now known. This places us in a new "universe," where everything is conditioned on the realized value of Y. Based on the knowledge of Y, the company constructs a revised forecast of yearly sales, which is $\mathbf{E}[X \mid Y]$.

We view $\mathbf{E}[X \mid Y] - \mathbf{E}[X]$ as the forecast revision, in light of the mid-year information. The law of iterated expectations implies that

$$\mathbf{E}\big[\mathbf{E}[X \mid Y] - \mathbf{E}[X]\big] = \mathbf{E}\big[\mathbf{E}[X \mid Y]\big] - \mathbf{E}[X] = \mathbf{E}[X] - \mathbf{E}[X] = 0.$$

This indicates that while the actual revision will usually be nonzero, in the beginning of the year we expect the revision to be zero, on the average. This is quite intuitive. Indeed, if the expected revision were positive, the original forecast should have been higher in the first place.

We finally note an important property: for any function g, we have

$$\mathbf{E}\big[Xg(Y) \mid Y\big] = g(Y)\,\mathbf{E}[X \mid Y].$$

This is because given the value of Y, $g(Y)$ is a constant and can be pulled outside the expectation; see also Problem 25.

The Conditional Expectation as an Estimator

If we view Y as an observation that provides information about X, it is natural to view the conditional expectation, denoted

$$\hat{X} = \mathbf{E}[X \mid Y],$$

as an **estimator** of X given Y. The **estimation error**

$$\tilde{X} = \hat{X} - X,$$

is a random variable satisfying

$$\mathbf{E}[\tilde{X} \mid Y] = \mathbf{E}[\hat{X} - X \mid Y] = \mathbf{E}[\hat{X} \mid Y] - \mathbf{E}[X \mid Y] = \hat{X} - \hat{X} = 0.$$

Thus, the random variable $\mathbf{E}[\tilde{X} \mid Y]$ is identically zero: $\mathbf{E}[\tilde{X} \mid Y = y] = 0$ for all values of y. By using the law of iterated expectations, we also have

$$\mathbf{E}[\tilde{X}] = \mathbf{E}\big[\mathbf{E}[\tilde{X} \mid Y]\big] = 0.$$

This property is reassuring, as it indicates that the estimation error does not have a systematic upward or downward bias.

We will now show that $\hat{X}$ has another interesting property: **it is uncorrelated with the estimation error** $\tilde{X}$. Indeed, using the law of iterated expectations, we have

$$\mathbf{E}[\hat{X}\tilde{X}] = \mathbf{E}\big[\mathbf{E}[\hat{X}\tilde{X} \,|\, Y]\big] = \mathbf{E}\big[\hat{X}\mathbf{E}[\tilde{X} \,|\, Y]\big] = 0,$$

where the last two equalities follow from the fact that $\hat{X}$ is completely determined by Y, so that

$$\mathbf{E}[\hat{X}\tilde{X} \,|\, Y] = \hat{X}\mathbf{E}[\tilde{X} \,|\, Y] = 0.$$

It follows that

$$\text{cov}(\hat{X}, \tilde{X}) = \mathbf{E}[\hat{X}\tilde{X}] - \mathbf{E}[\hat{X}]\,\mathbf{E}[\tilde{X}] = 0 - \mathbf{E}[X] \cdot 0 = 0,$$

and $\hat{X}$ and $\tilde{X}$ are uncorrelated.

An important consequence of the fact $\text{cov}(\hat{X}, \tilde{X}) = 0$ is that by considering the variance of both sides in the equation $X = \hat{X} + \tilde{X}$, we obtain

$$\text{var}(X) = \text{var}(\tilde{X}) + \text{var}(\hat{X}).$$

This relation can be written in the form of a useful law, as we now discuss.

The Conditional Variance

We introduce the random variable

$$\text{var}(X \,|\, Y) = \mathbf{E}\Big[\big(X - \mathbf{E}[X \,|\, Y]\big)^2 \,\Big|\, Y\Big] = \mathbf{E}\big[\tilde{X}^2 \,|\, Y\big].$$

This is the function of Y whose value is the conditional variance of X when Y takes the value y:

$$\text{var}(X \,|\, Y = y) = \mathbf{E}\big[\tilde{X}^2 \,|\, Y = y\big].$$

Using the fact $\mathbf{E}[\tilde{X}] = 0$ and the law of iterated expectations, we can write the variance of the estimation error as

$$\text{var}(\tilde{X}) = \mathbf{E}\big[\tilde{X}^2\big] = \mathbf{E}\big[\mathbf{E}[\tilde{X}^2 \,|\, Y]\big] = \mathbf{E}\big[\text{var}(X \,|\, Y)\big],$$

and rewrite the equation $\text{var}(X) = \text{var}(\tilde{X}) + \text{var}(\hat{X})$ as follows.

Law of Total Variance: $\quad \text{var}(X) = \mathbf{E}\big[\text{var}(X \,|\, Y)\big] + \text{var}\big(\mathbf{E}[X \,|\, Y]\big).$

The law of total variance is helpful in calculating variances of random variables by using conditioning, as illustrated by the following examples.

Example 4.16 (continued). We consider n independent tosses of a biased coin whose probability of heads, Y, is uniformly distributed over the interval $[0, 1]$. With X being the number of heads obtained, we have $\mathbf{E}[X \mid Y] = nY$ and $\text{var}(X \mid Y) = nY(1 - Y)$. Thus,

$$\begin{aligned}\mathbf{E}\big[\text{var}(X \mid Y)\big] &= \mathbf{E}\big[nY(1-Y)\big] = n\big(\mathbf{E}[Y] - \mathbf{E}[Y^2]\big) \\ &= n\big(\mathbf{E}[Y] - \text{var}(Y) - (\mathbf{E}[Y])^2\big) = n\left(\frac{1}{2} - \frac{1}{12} - \frac{1}{4}\right) = \frac{n}{6}.\end{aligned}$$

Furthermore,

$$\text{var}\big(\mathbf{E}[X \mid Y]\big) = \text{var}(nY) = \frac{n^2}{12}.$$

Therefore, by the law of total variance, we have

$$\text{var}(X) = \mathbf{E}\big[\text{var}(X \mid Y)\big] + \text{var}\big(\mathbf{E}[X \mid Y]\big) = \frac{n}{6} + \frac{n^2}{12}.$$

Example 4.17 (continued). Consider again the problem where we break twice a stick of length ℓ at randomly chosen points. Here Y is the length of the piece left after the first break and X is the length after the second break. We calculated the mean of X as $\ell/4$. We will now use the law of total variance to calculate $\text{var}(X)$.

Since X is uniformly distributed between 0 and Y, we have

$$\text{var}(X \mid Y) = \frac{Y^2}{12}.$$

Thus, since Y is uniformly distributed between 0 and ℓ, we have

$$\mathbf{E}\big[\text{var}(X \mid Y)\big] = \frac{1}{12}\int_0^\ell \frac{1}{\ell}y^2\,dy = \frac{1}{12}\cdot\frac{1}{3\ell}y^3\bigg|_0^\ell = \frac{\ell^2}{36}.$$

We also have $\mathbf{E}[X \mid Y] = Y/2$, so

$$\text{var}\big(\mathbf{E}[X \mid Y]\big) = \text{var}(Y/2) = \frac{1}{4}\text{var}(Y) = \frac{1}{4}\cdot\frac{\ell^2}{12} = \frac{\ell^2}{48}.$$

Using now the law of total variance, we obtain

$$\text{var}(X) = \mathbf{E}\big[\text{var}(X \mid Y)\big] + \text{var}\big(\mathbf{E}[X \mid Y]\big) = \frac{\ell^2}{36} + \frac{\ell^2}{48} = \frac{7\ell^2}{144}.$$

Example 4.20. Averaging Quiz Scores by Section – Variance. The setting is the same as in Example 4.18 and we consider again the random variables

$$\begin{aligned}X &= \text{quiz score of a student,} \\ Y &= \text{section of a student,} \quad \big(Y \in \{1, \ldots, k\}\big).\end{aligned}$$

Let n_s be the number of students in section s, and let n be the total number of students. We interpret the different quantities in the formula

$$\mathrm{var}(X) = \mathbf{E}\big[\mathrm{var}(X \mid Y)\big] + \mathrm{var}\big(\mathbf{E}[X \mid Y]\big).$$

Here, $\mathrm{var}(X \mid Y = s)$ is the variance of the quiz scores within section s. Thus,

$$\mathbf{E}\big[\mathrm{var}(X \mid Y)\big] = \sum_{s=1}^{k} \mathbf{P}(Y = s)\mathrm{var}(X \mid Y = s) = \sum_{s=1}^{k} \frac{n_s}{n} \mathrm{var}(X \mid Y = s),$$

so that $\mathbf{E}\big[\mathrm{var}(X \mid Y)\big]$ is the weighted average of the section variances, where each section is weighted in proportion to its size.

Recall that $\mathbf{E}[X \mid Y = s]$ is the average score in section s. Then, $\mathrm{var}\big(\mathbf{E}[X \mid Y]\big)$ is a measure of the variability of the averages of the different sections. The law of total variance states that the total quiz score variance can be broken into two parts:

(a) The average score variability $\mathbf{E}\big[\mathrm{var}(X \mid Y)\big]$ *within* individual sections.

(b) The variability $\mathrm{var}\big(\mathbf{E}[X \mid Y]\big)$ *between* sections.

We have seen earlier that the law of iterated expectations can be used to break down complicated expectation calculations, by considering different cases. A similar method applies to variance calculations.

Example 4.21. Computing Variances by Conditioning. Consider a continuous random variable X with the PDF given in Fig. 4.13. We define an auxiliary random variable Y as follows:

$$Y = \begin{cases} 1, & \text{if } x < 1, \\ 2, & \text{if } x \geq 1. \end{cases}$$

Here, $\mathbf{E}[X \mid Y]$ takes the values 1/2 and 2, each with probability 1/2. Thus, the mean of $\mathbf{E}[X \mid Y]$ is 5/4. It follows that

$$\mathrm{var}\big(\mathbf{E}[X \mid Y]\big) = \frac{1}{2}\left(\frac{1}{2} - \frac{5}{4}\right)^2 + \frac{1}{2}\left(2 - \frac{5}{4}\right)^2 = \frac{9}{16}.$$

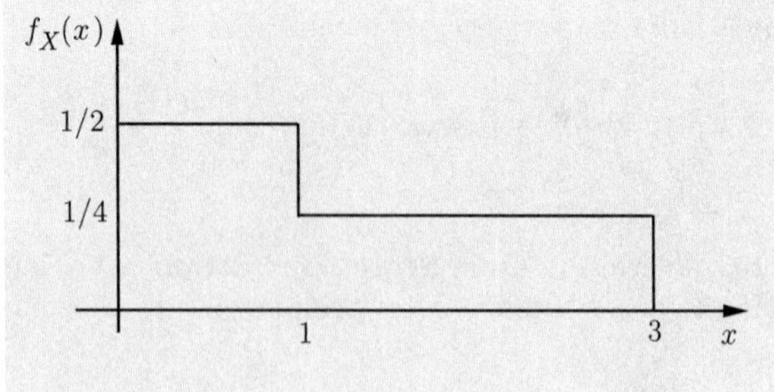

Figure 4.13: The PDF in Example 4.21.

Conditioned on $Y = 1$ or $Y = 2$, X is uniformly distributed on an interval of length 1 or 2, respectively. Therefore,

$$\text{var}(X \mid Y = 1) = \frac{1}{12}, \qquad \text{var}(X \mid Y = 2) = \frac{4}{12},$$

and

$$\mathbf{E}\big[\text{var}(X \mid Y)\big] = \frac{1}{2} \cdot \frac{1}{12} + \frac{1}{2} \cdot \frac{4}{12} = \frac{5}{24}.$$

Putting everything together, we obtain

$$\text{var}(X) = \mathbf{E}\big[\text{var}(X \mid Y)\big] + \text{var}\big(\mathbf{E}[X \mid Y]\big) = \frac{5}{24} + \frac{9}{16} = \frac{37}{48}.$$

We summarize the main points in this section.

Properties of the Conditional Expectation and Variance

- $\mathbf{E}[X \mid Y = y]$ is a number whose value depends on y.
- $\mathbf{E}[X \mid Y]$ is a function of the random variable Y, hence a random variable. Its value is $\mathbf{E}[X \mid Y = y]$ whenever the value of Y is y.
- $\mathbf{E}\big[\mathbf{E}[X \mid Y]\big] = \mathbf{E}[X]$ (**law of iterated expectations**).
- $\mathbf{E}[X \mid Y = y]$ may be viewed as an estimate of X given $Y = y$. The corresponding error $\mathbf{E}[X \mid Y] - X$ is a zero mean random variable that is uncorrelated with $\mathbf{E}[X \mid Y]$.
- $\text{var}(X \mid Y)$ is a random variable whose value is $\text{var}(X \mid Y = y)$ whenever the value of Y is y.
- $\text{var}(X) = \mathbf{E}\big[\text{var}(X \mid Y)\big] + \text{var}\big(\mathbf{E}[X \mid Y]\big)$ (**law of total variance**).

4.4 TRANSFORMS

In this section, we introduce the transform associated with a random variable. The transform provides us with an alternative representation of a probability law. It is not particularly intuitive, but it is often convenient for certain types of mathematical manipulations.

The **transform** associated with a random variable X (also referred to as the associated **moment generating function**) is a function $M_X(s)$ of a scalar parameter s, defined by

$$M_X(s) = \mathbf{E}[e^{sX}].$$

The simpler notation $M(s)$ can also be used whenever the underlying random variable X is clear from the context. In more detail, when X is a discrete random variable, the corresponding transform is given by

$$M(s) = \sum_x e^{sx} p_X(x),$$

while in the continuous case it is given by†

$$M(s) = \int_{-\infty}^{\infty} e^{sx} f_X(x)\, dx.$$

Let us now provide some examples of transforms.

Example 4.22. Let

$$p_X(x) = \begin{cases} 1/2, & \text{if } x = 2, \\ 1/6, & \text{if } x = 3, \\ 1/3, & \text{if } x = 5. \end{cases}$$

The corresponding transform is

$$M(s) = \mathbf{E}[s^{sx}] = \frac{1}{2}e^{2s} + \frac{1}{6}e^{3s} + \frac{1}{3}e^{5s}.$$

Example 4.23. The Transform Associated with a Poisson Random Variable. Let X be a Poisson random variable with parameter λ:

$$p_X(x) = \frac{\lambda^x e^{-\lambda}}{x!}, \qquad x = 0, 1, 2, \ldots$$

The corresponding transform is

$$M(s) = \sum_{x=0}^{\infty} e^{sx} \frac{\lambda^x e^{-\lambda}}{x!}.$$

---

† The reader who is familiar with Laplace transforms may recognize that the transform associated with a continuous random variable is essentially the same as the Laplace transform of its PDF, the only difference being that Laplace transforms usually involve e^{-sx} rather than e^{sx}. For the discrete case, a variable z is sometimes used in place of e^s and the resulting transform

$$M(z) = \sum_x z^x p_X(x)$$

is known as the **z-transform**. However, we will not be using z-transforms in this book.

We let $a = e^s\lambda$ and obtain

$$M(s) = e^{-\lambda}\sum_{x=0}^{\infty}\frac{a^x}{x!} = e^{-\lambda}e^a = e^{a-\lambda} = e^{\lambda(e^s-1)}.$$

Example 4.24. The Transform Associated with an Exponential Random Variable. Let X be an exponential random variable with parameter λ:

$$f_X(x) = \lambda e^{-\lambda x}, \qquad x \geq 0.$$

Then,

$$\begin{aligned} M(s) &= \lambda\int_0^\infty e^{sx}e^{-\lambda x}\,dx \\ &= \lambda\int_0^\infty e^{(s-\lambda)x}\,dx \\ &= \lambda\frac{e^{(s-\lambda)x}}{s-\lambda}\bigg|_0^\infty \qquad \text{(if } s<\lambda\text{)} \\ &= \frac{\lambda}{\lambda-s}. \end{aligned}$$

The above calculation and the formula for $M(s)$ is correct only if the integrand $e^{(s-\lambda)x}$ decays as x increases, which is the case if and only if $s < \lambda$; otherwise, the integral is infinite.

It is important to realize that the transform is not a number but rather a *function* of a parameter s. Thus, we are dealing with a transformation that starts with a function, e.g., a PDF, and results in a new function. Strictly speaking, $M(s)$ is only defined for those values of s for which $\mathbf{E}[e^{sX}]$ is finite, as noted in the preceding example.

Example 4.25. The Transform Associated with a Linear Function of a Random Variable. Let $M_X(s)$ be the transform associated with a random variable X. Consider a new random variable $Y = aX + b$. We then have

$$M_Y(s) = \mathbf{E}[e^{s(aX+b)}] = e^{sb}\mathbf{E}[e^{saX}] = e^{sb}M_X(sa).$$

For example, if X is exponential with parameter $\lambda = 1$, so that $M_X(s) = 1/(1-s)$, and if $Y = 2X + 3$, then

$$M_Y(s) = e^{3s}\frac{1}{1-2s}.$$

Example 4.26. The Transform Associated with a Normal Random Variable. Let X be a normal random variable with mean μ and variance σ^2. To calculate the corresponding transform, we first consider the special case of the standard normal random variable Y, where $\mu = 0$ and $\sigma^2 = 1$, and then use the formula derived in the preceding example. The PDF of the standard normal is

$$f_Y(y) = \frac{1}{\sqrt{2\pi}} e^{-y^2/2},$$

and the associated transform is

$$\begin{aligned}
M_Y(s) &= \int_{-\infty}^{\infty} \frac{1}{\sqrt{2\pi}} e^{-y^2/2} e^{sy} \, dy \\
&= \frac{1}{\sqrt{2\pi}} \int_{-\infty}^{\infty} e^{-(y^2/2)+sy} dy \\
&= e^{s^2/2} \frac{1}{\sqrt{2\pi}} \int_{-\infty}^{\infty} e^{-(y^2/2)+sy-(s^2/2)} dy \\
&= e^{s^2/2} \frac{1}{\sqrt{2\pi}} \int_{-\infty}^{\infty} e^{-(y-s)^2/2} dy \\
&= e^{s^2/2},
\end{aligned}$$

where the last equality follows by using the normalization property of a normal PDF with mean s and unit variance.

A general normal random variable with mean μ and variance σ^2 is obtained from the standard normal via the linear transformation

$$X = \sigma Y + \mu.$$

The transform associated with the standard normal is $M_Y(s) = e^{s^2/2}$, as verified above. By applying the formula of Example 4.25, we obtain

$$M_X(s) = e^{s\mu} M_Y(s\sigma) = e^{(\sigma^2 s^2/2)+\mu s}.$$

From Transforms to Moments

The reason behind the alternative name "moment generating function" is that the moments of a random variable are easily computed once a formula for the associated transform is available. To see this, let us consider a continuous random variable X, and let us take the derivative of both sides of the definition

$$M(s) = \int_{-\infty}^{\infty} e^{sx} f_X(x) \, dx,$$

with respect to s. We obtain

$$\begin{aligned}
\frac{d}{ds} M(s) &= \frac{d}{ds} \int_{-\infty}^{\infty} e^{sx} f_X(x) \, dx \\
&= \int_{-\infty}^{\infty} \frac{d}{ds} e^{sx} f_X(x) \, dx \\
&= \int_{-\infty}^{\infty} x e^{sx} f_X(x) \, dx.
\end{aligned}$$

This equality holds for all values of s.† By considering the special case where $s = 0$, we obtain

$$\frac{d}{ds}M(s)\bigg|_{s=0} = \int_{-\infty}^{\infty} x f_X(x)\,dx = \mathbf{E}[X].$$

More generally, if we differentiate n times the function $M(s)$ with respect to s, a similar calculation yields

$$\frac{d^n}{ds^n}M(s)\bigg|_{s=0} = \int_{-\infty}^{\infty} x^n f_X(x)\,dx = \mathbf{E}[X^n].$$

Example 4.27. We saw earlier (Example 4.22) that the PMF

$$p_X(x) = \begin{cases} 1/2, & \text{if } x = 2, \\ 1/6, & \text{if } x = 3, \\ 1/3, & \text{if } x = 5, \end{cases}$$

is associated with the transform

$$M(s) = \frac{1}{2}e^{2s} + \frac{1}{6}e^{3s} + \frac{1}{3}e^{5s}.$$

Thus,

$$\begin{aligned} \mathbf{E}[X] &= \frac{d}{ds}M(s)\bigg|_{s=0} \\ &= \frac{1}{2}\cdot 2e^{2s} + \frac{1}{6}\cdot 3e^{3s} + \frac{1}{3}\cdot 5e^{5s}\bigg|_{s=0} \\ &= \frac{1}{2}\cdot 2 + \frac{1}{6}\cdot 3 + \frac{1}{3}\cdot 5 \\ &= \frac{19}{6}. \end{aligned}$$

Also,

$$\begin{aligned} \mathbf{E}[X^2] &= \frac{d^2}{ds^2}M(s)\bigg|_{s=0} \\ &= \frac{1}{2}\cdot 4e^{2s} + \frac{1}{6}\cdot 9e^{3s} + \frac{1}{3}\cdot 25e^{5s}\bigg|_{s=0} \\ &= \frac{1}{2}\cdot 4 + \frac{1}{6}\cdot 9 + \frac{1}{3}\cdot 25 \\ &= \frac{71}{6}. \end{aligned}$$

---

† This derivation involves an interchange of differentiation and integration. The interchange turns out to be justified for all of the applications to be considered in this book. Furthermore, the derivation remains valid for general random variables, including discrete ones. In fact, it could be carried out more abstractly, in the form

$$\frac{d}{ds}M(s) = \frac{d}{ds}\mathbf{E}[e^{sX}] = \mathbf{E}\left[\frac{d}{ds}e^{sX}\right] = \mathbf{E}[Xe^{sX}],$$

leading to the same conclusion.

For an exponential random variable with PDF

$$f_X(x) = \lambda e^{-\lambda x}, \qquad x \geq 0,$$

we found earlier (Example 4.24) that

$$M(s) = \frac{\lambda}{\lambda - s}.$$

Thus,

$$\frac{d}{ds}M(s) = \frac{\lambda}{(\lambda - s)^2}, \qquad \frac{d^2}{ds^2}M(s) = \frac{2\lambda}{(\lambda - s)^3}.$$

By setting $s = 0$, we obtain

$$\mathbf{E}[X] = \frac{1}{\lambda}, \qquad \mathbf{E}[X^2] = \frac{2}{\lambda^2},$$

which agrees with the formulas derived in Chapter 3.

We close by noting two more useful and generic properties of transforms. For any random variable X, we have

$$M_X(0) = \mathbf{E}[e^{0X}] = \mathbf{E}[1] = 1,$$

and if X takes only nonnegative integer values, then

$$\lim_{s \to -\infty} M_X(s) = \mathbf{P}(X = 0)$$

(see the end-of-chapter problems).

Inversion of Transforms

A very important property of the transform $M_X(s)$ is that it can be inverted, i.e., it can be used to determine the probability law of the random variable X. Some appropriate mathematical conditions are required, which are satisfied in all of our examples that make use of the inversion property. The following is a more precise statement. Its proof is beyond our scope.

> **Inversion Property**
>
> The transform $M_X(s)$ associated with a random variable X uniquely determines the CDF of X, assuming that $M_X(s)$ is finite for all s in some interval $[-a, a]$, where a is a positive number.

There exist explicit formulas that allow us to recover the PMF or PDF of a random variable starting from the associated transform, but they are quite difficult to use. In practice, transforms are usually inverted by "pattern matching," based on tables of known distribution-transform pairs. We will see a number of such examples shortly.

Example 4.28. We are told that the transform associated with a random variable X is

$$M(s) = \frac{1}{4}e^{-s} + \frac{1}{2} + \frac{1}{8}e^{4s} + \frac{1}{8}e^{5s}.$$

Since $M(s)$ is a sum of terms of the form e^{sx}, we can compare with the general formula

$$M(s) = \sum_x e^{sx} p_X(x),$$

and infer that X is a discrete random variable. The different values that X can take can be read from the corresponding exponents, and are -1, 0, 4, and 5. The probability of each value x is given by the coefficient multiplying the corresponding e^{sx} term. In our case,

$$\mathbf{P}(X=-1) = \frac{1}{4}, \quad \mathbf{P}(X=0) = \frac{1}{2}, \quad \mathbf{P}(X=4) = \frac{1}{8}, \quad \mathbf{P}(X=5) = \frac{1}{8}.$$

Generalizing from the last example, the distribution of a finite-valued discrete random variable can be always found by inspection of the corresponding transform. The same procedure also works for discrete random variables with an infinite range, as in the example that follows.

Example 4.29. The Transform Associated with a Geometric Random Variable. We are told that the transform associated with a random variable X is of the form

$$M(s) = \frac{pe^s}{1-(1-p)e^s},$$

where p is a constant in the range $0 < p \leq 1$. We wish to find the distribution of X. We recall the formula for the geometric series:

$$\frac{1}{1-\alpha} = 1 + \alpha + \alpha^2 + \cdots,$$

which is valid whenever $|\alpha| < 1$. We use this formula with $\alpha = (1-p)e^s$, and for s sufficiently close to zero so that $(1-p)e^s < 1$. We obtain

$$M(s) = pe^s\big(1 + (1-p)e^s + (1-p)^2e^{2s} + (1-p)^3e^{3s} + \cdots\big).$$

As in the previous example, we infer that this is a discrete random variable that takes positive integer values. The probability $\mathbf{P}(X=k)$ is found by reading the coefficient of the term e^{ks}. In particular, $\mathbf{P}(X=1) = p$, $\mathbf{P}(X=2) = p(1-p)$, and

$$\mathbf{P}(X=k) = p(1-p)^{k-1}, \qquad k = 1, 2, \ldots$$

We recognize this as the geometric distribution with parameter p.

Note that

$$\frac{d}{ds}M(s) = \frac{pe^s}{1-(1-p)e^s} + \frac{(1-p)pe^{2s}}{(1-(1-p)e^s)^2}.$$

For $s = 0$, the right-hand side is equal to $1/p$, which agrees with the formula for $\mathbf{E}[X]$ derived in Chapter 2.

Example 4.30. The Transform Associated with a Mixture of Two Distributions. The neighborhood bank has three tellers, two of them fast, one slow. The time to assist a customer is exponentially distributed with parameter $\lambda = 6$ at the fast tellers, and $\lambda = 4$ at the slow teller. Jane enters the bank and chooses a teller at random, each one with probability 1/3. Find the PDF of the time it takes to assist Jane and the associated transform.

We have

$$f_X(x) = \frac{2}{3}\cdot 6e^{-6x} + \frac{1}{3}\cdot 4e^{-4x}, \qquad x \geq 0.$$

Then,

$$\begin{aligned} M(s) &= \int_0^\infty e^{sx}\left(\frac{2}{3}6e^{-6x} + \frac{1}{3}4e^{-4x}\right) dx \\ &= \frac{2}{3}\int_0^\infty e^{sx}6e^{-6x}\,dx + \frac{1}{3}\int_0^\infty e^{sx}4e^{-4x}\,dx \\ &= \frac{2}{3}\cdot\frac{6}{6-s} + \frac{1}{3}\cdot\frac{4}{4-s} \qquad\qquad \text{(for } s < 4\text{)}. \end{aligned}$$

More generally, let $X_1, \ldots, X_n$ be continuous random variables with PDFs $f_{X_1}, \ldots, f_{X_n}$. The value y of a random variable Y is generated as follows: an index i is chosen with a corresponding probability p_i, and y is taken to be equal to the value of X_i. Then,

$$f_Y(y) = p_1 f_{X_1}(y) + \cdots + p_n f_{X_n}(y),$$

and

$$M_Y(s) = p_1 M_{X_1}(s) + \cdots + p_n M_{X_n}(s).$$

The steps in this problem can be reversed. For example, we may be given that the transform associated with a random variable Y is of the form

$$\frac{1}{2}\cdot\frac{1}{2-s} + \frac{3}{4}\cdot\frac{1}{1-s}.$$

We can then rewrite it as

$$\frac{1}{4}\cdot\frac{2}{2-s} + \frac{3}{4}\cdot\frac{1}{1-s},$$

and recognize that Y is the mixture of two exponential random variables with parameters 2 and 1, which are selected with probabilities 1/4 and 3/4, respectively.

Sums of Independent Random Variables

Transform methods are particularly convenient when dealing with a sum of random variables. The reason is that *addition of independent random variables corresponds to multiplication of transforms*, as we will proceed to show. This provides an often convenient alternative to the convolution formula.

Let X and Y be independent random variables, and let $Z = X + Y$. The transform associated with Z is, by definition,

$$M_Z(s) = \mathbf{E}[e^{sZ}] = \mathbf{E}[e^{s(X+Y)}] = \mathbf{E}[e^{sX}e^{sY}].$$

Since X and Y are independent, e^{sX} and e^{sY} are independent random variables, for any fixed value of s. Hence, the expectation of their product is the product of the expectations, and

$$M_Z(s) = \mathbf{E}[e^{sX}]\mathbf{E}[e^{sY}] = M_X(s)M_Y(s).$$

By the same argument, if $X_1, \ldots, X_n$ is a collection of independent random variables, and

$$Z = X_1 + \cdots + X_n,$$

then

$$M_Z(s) = M_{X_1}(s) \cdots M_{X_n}(s).$$

Example 4.31. The Transform Associated with the Binomial. Let $X_1, \ldots, X_n$ be independent Bernoulli random variables with a common parameter p. Then,

$$M_{X_i}(s) = (1-p)e^{0s} + pe^{1s} = 1 - p + pe^s, \qquad \text{for all } i.$$

The random variable $Z = X_1 + \cdots + X_n$ is binomial with parameters n and p. The corresponding transform is given by

$$M_Z(s) = \left(1 - p + pe^s\right)^n.$$

Example 4.32. The Sum of Independent Poisson Random Variables is Poisson. Let X and Y be independent Poisson random variables with means λ and μ, respectively, and let $Z = X + Y$. Then,

$$M_X(s) = e^{\lambda(e^s-1)}, \qquad M_Y(s) = e^{\mu(e^s-1)},$$

and

$$M_Z(s) = M_X(s)M_Y(s) = e^{\lambda(e^s-1)}e^{\mu(e^s-1)} = e^{(\lambda+\mu)(e^s-1)}.$$

Thus, the transform associated with Z is the same as the transform associated with a Poisson random variable with mean $\lambda + \mu$. By the uniqueness property of transforms, Z is Poisson with mean $\lambda + \mu$.

Example 4.33. The Sum of Independent Normal Random Variables is Normal. Let X and Y be independent normal random variables with means μ_x, μ_y, and variances σ_x^2, σ_y^2, respectively, and let $Z = X + Y$. Then,

$$M_X(s) = \exp\left\{\frac{\sigma_x^2 s^2}{2} + \mu_x s\right\}, \qquad M_Y(s) = \exp\left\{\frac{\sigma_y^2 s^2}{2} + \mu_y s\right\},$$

and

$$M_W(s) = \exp\left\{\frac{(\sigma_x^2 + \sigma_y^2)s^2}{2} + (\mu_x + \mu_y)s\right\}.$$

We observe that the transform associated with Z is the same as the transform associated with a normal random variable with mean $\mu_x + \mu_y$ and variance $\sigma_x^2 + \sigma_y^2$. By the uniqueness property of transforms, Z is normal with these parameters, thus providing an alternative to the derivation described in Section 4.1, based on the convolution formula.

Summary of Transforms and their Properties

- The transform associated with a random variable X is given by

$$M_X(s) = \mathbf{E}[e^{sX}] = \begin{cases} \sum_x e^{sx} p_X(x), & X \text{ discrete}, \\ \int_{-\infty}^{\infty} e^{sx} f_X(x)\, dx, & X \text{ continuous}. \end{cases}$$

- The distribution of a random variable is completely determined by the corresponding transform.
- Moment generating properties:

$$M_X(0) = 1, \qquad \frac{d}{ds} M_X(s)\bigg|_{s=0} = \mathbf{E}[X], \qquad \frac{d^n}{ds^n} M_X(s)\bigg|_{s=0} = \mathbf{E}[X^n].$$

- If $Y = aX + b$, then $M_Y(s) = e^{sb} M_X(as)$.
- If X and Y are independent, then $M_{X+Y}(s) = M_X(s) M_Y(s)$.

We have obtained formulas for the transforms associated with a few common random variables. We can derive such formulas with a moderate amount of algebra for many other distributions (see the end-of-chapter problems for the case of the uniform distribution). We summarize the most useful ones in the tables that follow.

Transforms for Common Discrete Random Variables

Bernoulli(p) $(k = 0, 1)$

$$p_X(k) = \begin{cases} p, & \text{if } k = 1, \\ 1 - p, & \text{if } k = 0, \end{cases} \qquad M_X(s) = 1 - p + pe^s.$$

Binomial(n, p) $(k = 0, 1, \ldots, n)$

$$p_X(k) = \binom{n}{k} p^k (1-p)^{n-k}, \qquad M_X(s) = (1 - p + pe^s)^n.$$

Geometric(p) $(k = 1, 2, \ldots)$

$$p_X(k) = p(1-p)^{k-1}, \qquad M_X(s) = \frac{pe^s}{1 - (1-p)e^s}.$$

Poisson(λ) $(k = 0, 1, \ldots)$

$$p_X(k) = \frac{e^{-\lambda}\lambda^k}{k!}, \qquad M_X(s) = e^{\lambda(e^s - 1)}.$$

Uniform(a, b) $(k = a, a+1, \ldots, b)$

$$p_X(k) = \frac{1}{b - a + 1}, \qquad M_X(s) = \frac{e^{sa}\left(e^{s(b-a+1)} - 1\right)}{(b - a + 1)(e^s - 1)}.$$

Transforms for Common Continuous Random Variables

Uniform(a, b) $(a \leq x \leq b)$

$$f_X(x) = \frac{1}{b - a}, \qquad M_X(s) = \frac{e^{sb} - e^{sa}}{s(b - a)}.$$

Exponential(λ) $(x \geq 0)$

$$f_X(x) = \lambda e^{-\lambda x}, \qquad M_X(s) = \frac{\lambda}{\lambda - s}, \qquad (s < \lambda).$$

Normal(μ, σ^2) $(-\infty < x < \infty)$

$$f_X(x) = \frac{1}{\sqrt{2\pi}\,\sigma} e^{-(x-\mu)^2/2\sigma^2}, \qquad M_X(s) = e^{(\sigma^2 s^2/2) + \mu s}.$$

Transforms Associated with Joint Distributions

If two random variables X and Y are described by some joint distribution (e.g., a joint PDF), then each one is associated with a transform $M_X(s)$ or $M_Y(s)$. These are the transforms of the marginal distributions and do not convey information on the dependence between the two random variables. Such information is contained in a multivariate transform, which we now define.

Consider n random variables $X_1, \ldots, X_n$ related to the same experiment. Let $s_1, \ldots, s_n$ be scalar free parameters. The associated **multivariate transform** is a function of these n parameters and is defined by

$$M_{X_1,\ldots,X_n}(s_1, \ldots, s_n) = \mathbf{E}\left[e^{s_1X_1+\cdots+s_nX_n}\right].$$

The inversion property of transforms discussed earlier extends to the multivariate case. In particular, if $Y_1, \ldots, Y_n$ is another set of random variables and if $M_{X_1,\ldots,X_n}(s_1, \ldots, s_n) = M_{Y_1,\ldots,Y_n}(s_1, \ldots, s_n)$ for all $(s_1, \ldots, s_n)$ belonging to some n-dimensional cube with positive volume, then the joint distribution of $X_1, \ldots, X_n$ is the same as the joint distribution of $Y_1, \ldots, Y_n$.

4.5 SUM OF A RANDOM NUMBER OF INDEPENDENT RANDOM VARIABLES

In our discussion so far of sums of random variables, we have always assumed that the number of variables in the sum is known and fixed. In this section, we will consider the case where the number of random variables being added is itself random. In particular, we consider the sum

$$Y = X_1 + \cdots + X_N,$$

where N is a random variable that takes nonnegative integer values, and $X_1, X_2, \ldots$ are identically distributed random variables. (If $N = 0$, we let $Y = 0$.) We assume that $N, X_1, X_2, \ldots$ are independent, meaning that any finite subcollection of these random variables are independent.

Let us denote by $\mathbf{E}[X]$ and $\mathrm{var}(X)$ the common mean and variance, respectively, of the X_i. We wish to derive formulas for the mean, variance, and the transform of Y. The method that we follow is to first condition on the event $\{N = n\}$, which brings us to the more familiar case of a *fixed* number of random variables.

Fix a nonnegative integer n. The random variable $X_1 + \cdots + X_n$ is independent of N and, therefore, independent of $\{N = n\}$. Hence,

$$\begin{aligned}
\mathbf{E}[Y \mid N = n] &= \mathbf{E}[X_1 + \cdots + X_N \mid N = n] \\
&= \mathbf{E}[X_1 + \cdots + X_n \mid N = n] \\
&= \mathbf{E}[X_1 + \cdots + X_n] \\
&= n\mathbf{E}[X].
\end{aligned}$$

This is true for every nonnegative integer n, so

$$\mathbf{E}[Y \mid N] = N\mathbf{E}[X].$$

Using the law of iterated expectations, we obtain

$$\mathbf{E}[Y] = \mathbf{E}\big[\mathbf{E}[Y \mid N]\big] = \mathbf{E}\big[N\,\mathbf{E}[X]\big] = \mathbf{E}[N]\,\mathbf{E}[X].$$

Similarly,

$$\begin{aligned}\text{var}(Y \mid N = n) &= \text{var}(X_1 + \cdots + X_N \mid N = n)\\ &= \text{var}(X_1 + \cdots + X_n)\\ &= n\,\text{var}(X).\end{aligned}$$

Since this is true for every nonnegative integer n, the random variable $\text{var}(Y \mid N)$ is equal to $N\text{var}(X)$. We now use the law of total variance to obtain

$$\begin{aligned}\text{var}(Y) &= \mathbf{E}\big[\text{var}(Y \mid N)\big] + \text{var}\big(\mathbf{E}[Y \mid N]\big)\\ &= \mathbf{E}\big[N\text{var}(X)\big] + \text{var}\big(N\,\mathbf{E}[X]\big)\\ &= \mathbf{E}[N]\text{var}(X) + \big(\mathbf{E}[X]\big)^2\text{var}(N).\end{aligned}$$

The calculation of the transform proceeds along similar lines. The transform associated with Y, conditional on $N = n$, is $\mathbf{E}[e^{sY} \mid N = n]$. However, conditioned on $N = n$, Y is the sum of the independent random variables $X_1, \ldots, X_n$, and

$$\begin{aligned}\mathbf{E}[e^{sY} \mid N = n] &= \mathbf{E}\big[e^{sX_1} \cdots e^{sX_N} \mid N = n\big]\\ &= \mathbf{E}\big[e^{sX_1} \cdots e^{sX_n}\big]\\ &= \mathbf{E}[e^{sX_1}] \cdots \mathbf{E}[e^{sX_n}]\\ &= \big(M_X(s)\big)^n,\end{aligned}$$

where $M_X(s)$ is the transform associated with X_i, for each i. Using the law of iterated expectations, the (unconditional) transform associated with Y is

$$M_Y(s) = \mathbf{E}[e^{sY}] = \mathbf{E}\big[\mathbf{E}[e^{sY} \mid N]\big] = \mathbf{E}\big[\big(M_X(s)\big)^N\big] = \sum_{n=0}^{\infty} \big(M_X(s)\big)^n p_N(n).$$

Using the observation

$$\big(M_X(s)\big)^n = e^{\log(M_X(s))^n} = e^{n \log M_X(s)},$$

we have

$$M_Y(s) = \sum_{n=0}^{\infty} e^{n \log M_X(s)} p_N(n).$$

Comparing with the formula

$$M_N(s) = \mathbf{E}[e^{sN}] = \sum_{n=0}^{\infty} e^{sn} p_N(n),$$

we see that $M_Y(s) = M_N\big(\log M_X(s)\big)$, i.e., $M_Y(s)$ is obtained from the formula for $M_N(s)$, with s replaced by $\log M_X(s)$ or, equivalently, with e^s replaced by $M_X(s)$.

Let us summarize the properties derived so far.

Properties of the Sum of a Random Number of Independent Random Variables

Let $X_1, X_2, \ldots$ be identically distributed random variables with mean $\mathbf{E}[X]$ and variance $\text{var}(X)$. Let N be a random variable that takes nonnegative integer values. We assume that all of these random variables are independent, and we consider the sum

$$Y = X_1 + \cdots + X_N.$$

Then:

- $\mathbf{E}[Y] = \mathbf{E}[N]\,\mathbf{E}[X]$.
- $\text{var}(Y) = \mathbf{E}[N]\,\text{var}(X) + \big(\mathbf{E}[X]\big)^2 \text{var}(N)$.
- We have

  $$M_Y(s) = M_N\big(\log M_X(s)\big).$$

  Equivalently, the transform $M_Y(s)$ is found by starting with the transform $M_N(s)$ and replacing each occurrence of e^s with $M_X(s)$.

Example 4.34. A remote village has three gas stations. Each gas station is open on any given day with probability 1/2, independent of the others. The amount of gas available in each gas station is unknown and is uniformly distributed between 0 and 1000 gallons. We wish to characterize the probability law of the total amount of gas available at the gas stations that are open.

The number N of open gas stations is a binomial random variable with $p = 1/2$ and the corresponding transform is

$$M_N(s) = (1 - p + pe^s)^3 = \frac{1}{8}(1 + e^s)^3.$$

The transform $M_X(s)$ associated with the amount of gas available in an open gas station is

$$M_X(s) = \frac{e^{1000s} - 1}{1000s}.$$

The transform associated with the total amount Y available is the same as $M_N(s)$, except that each occurrence of e^s is replaced with $M_X(s)$, i.e.,

$$M_Y(s) = \frac{1}{8}\left(1 + \left(\frac{e^{1000s} - 1}{1000s}\right)\right)^3.$$

Example 4.35. Sum of a Geometric Number of Independent Exponential Random Variables. Jane visits a number of bookstores, looking for *Great Expectations.* Any given bookstore carries the book with probability p, independent of the others. In a typical bookstore visited, Jane spends a random amount of time, exponentially distributed with parameter λ, until she either finds the book or she determines that the bookstore does not carry it. We assume that Jane will keep visiting bookstores until she buys the book and that the time spent in each is independent of everything else. We wish to find the mean, variance, and PDF of the total time spent in bookstores.

The total number N of bookstores visited is geometrically distributed with parameter p. Hence, the total time Y spent in bookstores is the sum of a geometrically distributed number N of independent exponential random variables $X_1, X_2, \ldots$. We have

$$\mathbf{E}[Y] = \mathbf{E}[N]\,\mathbf{E}[X] = \frac{1}{p} \cdot \frac{1}{\lambda}.$$

Using the formulas for the variance of geometric and exponential random variables, we also obtain

$$\text{var}(Y) = \mathbf{E}[N]\,\text{var}(X) + \big(\mathbf{E}[X]\big)^2 \text{var}(N) = \frac{1}{p} \cdot \frac{1}{\lambda^2} + \frac{1}{\lambda^2} \cdot \frac{1-p}{p^2} = \frac{1}{\lambda^2 p^2}.$$

In order to find the transform $M_Y(s)$, let us recall that

$$M_X(s) = \frac{\lambda}{\lambda - s}, \qquad M_N(s) = \frac{pe^s}{1-(1-p)e^s}.$$

Then, $M_Y(s)$ is found by starting with $M_N(s)$ and replacing each occurrence of e^s with $M_X(s)$. This yields

$$M_Y(s) = \frac{pM_X(s)}{1-(1-p)M_X(s)} = \frac{\dfrac{p\lambda}{\lambda - s}}{1-(1-p)\dfrac{\lambda}{\lambda - s}},$$

which simplifies to

$$M_Y(s) = \frac{p\lambda}{p\lambda - s}.$$

We recognize this as the transform associated with an exponentially distributed random variable with parameter $p\lambda$, and therefore,

$$f_Y(y) = p\lambda e^{-p\lambda y}, \qquad y \geq 0.$$

This result can be surprising because the sum of a *fixed* number n of independent exponential random variables is not exponentially distributed. For example, if $n = 2$, the transform associated with the sum is $\big(\lambda/(\lambda - s)\big)^2$, which does not correspond to an exponential distribution.

Example 4.36. Sum of a Geometric Number of Independent Geometric Random Variables. This example is a discrete counterpart of the preceding one. We let N be geometrically distributed with parameter p. We also let each random variable X_i be geometrically distributed with parameter q. We assume that all of these random variables are independent. Let $Y = X_1 + \cdots + X_N$. We have

$$M_N(s) = \frac{pe^s}{1-(1-p)e^s}, \qquad M_X(s) = \frac{qe^s}{1-(1-q)e^s}.$$

To determine $M_Y(s)$, we start with the formula for $M_N(s)$ and replace each occurrence of e^s with $M_X(s)$. This yields

$$M_Y(s) = \frac{pM_X(s)}{1-(1-p)M_X(s)},$$

and, after some algebra,

$$M_Y(s) = \frac{pqe^s}{1-(1-pq)e^s}.$$

We conclude that Y is geometrically distributed, with parameter pq.

4.6 SUMMARY AND DISCUSSION

In this chapter, we have studied a number of advanced topics. We discuss here some of the highlights.

In Section 4.1, we addressed the problem of calculating the PDF of a function $g(X)$ of a continuous random variable X. The concept of a CDF is very useful here. In particular, the PDF of $g(X)$ is typically obtained by calculating and differentiating the corresponding CDF. In some cases, such as when the function g is strictly monotonic, the calculation is facilitated through the use of special formulas. We also considered some examples involving a function $g(X, Y)$ of two continuous random variables. In particular, we derived the convolution formula for the probability law of the sum of two independent random variables.

In Section 4.2, we introduced covariance and correlation, both of which are important qualitative indicators of the relationship between two random variables. The covariance and its scaled version, the correlation coefficient, are involved in determining the variance of the sum of dependent random variables. They also play an important role in the linear least mean squares estimation methodology of Section 8.4.

In Section 4.3, we reconsidered the subject of conditioning, with the aim of developing tools for computing expected values and variances. We took a closer look at the conditional expectation and indicated that it can be viewed as a random variable, with an expectation and variance of its own. We derived some related properties, including the law of iterated expectations, and the law of total variance.

In Section 4.4, we introduced the transform associated with a random variable, and saw how such a transform can be computed. Conversely, we indicated that given a transform, the distribution of an associated random variable is uniquely determined. It can be found, for example, using tables of commonly occurring transforms. We have found transforms useful for a variety of purposes, such as the following.

(a) Knowledge of the transform associated with a random variable provides a shortcut for calculating the moments of the random variable.

(b) The transform associated with the sum of two independent random variables is equal to the product of the transforms associated with each one of them. This property was used to show that the sum of two independent normal (respectively, Poisson) random variables is normal (respectively, Poisson).

(c) Transforms can be used to characterize the distribution of the sum of a random number of random variables (Section 4.5), something which is often impossible by other means.

Finally, in Section 4.5, we derived formulas for the mean, the variance, and the transform of the sum of a random number of random variables, by combining the methodology of Sections 4.3 and 4.4.

PROBLEMS

SECTION 4.1. Derived Distributions

Problem 1. If X is a random variable that is uniformly distributed between -1 and 1, find the PDF of $\sqrt{|X|}$ and the PDF of $-\ln|X|$.

Problem 2. Find the PDF of e^X in terms of the PDF of X. Specialize the answer to the case where X is uniformly distributed between 0 and 1.

Problem 3. Find the PDFs of $|X|^{1/3}$ and $|X|^{1/4}$ in terms of the PDF of X.

Problem 4. The metro train arrives at the station near your home every quarter hour starting at 6:00 a.m. You walk into the station every morning between 7:10 and 7:30 a.m., with the time in this interval being a random variable with given PDF (cf. Example 3.14, in Chapter 3). Let X be the elapsed time, in minutes, between 7:10 and the time of your arrival. Let Y be the time that you have to wait until you board a train. Calculate the CDF of Y in terms of the CDF of X and differentiate to obtain a formula for the PDF of Y.

Problem 5. Let X and Y be independent random variables, uniformly distributed in the interval $[0, 1]$. Find the CDF and the PDF of $|X - Y|$.

Problem 6. Let X and Y be the Cartesian coordinates of a randomly chosen point (according to a uniform PDF) in the triangle with vertices at $(0, 1)$, $(0, -1)$, and $(1, 0)$. Find the CDF and the PDF of $|X - Y|$.

Problem 7. Two points are chosen randomly and independently from the interval $[0, 1]$ according to a uniform distribution. Show that the expected distance between the two points is $1/3$.

Problem 8. Find the PDF of $Z = X+Y$, when X and Y are independent exponential random variables with common parameter λ.

Problem 9. Consider the same problem as in Example 4.9, but assume that the random variables X and Y are independent and exponentially distributed with different parameters λ and μ, respectively. Find the PDF of $X - Y$.

Problem 10. Let X and Y be independent random variables with PMFs

$$p_X(x) = \begin{cases} 1/3, & \text{if } x = 1, 2, 3, \\ 0, & \text{otherwise,} \end{cases} \qquad p_Y(y) = \begin{cases} 1/2, & \text{if } y = 0, \\ 1/3, & \text{if } y = 1, \\ 1/6, & \text{if } y = 2, \\ 0, & \text{otherwise.} \end{cases}$$

Find the PMF of $Z = X + Y$, using the convolution formula.

Problem 11. Use the convolution formula to establish that the sum of two independent Poisson random variables with parameters λ and μ, respectively, is Poisson with parameter $\lambda + \mu$.

Problem 12. The random variables X, Y, and Z are independent and uniformly distributed between zero and one. Find the PDF of $X + Y + Z$.

Problem 13. Consider a PDF that is positive only within an interval $[a, b]$ and is symmetric around the mean $(a + b)/2$. Let X and Y be independent random variables that both have this PDF. Suppose that you have calculated the PDF of $X + Y$. How can you easily obtain the PDF of $X - Y$?

Problem 14. **Competing exponentials.** The lifetimes of two light bulbs are modeled as independent and exponential random variables X and Y, with parameters λ and μ, respectively. The time at which a light bulb first burns out is

$$Z = \min\{X, Y\}.$$

Show that Z is an exponential random variable with parameter $\lambda + \mu$.

Problem 15.* Cauchy random variable.

(a) Let X be a random variable that is uniformly distributed between $-1/2$ and $1/2$. Show that the PDF of $Y = \tan(\pi X)$ is

$$f_Y(y) = \frac{1}{\pi(1 + y^2)}, \qquad -\infty < y < \infty.$$

(Y is called a *Cauchy random variable*.)

(b) Let Y be a Cauchy random variable. Find the PDF of the random variable X, which is equal to the angle between $-\pi/2$ and $\pi/2$ whose tangent is Y.

Solution. (a) We first note that Y is a continuous, strictly monotonically increasing function of X, which takes values between $-\infty$ and ∞, as X ranges over the interval $[-1/2, 1/2]$. Therefore, we have for all scalars y,

$$F_Y(y) = \mathbf{P}(Y \le y) = \mathbf{P}\big(\tan(\pi X) \le y\big) = \mathbf{P}\big(\pi X \le \tan^{-1} y\big) = \frac{1}{\pi}\tan^{-1} y + \frac{1}{2},$$

where the last equality follows using the CDF of X, which is uniformly distributed in the interval $[-1/2, 1/2]$. Therefore, by differentiation, using the formula $d/dy\big(\tan^{-1} y\big) = 1/(1 + y^2)$, we have for all y,

$$f_Y(y) = \frac{1}{\pi(1 + y^2)}.$$

(b) We first compute the CDF of X and then differentiate to obtain its PDF. We have for $-\pi/2 \le x \le \pi/2$,

$$\begin{aligned} \mathbf{P}(X \le x) &= \mathbf{P}(\tan^{-1} Y \le x) \\ &= \mathbf{P}(Y \le \tan x) \end{aligned}$$

$$= \frac{1}{\pi} \int_{-\infty}^{\tan x} \frac{1}{1+y^2}\, dy$$
$$= \frac{1}{\pi} \tan^{-1} y \Big|_{-\infty}^{\tan x}$$
$$= \frac{1}{\pi}\left(x + \frac{\pi}{2}\right).$$

For $x < -\pi/2$, we have $\mathbf{P}(X \leq x) = 0$, and for $\pi/2 < x$, we have $\mathbf{P}(X \leq x) = 1$. Taking the derivative of the CDF $\mathbf{P}(X \leq x)$, we find that X is uniformly distributed on the interval $[-\pi/2, \pi/2]$.

Note: An interesting property of the Cauchy random variable is that it satisfies

$$\int_0^{\infty} \frac{y}{\pi(1+y^2)}\, dy = -\int_{-\infty}^{0} \frac{y}{\pi(1+y^2)}\, dy = \infty,$$

as can be easily verified. As a result, the Cauchy random variable does not have a well-defined expected value, despite the symmetry of its PDF around 0; see the footnote in Section 3.1 on the definition of the expected value of a continuous random variable.

Problem 16.* The polar coordinates of two independent normal random variables. Let X and Y be independent standard normal random variables. The pair (X, Y) can be described in polar coordinates in terms of random variables $R \geq 0$ and $\Theta \in [0, 2\pi]$, so that

$$X = R\cos\Theta, \qquad Y = R\sin\Theta.$$

(a) Show that Θ is uniformly distributed in $[0, 2\pi]$, that R has the PDF

$$f_R(r) = re^{-r^2/2}, \qquad r \geq 0,$$

and that R and Θ are independent. (The random variable R is said to have a **Rayleigh** distribution.)

(b) Show that R^2 has an exponential distribution with parameter 1/2.

Note: Using the results in this problem, we see that samples of a normal random variable can be generated using samples of independent uniform and exponential random variables.

Solution. (a) The joint PDF of X and Y is

$$f_{X,Y}(x, y) = f_X(x)f_Y(y) = \frac{1}{2\pi}e^{-(x^2+y^2)/2}.$$

We first find the joint CDF of R and Θ. Fix some $r > 0$ and some $\theta \in [0, 2\pi]$, and let A be the set of points (x, y) whose polar coordinates $(\overline{r}, \overline{\theta})$ satisfy $0 \leq \overline{r} \leq r$ and $0 \leq \overline{\theta} \leq \theta$; note that the set A is a sector of a circle of radius r, with angle θ. We have

$$F_{R,\Theta}(r, \theta) = \mathbf{P}(R \leq r,\ \Theta \leq \theta) = \mathbf{P}\big((X, Y) \in A\big)$$
$$= \frac{1}{2\pi} \iint_{(x,y)\in A} e^{-(x^2+y^2)/2}\, dx\, dy = \frac{1}{2\pi} \int_0^{\theta} \int_0^{r} e^{-\overline{r}^2/2}\overline{r}\, d\overline{r}\, d\overline{\theta},$$

where the last equality is obtained by transforming to polar coordinates. We then differentiate, to find that

$$f_{R,\Theta}(r,\theta) = \frac{\partial^2 F_{R,\Theta}}{\partial r \partial \theta}(r,\theta) = \frac{r}{2\pi} e^{-r^2/2}, \qquad r \geq 0, \ \theta \in [0, 2\pi].$$

Thus,

$$f_R(r) = \int_0^{2\pi} f_{R,\Theta}(r,\theta)\, d\theta = r\, e^{-r^2/2}, \qquad r \geq 0.$$

Furthermore,

$$f_{\Theta|R}(\theta \,|\, r) = \frac{f_{R,\Theta}(r,\theta)}{f_R(r)} = \frac{1}{2\pi}, \qquad \theta \in [0, 2\pi].$$

Since the conditional PDF $f_{\Theta|R}$ of Θ is unaffected by the value of the conditioning variable R, it follows that it is also equal to the unconditional PDF f_Θ. In particular, $f_{R,\Theta}(r,\theta) = f_R(r) f_\Theta(\theta)$, so that R and Θ are independent.

(b) Let $t \geq 0$. We have

$$\mathbf{P}(R^2 \geq t) = \mathbf{P}\big(R \geq \sqrt{t}\big) = \int_{\sqrt{t}}^{\infty} re^{-r^2/2}\, dr = \int_{t/2}^{\infty} e^{-u}\, du = e^{-t/2},$$

where we have used the change of variables $u = r^2/2$. By differentiating, we obtain

$$f_{R^2}(t) = \frac{1}{2} e^{-t/2}, \qquad t \geq 0.$$

SECTION 4.2. Covariance and Correlation

Problem 17. Suppose that X and Y are random variables with the same variance. Show that $X - Y$ and $X + Y$ are uncorrelated.

Problem 18. Consider four random variables, W, X, Y, Z, with

$$\mathbf{E}[W] = \mathbf{E}[X] = \mathbf{E}[Y] = \mathbf{E}[Z] = 0,$$

$$\text{var}(W) = \text{var}(X) = \text{var}(Y) = \text{var}(Z) = 1,$$

and assume that W, X, Y, Z are pairwise uncorrelated. Find the correlation coefficients $\rho(R,S)$ and $\rho(R,T)$, where $R = W + X$, $S = X + Y$, and $T = Y + Z$.

Problem 19. Suppose that a random variable X satisfies

$$\mathbf{E}[X] = 0, \quad \mathbf{E}[X^2] = 1, \quad \mathbf{E}[X^3] = 0, \quad \mathbf{E}[X^4] = 3,$$

and let

$$Y = a + bX + cX^2.$$

Find the correlation coefficient $\rho(X,Y)$.

Problem 20.* Schwarz inequality. Show that for any random variables X and Y, we have

$$\big(\mathbf{E}[XY]\big)^2 \le \mathbf{E}[X^2]\,\mathbf{E}[Y^2].$$

Solution. We may assume that $\mathbf{E}[Y^2] \neq 0$; otherwise, we have $Y = 0$ with probability 1, and hence $\mathbf{E}[XY] = 0$, so the inequality holds. We have

$$\begin{aligned}
0 &\le \mathbf{E}\left[\left(X - \frac{\mathbf{E}[XY]}{\mathbf{E}[Y^2]}Y\right)^2\right] \\
&= \mathbf{E}\left[X^2 - 2\frac{\mathbf{E}[XY]}{\mathbf{E}[Y^2]}XY + \frac{\big(\mathbf{E}[XY]\big)^2}{\big(\mathbf{E}[Y^2]\big)^2}Y^2\right] \\
&= \mathbf{E}[X^2] - 2\frac{\mathbf{E}[XY]}{\mathbf{E}[Y^2]}\mathbf{E}[XY] + \frac{\big(\mathbf{E}[XY]\big)^2}{\big(\mathbf{E}[Y^2]\big)^2}\mathbf{E}[Y^2] \\
&= \mathbf{E}[X^2] - \frac{\big(\mathbf{E}[XY]\big)^2}{\mathbf{E}[Y^2]},
\end{aligned}$$

i.e., $\big(\mathbf{E}[XY]\big)^2 \le \mathbf{E}[X^2]\,\mathbf{E}[Y^2]$.

Problem 21.* Correlation coefficient. Consider the correlation coefficient

$$\rho(X,Y) = \frac{\text{cov}(X,Y)}{\sqrt{\text{var}(X)\text{var}(Y)}}$$

of two random variables X and Y that have positive variances. Show that:

(a) $|\rho(X,Y)| \le 1$. *Hint*: Use the Schwarz inequality from the preceding problem.

(b) If $Y - \mathbf{E}[Y]$ is a positive (or negative) multiple of $X - \mathbf{E}[X]$, then $\rho(X,Y) = 1$ [or $\rho(X,Y) = -1$, respectively].

(c) If $\rho(X,Y) = 1$ [or $\rho(X,Y) = -1$], then, with probability 1, $Y - \mathbf{E}[Y]$ is a positive (or negative, respectively) multiple of $X - \mathbf{E}[X]$.

Solution. (a) Let $\tilde{X} = X - \mathbf{E}[X]$ and $\tilde{Y} = Y - \mathbf{E}[Y]$. Using the Schwarz inequality, we get

$$\big(\rho(X,Y)\big)^2 = \frac{\big(\mathbf{E}[\tilde{X}\tilde{Y}]\big)^2}{\mathbf{E}[\tilde{X}^2]\,\mathbf{E}[\tilde{Y}^2]} \le 1,$$

and hence $|\rho(X,Y)| \le 1$.

(b) If $\tilde{Y} = a\tilde{X}$, then

$$\rho(X,Y) = \frac{\mathbf{E}[\tilde{X}a\tilde{X}]}{\sqrt{\mathbf{E}[\tilde{X}^2]\,\mathbf{E}[(a\tilde{X})^2]}} = \frac{a}{|a|}.$$

(c) If $\big(\rho(X,Y)\big)^2 = 1$, the calculation in the solution of Problem 20 yields

$$\begin{aligned}\mathbf{E}\left[\left(\tilde{X} - \frac{\mathbf{E}[\tilde{X}\tilde{Y}]}{\mathbf{E}[\tilde{Y}^2]}\tilde{Y}\right)^2\right] &= \mathbf{E}[\tilde{X}^2] - \frac{\big(\mathbf{E}[\tilde{X}\tilde{Y}]\big)^2}{\mathbf{E}[\tilde{Y}^2]} \\ &= \mathbf{E}[\tilde{X}^2]\Big(1 - \big(\rho(X,Y)\big)^2\Big) \\ &= 0.\end{aligned}$$

Thus, with probability 1, the random variable

$$\tilde{X} - \frac{\mathbf{E}[\tilde{X}\tilde{Y}]}{\mathbf{E}[\tilde{Y}^2]}Y$$

is equal to zero. It follows that, with probability 1,

$$\tilde{X} = \frac{\mathbf{E}[\tilde{X}\tilde{Y}]}{\mathbf{E}[\tilde{Y}^2]}\tilde{Y} = \sqrt{\frac{\mathbf{E}[\tilde{X}^2]}{\mathbf{E}[\tilde{Y}^2]}}\rho(X,Y)\tilde{Y},$$

i.e., the sign of the constant ratio of $\tilde{X}$ and $\tilde{Y}$ is determined by the sign of $\rho(X,Y)$.

SECTION 4.3. Conditional Expectation and Variance Revisited

Problem 22. Consider a gambler who at each gamble either wins or loses his bet with probabilities p and $1-p$, independent of earlier gambles. When $p > 1/2$, a popular gambling system, known as the Kelly strategy, is to always bet the fraction $2p-1$ of the current fortune. Compute the expected fortune after n gambles, starting with x units and employing the Kelly strategy.

Problem 23. Pat and Nat are dating, and all of their dates are scheduled to start at 9 p.m. Nat always arrives promptly at 9 p.m. Pat is highly disorganized and arrives at a time that is uniformly distributed between 8 p.m. and 10 p.m. Let X be the time in hours between 8 p.m. and the time when Pat arrives. If Pat arrives before 9 p.m., their date will last exactly 3 hours. If Pat arrives after 9 p.m., their date will last for a time that is uniformly distributed between 0 and $3-X$ hours. The date starts at the time they meet. Nat gets irritated when Pat is late and will end the relationship after the second date on which Pat is late by more than 45 minutes. All dates are independent of any other dates.

(a) What is the expected number of hours Nat waits for Pat to arrive?

(b) What is the expected duration of any particular date?

(c) What is the expected number of dates they will have before breaking up?

Problem 24. A retired professor comes to the office at a time which is uniformly distributed between 9 a.m. and 1 p.m., performs a single task, and leaves when the task is completed. The duration of the task is exponentially distributed with parameter $\lambda(y) = 1/(5-y)$, where y is the length of the time interval between 9 a.m. and the time of his arrival.

(a) What is the expected amount of time that the professor devotes to the task?

(b) What is the expected time at which the task is completed?

(c) The professor has a Ph.D. student who on a given day comes to see him at a time that is uniformly distributed between 9 a.m. and 5 p.m. If the student does not find the professor, he leaves and does not return. If he finds the professor, he spends an amount of time that is uniformly distributed between 0 and 1 hour. The professor will spend the same total amount of time on his task regardless of whether he is interrupted by the student. What is the expected amount of time that the professor will spend with the student and what is the expected time at which he will leave his office?

Problem 25.* Show that for a discrete or continuous random variable X, and any function $g(Y)$ of another random variable Y, we have $\mathbf{E}[Xg(Y) \,|\, Y] = g(Y)\,\mathbf{E}[X \,|\, Y]$.

Solution. Assume that X is continuous. From a version of the expected value rule for conditional expectations given in Chapter 3, we have

$$\begin{aligned}\mathbf{E}[Xg(Y) \,|\, Y = y] &= \int_{-\infty}^{\infty} xg(y) f_{X|Y}(x \,|\, y)\, dx \\ &= g(y) \int_{-\infty}^{\infty} x f_{X|Y}(x \,|\, y)\, dx \\ &= g(y)\,\mathbf{E}[X \,|\, Y = y].\end{aligned}$$

This shows that the realized values $\mathbf{E}[Xg(Y) \,|\, Y = y]$ and $g(y)\mathbf{E}[X \,|\, Y = y]$ of the random variables $\mathbf{E}[Xg(Y) \,|\, Y]$ and $g(Y)\mathbf{E}[X \,|\, Y]$ are always equal. Hence these two random variables are equal. The proof is similar if X is discrete.

Problem 26.* Let X and Y be independent random variables. Use the law of total variance to show that

$$\text{var}(XY) = \big(\mathbf{E}[X]\big)^2 \text{var}(Y) + \big(\mathbf{E}[Y]\big)^2 \text{var}(X) + \text{var}(X)\text{var}(Y).$$

Solution. Let $Z = XY$. The law of total variance yields

$$\text{var}(Z) = \text{var}\big(\mathbf{E}[Z \,|\, X]\big) + \mathbf{E}\big[\text{var}(Z \,|\, X)\big].$$

We have

$$\mathbf{E}[Z \,|\, X] = \mathbf{E}[XY \,|\, X] = X\mathbf{E}[Y],$$

so that

$$\text{var}\big(\mathbf{E}[Z \,|\, X]\big) = \text{var}\big(X\mathbf{E}[Y]\big) = \big(\mathbf{E}[Y]\big)^2 \text{var}(X).$$

Furthermore,

$$\text{var}(Z \,|\, X) = \text{var}(XY \,|\, X) = X^2 \text{var}(Y \,|\, X) = X^2 \text{var}(Y),$$

so that

$$\mathbf{E}\big[\text{var}(Z \,|\, X)\big] = \mathbf{E}[X^2]\text{var}(Y) = \big(\mathbf{E}[X]\big)^2 \text{var}(Y) + \text{var}(X)\text{var}(Y).$$

Combining the preceding relations, we obtain

$$\text{var}(XY) = \big(\mathbf{E}[X]\big)^2 \text{var}(Y) + \big(\mathbf{E}[Y]\big)^2 \text{var}(X) + \text{var}(X)\text{var}(Y).$$

Problem 27.* We toss n times a biased coin whose probability of heads, denoted by q, is the value of a random variable Q with given mean μ and positive variance σ^2. Let X_i be a Bernoulli random variable that models the outcome of the ith toss (i.e., $X_i = 1$ if the ith toss is a head). We assume that $X_1, \ldots, X_n$ are conditionally independent, given $Q = q$. Let X be the number of heads obtained in the n tosses.

(a) Use the law of iterated expectations to find $\mathbf{E}[X_i]$ and $\mathbf{E}[X]$.

(b) Find $\text{cov}(X_i, X_j)$. Are $X_1, \ldots, X_n$ independent?

(c) Use the law of total variance to find $\text{var}(X)$. Verify your answer using the covariance result of part (b).

Solution. (a) We have, from the law of iterated expectations and the fact $\mathbf{E}[X_i \,|\, Q] = Q$,

$$\mathbf{E}[X_i] = \mathbf{E}\big[\mathbf{E}[X_i \,|\, Q]\big] = \mathbf{E}[Q] = \mu.$$

Since $X = X_1 + \cdots + X_n$, it follows that

$$\mathbf{E}[X] = \mathbf{E}[X_1] + \cdots + \mathbf{E}[X_n] = n\mu.$$

(b) We have, for $i \neq j$, using the conditional independence assumption,

$$\mathbf{E}[X_i X_j \,|\, Q] = \mathbf{E}[X_i \,|\, Q]\, \mathbf{E}[X_j \,|\, Q] = Q^2,$$

and

$$\mathbf{E}[X_i X_j] = \mathbf{E}\big[\mathbf{E}[X_i X_j \,|\, Q]\big] = \mathbf{E}[Q^2].$$

Thus,

$$\text{cov}(X_i, X_j) = \mathbf{E}[X_i X_j] - \mathbf{E}[X_i]\, \mathbf{E}[X_j] = \mathbf{E}[Q^2] - \mu^2 = \sigma^2.$$

Since $\text{cov}(X_i, X_j) > 0$, $X_1, \ldots, X_n$ are not independent.

Also, for $i = j$, using the observation that $X_i^2 = X_i$,

$$\begin{aligned}
\text{var}(X_i) &= \mathbf{E}[X_i^2] - \big(\mathbf{E}[X_i]\big)^2 \\
&= \mathbf{E}[X_i] - \big(\mathbf{E}[X_i]\big)^2 \\
&= \mu - \mu^2.
\end{aligned}$$

(c) Using the law of total variance, and the conditional independence of $X_1, \ldots, X_n$, we have

$$\begin{aligned}
\text{var}(X) &= \mathbf{E}\big[\text{var}(X \mid Q)\big] + \text{var}\big(\mathbf{E}[X \mid Q]\big) \\
&= \mathbf{E}\big[\text{var}(X_1 + \cdots + X_n \mid Q)\big] + \text{var}\big(\mathbf{E}[X_1 + \cdots + X_n \mid Q]\big) \\
&= \mathbf{E}\big[nQ(1-Q)\big] + \text{var}(nQ) \\
&= n\mathbf{E}[Q - Q^2] + n^2\text{var}(Q) \\
&= n(\mu - \mu^2 - \sigma^2) + n^2\sigma^2 \\
&= n(\mu - \mu^2) + n(n-1)\sigma^2.
\end{aligned}$$

To verify the result using the covariance formulas of part (b), we write

$$\begin{aligned}
\text{var}(X) &= \text{var}(X_1 + \cdots + X_n) \\
&= \sum_{i=1}^{n} \text{var}(X_i) + \sum_{\{(i,j) \mid i \neq j\}} \text{cov}(X_i, X_j) \\
&= n\text{var}(X_1) + n(n-1)\text{cov}(X_1, X_2) \\
&= n(\mu - \mu^2) + n(n-1)\sigma^2.
\end{aligned}$$

Problem 28.* **The Bivariate Normal PDF**. The (zero mean) bivariate normal PDF is of the form

$$f_{X,Y}(x, y) = ce^{-q(x,y)},$$

where the exponent term $q(x, y)$ is a quadratic function of x and y,

$$q(x, y) = \frac{\dfrac{x^2}{\sigma_x^2} - 2\rho\dfrac{xy}{\sigma_x\sigma_y} + \dfrac{y^2}{\sigma_y^2}}{2(1-\rho^2)},$$

σ_x and σ_y are positive constants, ρ is a constant that satisfies $-1 < \rho < 1$, and c is a normalizing constant.

(a) By completing the square, rewrite $q(x, y)$ in the form $(\alpha x - \beta y)^2 + \gamma y^2$, for some constants α, β, and γ.

(b) Show that X and Y are zero mean normal random variables with variance σ_x^2 and σ_y^2, respectively.

(c) Find the normalizing constant c.

(d) Show that the conditional PDF of X given that $Y = y$ is normal, and identify its conditional mean and variance.

(e) Show that the correlation coefficient of X and Y is equal to ρ.

(f) Show that X and Y are independent if and only if they are uncorrelated.

(g) Show that the estimation error $\mathbf{E}[X \mid Y] - X$ is normal with mean zero and variance $(1 - \rho^2)\sigma_x^2$, and is independent from Y.

Solution. (a) We can rewrite $q(x,y)$ in the form

$$q(x,y) = q_1(x,y) + q_2(y),$$

where

$$q_1(x,y) = \frac{1}{2(1-\rho^2)}\left(\frac{x}{\sigma_x} - \rho\frac{y}{\sigma_y}\right)^2, \qquad \text{and} \qquad q_2(y) = \frac{y^2}{2\sigma_y^2}.$$

(b) We have

$$f_Y(y) = c\int_{-\infty}^{\infty} e^{-q_1(x,y)}e^{-q_2(y)}\,dx = ce^{-q_2(y)}\int_{-\infty}^{\infty} e^{-q_1(x,y)}\,dx.$$

Using the change of variables

$$u = \frac{x/\sigma_x - \rho y/\sigma_y}{\sqrt{1-\rho^2}},$$

we obtain

$$\int_{-\infty}^{\infty} e^{-q_1(x,y)}\,dx = \sigma_x\sqrt{1-\rho^2}\int_{-\infty}^{\infty} e^{-u^2/2}\,du = \sigma_x\sqrt{1-\rho^2}\,\sqrt{2\pi}.$$

Thus,

$$f_Y(y) = c\sigma_x\sqrt{1-\rho^2}\,\sqrt{2\pi}\,e^{-y^2/2\sigma_y^2}.$$

We recognize this as a normal PDF with mean zero and variance σ_y^2. The result for the random variable X follows by symmetry.

(c) The normalizing constant for the PDF of Y must be equal to $1/(\sqrt{2\pi}\,\sigma_y)$. It follows that

$$c\sigma_x\sqrt{1-\rho^2}\,\sqrt{2\pi} = \frac{1}{\sqrt{2\pi}\,\sigma_y},$$

which implies that

$$c = \frac{1}{2\pi\sigma_x\sigma_y\sqrt{1-\rho^2}}.$$

(d) Since

$$f_{X,Y}(x,y) = \frac{1}{2\pi\sigma_x\sigma_y\sqrt{1-\rho^2}}e^{-q_1(x,y)}e^{-q_2(y)},$$

and

$$f_Y(y) = \frac{1}{\sqrt{2\pi}\,\sigma_y}e^{-q_2(y)},$$

we obtain

$$f_{X|Y}(x\,|\,y) = \frac{f_{X,Y}(x,y)}{f_Y(y)} = \frac{1}{\sqrt{2\pi}\,\sigma_x\sqrt{1-\rho^2}}\exp\left\{-\frac{(x-\rho\sigma_x y/\sigma_y)^2}{2\sigma_x^2(1-\rho^2)}\right\}.$$

For any fixed y, we recognize this as a normal PDF with mean $(\rho\sigma_x/\sigma_y)y$, and variance $\sigma_x^2(1-\rho^2)$. In particular, $\mathbf{E}[X \mid Y=y] = (\rho\sigma_x/\sigma_y)y$, and $\mathbf{E}[X \mid Y] = (\rho\sigma_x/\sigma_y)Y$.

(e) Using the expected value rule and the law of iterated expectations, we have

$$\begin{aligned}
\mathbf{E}[XY] &= \mathbf{E}\big[\mathbf{E}[XY \mid Y]\big] \\
&= \mathbf{E}\big[Y\,\mathbf{E}[X \mid Y]\big] \\
&= \mathbf{E}\big[Y(\rho\sigma_x/\sigma_y)Y\big] \\
&= \rho\frac{\sigma_x}{\sigma_y}\mathbf{E}[Y^2] \\
&= \rho\sigma_x\sigma_y.
\end{aligned}$$

Thus, the correlation coefficient $\rho(X,Y)$ is equal to

$$\rho(X,Y) = \frac{\mathrm{cov}(X,Y)}{\sigma_x\sigma_y} = \frac{\mathbf{E}[XY]}{\sigma_x\sigma_y} = \rho.$$

(f) If X and Y are uncorrelated, then $\rho = 0$, and the joint PDF satisfies $f_{X,Y}(x,y) = f_X(x)f_Y(y)$, so that X and Y are independent. Conversely, if X and Y are independent, then they are automatically uncorrelated.

(g) From part (d), we know that conditioned on $Y = y$, X is normal with mean $\mathbf{E}[X \mid Y=y]$ and variance $(1-\rho^2)\sigma_x^2$. Therefore, conditioned on $Y = y$, the estimation error $\tilde{X} = \mathbf{E}[X \mid Y=y] - X$ is normal with mean zero and variance $(1-\rho^2)\sigma_x^2$, i.e.,

$$f_{\tilde{X}|Y}(\tilde{x} \mid y) = \frac{1}{\sqrt{2\pi(1-\rho^2)\sigma_x^2}} \exp\left\{ -\frac{\tilde{x}^2}{2(1-\rho^2)\sigma_x^2} \right\}.$$

Since the conditional PDF of $\tilde{X}$ does not depend on the value y of Y, it follows that $\tilde{X}$ is independent of Y, and the above conditional PDF is also the unconditional PDF of $\tilde{X}$.

SECTION 4.4. Transforms

Problem 29. Let X be a random variable that takes the values 1, 2, and 3, with the following probabilities:

$$\mathbf{P}(X=1) = \frac{1}{2}, \qquad \mathbf{P}(X=2) = \frac{1}{4}, \qquad \mathbf{P}(X=3) = \frac{1}{4}.$$

Find the transform associated with X and use it to obtain the first three moments, $\mathbf{E}[X]$, $\mathbf{E}[X^2]$, $\mathbf{E}[X^3]$.

Problem 30. Calculate $\mathbf{E}[X^3]$ and $\mathbf{E}[X^4]$ for a standard normal random variable X.

Problem 31. Find the third, fourth, and fifth moments of an exponential random variable with parameter λ.

Problem 32. A nonnegative integer-valued random variable X has one of the following two expressions as its transform:

1. $M(s) = e^{2(e^{e^s-1}-1)}$.
2. $M(s) = e^{2(e^{e^s}-1)}$.

(a) Explain why one of the two cannot possibly be the transform.

(b) Use the true transform to find $\mathbf{P}(X = 0)$.

Problem 33. Find the PDF of the continuous random variable X associated with the transform

$$M(s) = \frac{1}{3} \cdot \frac{2}{2-s} + \frac{2}{3} \cdot \frac{3}{3-s}.$$

Problem 34. A soccer team has three designated players who take turns striking penalty shots. The ith player has probability of success p_i, independent of the successes of the other players. Let X be the number of successful penalty shots after each player has had one turn. Use convolution to calculate the PMF of X. Confirm your answer by first calculating the transform associated with X and then obtaining the PMF from the transform.

Problem 35. Let X be a random variable that takes nonnegative integer values, and is associated with a transform of the form

$$M_X(s) = c \cdot \frac{3 + 4e^{2s} + 2e^{3s}}{3 - e^s},$$

where c is some scalar. Find $\mathbf{E}[X]$, $p_X(1)$, and $\mathbf{E}[X \mid X \neq 0]$.

Problem 36. Let X, Y, and Z be independent random variables, where X is Bernoulli with parameter $1/3$, Y is exponential with parameter 2, and Z is Poisson with parameter 3.

(a) Consider the new random variable $U = XY + (1 - X)Z$. Find the transform associated with U.

(b) Find the transform associated with $2Z + 3$.

(c) Find the transform associated with $Y + Z$.

Problem 37. A pizza parlor serves n different types of pizza, and is visited by a number K of customers in a given period of time, where K is a nonnegative integer random variable with a known associated transform $M_K(s) = \mathbf{E}[e^{sK}]$. Each customer orders a single pizza, with all types of pizza being equally likely, independent of the number of other customers and the types of pizza they order. Give a formula, in terms of $M_K(\cdot)$, for the expected number of different types of pizzas ordered.

Problem 38.* Let X be a discrete random variable taking nonnegative integer values. Let $M(s)$ be the transform associated with X.

(a) Show that

$$\mathbf{P}(X = 0) = \lim_{s \to -\infty} M(s).$$

(b) Use part (a) to verify that if X is a binomial random variable with parameters n and p, we have $\mathbf{P}(X = 0) = (1 - p)^n$. Furthermore, if X is a Poisson random variable with parameter λ, we have $\mathbf{P}(X = 0) = e^{-\lambda}$.

(c) Suppose that X is instead known to take only integer values that are greater than or equal to a given integer $\overline{k}$. How can we calculate $P(X = \overline{k})$ using the transform associated with X?

Solution. (a) We have

$$M(s) = \sum_{k=0}^{\infty} \mathbf{P}(X = k)e^{ks}.$$

As $s \to -\infty$, all the terms e^{ks} with $k > 0$ tend to 0, so we obtain $\lim_{s\to-\infty} M(s) = \mathbf{P}(X = 0)$.

(b) In the case of the binomial, we have from the transform tables

$$M(s) = (1 - p + pe^s)^n,$$

so that $\lim_{s\to-\infty} M(s) = (1 - p)^n$. In the case of the Poisson, we have

$$M(s) = e^{\lambda(e^s - 1)},$$

so that $\lim_{s\to-\infty} M(s) = e^{-\lambda}$.

(c) The random variable $Y = X - \overline{k}$ takes only nonnegative integer values and the associated transform is $M_Y(s) = e^{-s\overline{k}}M(s)$ (cf. Example 4.25). Since $\mathbf{P}(Y = 0) = \mathbf{P}(X = \overline{k})$, we have from part (a),

$$\mathbf{P}(X = \overline{k}) = \lim_{s\to-\infty} e^{-s\overline{k}}M(s).$$

Problem 39.* Transforms associated with uniform random variables.

(a) Find the transform associated with an integer-valued random variable X that is uniformly distributed in the range $\{a, a + 1, \ldots, b\}$.

(b) Find the transform associated with a continuous random variable X that is uniformly distributed in the range $[a, b]$.

Solution. (a) The PMF of X is

$$p_X(k) = \begin{cases} \dfrac{1}{b - a + 1}, & \text{if } k = a, a + 1, \ldots, b, \\ 0, & \text{otherwise.} \end{cases}$$

The transform is

$$M(s) = \sum_{k=-\infty}^{\infty} e^{sk}\mathbf{P}(X = k)$$

$$= \sum_{k=a}^{b} \frac{1}{b-a+1} e^{sk}$$

$$= \frac{e^{sa}}{b-a+1} \sum_{k=0}^{b-a} e^{sk}$$

$$= \frac{e^{sa}}{b-a+1} \cdot \frac{e^{s(b-a+1)} - 1}{e^s - 1}.$$

(b) We have

$$M(s) = \mathbf{E}[e^{sX}] = \int_a^b \frac{e^{sx}}{b-a}\, dx = \frac{e^{sb} - e^{sa}}{s(b-a)}.$$

Problem 40.* Suppose that the transform associated with a discrete random variable X has the form

$$M(s) = \frac{A(e^s)}{B(e^s)},$$

where $A(t)$ and $B(t)$ are polynomials of the generic variable t. Assume that $A(t)$ and $B(t)$ have no common roots and that the degree of $A(t)$ is smaller than the degree of $B(t)$. Assume also that $B(t)$ has distinct, real, and nonzero roots that have absolute value greater than 1. Then it can be seen that $M(s)$ can be written in the form

$$M(s) = \frac{a_1}{1 - r_1 e^s} + \cdots + \frac{a_m}{1 - r_m e^s},$$

where $1/r_1, \ldots, 1/r_m$ are the roots of $B(t)$ and the a_i are constants that are equal to $\lim_{e^s \to \frac{1}{r_i}} (1 - r_i e^s) M(s)$, $i = 1, \ldots, m$.

(a) Show that the PMF of X has the form

$$\mathbf{P}(X = k) = \begin{cases} \displaystyle\sum_{i=1}^{m} a_i r_i^k, & \text{if } k = 0, 1, \ldots, \\ 0, & \text{otherwise.} \end{cases}$$

Note: For large k, the PMF of X can be approximated by $a_{\bar{i}} r_{\bar{i}}^k$, where $\bar{i}$ is the index corresponding to the largest $|r_i|$ (assuming $\bar{i}$ is unique).

(b) Extend the result of part (a) to the case where $M(s) = e^{bs} A(e^s)/B(e^s)$ and b is an integer.

Solution. (a) We have for all s such that $|r_i| e^s < 1$

$$\frac{1}{1 - r_i e^s} = 1 + r_i e^s + r_i^2 e^{2s} + \cdots.$$

Therefore,

$$M(s) = \sum_{i=1}^{m} a_i + \left(\sum_{i=1}^{m} a_i r_i \right) e^s + \left(\sum_{i=1}^{m} a_i r_i^2 \right) e^{2s} + \cdots,$$

and by inverting this transform, we see that

$$\mathbf{P}(X = k) = \sum_{i=1}^{m} a_i r_i^k$$

for $k \geq 0$, and $\mathbf{P}(X = k) = 0$ for $k < 0$. Note that if the coefficients a_i are nonnegative, this PMF is a mixture of geometric PMFs.

(b) In this case, $M(s)$ corresponds to the translation by b of a random variable whose transform is $A(e^s)/B(e^s)$ (cf. Example 4.25), so we have

$$\mathbf{P}(X = k) = \begin{cases} \displaystyle\sum_{i=1}^{m} a_i r_i^{(k-b)}, & \text{if } k = b, b+1, \ldots, \\ 0, & \text{otherwise.} \end{cases}$$

SECTION 4.4. Sum of a Random Number of Independent Random Variables

Problem 41. At a certain time, the number of people that enter an elevator is a Poisson random variable with parameter λ. The weight of each person is independent of every other person's weight, and is uniformly distributed between 100 and 200 lbs. Let X_i be the fraction of 100 by which the ith person exceeds 100 lbs, e.g., if the 7th person weighs 175 lbs., then $X_7 = 0.75$. Let Y be the sum of the X_i.

(a) Find the transform associated with Y.

(b) Use the transform to compute the expected value of Y.

(c) Verify your answer to part (b) by using the law of iterated expectations.

Problem 42. Construct an example to show that the sum of a random number of independent normal random variables is not normal (even though a fixed sum is).

Problem 43. A motorist goes through 4 lights, each of which is found to be red with probability 1/2. The waiting times at each light are modeled as independent normal random variables with mean 1 minute and standard deviation 1/2 minute. Let X be the total waiting time at the red lights.

(a) Use the total probability theorem to find the PDF and the transform associated with X, and the probability that X exceeds 4 minutes. Is X normal?

(b) Find the transform associated with X by viewing X as a sum of a random number of random variables.

Problem 44. Consider the calculation of the mean and variance of a sum

$$Y = X_1 + \cdots + X_N,$$

where N is itself a sum of integer-valued random variables, i.e.,

$$N = K_1 + \cdots + K_M.$$

Here N, M, $K_1, K_2, \ldots$, $X_1, X_2, \ldots$ are independent random variables, N, M, $K_1, K_2, \ldots$ are integer-valued and nonnegative, $K_1, K_2, \ldots$ are identically distributed with common mean and variance denoted $\mathbf{E}[K]$ and $\text{var}(K)$, and $X_1, X_2, \ldots$ are identically distributed with common mean and variance denoted $\mathbf{E}[X]$ and $\text{var}(X)$.

(a) Derive formulas for $\mathbf{E}[N]$ and $\text{var}(N)$ in terms of $\mathbf{E}[M]$, $\text{var}(M)$, $\mathbf{E}[K]$, $\text{var}(K)$.

(b) Derive formulas for $\mathbf{E}[Y]$ and $\text{var}(Y)$ in terms of $\mathbf{E}[M]$, $\text{var}(M)$, $\mathbf{E}[K]$, $\text{var}(K)$, $\mathbf{E}[X]$, $\text{var}(X)$.

(c) A crate contains M cartons, where M is geometrically distributed with parameter p. The ith carton contains K_i widgets, where K_i is Poisson-distributed with parameter μ. The weight of each widget is exponentially distributed with parameter λ. All these random variables are independent. Find the expected value and variance of the total weight of a crate.

Problem 45.* Use transforms to show that the sum of a Poisson-distributed number of independent, identically distributed Bernoulli random variables is Poisson.

Solution. Let N be a Poisson-distributed random variable with parameter λ. Let X_i, $i = 1, \ldots, N$, be independent Bernoulli random variables with parameter p, and let

$$L = X_1 + \cdots + X_N$$

be the corresponding sum. The transform associated with L is found by starting with the transform associated with N, which is

$$M_N(s) = e^{\lambda(e^s - 1)},$$

and replacing each occurrence of e^s by the transform associated with X_i, which is

$$M_X(s) = 1 - p + pe^s.$$

We obtain

$$M_L(s) = e^{\lambda(1-p+pe^s-1)} = e^{\lambda p(e^s-1)}.$$

This is the transform associated with a Poisson random variable with parameter λp.

5

Limit Theorems

Contents

In this chapter, we discuss some fundamental issues related to the asymptotic behavior of sequences of random variables. Our principal context involves a sequence $X_1, X_2, \ldots$ of independent identically distributed random variables with mean μ and variance σ^2. Let

$$S_n = X_1 + \cdots + X_n$$

be the sum of the first n of them. Limit theorems are mostly concerned with the properties of S_n and related random variables as n becomes very large.

Because of independence, we have

$$\text{var}(S_n) = \text{var}(X_1) + \cdots + \text{var}(X_n) = n\sigma^2.$$

Thus, the distribution of S_n spreads out as n increases, and cannot have a meaningful limit. The situation is different if we consider the **sample mean**

$$M_n = \frac{X_1 + \cdots + X_n}{n} = \frac{S_n}{n}.$$

A quick calculation yields

$$\mathbf{E}[M_n] = \mu, \qquad \text{var}(M_n) = \frac{\sigma^2}{n}.$$

In particular, the variance of M_n decreases to zero as n increases, and the bulk of the distribution of M_n must be very close to the mean μ. This phenomenon is the subject of certain laws of large numbers, which generally assert that the sample mean M_n (a random variable) converges to the true mean μ (a number), in a precise sense. These laws provide a mathematical basis for the loose interpretation of an expectation $\mathbf{E}[X] = \mu$ as the average of a large number of independent samples drawn from the distribution of X.

We will also consider a quantity which is intermediate between S_n and M_n. We first subtract $n\mu$ from S_n, to obtain the zero-mean random variable $S_n - n\mu$ and then divide by $\sigma\sqrt{n}$, to form the random variable

$$Z_n = \frac{S_n - n\mu}{\sigma\sqrt{n}}.$$

It can be seen that

$$\mathbf{E}[Z_n] = 0, \qquad \text{var}(Z_n) = 1.$$

Since the mean and the variance of Z_n remain unchanged as n increases, its distribution neither spreads, nor shrinks to a point. The **central limit theorem** is concerned with the asymptotic shape of the distribution of Z_n and asserts that it becomes the standard normal distribution.

Limit theorems are useful for several reasons:

(a) Conceptually, they provide an interpretation of expectations (as well as probabilities) in terms of a long sequence of identical independent experiments.

(b) They allow for an approximate analysis of the properties of random variables such as S_n. This is to be contrasted with an exact analysis which would require a formula for the PMF or PDF of S_n, a complicated and tedious task when n is large.

(c) They play a major role in inference and statistics, in the presence of large data sets.

5.1 MARKOV AND CHEBYSHEV INEQUALITIES

In this section, we derive some important inequalities. These inequalities use the mean and possibly the variance of a random variable to draw conclusions on the probabilities of certain events. They are primarily useful in situations where exact values or bounds for the mean and variance of a random variable X are easily computable, but the distribution of X is either unavailable or hard to calculate.

We first present the **Markov inequality**. Loosely speaking, it asserts that if a *nonnegative* random variable has a small mean, then the probability that it takes a large value must also be small.

Markov Inequality

If a random variable X can only take nonnegative values, then

$$\mathbf{P}(X \geq a) \leq \frac{\mathbf{E}[X]}{a}, \qquad \text{for all } a > 0.$$

To justify the Markov inequality, let us fix a positive number a and consider the random variable Y_a defined by

$$Y_a = \begin{cases} 0, & \text{if } X < a, \\ a, & \text{if } X \geq a. \end{cases}$$

It is seen that the relation

$$Y_a \leq X$$

always holds and therefore,

$$\mathbf{E}[Y_a] \leq \mathbf{E}[X].$$

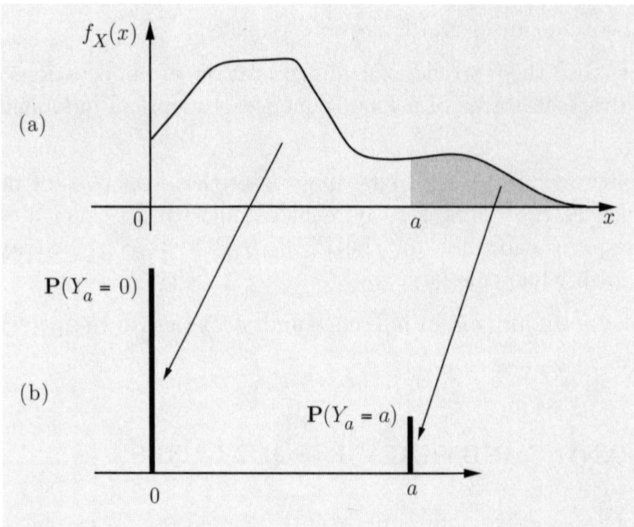

Figure 5.1: Illustration of the derivation of the Markov inequality. Part (a) of the figure shows the PDF of a nonnegative random variable X. Part (b) shows the PMF of a related random variable Y_a, which is constructed as follows. All of the probability mass in the PDF of X that lies between 0 and a is assigned to 0, and all of the mass that lies above a is assigned to a. Since mass is shifted to the left, the expectation can only decrease and, therefore,

$$\mathbf{E}[X] \geq \mathbf{E}[Y_a] = a\mathbf{P}(Y_a = a) = a\mathbf{P}(X \geq a).$$

On the other hand,

$$\mathbf{E}[Y_a] = a\mathbf{P}(Y_a = a) = a\mathbf{P}(X \geq a),$$

from which we obtain

$$a\mathbf{P}(X \geq a) \leq \mathbf{E}[X];$$

see Fig. 5.1 for an illustration.

Example 5.1. Let X be uniformly distributed in the interval $[0, 4]$ and note that $\mathbf{E}[X] = 2$. Then, the Markov inequality asserts that

$$\mathbf{P}(X \geq 2) \leq \frac{2}{2} = 1, \qquad \mathbf{P}(X \geq 3) \leq \frac{2}{3} = 0.67, \qquad \mathbf{P}(X \geq 4) \leq \frac{2}{4} = 0.5.$$

By comparing with the exact probabilities

$$\mathbf{P}(X \geq 2) = 0.5, \qquad \mathbf{P}(X \geq 3) = 0.25, \qquad \mathbf{P}(X \geq 4) = 0,$$

we see that the bounds provided by the Markov inequality can be quite loose.

We continue with the **Chebyshev inequality**. Loosely speaking, it asserts that if a random variable has small variance, then the probability that it takes a value far from its mean is also small. Note that the Chebyshev inequality does not require the random variable to be nonnegative.

Chebyshev Inequality

If X is a random variable with mean μ and variance σ^2, then

$$\mathbf{P}\big(|X-\mu| \geq c\big) \leq \frac{\sigma^2}{c^2}, \qquad \text{for all } c > 0.$$

To justify the Chebyshev inequality, we consider the nonnegative random variable $(X-\mu)^2$ and apply the Markov inequality with $a = c^2$. We obtain

$$\mathbf{P}\big((X-\mu)^2 \geq c^2\big) \leq \frac{\mathbf{E}\big[(X-\mu)^2\big]}{c^2} = \frac{\sigma^2}{c^2}.$$

We complete the derivation by observing that the event $(X-\mu)^2 \geq c^2$ is identical to the event $|X-\mu| \geq c$, so that

$$\mathbf{P}\big(|X-\mu| \geq c\big) = \mathbf{P}\big((X-\mu)^2 \geq c^2\big) \leq \frac{\sigma^2}{c^2}.$$

For a similar derivation that bypasses the Markov inequality, assume for simplicity that X is a continuous random variable, introduce the function

$$g(x) = \begin{cases} 0, & \text{if } |x-\mu| < c, \\ c^2, & \text{if } |x-\mu| \geq c, \end{cases}$$

note that $(x-\mu)^2 \geq g(x)$ for all x, and write

$$\sigma^2 = \int_{-\infty}^{\infty} (x-\mu)^2 f_X(x)\,dx \geq \int_{-\infty}^{\infty} g(x)\, f_X(x)\,dx = c^2\,\mathbf{P}\big(|X-\mu| \geq c\big),$$

which is the Chebyshev inequality.

An alternative form of the Chebyshev inequality is obtained by letting $c = k\sigma$, where k is positive, which yields

$$\mathbf{P}\big(|X-\mu| \geq k\sigma\big) \leq \frac{\sigma^2}{k^2\sigma^2} = \frac{1}{k^2}.$$

Thus, the probability that a random variable takes a value more than k standard deviations away from its mean is at most $1/k^2$.

The Chebyshev inequality tends to be more powerful than the Markov inequality (the bounds that it provides are more accurate), because it also uses information on the variance of X. Still, the mean and the variance of a random variable are only a rough summary of its properties, and we cannot expect the bounds to be close approximations of the exact probabilities.

Example 5.2. As in Example 5.1, let X be uniformly distributed in $[0,4]$. Let us use the Chebyshev inequality to bound the probability that $|X-2| \geq 1$. We have $\sigma^2 = 16/12 = 4/3$, and

$$\mathbf{P}\big(|X-2| \geq 1\big) \leq \frac{4}{3},$$

which is uninformative.

For another example, let X be exponentially distributed with parameter $\lambda = 1$, so that $\mathbf{E}[X] = \text{var}(X) = 1$. For $c > 1$, using the Chebyshev inequality, we obtain

$$\mathbf{P}(X \geq c) = \mathbf{P}(X - 1 \geq c - 1) \leq \mathbf{P}\big(|X-1| \geq c-1\big) \leq \frac{1}{(c-1)^2}.$$

This is again conservative compared to the exact answer $\mathbf{P}(X \geq c) = e^{-c}$.

Example 5.3. Upper Bounds in the Chebyshev Inequality. When X is known to take values in a range $[a,b]$, we claim that $\sigma^2 \leq (b-a)^2/4$. Thus, if σ^2 is unknown, we may use the bound $(b-a)^2/4$ in place of σ^2 in the Chebyshev inequality, and obtain

$$\mathbf{P}\big(|X-\mu| \geq c\big) \leq \frac{(b-a)^2}{4c^2}, \qquad \text{for all } c > 0.$$

To verify our claim, note that for any constant γ, we have

$$\mathbf{E}\big[(X-\gamma)^2\big] = \mathbf{E}[X^2] - 2\mathbf{E}[X]\gamma + \gamma^2,$$

and the above quadratic is minimized when $\gamma = \mathbf{E}[X]$. It follows that

$$\sigma^2 = \mathbf{E}\Big[\big(X - \mathbf{E}[X]\big)^2\Big] \leq \mathbf{E}\big[(X-\gamma)^2\big], \qquad \text{for all } \gamma.$$

By letting $\gamma = (a+b)/2$, we obtain

$$\sigma^2 \leq \mathbf{E}\left[\left(X - \frac{a+b}{2}\right)^2\right] = \mathbf{E}\big[(X-a)(X-b)\big] + \frac{(b-a)^2}{4} \leq \frac{(b-a)^2}{4},$$

where the equality above is verified by straightforward calculation, and the last inequality follows from the fact

$$(x-a)(x-b) \leq 0$$

for all x in the range $[a,b]$.

The bound $\sigma^2 \leq (b-a)^2/4$ may be quite conservative, but in the absence of further information about X, it cannot be improved. It is satisfied with equality when X is the random variable that takes the two extreme values a and b with equal probability $1/2$.

5.2 THE WEAK LAW OF LARGE NUMBERS

The weak law of large numbers asserts that the sample mean of a large number of independent identically distributed random variables is very close to the true mean, with high probability.

As in the introduction to this chapter, we consider a sequence $X_1, X_2, \ldots$ of independent identically distributed random variables with mean μ and variance σ^2, and define the sample mean by

$$M_n = \frac{X_1 + \cdots + X_n}{n}.$$

We have

$$\mathbf{E}[M_n] = \frac{\mathbf{E}[X_1] + \cdots + \mathbf{E}[X_n]}{n} = \frac{n\mu}{n} = \mu,$$

and, using independence,

$$\text{var}(M_n) = \frac{\text{var}(X_1 + \cdots + X_n)}{n^2} = \frac{\text{var}(X_1) + \cdots + \text{var}(X_n)}{n^2} = \frac{n\sigma^2}{n^2} = \frac{\sigma^2}{n}.$$

We apply the Chebyshev inequality and obtain

$$\mathbf{P}\big(|M_n - \mu| \geq \epsilon\big) \leq \frac{\sigma^2}{n\epsilon^2}, \qquad \text{for any } \epsilon > 0.$$

We observe that for any fixed $\epsilon > 0$, the right-hand side of this inequality goes to zero as n increases. As a consequence, we obtain the weak law of large numbers, which is stated below. It turns out that this law remains true even if the X_i have infinite variance, but a much more elaborate argument is needed, which we omit. The only assumption needed is that $\mathbf{E}[X_i]$ is well-defined.

The Weak Law of Large Numbers

Let $X_1, X_2, \ldots$ be independent identically distributed random variables with mean μ. For every $\epsilon > 0$, we have

$$\mathbf{P}\big(|M_n - \mu| \geq \epsilon\big) = \mathbf{P}\left(\left|\frac{X_1 + \cdots + X_n}{n} - \mu\right| \geq \epsilon\right) \to 0, \qquad \text{as } n \to \infty.$$

The weak law of large numbers states that for large n, the bulk of the distribution of M_n is concentrated near μ. That is, if we consider a positive length interval $[\mu - \epsilon, \mu + \epsilon]$ around μ, then there is high probability that M_n will fall in that interval; as $n \to \infty$, this probability converges to 1. Of course, if ϵ is very small, we may have to wait longer (i.e., need a larger value of n) before we can assert that M_n is highly likely to fall in that interval.

Example 5.4. Probabilities and Frequencies. Consider an event A defined in the context of some probabilistic experiment. Let $p = \mathbf{P}(A)$ be the probability of this event. We consider n independent repetitions of the experiment, and let M_n be the fraction of time that event A occurs; in this context, M_n is often called the **empirical frequency** of A. Note that

$$M_n = \frac{X_1 + \cdots + X_n}{n},$$

where X_i is 1 whenever A occurs, and 0 otherwise; in particular, $\mathbf{E}[X_i] = p$. The weak law applies and shows that when n is large, the empirical frequency is most likely to be within ϵ of p. Loosely speaking, this allows us to conclude that empirical frequencies are faithful estimates of p. Alternatively, this is a step towards interpreting the probability p as the frequency of occurrence of A.

Example 5.5. Polling. Let p be the fraction of voters who support a particular candidate for office. We interview n "randomly selected" voters and record M_n, the fraction of them that support the candidate. We view M_n as our estimate of p and would like to investigate its properties.

We interpret "randomly selected" to mean that the n voters are chosen independently and uniformly from the given population. Thus, the reply of each person interviewed can be viewed as an independent Bernoulli random variable X_i with success probability p and variance $\sigma^2 = p(1-p)$. The Chebyshev inequality yields

$$\mathbf{P}\big(|M_n - p| \geq \epsilon\big) \leq \frac{p(1-p)}{n\epsilon^2}.$$

The true value of the parameter p is assumed to be unknown. On the other hand, it may be verified that $p(1-p) \leq 1/4$ (cf. Example 5.3), which yields

$$\mathbf{P}\big(|M_n - p| \geq \epsilon\big) \leq \frac{1}{4n\epsilon^2}.$$

For example, if $\epsilon = 0.1$ and $n = 100$, we obtain

$$\mathbf{P}\big(|M_{100} - p| \geq 0.1\big) \leq \frac{1}{4 \cdot 100 \cdot (0.1)^2} = 0.25.$$

In words, with a sample size of $n = 100$, the probability that our estimate is incorrect by more than 0.1 is no larger than 0.25.

Suppose now that we impose some tight specifications on our poll. We would like to have high confidence (probability at least 95%) that our estimate will be very accurate (within .01 of p). How many voters should be sampled?

The only guarantee that we have at this point is the inequality

$$\mathbf{P}\big(|M_n - p| \geq 0.01\big) \leq \frac{1}{4n(0.01)^2}.$$

We will be sure to satisfy the above specifications if we choose n large enough so that

$$\frac{1}{4n(0.01)^2} \leq 1 - 0.95 = 0.05,$$

which yields $n \geq 50{,}000$. This choice of n satisfies our specifications, but turns out to be fairly conservative, because it is based on the rather loose Chebyshev inequality. A refinement will be considered in Section 5.4.

5.3 CONVERGENCE IN PROBABILITY

We can interpret the weak law of large numbers as stating that "M_n converges to μ." However, since $M_1, M_2, \ldots$ is a sequence of random variables, not a sequence of numbers, the meaning of convergence has to be made precise. A particular definition is provided below. To facilitate the comparison with the ordinary notion of convergence, we also include the definition of the latter.

Convergence of a Deterministic Sequence

Let $a_1, a_2, \ldots$ be a sequence of real numbers, and let a be another real number. We say that the sequence a_n converges to a, or $\lim_{n\to\infty} a_n = a$, if for every $\epsilon > 0$ there exists some n_0 such that

$$|a_n - a| \le \epsilon, \qquad \text{for all } n \ge n_0.$$

Intuitively, if $\lim_{n\to\infty} a_n = a$, then for any given accuracy level ϵ, a_n must be within ϵ of a, when n is large enough.

Convergence in Probability

Let $Y_1, Y_2, \ldots$ be a sequence of random variables (not necessarily independent), and let a be a real number. We say that the sequence Y_n **converges to a in probability**, if for every $\epsilon > 0$, we have

$$\lim_{n\to\infty} \mathbf{P}\big(|Y_n - a| \ge \epsilon\big) = 0.$$

Given this definition, the weak law of large numbers simply states that the sample mean converges in probability to the true mean μ. More generally, the Chebyshev inequality implies that if all Y_n have the same mean μ, and $\text{var}(Y_n)$ converges to 0, then Y_n converges to μ in probability.

If the random variables $Y_1, Y_2, \ldots$ have a PMF or a PDF and converge in probability to a, then according to the above definition, "almost all" of the PMF or PDF of Y_n is concentrated within ϵ of a for large values of n. It is also instructive to rephrase the above definition as follows: for every $\epsilon > 0$, and for every $\delta > 0$, there exists some n_0 such that

$$\mathbf{P}\big(|Y_n - a| \ge \epsilon\big) \le \delta, \qquad \text{for all } n \ge n_0.$$

If we refer to ϵ as the *accuracy* level, and δ as the *confidence* level, the definition takes the following intuitive form: for any given level of accuracy and confidence,

Y_n will be equal to a, within these levels of accuracy and confidence, provided that n is large enough.

Example 5.6. Consider a sequence of independent random variables X_n that are uniformly distributed in the interval $[0, 1]$, and let

$$Y_n = \min\{X_1, \ldots, X_n\}.$$

The sequence of values of Y_n cannot increase as n increases, and it will occasionally decrease (whenever a value of X_n that is smaller than the preceding values is obtained). Thus, we intuitively expect that Y_n converges to zero. Indeed, for $\epsilon > 0$, we have using the independence of the X_n,

$$\begin{aligned} \mathbf{P}\big(|Y_n - 0| \geq \epsilon\big) &= \mathbf{P}(X_1 \geq \epsilon, \ldots, X_n \geq \epsilon) \\ &= \mathbf{P}(X_1 \geq \epsilon) \cdots \mathbf{P}(X_n \geq \epsilon) \\ &= (1 - \epsilon)^n. \end{aligned}$$

In particular,

$$\lim_{n \to \infty} \mathbf{P}\big(|Y_n - 0| \geq \epsilon\big) = \lim_{n \to \infty} (1 - \epsilon)^n = 0.$$

Since this is true for every $\epsilon > 0$, we conclude that Y_n converges to zero, in probability.

Example 5.7. Let Y be an exponentially distributed random variable with parameter $\lambda = 1$. For any positive integer n, let $Y_n = Y/n$. (Note that these random variables are dependent.) We wish to investigate whether the sequence Y_n converges to zero.

For $\epsilon > 0$, we have

$$\mathbf{P}\big(|Y_n - 0| \geq \epsilon\big) = \mathbf{P}(Y_n \geq \epsilon) = \mathbf{P}(Y \geq n\epsilon) = e^{-n\epsilon}.$$

In particular,

$$\lim_{n \to \infty} \mathbf{P}\big(|Y_n - 0| \geq \epsilon\big) = \lim_{n \to \infty} e^{-n\epsilon} = 0.$$

Since this is the case for every $\epsilon > 0$, Y_n converges to zero, in probability.

One might be tempted to believe that if a sequence Y_n converges to a number a, then $\mathbf{E}[Y_n]$ must also converge to a. The following example shows that this need not be the case, and illustrates some of the limitations of the notion of convergence in probability.

Example 5.8. Consider a sequence of discrete random variables Y_n with the following distribution:

$$\mathbf{P}(Y_n = y) = \begin{cases} 1 - \dfrac{1}{n}, & \text{for } y = 0, \\ \dfrac{1}{n}, & \text{for } y = n^2, \\ 0, & \text{elsewhere;} \end{cases}$$

see Fig. 5.2 for an illustration. For every $\epsilon > 0$, we have

$$\lim_{n\to\infty} \mathbf{P}\big(|Y_n| \geq \epsilon\big) = \lim_{n\to\infty} \frac{1}{n} = 0,$$

and Y_n converges to zero in probability. On the other hand, $\mathbf{E}[Y_n] = n^2/n = n$, which goes to infinity as n increases.

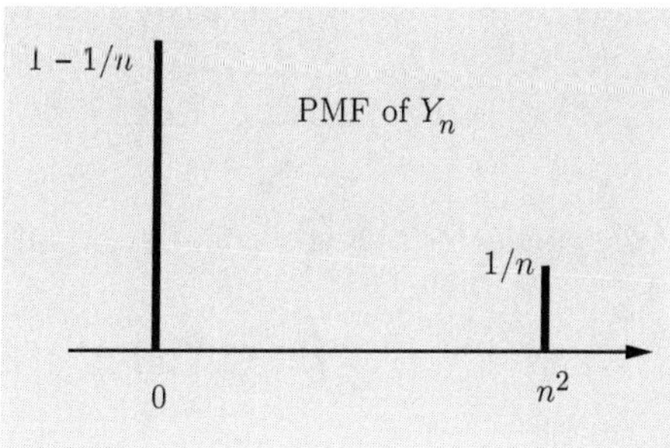

Figure 5.2: The PMF of the random variable Y_n in Example 5.8.

5.4 THE CENTRAL LIMIT THEOREM

According to the weak law of large numbers, the distribution of the sample mean $M_n = (X_1 + \cdots + X_n)/n$ is increasingly concentrated in the near vicinity of the true mean μ. In particular, its variance tends to zero. On the other hand, the variance of the sum

$$S_n = X_1 + \cdots + X_n = nM_n$$

increases to infinity, and the distribution of S_n cannot be said to converge to anything meaningful. An intermediate view is obtained by considering the deviation $S_n - n\mu$ of S_n from its mean $n\mu$, and scaling it by a factor proportional to $1/\sqrt{n}$. What is special about this particular scaling is that it keeps the variance at a constant level. The central limit theorem asserts that the distribution of this scaled random variable approaches a normal distribution.

More precisely, let $X_1, X_2, \ldots$ be a sequence of independent identically distributed random variables with mean μ and variance σ^2. We define

$$Z_n = \frac{S_n - n\mu}{\sigma\sqrt{n}} = \frac{X_1 + \cdots + X_n - n\mu}{\sigma\sqrt{n}}.$$

An easy calculation yields

$$\mathbf{E}[Z_n] = \frac{\mathbf{E}[X_1 + \cdots + X_n] - n\mu}{\sigma\sqrt{n}} = 0,$$

and

$$\text{var}(Z_n) = \frac{\text{var}(X_1 + \cdots + X_n)}{\sigma^2 n} = \frac{\text{var}(X_1) + \cdots + \text{var}(X_n)}{\sigma^2 n} = \frac{n\sigma^2}{n\sigma^2} = 1.$$

The Central Limit Theorem

Let $X_1, X_2, \ldots$ be a sequence of independent identically distributed random variables with common mean μ and variance σ^2, and define

$$Z_n = \frac{X_1 + \cdots + X_n - n\mu}{\sigma\sqrt{n}}.$$

Then, the CDF of Z_n converges to the standard normal CDF

$$\Phi(z) = \frac{1}{\sqrt{2\pi}} \int_{-\infty}^{z} e^{-x^2/2}\, dx,$$

in the sense that

$$\lim_{n\to\infty} \mathbf{P}(Z_n \leq z) = \Phi(z), \qquad \text{for every } z.$$

The central limit theorem is surprisingly general. Besides independence, and the implicit assumption that the mean and variance are finite, it places no other requirement on the distribution of the X_i, which could be discrete, continuous, or mixed; see the end-of-chapter problems for an outline of its proof.

This theorem is of tremendous importance for several reasons, both conceptual and practical. On the conceptual side, it indicates that the sum of a large number of independent random variables is approximately normal. As such, it applies to many situations in which a random effect is the sum of a large number of small but independent random factors. Noise in many natural or engineered systems has this property. In a wide array of contexts, it has been found empirically that the statistics of noise are well-described by normal distributions, and the central limit theorem provides a convincing explanation for this phenomenon.

On the practical side, the central limit theorem eliminates the need for detailed probabilistic models, and for tedious manipulations of PMFs and PDFs. Rather, it allows the calculation of certain probabilities by simply referring to the normal CDF table. Furthermore, these calculations only require the knowledge of means and variances.

Approximations Based on the Central Limit Theorem

The central limit theorem allows us to calculate probabilities related to Z_n as if Z_n were normal. Since normality is preserved under linear transformations, this is equivalent to treating S_n as a normal random variable with mean $n\mu$ and variance $n\sigma^2$.

Normal Approximation Based on the Central Limit Theorem

Let $S_n = X_1 + \cdots + X_n$, where the X_i are independent identically distributed random variables with mean μ and variance σ^2. If n is large, the probability $\mathbf{P}(S_n \leq c)$ can be approximated by treating S_n as if it were normal, according to the following procedure.

1. Calculate the mean $n\mu$ and the variance $n\sigma^2$ of S_n.
2. Calculate the normalized value $z = (c - n\mu)/\sigma\sqrt{n}$.
3. Use the approximation

$$\mathbf{P}(S_n \leq c) \approx \Phi(z),$$

where $\Phi(z)$ is available from standard normal CDF tables.

Example 5.9. We load on a plane 100 packages whose weights are independent random variables that are uniformly distributed between 5 and 50 pounds. What is the probability that the total weight will exceed 3000 pounds? It is not easy to calculate the CDF of the total weight and the desired probability, but an approximate answer can be quickly obtained using the central limit theorem.

We want to calculate $\mathbf{P}(S_{100} > 3000)$, where S_{100} is the sum of the weights of 100 packages. The mean and the variance of the weight of a single package are

$$\mu = \frac{5 + 50}{2} = 27.5, \qquad \sigma^2 = \frac{(50 - 5)^2}{12} = 168.75,$$

based on the formulas for the mean and variance of the uniform PDF. We calculate the normalized value

$$z = \frac{3000 - 100 \cdot 27.5}{\sqrt{168.75 \cdot 100}} = \frac{250}{129.9} = 1.92,$$

and use the standard normal tables to obtain the approximation

$$\mathbf{P}(S_{100} \leq 3000) \approx \Phi(1.92) = 0.9726.$$

Thus, the desired probability is

$$\mathbf{P}(S_{100} > 3000) = 1 - \mathbf{P}(S_{100} \leq 3000) \approx 1 - 0.9726 = 0.0274.$$

Example 5.10. A machine processes parts, one at a time. The processing times of different parts are independent random variables, uniformly distributed in $[1, 5]$.

We wish to approximate the probability that the number of parts processed within 320 time units, denoted by N_{320}, is at least 100.

There is no obvious way of expressing the random variable N_{320} as a sum of independent random variables, but we can proceed differently. Let X_i be the processing time of the ith part, and let $S_{100} = X_1 + \cdots + X_{100}$ be the total processing time of the first 100 parts. The event $\{N_{320} \geq 100\}$ is the same as the event $\{S_{100} \leq 320\}$, and we can now use a normal approximation to the distribution of S_{100}. Note that $\mu = \mathbf{E}[X_i] = 3$ and $\sigma^2 = \text{var}(X_i) = 16/12 = 4/3$. We calculate the normalized value

$$z = \frac{320 - n\mu}{\sigma\sqrt{n}} = \frac{320 - 300}{\sqrt{100 \cdot 4/3}} = 1.73,$$

and use the approximation

$$\mathbf{P}(S_{100} \leq 320) \approx \Phi(1.73) = 0.9582.$$

If the variance of the X_i is unknown, but an upper bound is available, the normal approximation can be used to obtain bounds on the probabilities of interest.

Example 5.11. Polling. Let us revisit the polling problem in Example 5.5. We poll n voters and record the fraction M_n of those polled who are in favor of a particular candidate. If p is the fraction of the entire voter population that supports this candidate, then

$$M_n = \frac{X_1 + \cdots + X_n}{n},$$

where the X_i are independent Bernoulli random variables with parameter p. In particular, M_n has mean p and variance $p(1-p)/n$. By the normal approximation, $X_1 + \cdots + X_n$ is approximately normal, and therefore M_n is also approximately normal.

We are interested in the probability $\mathbf{P}(|M_n - p| \geq \epsilon)$ that the polling error is larger than some desired accuracy ϵ. Because of the symmetry of the normal PDF around the mean, we have

$$\mathbf{P}\big(|M_n - p| \geq \epsilon\big) \approx 2\mathbf{P}(M_n - p \geq \epsilon).$$

The variance $p(1-p)/n$ of $M_n - p$ depends on p and is therefore unknown. We note that the probability of a large deviation from the mean increases with the variance. Thus, we can obtain an upper bound on $\mathbf{P}(M_n - p \geq \epsilon)$ by assuming that $M_n - p$ has the largest possible variance, namely, $1/(4n)$ which corresponds to $p = 1/2$. To calculate this upper bound, we evaluate the standardized value

$$z = \frac{\epsilon}{1/(2\sqrt{n})},$$

and use the normal approximation

$$\mathbf{P}\big(M_n - p \geq \epsilon\big) \leq 1 - \Phi(z) = 1 - \Phi\big(2\epsilon\sqrt{n}\big).$$

For instance, consider the case where $n = 100$ and $\epsilon = 0.1$. Assuming the worst-case variance, and treating M_n as if it were normal, we obtain

$$\begin{aligned}\mathbf{P}\big(|M_{100} - p| \geq 0.1\big) &\approx 2\mathbf{P}(M_n - p \geq 0.1) \\ &\leq 2 - 2\Phi\big(2 \cdot 0.1 \cdot \sqrt{100}\big) = 2 - 2\Phi(2) = 2 - 2 \cdot 0.977 = 0.046.\end{aligned}$$

This is much smaller (and more accurate) than the estimate of 0.25 that was obtained in Example 5.5 using the Chebyshev inequality.

We now consider a reverse problem. How large a sample size n is needed if we wish our estimate M_n to be within 0.01 of p with probability at least 0.95? Assuming again the worst possible variance, we are led to the condition

$$2 - 2\Phi\big(2 \cdot 0.01 \cdot \sqrt{n}\big) \leq 0.05,$$

or

$$\Phi\big(2 \cdot 0.01 \cdot \sqrt{n}\big) \geq 0.975.$$

From the normal tables, we see that $\Phi(1.96) = 0.975$, which leads to

$$2 \cdot 0.01 \cdot \sqrt{n} \geq 1.96,$$

or

$$n \geq \frac{(1.96)^2}{4 \cdot (0.01)^2} = 9604.$$

This is significantly better than the sample size of 50,000 that we found using Chebyshev's inequality.

The normal approximation is increasingly accurate as n tends to infinity, but in practice we are generally faced with specific and finite values of n. It would be useful to know how large n should be before the approximation can be trusted, but there are no simple and general guidelines. Much depends on whether the distribution of the X_i is close to normal and, in particular, whether it is symmetric. For example, if the X_i are uniform, then S_8 is already very close to normal. But if the X_i are, say, exponential, a significantly larger n will be needed before the distribution of S_n is close to a normal one. Furthermore, the normal approximation to $\mathbf{P}(S_n \leq c)$ tends to be more faithful when c is in the vicinity of the mean of S_n.

De Moivre-Laplace Approximation to the Binomial

A binomial random variable S_n with parameters n and p can be viewed as the sum of n independent Bernoulli random variables $X_1, \ldots, X_n$, with common parameter p:

$$S_n = X_1 + \cdots + X_n.$$

Recall that

$$\mu = \mathbf{E}[X_i] = p, \qquad \sigma = \sqrt{\text{var}(X_i)} = \sqrt{p(1-p)},$$

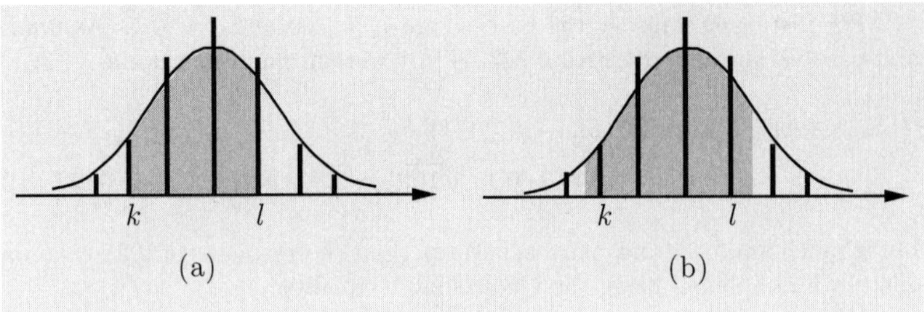

Figure 5.3: The central limit approximation treats a binomial random variable S_n as if it were normal with mean np and variance $np(1-p)$. This figure shows a binomial PMF together with the approximating normal PDF. (a) A first approximation of a binomial probability $\mathbf{P}(k \le S_n \le l)$ is obtained by integrating the area under the normal PDF from k to l, which is the shaded area in the figure. With this approach, if we have $k = l$, the probability $\mathbf{P}(S_n = k)$ will be approximated by zero. (b) A possible remedy is to use the normal probability between $k - \frac{1}{2}$ and $k + \frac{1}{2}$ to approximate $\mathbf{P}(S_n = k)$. By extending this idea, $\mathbf{P}(k \le S_n \le l)$ can be approximated by using the area under the normal PDF from $k - \frac{1}{2}$ to $l + \frac{1}{2}$, which corresponds to the shaded area.

We will now use the approximation suggested by the central limit theorem to provide an approximation for the probability of the event $\{k \le S_n \le l\}$, where k and l are given integers. We express the event of interest in terms of a standardized random variable, using the equivalence

$$k \le S_n \le l \qquad \Longleftrightarrow \qquad \frac{k - np}{\sqrt{np(1-p)}} \le \frac{S_n - np}{\sqrt{np(1-p)}} \le \frac{l - np}{\sqrt{np(1-p)}}.$$

By the central limit theorem, $(S_n - np)/\sqrt{np(1-p)}$ has approximately a standard normal distribution, and we obtain

$$\begin{aligned} \mathbf{P}(k \le S_n \le l) &= \mathbf{P}\left(\frac{k - np}{\sqrt{np(1-p)}} \le \frac{S_n - np}{\sqrt{np(1-p)}} \le \frac{l - np}{\sqrt{np(1-p)}}\right) \\ &\approx \Phi\left(\frac{l - np}{\sqrt{np(1-p)}}\right) - \Phi\left(\frac{k - np}{\sqrt{np(1-p)}}\right). \end{aligned}$$

An approximation of this form is equivalent to treating S_n as a normal random variable with mean np and variance $np(1-p)$. Figure 5.3 provides an illustration and indicates that a more accurate approximation may be possible if we replace k and l by $k - \frac{1}{2}$ and $l + \frac{1}{2}$, respectively. The corresponding formula is given below.

De Moivre-Laplace Approximation to the Binomial

If S_n is a binomial random variable with parameters n and p, n is large, and k, l are nonnegative integers, then

$$\mathbf{P}(k \le S_n \le l) \approx \Phi\left(\frac{l + \frac{1}{2} - np}{\sqrt{np(1-p)}}\right) - \Phi\left(\frac{k - \frac{1}{2} - np}{\sqrt{np(1-p)}}\right).$$

When p is close to $1/2$, in which case the PMF of the X_i is symmetric, the above formula yields a very good approximation for n as low as 40 or 50. When p is near 1 or near 0, the quality of the approximation drops, and a larger value of n is needed to maintain the same accuracy.

Example 5.12. Let S_n be a binomial random variable with parameters $n = 36$ and $p = 0.5$. An exact calculation yields

$$\mathbf{P}(S_n \le 21) = \sum_{k=0}^{21} \binom{36}{k} (0.5)^{36} = 0.8785.$$

The central limit theorem approximation, without the above discussed refinement, yields

$$\mathbf{P}(S_n \le 21) \approx \Phi\left(\frac{21 - np}{\sqrt{np(1-p)}}\right) = \Phi\left(\frac{21 - 18}{3}\right) = \Phi(1) = 0.8413.$$

Using the proposed refinement, we have

$$\mathbf{P}(S_n \le 21) \approx \Phi\left(\frac{21.5 - np}{\sqrt{np(1-p)}}\right) = \Phi\left(\frac{21.5 - 18}{3}\right) = \Phi(1.17) = 0.879,$$

which is much closer to the exact value.

The de Moivre-Laplace formula also allows us to approximate the probability of a single value. For example,

$$\mathbf{P}(S_n = 19) \approx \Phi\left(\frac{19.5 - 18}{3}\right) - \Phi\left(\frac{18.5 - 18}{3}\right) = 0.6915 - 05675 = 0.124.$$

This is very close to the exact value which is

$$\binom{36}{19} (0.5)^{36} = 0.1251.$$

5.5 THE STRONG LAW OF LARGE NUMBERS

The strong law of large numbers is similar to the weak law in that it also deals with the convergence of the sample mean to the true mean. It is different, however, because it refers to another type of convergence.

The following is a general statement of the strong law of large numbers. A proof of the strong law, under the mildly restrictive assumption that the X_i have finite fourth moments is developed in the end-of-chapter problems.

The Strong Law of Large Numbers

Let $X_1, X_2, \ldots$ be a sequence of independent identically distributed random variables with mean μ. Then, the sequence of sample means $M_n = (X_1 + \cdots + X_n)/n$ converges to μ, **with probability 1**, in the sense that

$$\mathbf{P}\left(\lim_{n\to\infty} \frac{X_1 + \cdots + X_n}{n} = \mu\right) = 1.$$

In order to interpret the strong law of large numbers, we need to go back to our original description of probabilistic models in terms of sample spaces. The contemplated experiment is infinitely long and generates a sequence of values, one value for each one of the random variables in the sequence $X_1, X_2, \ldots$. Thus, it is best to think of the sample space as a set of infinite sequences $(x_1, x_2, \ldots)$ of real numbers: any such sequence is a possible outcome of the experiment. Let us now consider the set A consisting of those sequences $(x_1, x_2, \ldots)$ whose long-term average is μ, i.e.,

$$(x_1, x_2, \ldots) \in A \qquad \Longleftrightarrow \qquad \lim_{n\to\infty} \frac{x_1 + \cdots + x_n}{n} = \mu.$$

The strong law of large numbers states that all of the probability is concentrated on this particular subset of the sample space. Equivalently, the collection of outcomes that do not belong to A (infinite sequences whose long-term average is not μ) has probability zero.

The difference between the weak and the strong law is subtle and deserves close scrutiny. The weak law states that the probability $\mathbf{P}\big(|M_n - \mu| \geq \epsilon\big)$ of a significant deviation of M_n from μ goes to zero as $n \to \infty$. Still, for any finite n, this probability can be positive and it is conceivable that once in a while, even if infrequently, M_n deviates significantly from μ. The weak law provides no conclusive information on the number of such deviations, but the strong law does. According to the strong law, and with probability 1, M_n converges to μ. This implies that for any given $\epsilon > 0$, the probability that the difference $|M_n - \mu|$ will exceed ϵ an infinite number of times is equal to zero.

Example 5.13. Probabilities and Frequencies. As in Example 5.4, consider an event A defined in terms of some probabilistic experiment. We consider a sequence of independent repetitions of the same experiment, and let M_n be the fraction of the first n repetitions in which A occurs. The strong law of large numbers asserts that M_n converges to $\mathbf{P}(A)$, with probability 1. In contrast, the weak law of large numbers asserts that M_n converges to $\mathbf{P}(A)$ in probability (cf. Example 5.4).

We have often talked intuitively about the probability of an event A as the frequency with which it occurs in an infinitely long sequence of independent trials. The strong law backs this intuition and establishes that the long-term frequency of occurrence of A is indeed equal to $\mathbf{P}(A)$, with essential certainty (the probability of this happening is 1).

Convergence with Probability 1

The convergence concept behind the strong law is different than the notion employed in the weak law. We provide here a definition and some discussion of this new convergence concept.

Convergence with Probability 1

Let $Y_1, Y_2, \ldots$ be a sequence of random variables (not necessarily independent). Let c be a real number. We say that Y_n converges to c **with probability 1** (or **almost surely**) if

$$\mathbf{P}\Big(\lim_{n\to\infty} Y_n = c\Big) = 1.$$

Similar to our earlier discussion, a proper interpretation of this type of convergence involves a sample space consisting of infinite sequences: all of the probability is concentrated on those sequences that converge to c. This does not mean that other sequences are impossible, only that they are extremely unlikely, in the sense that their total probability is zero.

Example 5.14. Let $X_1, X_2, \ldots$ be a sequence of independent random variables that are uniformly distributed in $[0, 1]$, and let $Y_n = \min\{X_1, \ldots, X_n\}$. We wish to show that Y_n converges to 0, with probability 1.

In any execution of the experiment, the sequence Y_n is nonincreasing, i.e., $Y_{n+1} \leq Y_n$ for all n. Since this sequence is bounded below by zero, it must have a limit, which we denote by Y. Let us fix some $\epsilon > 0$. We have $Y \geq \epsilon$ if and only if $X_i \geq \epsilon$ for all i, which implies that

$$\mathbf{P}(Y \geq \epsilon) = \mathbf{P}(X_1 \geq \epsilon, \ldots, X_n \geq \epsilon) = (1-\epsilon)^n.$$

Since this is true for all n, we must have

$$\mathbf{P}(Y \geq \epsilon) \leq \lim_{n\to\infty} (1-\epsilon)^n = 0.$$

This shows that $\mathbf{P}(Y \geq \epsilon) = 0$, for any positive ϵ. We conclude that $\mathbf{P}(Y > 0) = 0$, which implies that $\mathbf{P}(Y = 0) = 1$. Since Y is the limit of Y_n, we see that Y_n converges to zero with probability 1.

Convergence with probability 1 implies convergence in probability (see the end-of-chapter problems), but the converse is not necessarily true. Our last example illustrates the difference between convergence in probability and convergence with probability 1.

Example 5.15. Consider a discrete-time arrival process. The set of times is partitioned into consecutive intervals of the form $I_k = \{2^k, 2^k + 1, \ldots, 2^{k+1} - 1\}$. Note that the length of I_k is 2^k, which increases with k. During each interval I_k, there is exactly one arrival, and all times within an interval are equally likely. The arrival times within different intervals are assumed to be independent. Let us define $Y_n = 1$ if there is an arrival at time n, and $Y_n = 0$ if there is no arrival.

We have $\mathbf{P}(Y_n \neq 0) = 1/2^k$, if $n \in I_k$. Note that as n increases, it belongs to intervals I_k with increasingly large indices k. Consequently,

$$\lim_{n\to\infty} \mathbf{P}(Y_n \neq 0) = \lim_{k\to\infty} \frac{1}{2^k} = 0,$$

and we conclude that Y_n converges to 0 in probability. However, when we carry out the experiment, the total number of arrivals is infinite (one arrival during each interval I_k). Therefore, Y_n is unity for infinitely many values of n, the event $\{\lim_{n\to\infty} Y_n = 0\}$ has zero probability, and we do not have convergence with probability 1.

Intuitively, the following is happening. At any given time, there is only a small, and diminishing with n, probability of a substantial deviation from 0, which implies convergence in probability. On the other hand, given enough time, a substantial deviation from 0 is certain to occur and for this reason, we do not have convergence with probability 1.

5.6 SUMMARY AND DISCUSSION

In this chapter, we explored some fundamental aspects of probability theory that have major conceptual and practical implications. On the conceptual side, they put on a firm ground the interpretation of probability as relative frequency in a large number of independent trials. On the practical side, they allow the approximate calculation of probabilities in models that involve sums of independent random variables and that would be too hard to compute with other means. We will see a wealth of applications in the chapter on statistical inference.

We discussed three major laws that take the form of limit theorems.

(a) The first one, the weak law of large numbers, indicates that the sample mean is very likely to be close to the true mean, as the sample size increases. It is based on the Chebyshev inequality, which is of independent interest and is representative of a large collection of useful inequalities that permeate probability theory.

(b) The second one, the central limit theorem, is one of the most remarkable results of probability theory, and asserts that the sum of a large number of independent random variables is approximately normal. The central limit theorem finds many applications: it is one of the principal tools of statistical analysis and also justifies the use of normal random variables in modeling a wide array of situations.

(c) The third one, the strong law of large numbers, makes a more emphatic connection of probabilities and relative frequencies, and is often an important tool in theoretical studies.

While developing the various limit theorems, we introduced a number of convergence concepts (convergence in probability and convergence with probability 1), which provide a precise language for discussing convergence in probabilistic models. The limit theorems and the convergence concepts discussed in this chapter underlie several more advanced topics in the study of probabilistic models and stochastic processes.

PROBLEMS

SECTION 5.1. Some Useful Inequalities

Problem 1. A statistician wants to estimate the mean height h (in meters) of a population, based on n independent samples $X_1, \ldots, X_n$, chosen uniformly from the entire population. He uses the sample mean $M_n = (X_1 + \cdots + X_n)/n$ as the estimate of h, and a rough guess of 1.0 meters for the standard deviation of the samples X_i.

(a) How large should n be so that the standard deviation of M_n is at most 1 centimeter?

(b) How large should n be so that Chebyshev's inequality guarantees that the estimate is within 5 centimeters from h, with probability at least 0.99?

(c) The statistician realizes that all persons in the population have heights between 1.4 and 2.0 meters, and revises the standard deviation figure that he uses based on the bound of Example 5.3. How should the values of n obtained in parts (a) and (b) be revised?

Problem 2.* The Chernoff bound. The Chernoff bound is a powerful tool that relies on the transform associated with a random variable, and provides bounds on the probabilities of certain tail events.

(a) Show that the inequality

$$\mathbf{P}(X \geq a) \leq e^{-sa} M(s)$$

holds for every a and every $s \geq 0$, where $M(s) = \mathbf{E}[e^{sX}]$ is the transform associated with the random variable X, assumed to be finite in a small open interval containing $s = 0$.

(b) Show that the inequality

$$\mathbf{P}(X \leq a) \leq e^{-sa} M(s)$$

holds for every a and every $s \leq 0$.

(c) Show that the inequality

$$\mathbf{P}(X \geq a) \leq e^{-\phi(a)}$$

holds for every a, where

$$\phi(a) = \max_{s \geq 0} \big(sa - \ln M(s)\big).$$

(d) Show that if $a > \mathbf{E}[X]$, then $\phi(a) > 0$.

(e) Apply the result of part (c) to obtain a bound for $\mathbf{P}(X \geq a)$, for the case where X is a standard normal random variable and $a > 0$.

(f) Let $X_1, X_2, \ldots$ be independent random variables with the same distribution as X. Show that for any $a > \mathbf{E}[X]$, we have

$$\mathbf{P}\left(\frac{1}{n}\sum_{i=1}^{n} X_i \geq a\right) \leq e^{-n\phi(a)},$$

so that the probability that the sample mean exceeds the mean by a certain amount decreases exponentially with n.

Solution. (a) Given some a and $s \geq 0$, consider the random variable Y_a defined by

$$Y_a = \begin{cases} 0, & \text{if } X < a, \\ e^{sa}, & \text{if } X \geq a. \end{cases}$$

It is seen that the relation

$$Y_a \leq e^{sX}$$

always holds and therefore,

$$\mathbf{E}[Y_a] \leq \mathbf{E}[e^{sX}] = M(s).$$

On the other hand,

$$\mathbf{E}[Y_a] = e^{sa}\mathbf{P}(Y_a = e^{sa}) = e^{sa}\mathbf{P}(X \geq a),$$

from which we obtain

$$\mathbf{P}(X \geq a) \leq e^{-sa}M(s).$$

(b) The argument is similar to the one for part (a). We define Y_a by

$$Y_a = \begin{cases} e^{sa}, & \text{if } X \leq a, \\ 0, & \text{if } X > a. \end{cases}$$

Since $s \leq 0$, the relation

$$Y_a \leq e^{sX}$$

always holds and therefore,

$$\mathbf{E}[Y_a] \leq \mathbf{E}[e^{sX}] = M(s).$$

On the other hand,

$$\mathbf{E}[Y_a] = e^{sa}\mathbf{P}(Y_a = e^{sa}) = e^{sa}\mathbf{P}(X \leq a),$$

from which we obtain

$$\mathbf{P}(X \leq a) \leq e^{-sa}M(s).$$

(c) Since the inequality from part (a) is valid for every $s \geq 0$, we obtain

$$\begin{aligned}\mathbf{P}(X \geq a) &\leq \min_{s\geq 0} \big(e^{-sa} M(s)\big) \\ &= \min_{s\geq 0} e^{-\big(sa - \ln M(s)\big)} \\ &= e^{-\max_{s\geq 0}\big(sa - \ln M(s)\big)} \\ &= e^{-\phi(a)}.\end{aligned}$$

(d) For $s = 0$, we have

$$sa - \ln M(s) = 0 - \ln 1 = 0,$$

where we have used the generic property $M(0) = 1$ of transforms. Furthermore,

$$\frac{d}{ds}\big(sa - \ln M(s)\big)\Big|_{s=0} = a - \frac{1}{M(s)} \cdot \frac{d}{ds} M(s)\Big|_{s=0} = a - 1 \cdot \mathbf{E}[X] > 0.$$

Since the function $sa - \ln M(s)$ is zero and has a positive derivative at $s = 0$, it must be positive when s is positive and small. It follows that the maximum $\phi(a)$ of the function $sa - \ln M(s)$ over all $s \geq 0$ is also positive.

(e) For a standard normal random variable X, we have $M(s) = e^{s^2/2}$. Therefore, $sa - \ln M(s) = sa - s^2/2$. To maximize this expression over all $s \geq 0$, we form the derivative, which is $a - s$, and set it to zero, resulting in $s = a$. Thus, $\phi(a) = a^2/2$, which leads to the bound

$$\mathbf{P}(X \geq a) \leq e^{-a^2/2}.$$

Note: In the case where $a \leq \mathbf{E}[X]$, the maximizing value of s turns out to be $s = 0$, resulting in $\phi(a) = 0$ and in the uninteresting bound

$$\mathbf{P}(X \geq a) \leq 1.$$

(f) Let $Y = X_1 + \cdots + X_n$. Using the result of part (c), we have

$$\mathbf{P}\left(\frac{1}{n}\sum_{i=1}^{n} X_i \geq a\right) = \mathbf{P}(Y \geq na) \leq e^{-\phi_Y(na)},$$

where

$$\phi_Y(na) = \max_{s\geq 0}\big(nsa - \ln M_Y(s)\big),$$

and

$$M_Y(s) = \big(M(s)\big)^n$$

is the transform associated with Y. We have $\ln M_Y(s) = n \ln M(s)$, from which we obtain

$$\phi_Y(na) = n \cdot \max_{s\geq 0}\big(sa - \ln M(s)\big) = n\phi(a),$$

and

$$\mathbf{P}\left(\frac{1}{n}\sum_{i=1}^{n} X_i \geq a\right) \leq e^{-n\phi(a)}.$$

Note that when $a > \mathbf{E}[X]$, part (d) asserts that $\phi(a) > 0$, so the probability of interest decreases exponentially with n.

Problem 3.* Jensen inequality. A twice differentiable real-valued function f of a single variable is called **convex** if its second derivative $(d^2f/dx^2)(x)$ is nonnegative for all x in its domain of definition.

(a) Show that the functions $f(x) = e^{\alpha x}$, $f(x) = -\ln x$, and $f(x) = x^4$ are all convex.

(b) Show that if f is twice differentiable and convex, then the first order Taylor approximation of f is an underestimate of the function, that is,

$$f(a) + (x - a)\frac{df}{dx}(a) \leq f(x),$$

for every a and x.

(c) Show that if f has the property in part (b), and if X is a random variable, then

$$f\big(\mathbf{E}[X]\big) \leq \mathbf{E}\big[f(X)\big].$$

Solution. (a) We have

$$\frac{d^2}{dx^2}e^{\alpha x} = \alpha^2 e^{\alpha x} > 0, \qquad \frac{d^2}{dx^2}(-\ln x) = \frac{1}{x^2} > 0, \qquad \frac{d^2}{dx^2}x^4 = 4\cdot 3\cdot x^2 \geq 0.$$

(b) Since the second derivative of f is nonnegative, its first derivative must be nondecreasing. Using the fundamental theorem of calculus, we obtain

$$f(x) = f(a) + \int_a^x \frac{df}{dt}(t)\,dt \geq f(a) + \int_a^x \frac{df}{dt}(a)\,dt = f(a) + (x-a)\frac{df}{dx}(a).$$

(c) Since the inequality from part (b) is assumed valid for every possible value x of the random variable X, we obtain

$$f(a) + (X - a)\frac{df}{dx}(a) \leq f(X).$$

We now choose $a = \mathbf{E}[X]$ and take expectations, to obtain

$$f\big(\mathbf{E}[X]\big) + \big(\mathbf{E}[X] - \mathbf{E}[X]\big)\frac{df}{dx}\big(\mathbf{E}[X]\big) \leq \mathbf{E}\big[f(X)\big],$$

or

$$f\big(\mathbf{E}[X]\big) \leq \mathbf{E}\big[f(X)\big].$$

SECTION 5.2. The Weak Law of Large Numbers

Problem 4. In order to estimate f, the true fraction of smokers in a large population, Alvin selects n people at random. His estimator M_n is obtained by dividing S_n, the number of smokers in his sample, by n, i.e., $M_n = S_n/n$. Alvin chooses the sample size n to be the smallest possible number for which the Chebyshev inequality yields a guarantee that

$$\mathbf{P}\big(|M_n - f| \geq \epsilon\big) \leq \delta,$$

where ϵ and δ are some prespecified tolerances. Determine how the value of n recommended by the Chebyshev inequality changes in the following cases.

(a) The value of ϵ is reduced to half its original value.

(b) The probability δ is reduced to half its original value.

SECTION 5.3. Convergence in Probability

Problem 5. Let $X_1, X_2, \ldots$ be independent random variables that are uniformly distributed over $[-1, 1]$. Show that the sequence $Y_1, Y_2, \ldots$ converges in probability to some limit, and identify the limit, for each of the following cases:

(a) $Y_n = X_n/n$.

(b) $Y_n = (X_n)^n$.

(c) $Y_n = X_1 \cdot X_2 \cdots X_n$.

(d) $Y_n = \max\{X_1, \ldots, X_n\}$.

Problem 6.* Consider two sequences of random variables $X_1, X_2, \ldots$ and $Y_1, Y_2, \ldots$, which converge in probability to some constants. Let c be another constant. Show that cX_n, X_n+Y_n, $\max\{0, X_n\}$, $|X_n|$, and X_nY_n all converge in probability to corresponding limits.

Solution. Let x and y be the limits of X_n and Y_n, respectively. Fix some $\epsilon > 0$ and a constant c. If $c = 0$, then cX_n equals zero for all n, and convergence trivially holds. If $c \neq 0$, we observe that $\mathbf{P}\big(|cX_n - cx| \geq \epsilon\big) = \mathbf{P}\big(|X_n - x| \geq \epsilon/|c|\big)$, which converges to zero, thus establishing convergence in probability of cX_n.

We will now show that $\mathbf{P}(|X_n + Y_n - x - y| \geq \epsilon)$ converges to zero, for any $\epsilon > 0$. To bound this probability, we note that for $|X_n + Y_n - x - y|$ to be as large as ϵ, we need either $|X_n - x|$ or $|Y_n - x|$ (or both) to be at least $\epsilon/2$. Therefore, in terms of events, we have

$$\big\{|X_n + Y_n - x - y| \geq \epsilon\big\} \subset \big\{|X_n - x| \geq \epsilon/2\big\} \cup \big\{|Y_n - y| \geq \epsilon/2\big\}.$$

This implies that

$$\mathbf{P}\big(|X_n + Y_n - x - y| \geq \epsilon\big) \leq \mathbf{P}\big(|X_n - x| \geq \epsilon/2\big) + \mathbf{P}\big(|Y_n - y| \geq \epsilon/2\big),$$

and

$$\lim_{n\to\infty} \mathbf{P}\big(|X_n + Y_n - x - y| \geq \epsilon\big) \leq \lim_{n\to\infty} \mathbf{P}\big(|X_n - x| \geq \epsilon/2\big) + \lim_{n\to\infty} \mathbf{P}\big(|Y_n - y| \geq \epsilon/2\big) = 0,$$

where the last equality follows since X_n and Y_n converge, in probability, to x and y, respectively.

By a similar argument, it is seen that the event $\{|\max\{0, X_n\} - \max\{0, x\}| \geq \epsilon\}$ is contained in the event $\{|X_n - x| \geq \epsilon\}$. Since $\lim_{n\to\infty} \mathbf{P}(|X_n - x| \geq \epsilon) = 0$, this implies that

$$\lim_{n\to\infty} \mathbf{P}\big(|\max\{0, X_n\} - \max\{0, x\}| \geq \epsilon\big) = 0.$$

Hence $\max\{0, X_n\}$ converges to $\max\{0, x\}$ in probability.

We have $|X_n| = \max\{0, X_n\} + \max\{0, -X_n\}$. Since $\max\{0, X_n\}$ and $\max\{0, -X_n\}$ converge, as shown earlier, it follows that their sum, $|X_n|$, converges to $\max\{0, x\} + \max\{0, -x\} = |x|$ in probability.

Finally, we have

$$\begin{aligned}\mathbf{P}\big(|X_nY_n - xy| \geq \epsilon\big) &= \mathbf{P}\Big(\big|(X_n - x)(Y_n - y) + xY_n + yX_n - 2xy\big| \geq \epsilon\Big)\\ &\leq \mathbf{P}\Big(\big|(X_n - x)(Y_n - y)\big| \geq \epsilon/2\Big) + \mathbf{P}\big(|xY_n + yX_n - 2xy| \geq \epsilon/2\big).\end{aligned}$$

Since xY_n and yX_n both converge to xy in probability, the last probability in the above expression converges to 0. It will thus suffice to show that

$$\lim_{n\to\infty} \mathbf{P}\Big(\big|(X_n - x)(Y_n - y)\big| \geq \epsilon/2\Big) = 0.$$

To bound this probability, we note that for $|(X_n - x)(Y_n - y)|$ to be as large as $\epsilon/2$, we need either $|X_n - x|$ or $|Y_n - x|$ (or both) to be at least $\sqrt{\epsilon/2}$. The rest of the proof is similar to the earlier proof that $X_n + Y_n$ converges in probability.

Problem 7.* A sequence X_n of random variables is said to converge to a number c **in the mean square**, if

$$\lim_{n\to\infty} \mathbf{E}\big[(X_n - c)^2\big] = 0.$$

(a) Show that convergence in the mean square implies convergence in probability.

(b) Give an example that shows that convergence in probability does not imply convergence in the mean square.

Solution. (a) Suppose that X_n converges to c in the mean square. Using the Markov inequality, we have

$$\mathbf{P}\big(|X_n - c| \geq \epsilon\big) = \mathbf{P}\big(|X_n - c|^2 \geq \epsilon^2\big) \leq \frac{\mathbf{E}\big[(X_n - c)^2\big]}{\epsilon^2}.$$

Taking the limit as $n \to \infty$, we obtain

$$\lim_{n\to\infty} \mathbf{P}\big(|X_n - c| \geq \epsilon\big) = 0,$$

which establishes convergence in probability.

(b) In Example 5.8, we have convergence in probability to 0 but $\mathbf{E}[Y_n^2] = n^3$, which diverges to infinity.

SECTION 5.4. The Central Limit Theorem

Problem 8. Before starting to play the roulette in a casino, you want to look for biases that you can exploit. You therefore watch 100 rounds that result in a number between 1 and 36, and count the number of rounds for which the result is odd. If the count exceeds 55, you decide that the roulette is not fair. Assuming that the roulette is fair, find an approximation for the probability that you will make the wrong decision.

Problem 9. During each day, the probability that your computer's operating system crashes at least once is 5%, independent of every other day. You are interested in the probability of at least 45 crash-free days out of the next 50 days.

(a) Find the probability of interest by using the normal approximation to the binomial.

(b) Repeat part (a), this time using the Poisson approximation to the binomial.

Problem 10. A factory produces X_n gadgets on day n, where the X_n are independent and identically distributed random variables, with mean 5 and variance 9.

(a) Find an approximation to the probability that the total number of gadgets produced in 100 days is less than 440.

(b) Find (approximately) the largest value of n such that

$$\mathbf{P}(X_1 + \cdots + X_n \geq 200 + 5n) \leq 0.05.$$

(c) Let N be the first day on which the total number of gadgets produced exceeds 1000. Calculate an approximation to the probability that $N \geq 220$.

Problem 11. Let $X_1, Y_1, X_2, Y_2, \ldots$ be independent random variables, uniformly distributed in the unit interval $[0, 1]$, and let

$$W = \frac{(X_1 + \cdots + X_{16}) - (Y_1 + \cdots + Y_{16})}{16}.$$

Find a numerical approximation to the quantity

$$\mathbf{P}\big(|W - \mathbf{E}[W]| < 0.001\big).$$

Problem 12.* Proof of the central limit theorem. Let $X_1, X_2, \ldots$ be a sequence of independent identically distributed zero-mean random variables with common variance σ^2, and associated transform $M_X(s)$. We assume that $M_X(s)$ is finite when $-d < s < d$, where d is some positive number. Let

$$Z_n = \frac{X_1 + \cdots + X_n}{\sigma\sqrt{n}}.$$

(a) Show that the transform associated with Z_n satisfies

$$M_{Z_n}(s) = \left(M_X\left(\frac{s}{\sigma\sqrt{n}}\right)\right)^n.$$

(b) Suppose that the transform $M_X(s)$ has a second order Taylor series expansion around $s = 0$, of the form

$$M_X(s) = a + bs + cs^2 + o(s^2),$$

where $o(s^2)$ is a function that satisfies $\lim_{s\to 0} o(s^2)/s^2 = 0$. Find a, b, and c in terms of σ^2.

(c) Combine the results of parts (a) and (b) to show that the transform $M_{Z_n}(s)$ converges to the transform associated with a standard normal random variable, that is,

$$\lim_{n\to\infty} M_{Z_n}(s) = e^{s^2/2}, \qquad \text{for all } s.$$

Note: The central limit theorem follows from the result of part (c), together with the fact (whose proof lies beyond the scope of this text) that if the transforms $M_{Z_n}(s)$ converge to the transform $M_Z(s)$ of a random variable Z whose CDF is continuous, then the CDFs F_{Z_n} converge to the CDF of Z. In our case, this implies that the CDF of Z_n converges to the CDF of a standard normal.

Solution. (a) We have, using the independence of the X_i,

$$\begin{aligned} M_{Z_n}(s) &= \mathbf{E}\big[e^{sZ_n}\big] \\ &= \mathbf{E}\left[\exp\left\{\frac{s}{\sigma\sqrt{n}}\sum_{i=1}^n X_i\right\}\right] \\ &= \prod_{i=1}^n \mathbf{E}\big[e^{sX_i/(\sigma\sqrt{n})}\big] \\ &= \left(M_X\left(\frac{s}{\sigma\sqrt{n}}\right)\right)^n. \end{aligned}$$

(b) Using the moment generating properties of the transform, we have

$$a = M_X(0) = 1, \qquad b = \frac{d}{ds}M_X(s)\bigg|_{s=0} = \mathbf{E}[X] = 0,$$

and

$$c = \frac{1}{2}\cdot\frac{d^2}{ds^2}M_X(s)\bigg|_{s=0} = \frac{\mathbf{E}[X^2]}{2} = \frac{\sigma^2}{2}.$$

(c) We combine the results of parts (a) and (b). We have

$$M_{Z_n}(s) = \left(M_X\left(\frac{s}{\sigma\sqrt{n}}\right)\right)^n = \left(a + \frac{bs}{\sigma\sqrt{n}} + \frac{cs^2}{\sigma^2 n} + o\left(\frac{s^2}{\sigma^2 n}\right)\right)^n,$$

and using the formulas for a, b, and c from part (b), it follows that

$$M_{Z_n}(s) = \left(1 + \frac{s^2}{2n} + o\left(\frac{s^2}{\sigma^2 n}\right)\right)^n.$$

We now take the limit as $n \to \infty$, and use the identity

$$\lim_{n\to\infty} \left(1 + \frac{c}{n}\right)^n = e^c,$$

to obtain

$$\lim_{n\to\infty} M_{Z_n}(s) = e^{s^2/2}.$$

SECTION 5.5. The Strong Law of Large Numbers

Problem 13.* Consider two sequences of random variables $X_1, X_2, \ldots$ and $Y_1, Y_2, \ldots$. Suppose that X_n converges to a and Y_n converges to b, with probability 1. Show that X_n+Y_n converges to $a+b$, with probability 1. Also, assuming that the random variables Y_n cannot be equal to zero, show that X_n/Y_n converges to a/b, with probability 1.

Solution. Let A (respectively, B) be the event that the sequence of values of the random variables X_n (respectively, Y_n) does not converge to a (respectively, b). Let C be the event that the sequence of values of $X_n + Y_n$ does not converge to $a + b$ and notice that $C \subset A \cup B$.

Since X_n and Y_n converge to a and b, respectively, with probability 1, we have $\mathbf{P}(A) = 0$ and $\mathbf{P}(B) = 0$. Hence,

$$\mathbf{P}(C) \le \mathbf{P}(A \cup B) \le \mathbf{P}(A) + \mathbf{P}(B) = 0.$$

Therefore, $\mathbf{P}(C^c) = 1$, or equivalently, $X_n + Y_n$ converges to $a + b$ with probability 1. For the convergence of X_n/Y_n, the argument is similar.

Problem 14.* Let $X_1, X_2, \ldots$ be a sequence of independent identically distributed random variables. Let $Y_1, Y_2, \ldots$ be another sequence of independent identically distributed random variables. We assume that the X_i and Y_i have finite mean, and that $Y_1 + \cdots + Y_n$ cannot be equal to zero. Does the sequence

$$Z_n = \frac{X_1 + \cdots + X_n}{Y_1 + \cdots + Y_n}$$

converge with probability 1, and if so, what is the limit?

Solution. We have

$$Z_n = \frac{(X_1 + \cdots + X_n)/n}{(Y_1 + \cdots + Y_n)/n}.$$

By the strong law of large numbers, the numerator and denominator converge with probability 1 to $\mathbf{E}[X]$ and $\mathbf{E}[Y]$, respectively. It follows that Z_n converges to $\mathbf{E}[X]/\mathbf{E}[Y]$, with probability 1 (cf. the preceding problem).

Problem 15.* Suppose that a sequence $Y_1, Y_2, \ldots$ of random variables converges to a real number c, with probability 1. Show that the sequence also converges to c in probability.

Solution. Let C be the event that the sequence of values of the random variables Y_n converges to c. By assumption, we have $\mathbf{P}(C) = 1$. Fix some $\epsilon > 0$, and let A_k be the event that $|Y_n - c| < \epsilon$ for every $n \ge k$. If the sequence of values of the random variables Y_n converges to c, then there must exist some k such that for every $n \ge k$, this sequence of values is within less than ϵ from c. Therefore, every element of C belongs to A_k for some k, or

$$C \subset \bigcup_{k=1}^{\infty} A_k.$$

Note also that the sequence of events A_k is monotonically increasing, in the sense that $A_k \subset A_{k+1}$ for all k. Finally, note that the event A_k is a subset of the event $\{|Y_k - c| < \epsilon\}$. Therefore,

$$\lim_{k\to\infty} \mathbf{P}\big(|Y_k - c| < \epsilon\big) \geq \lim_{k\to\infty} \mathbf{P}(A_k) = \mathbf{P}(\cup_{k=1}^{\infty} A_k) \geq \mathbf{P}(C) = 1,$$

where the first equality uses the continuity property of probabilities (Problem 13 in Chapter 1). It follows that

$$\lim_{k\to\infty} \mathbf{P}\big(|Y_k - c| \geq \epsilon\big) = 0,$$

which establishes convergence in probability.

Problem 16.* Consider a sequence Y_n of nonnegative random variables and suppose that

$$\mathbf{E}\left[\sum_{n=1}^{\infty} Y_n\right] < \infty.$$

Show that Y_n converges to 0, with probability 1.

Note: This result provides a commonly used method for establishing convergence with probability 1. To evaluate the expectation of $\sum_{n=1}^{\infty} Y_n$, one typically uses the formula

$$\mathbf{E}\left[\sum_{n=1}^{\infty} Y_n\right] = \sum_{n=1}^{\infty} \mathbf{E}[Y_n].$$

The fact that the expectation and the infinite summation can be interchanged, for the case of nonnegative random variables, is known as the **monotone convergence theorem**, a fundamental result of probability theory, whose proof lies beyond the scope of this text.

Solution. We note that the infinite sum $\sum_{n=1}^{\infty} Y_n$ must be finite, with probability 1. Indeed, if it had a positive probability of being infinite, then its expectation would also be infinite. But if the sum of the values of the random variables Y_n is finite, the sequence of these values must converge to zero. Since the probability of this event is equal to 1, it follows that the sequence Y_n converges to zero, with probability 1.

Problem 17.* Consider a sequence of Bernoulli random variables X_n, and let $p_n = \mathbf{P}(X_n = 1)$ be the probability of success in the nth trial. Assuming that $\sum_{n=1}^{\infty} p_n < \infty$, show that the number of successes is finite, with probability 1. [Compare with Problem 48(b) in Chapter 1.]

Solution. Using the monotone convergence theorem (see above note), we have

$$\mathbf{E}\left[\sum_{n=1}^{\infty} X_n\right] = \sum_{n=1}^{\infty} \mathbf{E}[X_n] = \sum_{n=1}^{\infty} p_n < \infty.$$

This implies that

$$\sum_{n=1}^{\infty} X_n < \infty,$$

with probability 1. We then note that the event $\left\{\sum_{n=1}^{\infty} X_n < \infty\right\}$ is the same as the event that there is a finite number of successes.

Problem 18.* The strong law of large numbers. Let $X_1, X_2, \ldots$ be a sequence of independent identically distributed random variables and assume that $\mathbf{E}[X_i^4] < \infty$. Prove the strong law of large numbers.

Solution. We note that the assumption $\mathbf{E}[X_i^4] < \infty$ implies that the expected value of the X_i is finite. Indeed, using the inequality $|x| \le 1 + x^4$, we have

$$\mathbf{E}\big[|X_i|\big] \le 1 + \mathbf{E}[X_i^4] < \infty.$$

Let us assume first that $\mathbf{E}[X_i] = 0$. We will show that

$$\mathbf{E}\left[\sum_{n=1}^{\infty} \frac{(X_1 + \cdots + X_n)^4}{n^4}\right] < \infty.$$

We have

$$\mathbf{E}\left[\frac{(X_1 + \cdots + X_n)^4}{n^4}\right] = \frac{1}{n^4} \sum_{i_1=1}^{n} \sum_{i_2=1}^{n} \sum_{i_3=1}^{n} \sum_{i_4=1}^{n} \mathbf{E}[X_{i_1} X_{i_2} X_{i_3} X_{i_4}].$$

Let us consider the various terms in this sum. If one of the indices is different from all of the other indices, the corresponding term is equal to zero. For example, if i_1 is different from i_2, i_3, or i_4, the assumption $\mathbf{E}[X_i] = 0$ yields

$$\mathbf{E}[X_{i_1} X_{i_2} X_{i_3} X_{i_4}] = \mathbf{E}[X_{i_1}]\mathbf{E}[X_{i_2} X_{i_3} X_{i_4}] = 0.$$

Therefore, the nonzero terms in the above sum are either of the form $\mathbf{E}[X_i^4]$ (there are n such terms), or of the form $\mathbf{E}[X_i^2 X_j^2]$, with $i \ne j$. Let us count how many terms there are of this form. Such terms are obtained in three different ways: by setting $i_1 = i_2 \ne i_3 = i_4$, or by setting $i_1 = i_3 \ne i_2 = i_4$, or by setting $i_1 = i_4 \ne i_2 = i_3$. For each one of these three ways, we have n choices for the first pair of indices, and $n-1$ choices for the second pair. We conclude that there are $3n(n-1)$ terms of this type. Thus,

$$\mathbf{E}\big[(X_1 + \cdots + X_n)^4\big] = n\mathbf{E}[X_1^4] + 3n(n-1)\mathbf{E}[X_1^2 X_2^2].$$

Using the inequality $xy \le (x^2 + y^2)/2$, we obtain $\mathbf{E}[X_1^2 X_2^2] \le \mathbf{E}[X_1^4]$, and

$$\mathbf{E}\big[(X_1 + \cdots + X_n)^4\big] \le \big(n + 3n(n-1)\big)\mathbf{E}[X_1^4] \le 3n^2\mathbf{E}[X_1^4].$$

It follows that

$$\mathbf{E}\left[\sum_{n=1}^{\infty} \frac{(X_1 + \cdots + X_n)^4}{n^4}\right] = \sum_{n=1}^{\infty} \frac{1}{n^4}\mathbf{E}\big[(X_1 + \cdots + X_n)^4\big] \le \sum_{n=1}^{\infty} \frac{3}{n^2}\mathbf{E}[X_1^4] < \infty,$$

where the last step uses the well known property $\sum_{n=1}^{\infty} n^{-2} < \infty$. This implies that $(X_1 + \cdots + X_n)^4/n^4$ converges to zero with probability 1 (cf. Problem 16), and therefore, $(X_1 + \cdots + X_n)/n$ also converges to zero with probability 1, which is the strong law of large numbers.

For the more general case where the mean of the random variables X_i is nonzero, the preceding argument establishes that $\big(X_1 + \cdots + X_n - n\mathbf{E}[X_1]\big)/n$ converges to zero, which is the same as $(X_1 + \cdots + X_n)/n$ converging to $\mathbf{E}[X_1]$, with probability 1.

6

The Bernoulli and Poisson Processes

Contents

A stochastic process is a mathematical model of a probabilistic experiment that evolves in time and generates a sequence of numerical values. For example, a stochastic process can be used to model:

(a) the sequence of daily prices of a stock;

(b) the sequence of scores in a football game;

(c) the sequence of failure times of a machine;

(d) the sequence of hourly traffic loads at a node of a communication network;

(e) the sequence of radar measurements of the position of an airplane.

Each numerical value in the sequence is modeled by a random variable, so a stochastic process is simply a (finite or infinite) sequence of random variables and does not represent a major conceptual departure from our basic framework. We are still dealing with a single basic experiment that involves outcomes governed by a probability law, and random variables that inherit their probabilistic properties from that law.†

However, stochastic processes involve some change in emphasis over our earlier models. In particular:

(a) We tend to focus on the **dependencies** in the sequence of values generated by the process. For example, how do future prices of a stock depend on past values?

(b) We are often interested in **long-term averages** involving the entire sequence of generated values. For example, what is the fraction of time that a machine is idle?

(c) We sometimes wish to characterize the likelihood or frequency of certain **boundary events.** For example, what is the probability that within a given hour all circuits of some telephone system become simultaneously busy; or what is the frequency with which some buffer in a computer network overflows with data?

There is a wide variety of stochastic processes, but in this book we will only discuss two major categories.

(i) *Arrival-Type Processes*: Here, we are interested in occurrences that have the character of an "arrival," such as message receptions at a receiver, job completions in a manufacturing cell, customer purchases at a store, etc. We will focus on models in which the interarrival times (the times between successive arrivals) are independent random variables. In Section 6.1, we consider the case where arrivals occur in discrete time and the interarrival

---

† Let us emphasize that all of the random variables arising in a stochastic process refer to a single and common experiment, and are therefore defined on a common sample space. The corresponding probability law can be specified explicitly or implicitly (in terms of its properties), provided that it determines unambiguously the joint CDF of any subset of the random variables involved.

times are geometrically distributed – this is the *Bernoulli process*. In Section 6.2, we consider the case where arrivals occur in continuous time and the interarrival times are exponentially distributed – this is the *Poisson process*.

(ii) *Markov Processes*: Here, we are looking at experiments that evolve in time and in which the future evolution exhibits a probabilistic dependence on the past. As an example, the future daily prices of a stock are typically dependent on past prices. However, in a Markov process, we assume a very special type of dependence: the next value depends on past values only through the current value. There is a rich methodology that applies to such processes, and is the subject of Chapter 7.

6.1 THE BERNOULLI PROCESS

The Bernoulli process can be visualized as a sequence of independent coin tosses, where the probability of heads in each toss is a fixed number p in the range $0 < p < 1$. In general, the Bernoulli process consists of a sequence of Bernoulli trials. Each trial produces a 1 (a success) with probability p, and a 0 (a failure) with probability $1 - p$, independent of what happens in other trials.

Of course, coin tossing is just a paradigm for a broad range of contexts involving a sequence of independent binary outcomes. For example, a Bernoulli process is often used to model systems involving arrivals of customers or jobs at service centers. Here, time is discretized into periods, and a "success" at the kth trial is associated with the arrival of at least one customer at the service center during the kth period. We will often use the term "arrival" in place of "success" when this is justified by the context.

In a more formal description, we define the Bernoulli process as a sequence $X_1, X_2, \ldots$ of **independent** Bernoulli random variables X_i with

$$\mathbf{P}(X_i = 1) = \mathbf{P}(\text{success at the } i\text{th trial}) = p,$$
$$\mathbf{P}(X_i = 0) = \mathbf{P}(\text{failure at the } i\text{th trial}) = 1 - p,$$

for each i.†

Given an arrival process, one is often interested in random variables such as the number of arrivals within a certain time period, or the time until the first arrival. For the case of a Bernoulli process, some answers are already available from earlier chapters. Here is a summary of the main facts.

---

† Generalizing from the case of a finite number of random variables, the independence of an *infinite* sequence of random variables X_i is defined by the requirement that the random variables $X_1, \ldots, X_n$ be independent for any finite n. Intuitively, knowing the values of any finite subset of the random variables does not provide any new probabilistic information on the remaining random variables, and the conditional distribution of the latter is the same as the unconditional one.

Some Random Variables Associated with the Bernoulli Process and their Properties

- **The binomial with parameters p and n.** This is the number S of successes in n independent trials. Its PMF, mean, and variance are

$$p_S(k) = \binom{n}{k} p^k (1-p)^{n-k}, \qquad k = 0, 1, \ldots, n,$$

$$\mathbf{E}[S] = np, \qquad \mathrm{var}(S) = np(1-p).$$

- **The geometric with parameter p.** This is the number T of trials up to (and including) the first success. Its PMF, mean, and variance are

$$p_T(t) = (1-p)^{t-1} p, \qquad t = 1, 2, \ldots,$$

$$\mathbf{E}[T] = \frac{1}{p}, \qquad \mathrm{var}(T) = \frac{1-p}{p^2}.$$

Independence and Memorylessness

The independence assumption underlying the Bernoulli process has important implications, including a memorylessness property (whatever has happened in past trials provides no information on the outcomes of future trials). An appreciation and intuitive understanding of such properties is very useful, and allows the quick solution of many problems that would be difficult with a more formal approach. In this subsection, we aim at developing the necessary intuition.

Let us start by considering random variables that are defined in terms of what happened in a certain set of trials. For example, the random variable $Z = (X_1 + X_3)X_6X_7$ is defined in terms of the first, third, sixth, and seventh trial. If we have two random variables of this type and if the two sets of trials that define them have no common element, then these random variables are independent. This is a generalization of a fact first seen in Chapter 2: if two random variables U and V are independent, then any two functions of them, $g(U)$ and $h(V)$, are also independent.

Example 6.1.

(a) Let U be the number of successes in trials 1 to 5. Let V be the number of successes in trials 6 to 10. Then, U and V are independent. This is because $U = X_1 + \cdots + X_5$, $V = X_6 + \cdots + X_{10}$, and the two collections $\{X_1, \ldots, X_5\}$, $\{X_6, \ldots, X_{10}\}$ have no common elements.

(b) Let U (respectively, V) be the first odd (respectively, even) time i in which we have a success. Then, U is determined by the odd-time sequence $X_1, X_3, \ldots,$

whereas V is determined by the even-time sequence $X_2, X_4, \ldots$. Since these two sequences have no common elements, U and V are independent.

Suppose now that a Bernoulli process has been running for n time steps, and that we have observed the values of $X_1, X_2, \ldots, X_n$. We notice that the sequence of future trials $X_{n+1}, X_{n+2}, \ldots$ are independent Bernoulli trials and therefore form a Bernoulli process. In addition, these future trials are independent from the past ones. We conclude that starting from any given point in time, the future is also modeled by a Bernoulli process, which is independent of the past. We refer to this loosely as the **fresh-start** property of the Bernoulli process.

Let us now recall that the time T until the first success is a geometric random variable. Suppose that we have been watching the process for n time steps and no success has been recorded. What can we say about the number $T-n$ of remaining trials until the first success? Since the future of the process (after time n) is independent of the past and constitutes a fresh-starting Bernoulli process, the number of future trials until the first success is described by the same geometric PMF. Mathematically, we have

$$\mathbf{P}(T - n = t \,|\, T > n) = (1-p)^{t-1}p = \mathbf{P}(T = t), \qquad t = 1, 2, \ldots.$$

We refer to this as the **memorylessness** property. It can also be derived algebraically, using the definition of conditional probabilities, but the argument given here is certainly more intuitive.

Independence Properties of the Bernoulli Process

- For any given time n, the sequence of random variables $X_{n+1}, X_{n+2}, \ldots$ (the future of the process) is also a Bernoulli process, and is independent from $X_1, \ldots, X_n$ (the past of the process).
- Let n be a given time and let $\overline{T}$ be the time of the first success after time n. Then, $\overline{T} - n$ has a geometric distribution with parameter p, and is independent of the random variables $X_1, \ldots, X_n$.

Example 6.2. A computer executes two types of tasks, priority and nonpriority, and operates in discrete time units (*slots*). A priority task arrives with probability p at the beginning of each slot, independent of other slots, and requires one full slot to complete. A nonpriority task is always available and is executed at a given slot if no priority task is available. In this context, it may be important to know the probabilistic properties of the time intervals available for nonpriority tasks.

With this in mind, let us call a slot *busy* if within this slot, the computer executes a priority task, and otherwise let us call it *idle*. We call a string of idle (or busy) slots, flanked by busy (or idle, respectively) slots, an *idle period* (or *busy period*, respectively). Let us derive the PMF, mean, and variance of the following random variables (cf. Fig. 6.1):

(a) $T =$ the time index of the first idle slot;

(b) $B =$ the length (number of slots) of the first busy period;

(c) $I =$ the length of the first idle period.

(d) $Z =$ the number of slots after the first slot of the first busy period up to and including the first subsequent idle slot.

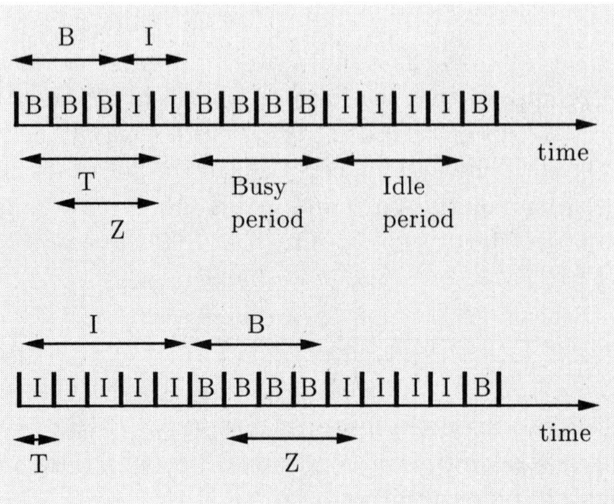

Figure 6.1: Illustration of random variables, and busy and idle periods in Example 6.2. In the top diagram, $T = 4$, $B = 3$, $I = 2$, and $Z = 3$. In the bottom diagram, $T = 1$, $I = 5$, $B = 4$, and $Z = 4$.

We recognize T as a geometrically distributed random variable with parameter $1 - p$. Its PMF is

$$p_T(k) = p^{k-1}(1-p), \qquad k = 1, 2, \ldots.$$

Its mean and variance are

$$\mathbf{E}[T] = \frac{1}{1-p}, \qquad \text{var}(T) = \frac{p}{(1-p)^2}.$$

Let us now consider the first busy period. It starts with the first busy slot, call it slot L. (In the top diagram in Fig. 6.1, $L = 1$; in the bottom diagram, $L = 6$.) The number Z of subsequent slots until (and including) the first subsequent idle slot has the same distribution as T, because the Bernoulli process starts fresh at time $L + 1$. We then notice that $Z = B$ and conclude that B has the same PMF as T.

If we reverse the roles of idle and busy slots, and interchange p with $1-p$, we see that the length I of the first idle period has the same PMF as the time index of the first busy slot, so that

$$p_I(k) = (1-p)^{k-1}p, \quad k = 1, 2, \ldots, \qquad \mathbf{E}[I] = \frac{1}{p}, \quad \text{var}(I) = \frac{1-p}{p^2}.$$

We finally note that the argument given here also works for the second, third, etc., busy (or idle) period. Thus, the PMFs calculated above apply to the ith busy and idle period, for any i.

If we start watching a Bernoulli process at a certain time n, what we see is indistinguishable from a Bernoulli process that has just started. It turns out that the same is true if we start watching the process at some *random* time N, as long as N is determined only by the past history of the process and does not convey any information on the future. Indeed, such a property was used in Example 6.2, when we stated that the process starts fresh at time $L+1$. For another example, consider a roulette wheel with each occurrence of red viewed as a success. The sequence generated starting at some fixed spin (say, the 25th spin) is probabilistically indistinguishable from the sequence generated starting immediately after red occurs in five consecutive spins. In either case, the process starts fresh (although one can certainly find gamblers with alternative theories). The next example makes a similar argument, but more formally.

Example 6.3. Fresh-Start at a Random Time. Let N be the first time that we have a success immediately following a previous success. (That is, N is the first i for which $X_{i-1} = X_i = 1$.) What is the probability $\mathbf{P}(X_{N+1} = X_{N+2} = 0)$ that there are no successes in the two trials that follow?

Intuitively, once the condition $X_{N-1} = X_N = 1$ is satisfied, and from then on, the future of the process consists of independent Bernoulli trials. Therefore, the probability of an event that refers to the future of the process is the same as in a fresh-starting Bernoulli process, so that $\mathbf{P}(X_{N+1} = X_{N+2} = 0) = (1-p)^2$.

To provide a rigorous justification of the above argument, we note that the time N is a random variable, and by conditioning on the possible values of N, we have

$$\begin{aligned}\mathbf{P}(X_{N+1} = X_{N+2} = 0) &= \sum_{n=1}^{\infty} \mathbf{P}(N = n)\mathbf{P}(X_{N+1} = X_{N+2} = 0 \,|\, N = n) \\ &= \sum_{n=1}^{\infty} \mathbf{P}(N = n)\mathbf{P}(X_{n+1} = X_{n+2} = 0 \,|\, N = n).\end{aligned}$$

Because of the way that N was defined, the event $\{N = n\}$ occurs if and only if the values of $X_1, \ldots, X_n$ satisfy a certain condition. But these random variables are independent of X_{n+1} and X_{n+2}. Therefore,

$$\mathbf{P}(X_{n+1} = X_{n+2} = 0 \,|\, N = n) = \mathbf{P}(X_{n+1} = X_{n+2} = 0) = (1-p)^2,$$

which leads to

$$\mathbf{P}(X_{N+1} = X_{N+2} = 0) = \sum_{n=1}^{\infty} \mathbf{P}(N = n)(1-p)^2 = (1-p)^2.$$

Interarrival Times

An important random variable associated with the Bernoulli process is the time of the kth success (or arrival), which we denote by Y_k. A related random variable is the kth interarrival time, denoted by T_k. It is defined by

$$T_1 = Y_1, \qquad T_k = Y_k - Y_{k-1}, \qquad k = 2, 3, \ldots,$$

and represents the number of trials following the $(k-1)$st success until the next success. See Fig. 6.2 for an illustration, and also note that

$$Y_k = T_1 + T_2 + \cdots + T_k.$$

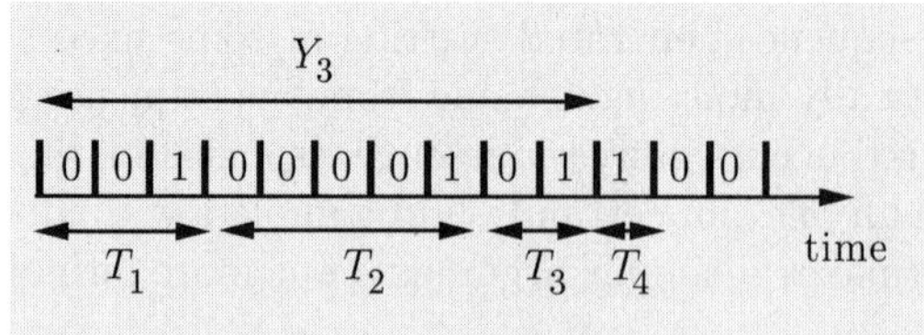

Figure 6.2: Illustration of interarrival times, where a 1 represents an arrival. In this example, $T_1 = 3$, $T_2 = 5$, $T_3 = 2$, $T_4 = 1$. Furthermore, $Y_1 = 3$, $Y_2 = 8$, $Y_3 = 10$, $Y_4 = 11$.

We have already seen that the time T_1 until the first success is a geometric random variable with parameter p. Having had a success at time T_1, the future is a new Bernoulli process, similar to the original: the number of trials T_2 until the next success has the same geometric PMF. Furthermore, past trials (up to and including time T_1) are independent of future trials (from time $T_1 + 1$ onward). Since T_2 is determined exclusively by what happens in these future trials, we see that T_2 is independent of T_1. Continuing similarly, we conclude that the random variables $T_1, T_2, T_3, \ldots$ are independent and all have the same geometric distribution.

This important observation leads to an alternative, but equivalent way of describing the Bernoulli process, which is sometimes more convenient.

Alternative Description of the Bernoulli Process

1. Start with a sequence of independent geometric random variables T_1, $T_2, \ldots$, with common parameter p, and let these stand for the interarrival times.
2. Record a success (or arrival) at times T_1, $T_1 + T_2$, $T_1 + T_2 + T_3$, etc.

Example 6.4. It has been observed that after a rainy day, the number of days until it rains again is geometrically distributed with parameter p, independent of the past. Find the probability that it rains on both the 5th and the 8th day of the month.

If we attempt to approach this problem by manipulating the geometric PMFs in the problem statement, the solution is quite tedious. However, if we view rainy days as "arrivals," we notice that the description of the weather conforms to the alternative description of the Bernoulli process given above. Therefore, any given day is rainy with probability p, independent of other days. In particular, the probability that days 5 and 8 are rainy is equal to p^2.

The kth Arrival Time

The time Y_k of the kth success (or arrival) is equal to the sum $Y_k = T_1 + T_2 + \cdots + T_k$ of k independent identically distributed geometric random variables. This allows us to derive formulas for the mean, variance, and PMF of Y_k, which are given in the table that follows.

Properties of the kth Arrival Time

- The kth arrival time is equal to the sum of the first k interarrival times

$$Y_k = T_1 + T_2 + \cdots + T_k,$$

and the latter are independent geometric random variables with common parameter p.

- The mean and variance of Y_k are given by

$$\mathbf{E}[Y_k] = \mathbf{E}[T_1] + \cdots + \mathbf{E}[T_k] = \frac{k}{p},$$

$$\text{var}(Y_k) = \text{var}(T_1) + \cdots + \text{var}(T_k) = \frac{k(1-p)}{p^2}.$$

- The PMF of Y_k is given by

$$p_{Y_k}(t) = \binom{t-1}{k-1} p^k (1-p)^{t-k}, \qquad t = k, k+1, \ldots,$$

and is known as the **Pascal PMF of order k.**

To verify the formula for the PMF of Y_k, we first note that Y_k cannot be smaller than k. For $t \geq k$, we observe that the event $\{Y_k = t\}$ (the kth success comes at time t) will occur if and only if both of the following two events A and B occur:

(a) event A: trial t is a success;

(b) event B: exactly $k-1$ successes occur in the first $t-1$ trials.

The probabilities of these two events are

$$\mathbf{P}(A) = p,$$

and

$$\mathbf{P}(B) = \binom{t-1}{k-1} p^{k-1}(1-p)^{t-k},$$

respectively. In addition, these two events are independent (whether trial t is a success or not is independent of what happened in the first $t-1$ trials). Therefore,

$$p_{Y_k}(t) = \mathbf{P}(Y_k = t) = \mathbf{P}(A \cap B) = \mathbf{P}(A)\mathbf{P}(B) = \binom{t-1}{k-1} p^k(1-p)^{t-k},$$

as claimed.

Example 6.5. In each minute of basketball play, Alicia commits a single foul with probability p and no foul with probability $1-p$. The number of fouls in different minutes are assumed to be independent. Alicia will foul out of the game once she commits her sixth foul, and will play 30 minutes if she does not foul out. What is the PMF of Alicia's playing time?

We model fouls as a Bernoulli process with parameter p. Alicia's playing time Z is equal to Y_6, the time until the sixth foul, except if Y_6 is larger than 30, in which case, her playing time is 30; that is, $Z = \min\{Y_6, 30\}$. The random variable Y_6 has a Pascal PMF of order 6, which is given by

$$p_{Y_6}(t) = \binom{t-1}{5} p^6(1-p)^{t-6}, \qquad t = 6, 7, \ldots$$

To determine the PMF $p_Z(z)$ of Z, we first consider the case where z is between 6 and 29. For z in this range, we have

$$p_Z(z) = \mathbf{P}(Z = z) = \mathbf{P}(Y_6 = z) = \binom{z-1}{5} p^6(1-p)^{z-6}, \qquad z = 6, 7, \ldots, 29.$$

The probability that $Z = 30$ is then determined from

$$p_Z(30) = 1 - \sum_{z=6}^{29} p_Z(z).$$

Splitting and Merging of Bernoulli Processes

Starting with a Bernoulli process in which there is a probability p of an arrival at each time, consider **splitting** it as follows. Whenever there is an arrival, we choose to either keep it (with probability q), or to discard it (with probability $1-q$); see Fig. 6.3. Assume that the decisions to keep or discard are independent for different arrivals. If we focus on the process of arrivals that are kept, we see that it is a Bernoulli process: in each time slot, there is a probability pq of a kept arrival, independent of what happens in other slots. For the same reason, the process of discarded arrivals is also a Bernoulli process, with a probability of a discarded arrival at each time slot equal to $p(1-q)$.

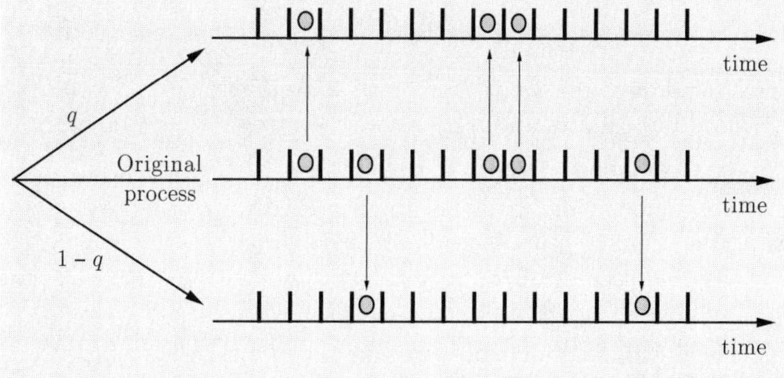

Figure 6.3: Splitting of a Bernoulli process.

In a reverse situation, we start with two *independent* Bernoulli processes (with parameters p and q, respectively) and **merge** them into a single process, as follows. An arrival is recorded in the merged process if and only if there is an arrival in at least one of the two original processes. This happens with probability $p+q-pq$ [one minus the probability $(1-p)(1-q)$ of no arrival in either process]. Since different time slots in either of the original processes are independent, different slots in the merged process are also independent. Thus, the merged process is Bernoulli, with success probability $p+q-pq$ at each time step; see Fig. 6.4.

Splitting and merging of Bernoulli (or other) arrival processes arises in many contexts. For example, a two-machine work center may see a stream of arriving parts to be processed and split them by sending each part to a randomly chosen machine. Conversely, a machine may be faced with arrivals of different types that can be merged into a single arrival stream.

The Poisson Approximation to the Binomial

The number of successes in n independent Bernoulli trials is a binomial random variable with parameters n and p, and its mean is np. In this subsection, we

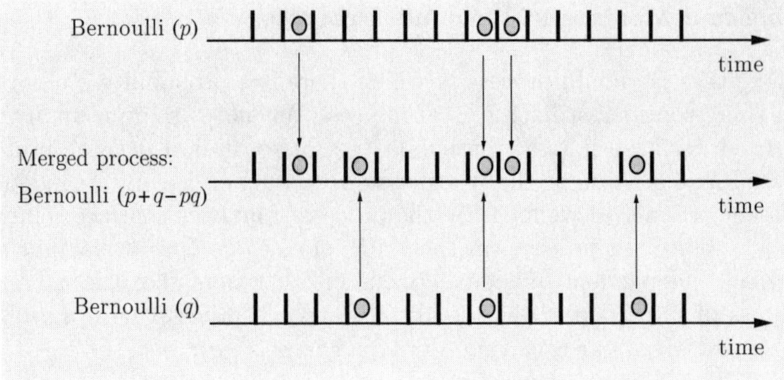

Figure 6.4: Merging of independent Bernoulli processes.

concentrate on the special case where n is large but p is small, so that the mean np has a moderate value. A situation of this type arises when one passes from discrete to continuous time, a theme to be picked up in the next section. For some examples, think of the number of airplane accidents on any given day: there is a large number n of trials (airplane flights), but each one has a very small probability p of being involved in an accident. Or think of counting the number of typos in a book: there is a large number of words, but a very small probability of misspelling any single one.

Mathematically, we can address situations of this kind, by letting n grow while simultaneously decreasing p, in a manner that keeps the product np at a constant value λ. In the limit, it turns out that the formula for the binomial PMF simplifies to the Poisson PMF. A precise statement is provided next, together with a reminder of some of the properties of the Poisson PMF that were derived in Chapter 2.

Poisson Approximation to the Binomial

- A Poisson random variable Z with parameter λ takes nonnegative integer values and is described by the PMF

$$p_Z(k) = e^{-\lambda}\frac{\lambda^k}{k!}, \qquad k = 0, 1, 2, \ldots.$$

Its mean and variance are given by

$$\mathbf{E}[Z] = \lambda, \qquad \text{var}(Z) = \lambda.$$

- For any fixed nonnegative integer k, the binomial probability

$$p_S(k) = \frac{n!}{(n-k)!\,k!} \cdot p^k(1-p)^{n-k}$$

converges to $p_Z(k)$, when we take the limit as $n \to \infty$ and $p = \lambda/n$, while keeping λ constant.

- In general, the Poisson PMF is a good approximation to the binomial as long as $\lambda = np$, n is very large, and p is very small.

To verify the validity of the Poisson approximation, we let $p = \lambda/n$ and note that

$$\begin{aligned} p_S(k) &= \frac{n!}{(n-k)!\,k!} \cdot p^k(1-p)^{n-k} \\ &= \frac{n(n-1)\cdots(n-k+1)}{k!} \cdot \frac{\lambda^k}{n^k} \cdot \left(1 - \frac{\lambda}{n}\right)^{n-k} \\ &= \frac{n}{n} \cdot \frac{(n-1)}{n} \cdots \frac{(n-k+1)}{n} \cdot \frac{\lambda^k}{k!} \cdot \left(1 - \frac{\lambda}{n}\right)^{n-k}. \end{aligned}$$

Let us focus on a fixed k and let $n \to \infty$. Each one of the ratios $(n-1)/n$, $(n-2)/n, \ldots, (n-k+1)/n$ converges to 1. Furthermore,†

$$\left(1 - \frac{\lambda}{n}\right)^{-k} \to 1, \qquad \left(1 - \frac{\lambda}{n}\right)^{n} \to e^{-\lambda}.$$

We conclude that for each fixed k, and as $n \to \infty$, we have

$$p_S(k) \to e^{-\lambda}\frac{\lambda^k}{k!}.$$

Example 6.6. As a rule of thumb, the Poisson/binomial approximation

$$e^{-\lambda}\frac{\lambda^k}{k!} \approx \frac{n!}{(n-k)!\,k!} \cdot p^k(1-p)^{n-k}, \qquad k = 0, 1, \ldots, n,$$

is valid to several decimal places if $n \geq 100$, $p \leq 0.01$, and $\lambda = np$. To check this, consider the following.

Gary Kasparov, a world chess champion, plays against 100 amateurs in a large simultaneous exhibition. It has been estimated from past experience that Kasparov

† We are using here, the well known formula $\lim_{x\to\infty}(1 - \frac{1}{x})^x = e^{-1}$. Letting $x = n/\lambda$, we have $\lim_{n\to\infty}(1 - \frac{\lambda}{n})^{n/\lambda} = e^{-1}$, and $\lim_{n\to\infty}(1 - \frac{\lambda}{n})^n = e^{-\lambda}$.

wins in such exhibitions 99% of his games on the average (in precise probabilistic terms, we assume that he wins each game with probability 0.99, independent of other games). What are the probabilities that he will win 100 games, 98 games, 95 games, and 90 games?

We model the number of games X that Kasparov does *not* win as a binomial random variable with parameters $n = 100$ and $p = 0.01$. Thus the probabilities that he will win 100 games, 98, 95 games, and 90 games are

$$p_X(0) = (1 - 0.01)^{100} = 0.366,$$

$$p_X(2) = \frac{100!}{98!\,2!} \cdot 0.01^2(1 - 0.01)^{98} = 0.185,$$

$$p_X(5) = \frac{100!}{95!\,5!} \cdot 0.01^5(1 - 0.01)^{95} = 0.00290,$$

$$p_X(10) = \frac{100!}{90!\,10!} \cdot 0.01^{10}(1 - 0.01)^{90} = 7.006 \cdot 10^{-8},$$

respectively. Now let us check the corresponding Poisson approximations with $\lambda = 100 \cdot 0.01 = 1$. They are:

$$p_Z(0) = e^{-1}\frac{1}{0!} = 0.368,$$

$$p_Z(2) = e^{-1}\frac{1}{2!} = 0.184,$$

$$p_Z(5) = e^{-1}\frac{1}{5!} = 0.00306,$$

$$p_Z(10) = e^{-1}\frac{1}{10!} = 1.001 \cdot 10^{-8}.$$

By comparing the binomial PMF values $p_X(k)$ with their Poisson approximations $p_Z(k)$, we see that there is close agreement.

Suppose now that Kasparov plays simultaneously against just 5 opponents, who are, however, stronger so that his probability of a win per game is 0.9. Here are the binomial probabilities $p_X(k)$ for $n = 5$ and $p = 0.1$, and the corresponding Poisson approximations $p_Z(k)$ for $\lambda = np = 0.5$:

k	0	1	2	3	4	5
$p_X(k)$	0.590	0.328	0.0729	0.0081	0.00045	0.00001
$p_Z(k)$	0.605	0.303	0.0758	0.0126	0.0016	0.00016

We see that the approximation, while not poor, is considerably less accurate than in the case where $n = 100$ and $p = 0.01$.

Example 6.7. A packet consisting of a string of n symbols is transmitted over a noisy channel. Each symbol has probability $p = 0.0001$ of being transmitted in error, independent of errors in the other symbols. How small should n be in order for the probability of incorrect transmission (at least one symbol in error) to be less than 0.001?

Each symbol transmission is viewed as an independent Bernoulli trial. Thus, the probability of a positive number S of errors in the packet is

$$1 - \mathbf{P}(S = 0) = 1 - (1 - p)^n.$$

For this probability to be less than 0.001, we must have $1 - (1 - 0.0001)^n < 0.001$ or

$$n < \frac{\ln 0.999}{\ln 0.9999} = 10.0045.$$

We can also use the Poisson approximation for $\mathbf{P}(S = 0)$, which is $e^{-\lambda}$ with $\lambda = np = 0.0001 \cdot n$, and obtain the condition $1 - e^{-0.0001 \cdot n} < 0.001$, which leads to

$$n < \frac{-\ln 0.999}{0.0001} = 10.005.$$

Given that n must be integer, both methods lead to the same conclusion that n can be at most 10.

6.2 THE POISSON PROCESS

The Poisson process is a continuous-time analog of the Bernoulli process and applies to situations where there is no natural way of dividing time into discrete periods.

To see the need for a continuous-time version of the Bernoulli process, let us consider a possible model of traffic accidents within a city. We can start by discretizing time into one-minute periods and record a "success" during every minute in which there is at least one traffic accident. Assuming the traffic intensity to be constant over time, the probability of an accident should be the same during each period. Under the additional (and quite plausible) assumption that different time periods are independent, the sequence of successes becomes a Bernoulli process. Note that in real life, two or more accidents during the same one-minute interval are certainly possible, but the Bernoulli process model does not keep track of the exact number of accidents. In particular, it does not allow us to calculate the expected number of accidents within a given period.

One way around this difficulty is to choose the length of a time period to be very small, so that the probability of two or more accidents becomes negligible. But how small should it be? A second? A millisecond? Instead of making an arbitrary choice, it is preferable to consider a limiting situation where the length of the time period becomes zero, and work with a continuous-time model.

We consider an arrival process that evolves in continuous time, in the sense that any real number t is a possible arrival time. We define

$$P(k, \tau) = \mathbf{P}(\text{there are exactly } k \text{ arrivals during an interval of length } \tau),$$

and assume that this probability is the same for all intervals of the same length τ. We also introduce a positive parameter λ, called the **arrival rate** or **intensity** of the process, for reasons that will soon become apparent.

Definition of the Poisson Process

An arrival process is called a Poisson process with rate λ if it has the following properties:

(a) **(Time-homogeneity)** The probability $P(k,\tau)$ of k arrivals is the same for all intervals of the same length τ.

(b) **(Independence)** The number of arrivals during a particular interval is independent of the history of arrivals outside this interval.

(c) **(Small interval probabilities)** The probabilities $P(k,\tau)$ satisfy

$$\begin{aligned} P(0,\tau) &= 1 - \lambda\tau + o(\tau), \\ P(1,\tau) &= \lambda\tau + o_1(\tau), \\ P(k,\tau) &= o_k(\tau), \qquad\qquad \text{for } k = 2, 3, \ldots \end{aligned}$$

Here, $o(\tau)$ and $o_k(\tau)$ are functions of τ that satisfy

$$\lim_{\tau \to 0} \frac{o(\tau)}{\tau} = 0, \qquad \lim_{\tau \to 0} \frac{o_k(\tau)}{\tau} = 0.$$

The first property states that arrivals are "equally likely" at all times. The arrivals during any time interval of length τ are statistically the same, i.e., they obey the same probability law. This is a counterpart to the assumption that the success probability p in a Bernoulli process is the same for all trials.

To interpret the second property, consider a particular interval $[t, t']$, of length $t' - t$. The unconditional probability of k arrivals during that interval is $P(k, t' - t)$. Suppose now that we are given complete or partial information on the arrivals outside this interval. Property (b) states that this information is irrelevant: the conditional probability of k arrivals during $[t, t']$ remains equal to the unconditional probability $P(k, t' - t)$. This property is analogous to the independence of trials in a Bernoulli process.

The third property is critical. The $o(\tau)$ and $o_k(\tau)$ terms are meant to be negligible in comparison to τ, when the interval length τ is very small. They can be thought of as the $O(\tau^2)$ terms in a Taylor series expansion of $P(k,\tau)$. Thus, for small τ, the probability of a single arrival is roughly $\lambda\tau$, plus a negligible term. Similarly, for small τ, the probability of zero arrivals is roughly $1 - \lambda\tau$. Finally, the probability of two or more arrivals is negligible in comparison to $P(1,\tau)$, as τ becomes smaller.

Number of Arrivals in an Interval

We will now derive some probability distributions associated with the arrivals in a Poisson process. We first use the connection with the Bernoulli process to obtain the PMF of the number of arrivals in a given time interval.

Let us consider a fixed time interval of length τ, and partition it into τ/δ periods of length δ, where δ is a very small number; see Fig. 6.5. The probability of more than two arrivals during any period can be neglected, because of property (c) and the preceding discussion. Different periods are independent, by property (b). Furthermore, each period has one arrival with probability approximately equal to $\lambda\delta$, or zero arrivals with probability approximately equal to $1 - \lambda\delta$. Therefore, the process being studied can be approximated by a Bernoulli process, with the approximation becoming more and more accurate as δ becomes smaller.

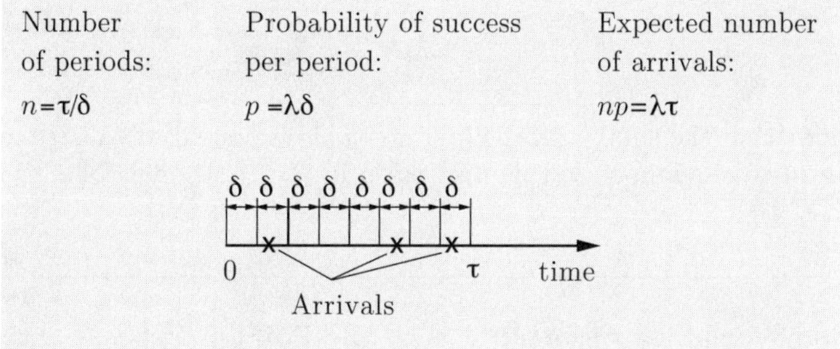

Figure 6.5: Bernoulli approximation of the Poisson process over an interval of length τ.

The probability $P(k, \tau)$ of k arrivals in time τ is approximately the same as the (binomial) probability of k successes in $n = \tau/\delta$ independent Bernoulli trials with success probability $p = \lambda\delta$ at each trial. While keeping the length τ of the interval fixed, we let the period length δ decrease to zero. We then note that the number n of periods goes to infinity, while the product np remains constant and equal to $\lambda\tau$. Under these circumstances, we saw in the previous section that the binomial PMF converges to a Poisson PMF with parameter $\lambda\tau$. We are then led to the important conclusion that

$$P(k, \tau) = e^{-\lambda\tau} \frac{(\lambda\tau)^k}{k!}, \quad k = 0, 1, \ldots.$$

Note that a Taylor series expansion of $e^{-\lambda\tau}$ yields

$$\begin{aligned} P(0, \tau) &= e^{-\lambda\tau} = 1 - \lambda\tau + o(\tau), \\ P(1, \tau) &= \lambda\tau e^{-\lambda\tau} = \lambda\tau - \lambda^2\tau^2 + O(\tau^3) = \lambda\tau + o_1(\tau), \end{aligned}$$

consistent with property (c).

Using our earlier formulas for the mean and variance of the Poisson PMF, we obtain

$$\mathbf{E}[N_\tau] = \lambda\tau, \qquad \text{var}(N_\tau) = \lambda\tau,$$

where N_τ denotes the number of arrivals during a time interval of length τ. These formulas are hardly surprising, since we are dealing with the limit of a binomial PMF with parameters $n = \tau/\delta$, $p = \lambda\delta$, mean $np = \lambda\tau$, and variance $np(1-p) \approx np = \lambda\tau$.

Let us now derive the probability law for the time T of the first arrival, assuming that the process starts at time zero. Note that we have $T > t$ if and only if there are no arrivals during the interval $[0, t]$. Therefore,

$$F_T(t) = \mathbf{P}(T \le t) = 1 - \mathbf{P}(T > t) = 1 - P(0, t) = 1 - e^{-\lambda t}, \qquad t \ge 0.$$

We then differentiate the CDF $F_T(t)$ of T, and obtain the PDF formula

$$f_T(t) = \lambda e^{-\lambda t}, \qquad t \ge 0,$$

which shows that the time until the first arrival is exponentially distributed with parameter λ. We summarize this discussion in the table that follows. See also Fig. 6.6.

Random Variables Associated with the Poisson Process and their Properties

- **The Poisson with parameter $\lambda\tau$.** This is the number N_τ of arrivals in a Poisson process with rate λ, over an interval of length τ. Its PMF, mean, and variance are

$$p_{N_\tau}(k) = P(k, \tau) = e^{-\lambda\tau}\frac{(\lambda\tau)^k}{k!}, \qquad k = 0, 1, \ldots,$$

$$\mathbf{E}[N_\tau] = \lambda\tau, \qquad \text{var}(N_\tau) = \lambda\tau.$$

- **The exponential with parameter λ.** This is the time T until the first arrival. Its PDF, mean, and variance are

$$f_T(t) = \lambda e^{-\lambda t}, \quad t \ge 0, \qquad \mathbf{E}[T] = \frac{1}{\lambda}, \qquad \text{var}(T) = \frac{1}{\lambda^2}.$$

Example 6.8. You get email according to a Poisson process at a rate of $\lambda = 0.2$ messages per hour. You check your email every hour. What is the probability of finding 0 and 1 new messages?

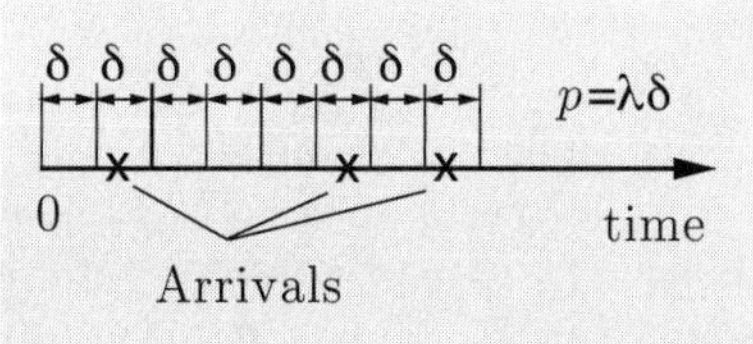

	Poisson	**Bernoulli**
Times of Arrival	Continuous	Discrete
PMF of # of Arrivals	Poisson	Binomial
Interarrival Time CDF	Exponential	Geometric
Arrival Rate	λ/unit time	p/trial

Figure 6.6: View of the Bernoulli process as the discrete-time version of the Poisson process. We discretize time in small intervals δ and associate each interval with a Bernoulli trial whose parameter is $p = \lambda\delta$. The table summarizes some of the basic correspondences.

These probabilities can be found using the Poisson PMF $e^{-\lambda\tau}(\lambda\tau)^k/k!$, with $\tau = 1$, and $k = 0$ or $k = 1$:

$$P(0, 1) = e^{-0.2} = 0.819, \qquad P(1, 1) = 0.2 \cdot e^{-0.2} = 0.164.$$

Suppose that you have not checked your email for a whole day. What is the probability of finding no new messages? We use again the Poisson PMF and obtain

$$P(0, 24) = e^{-0.2 \cdot 24} = 0.0083.$$

Alternatively, we can argue that the event of no messages in a 24-hour period is the intersection of the events of no messages during each of 24 hours. The latter events are independent and the probability of each is $P(0, 1) = e^{-0.2}$, so

$$P(0, 24) = \big(P(0, 1)\big)^{24} = \big(e^{-0.2}\big)^{24} = 0.0083,$$

which is consistent with the preceding calculation method.

Example 6.9. The Sum of Independent Poisson Random Variables is Poisson. Arrivals of customers at the local supermarket are modeled by a Poisson process with a rate of $\lambda = 10$ customers per minute. Let M be the number of customers arriving between 9:00 and 9:10. Also, let N be the number of customers arriving between 9:30 and 9:35. What is the distribution of $M + N$?

We notice that M is Poisson with parameter $\mu = 10\cdot 10 = 100$ and N is Poisson with parameter $\nu = 10\cdot 5 = 50$. Furthermore, M and N are independent. As shown in Section 4.4, using transforms, $M+N$ is Poisson with parameter $\mu+\nu = 150$ (see also Probem 11 in Chapter 4). We will now proceed to derive the same result in a more direct and intuitive manner.

Let $\tilde{N}$ be the number of customers that arrive between 9:10 and 9:15. Note that $\tilde{N}$ has the same distribution as N (Poisson with parameter 50). Furthermore, $\tilde{N}$ is also independent of M. Thus, the distribution of $M + N$ is the same as the distribution of $M + \tilde{N}$. But $M + \tilde{N}$ is the number of arrivals during an interval of length 15, and has therefore a Poisson distribution with parameter $10 \cdot 15 = 150$.

This example makes a point that is valid in general. The probability of k arrivals during a *set* of times of total length τ is always given by $P(k, \tau)$, even if that set is not an interval. (In this example, we dealt with the set [9:00,9:10]∪[9:30,9:35], of total length 15.)

Independence and Memorylessness

The Poisson process has several properties that parallel those of the Bernoulli process, including the independence of nonoverlapping time sets, and the memorylessness of the interarrival time distribution. Given that the Poisson process can be viewed as a limiting case of a Bernoulli process, the fact that it inherits the qualitative properties of the latter should be hardly surprising.

Independence Properties of the Poisson Process

- For any given time $t > 0$, the history of the process after time t is also a Poisson process, and is independent from the history of the process until time t.
- Let t be a given time and let $\overline{T}$ be the time of the first arrival after time t. Then, $\overline{T} - t$ has an exponential distribution with parameter λ, and is independent of the history of the process until time t.

The first property in the above table is established by observing that the portion of the process that starts at time t satisfies the properties required by the definition of the Poisson process. The independence of the future from the past is a direct consequence of the independence assumption in the definition of the Poisson process. Finally, the fact that $\overline{T} - t$ has the same exponential distribution can be verified by noting that

$$\mathbf{P}(\overline{T} - t > s) = \mathbf{P}\big(0 \text{ arrivals during } [t, t+s]\big) = P(0, s) = e^{-\lambda s}.$$

This is the memorylessness property, which is analogous to the one for the Bernoulli process. The following examples make use of this property.

Example 6.10. You and your partner go to a tennis court, and have to wait until the players occupying the court finish playing. Assume (somewhat unrealistically) that their playing time has an exponential PDF. Then, the PDF of your waiting time (equivalently, their remaining playing time) also has the same exponential PDF, regardless of when they started playing.

Example 6.11. When you enter the bank, you find that all three tellers are busy serving other customers, and there are no other customers in queue. Assume that the service times for you and for each of the customers being served are independent identically distributed exponential random variables. What is the probability that you will be the last to leave?

The answer is 1/3. To see this, focus at the moment when you start service with one of the tellers. Then, the remaining time of each of the other two customers being served, as well as your own remaining time, have the same PDF. Therefore, you and the other two customers have equal probability 1/3 of being the last to leave.

Interarrival Times

An important random variable associated with a Poisson process that starts at time 0, is the time of the kth arrival, which we denote by Y_k. A related random variable is the kth interarrival time, denoted by T_k. It is defined by

$$T_1 = Y_1, \qquad T_k = Y_k - Y_{k-1}, \qquad k = 2, 3, \ldots$$

and represents the amount of time between the $(k-1)$st and the kth arrival. Note that

$$Y_k = T_1 + T_2 + \cdots + T_k.$$

We have already seen that the time T_1 until the first arrival is an exponential random variable with parameter λ. Starting from the time T_1 of the first arrival, the future is a fresh-starting Poisson process.† Thus, the time until the next arrival has the same exponential PDF. Furthermore, the past of the process (up to time T_1) is independent of the future (after time T_1). Since T_2 is determined exclusively by what happens in the future, we see that T_2 is independent of T_1. Continuing similarly, we conclude that the random variables $T_1, T_2, T_3, \ldots$ are independent and all have the same exponential distribution.

This important observation leads to an alternative, but equivalent, way of describing the Poisson process.

† This statement is a bit stronger than the fact, discussed earlier, that starting from any given deterministic time t the process starts fresh, but is quite intuitive. It can be formally justified using an argument analogous to the one in Example 6.3, by conditioning on all possible values of the random variable T_1.

Alternative Description of the Poisson Process

1. Start with a sequence of independent exponential random variables $T_1, T_2, \ldots$, with common parameter λ, and let these represent the interarrival times.
2. Record an arrival at times T_1, $T_1 + T_2$, $T_1 + T_2 + T_3$, etc.

The kth Arrival Time

The time Y_k of the kth arrival is equal to the sum $Y_k = T_1 + T_2 + \cdots + T_k$ of k independent identically distributed exponential random variables. This allows us to derive formulas for the mean, variance, and PDF of Y_k, which are given in the table that follows.

Properties of the kth Arrival Time

- The kth arrival time is equal to the sum of the first k interarrival times

$$Y_k = T_1 + T_2 + \cdots + T_k,$$

  and the latter are independent exponential random variables with common parameter λ.
- The mean and variance of Y_k are given by

$$\mathbf{E}[Y_k] = \mathbf{E}[T_1] + \cdots + \mathbf{E}[T_k] = \frac{k}{\lambda},$$

$$\text{var}(Y_k) = \text{var}(T_1) + \cdots + \text{var}(T_k) = \frac{k}{\lambda^2}.$$

- The PDF of Y_k is given by

$$f_{Y_k}(y) = \frac{\lambda^k y^{k-1} e^{-\lambda y}}{(k-1)!}, \qquad y \geq 0,$$

  and is known as the **Erlang PDF of order k**.

To evaluate the PDF f_{Y_k} of Y_k, we argue that for a small δ, the product $\delta \cdot f_{Y_k}(y)$ approximates the probability that the kth arrival occurs between times

y and $y + \delta$.† When δ is very small, the probability of more than one arrival during the interval $[y, y + \delta]$ is negligible. Thus, the kth arrival occurs between y and $y + \delta$ if and only if the following two events A and B occur:

(a) event A: there is an arrival during the interval $[y, y + \delta]$;

(b) event B: there are exactly $k - 1$ arrivals before time y.

The probabilities of these two events are

$$\mathbf{P}(A) \approx \lambda\delta, \qquad \text{and} \qquad \mathbf{P}(B) = P(k-1, y) = \frac{\lambda^{k-1} y^{k-1} e^{-\lambda y}}{(k-1)!}.$$

Since A and B are independent, we have

$$\delta f_{Y_k}(y) \approx \mathbf{P}(y \le Y_k \le y + \delta) \approx \mathbf{P}(A \cap B) = \mathbf{P}(A)\,\mathbf{P}(B) \approx \lambda\delta \frac{\lambda^{k-1} y^{k-1} e^{-\lambda y}}{(k-1)!},$$

from which we obtain

$$f_{Y_k}(y) = \frac{\lambda^k y^{k-1} e^{-\lambda y}}{(k-1)!}, \qquad y \ge 0.$$

Example 6.12. You call the IRS hotline and you are told that you are the 56th person in line, excluding the person currently being served. Callers depart according to a Poisson process with a rate of $\lambda = 2$ per minute. How long will you have to wait on the average until your service starts, and what is the probability you will have to wait for more than 30 minutes?

By the memorylessness property, the remaining service time of the person currently being served is exponentially distributed with parameter 2. The service times of the 55 persons ahead of you are also exponential with the same parameter,

---

† For an alternative derivation that does not rely on approximation arguments, note that for a given $y \ge 0$, the event $\{Y_k \le y\}$ is the same as the event

$$\big\{\text{number of arrivals in the interval } [0, y] \text{ is at least } k\big\}.$$

Thus, the CDF of Y_k is given by

$$F_{Y_k}(y) = \mathbf{P}\big(Y_k \le y\big) = \sum_{n=k}^{\infty} P(n, y) = 1 - \sum_{n=0}^{k-1} P(n, y) = 1 - \sum_{n=0}^{k-1} \frac{(\lambda y)^n e^{-\lambda y}}{n!}.$$

The PDF of Y_k can be obtained by differentiating the above expression, which by straightforward calculation yields the Erlang PDF formula

$$f_{Y_k}(y) = \frac{d}{dy} F_{Y_k}(y) = \frac{\lambda^k y^{k-1} e^{-\lambda y}}{(k-1)!}.$$

and all of these random variables are independent. Thus, your waiting time in minutes, call it Y, is Erlang of order 56, and

$$\mathbf{E}[Y] = \frac{56}{\lambda} = 28.$$

The probability that you have to wait for more than 30 minutes is given by the formula

$$\mathbf{P}(Y \geq 30) = \int_{30}^{\infty} \frac{\lambda^{56} y^{55} e^{-\lambda y}}{55!} \, dy.$$

Computing this probability is quite tedious. On the other hand, since Y is the sum of a large number of independent identically distributed random variables, we can use an approximation based on the central limit theorem and the normal tables.

Splitting and Merging of Poisson Processes

Similar to the case of a Bernoulli process, we can start with a Poisson process with rate λ and **split** it, as follows: each arrival is kept with probability p and discarded with probability $1-p$, independent of what happens to other arrivals. In the Bernoulli case, we saw that the result of the splitting was also a Bernoulli process. In the present context, the result of the splitting turns out to be a Poisson process with rate λp.

Alternatively, we can start with two independent Poisson processes, with rates λ_1 and λ_2, and merge them by recording an arrival whenever an arrival occurs in either process. It turns out that the merged process is also Poisson with rate $\lambda_1 + \lambda_2$. Furthermore, any particular arrival of the merged process has probability $\lambda_1/(\lambda_1 + \lambda_2)$ of originating from the first process, and probability $\lambda_2/(\lambda_1 + \lambda_2)$ of originating from the second, independent of all other arrivals and their origins.

We discuss these properties in the context of some examples, and at the same time provide the arguments that establish their validity.

Example 6.13. Splitting of Poisson Processes. A packet that arrives at a node of a data network is either a local packet that is destined for that node (this happens with probability p), or else it is a transit packet that must be relayed to another node (this happens with probability $1 - p$). Packets arrive according to a Poisson process with rate λ, and each one is a local or transit packet independent of other packets and of the arrival times. As stated above, the process of *local* packet arrivals is Poisson with rate λp. Let us see why.

We verify that the process of local packet arrivals satisfies the defining properties of a Poisson process. Since λ and p are constant (do not change with time), the first property (time-homogeneity) clearly holds. Furthermore, there is no dependence between what happens in disjoint time intervals, verifying the second property. Finally, if we focus on a small interval of length δ, the probability of a local arrival is approximately the probability that there is a packet arrival, and that this turns out to be a local one, i.e., $\lambda\delta \cdot p$. In addition, the probability of two or more local arrivals is negligible in comparison to δ, and this verifies the third property.

We conclude that local packet arrivals form a Poisson process and, in particular, the number of such arrivals during an interval of length τ has a Poisson PMF with parameter $p\lambda\tau$. By a symmetrical argument, the process of transit packet arrivals is also Poisson, with rate $\lambda(1-p)$. A somewhat surprising fact in this context is that the two Poisson processes obtained by splitting an original Poisson process are independent; see the end-of-chapter problems.

Example 6.14. Merging of Poisson Processes. People with letters to mail arrive at the post office according to a Poisson process with rate λ_1, while people with packages to mail arrive according to an independent Poisson process with rate λ_2. As stated earlier the merged process, which includes arrivals of both types, is Poisson with rate $\lambda_1 + \lambda_2$. Let us see why.

First, it should be clear that the merged process satisfies the time-homogeneity property. Furthermore, since different intervals in each of the two arrival processes are independent, the same property holds for the merged process. Let us now focus on a small interval of length δ. Ignoring terms that are negligible compared to δ, we have

$$\mathbf{P}(0 \text{ arrivals in the merged process}) \approx (1-\lambda_1\delta)(1-\lambda_2\delta) \approx 1-(\lambda_1+\lambda_2)\delta,$$

$$\mathbf{P}(1 \text{ arrival in the merged process}) \approx \lambda_1\delta(1-\lambda_2\delta) + (1-\lambda_1\delta)\lambda_2\delta \approx (\lambda_1+\lambda_2)\delta,$$

and the third property has been verified.

Given that an arrival has just been recorded, what is the probability that it is an arrival of a person with a letter to mail? We focus again on a small interval of length δ around the current time, and we seek the probability

$$\mathbf{P}(1 \text{ arrival of person with a letter} \mid 1 \text{ arrival}).$$

Using the definition of conditional probabilities, and ignoring the negligible probability of more than one arrival, this is

$$\frac{\mathbf{P}(1 \text{ arrival of person with a letter})}{\mathbf{P}(1 \text{ arrival})} \approx \frac{\lambda_1\delta}{(\lambda_1+\lambda_2)\delta} = \frac{\lambda_1}{\lambda_1+\lambda_2}.$$

Generalizing this calculation, we let L_k be the event that the kth arrival corresponds to an arrival of a person with a letter to mail, and we have

$$\mathbf{P}(L_k) = \frac{\lambda_1}{\lambda_1+\lambda_2}.$$

Furthermore, since distinct arrivals happen at different times, and since, for Poisson processes, events at different times are independent, it follows that the random variables $L_1, L_2, \ldots$ are independent.

Example 6.15. Competing Exponentials. Two light bulbs have independent and exponentially distributed lifetimes T_a and T_b, with parameters λ_a and λ_b, respectively. What is the distribution of $Z = \min\{T_a, T_b\}$, the first time when a bulb burns out?

For all $z \geq 0$, we have,

$$\begin{aligned} F_Z(z) &= \mathbf{P}\big(\min\{T_a, T_b\} \leq z\big) \\ &= 1 - \mathbf{P}\big(\min\{T_a, T_b\} > z\big) \\ &= 1 - \mathbf{P}(T_a > z,\, T_b > z) \\ &= 1 - \mathbf{P}(T_a > z)\mathbf{P}(T_b > z) \\ &= 1 - e^{-\lambda_a z} e^{-\lambda_b z} \\ &= 1 - e^{-(\lambda_a + \lambda_b) z}. \end{aligned}$$

This is recognized as the exponential CDF with parameter $\lambda_a + \lambda_b$. Thus, the minimum of two independent exponentials with parameters λ_a and λ_b is an exponential with parameter $\lambda_a + \lambda_b$.

For a more intuitive explanation of this fact, let us think of T_a and T_b as the times of the first arrivals in two independent Poisson processes with rates λ_a and λ_b, respectively. If we merge these two processes, the first arrival time will be $\min\{T_a, T_b\}$. But we already know that the merged process is Poisson with rate $\lambda_a + \lambda_b$, and it follows that the first arrival time, $\min\{T_a, T_b\}$, is exponential with parameter $\lambda_a + \lambda_b$.

The preceding discussion can be generalized to the case of more than two processes. Thus, the total arrival process obtained by merging the arrivals of n independent Poisson processes with arrival rates $\lambda_1, \ldots, \lambda_n$ is Poisson with arrival rate equal to the sum $\lambda_1 + \cdots + \lambda_n$.

Example 6.16. More on Competing Exponentials. Three light bulbs have independent exponentially distributed lifetimes with a common parameter λ. What is the expected value of the time until the last bulb burns out?

We think of the times when each bulb burns out as the first arrival times in independent Poisson processes. In the beginning, we have three bulbs, and the merged process has rate 3λ. Thus, the time T_1 of the first burnout is exponential with parameter 3λ, and mean $1/3\lambda$. Once a bulb burns out, and because of the memorylessness property of the exponential distribution, the remaining lifetimes of the other two light bulbs are again independent exponential random variables with parameter λ. We thus have *two* Poisson processes running in parallel, and the remaining time T_2 until the first arrival in one of these two processes is now exponential with parameter 2λ and mean $1/2\lambda$. Finally, once a second bulb burns out, we are left with a single one. Using memorylessness once more, the remaining time T_3 until the last bulb burns out is exponential with parameter λ and mean $1/\lambda$. Thus, the expected value of the total time is

$$\mathbf{E}[T_1 + T_2 + T_3] = \frac{1}{3\lambda} + \frac{1}{2\lambda} + \frac{1}{\lambda}.$$

Note that the random variables T_1, T_2, T_3 are independent, because of memorylessness. This allows us to also compute the variance of the total time:

$$\text{var}(T_1 + T_2 + T_3) = \text{var}(T_1) + \text{var}(T_2) + \text{var}(T_3) = \frac{1}{9\lambda^2} + \frac{1}{4\lambda^2} + \frac{1}{\lambda^2}.$$

Bernoulli and Poisson Processes, and Sums of Random Variables

The insights obtained from splitting and merging of Bernoulli and Poisson processes can be used to provide simple explanations of some interesting properties involving sums of a random number of independent random variables. Alternative proofs, based for example on manipulating PMFs/PDFs, solving derived distribution problems, or using transforms, tend to be unintuitive. We collect these properties in the following table.

> **Properties of Sums of a Random Number of Random Variables**
>
> Let $N, X_1, X_2, \ldots$ be independent random variables, where N takes nonnegative integer values. Let $Y = X_1 + \cdots + X_N$ for positive values of N, and let $Y = 0$ when $N = 0$.
>
> - If X_i is Bernoulli with parameter p, and N is binomial with parameters m and q, then Y is binomial with parameters m and pq.
> - If X_i is Bernoulli with parameter p, and N is Poisson with parameter λ, then Y is Poisson with parameter λp.
> - If X_i is geometric with parameter p, and N is geometric with parameter q, then Y is geometric with parameter pq.
> - If X_i is exponential with parameter λ, and N is geometric with parameter q, then Y is exponential with parameter λq.

The first two properties are shown in Problem 22, the third property is shown in Problem 6, and the last property is shown in Problem 23. The last three properties were also shown in Chapter 4, by using transforms (see Section 4.4 and the last end-of-chapter problem of Chapter 4). One more related property is shown in Problem 24, namely that if N_t denotes the number of arrivals of a Poisson process with parameter λ within an interval of length t, and T is an interval with length that is exponentially distributed with parameter ν and is independent of the Poisson process, then $N_T + 1$ is geometrically distributed with parameter $\nu/(\lambda + \nu)$.

Let us also note a related and quite deep fact, namely that the sum of a *large* number of (*not* necessarily Poisson) independent arrival processes, can be approximated by a Poisson process with arrival rate equal to the sum of the individual arrival rates. The component processes must have a small rate relative to the total (so that none of them imposes its probabilistic character on the total arrival process) and they must also satisfy some technical mathematical assumptions. Further discussion of this fact is beyond our scope, but we note that it is in large measure responsible for the abundance of Poisson-like processes in practice. For example, the telephone traffic originating in a city consists of many component processes, each of which characterizes the phone calls placed by individual residents. The component processes need not be Poisson; some people for example tend to make calls in batches, and (usually) while in the process of talking, cannot initiate or receive a second call. However, the total telephone traffic is well-modeled by a Poisson process. For the same reasons, the process of auto accidents in a city, customer arrivals at a store, particle emissions from radioactive material, etc., tend to have the character of the Poisson process.

The Random Incidence Paradox

The arrivals of a Poisson process partition the time axis into a sequence of interarrival intervals; each interarrival interval starts with an arrival and ends at the time of the next arrival. We have seen that the lengths of these interarrival intervals are independent exponential random variables with parameter λ, where λ is the rate of the process. More precisely, for every k, the length of the kth interarrival interval has this exponential distribution. In this subsection, we look at these interarrival intervals from a different perspective.

Let us fix a time instant t^* and consider the length L of the interarrival interval that contains t^*. For a concrete context, think of a person who shows up at the bus station at some arbitrary time t^* and records the time from the previous bus arrival until the next bus arrival. The arrival of this person is often referred to as a "random incidence," but the reader should be aware that the term is misleading: t^* is just a particular time instance, not a random variable.

We assume that t^* is much larger than the starting time of the Poisson process so that we can be fairly certain that there has been an arrival prior to time t^*. To avoid the issue of how large t^* should be, we assume that the Poisson process has been running forever, so that we can be certain that there has been a prior arrival, and that L is well-defined. One might superficially argue that L is the length of a "typical" interarrival interval, and is exponentially distributed, but this turns out to be false. Instead, we will establish that L has an Erlang PDF of order two.

This is known as the *random incidence phenomenon* or *paradox*, and it can be explained with the help of Fig. 6.7. Let $[U, V]$ be the interarrival interval that contains t^*, so that $L = V - U$. In particular, U is the time of the first arrival prior to t^* and V is the time of the first arrival after t^*. We split L into two parts,

$$L = (t^* - U) + (V - t^*),$$

where $t^* - U$ is the elapsed time since the last arrival, and $V - t^*$ is the remaining time until the next arrival. Note that $t^* - U$ is determined by the past history of the process (before t^*), while $V - t^*$ is determined by the future of the process (after time t^*). By the independence properties of the Poisson process, the random variables $t^* - U$ and $V - t^*$ are independent. By the memorylessness property, the Poisson process starts fresh at time t^*, and therefore $V - t^*$ is exponential with parameter λ. The random variable $t^* - U$ is also exponential with parameter λ. The easiest way to see this is to realize that if we run a Poisson process backwards in time it remains Poisson; this is because the defining properties of a Poisson process make no reference to whether time moves forward or backward. A more formal argument is obtained by noting that

$$\mathbf{P}(t^* - U > x) = \mathbf{P}\big(\text{no arrivals during } [t^* - x, t^*]\big) = P(0, x) = e^{-\lambda x}, \qquad x \geq 0.$$

We have therefore established that L is the sum of two independent exponential random variables with parameter λ, i.e., Erlang of order two, with mean $2/\lambda$.

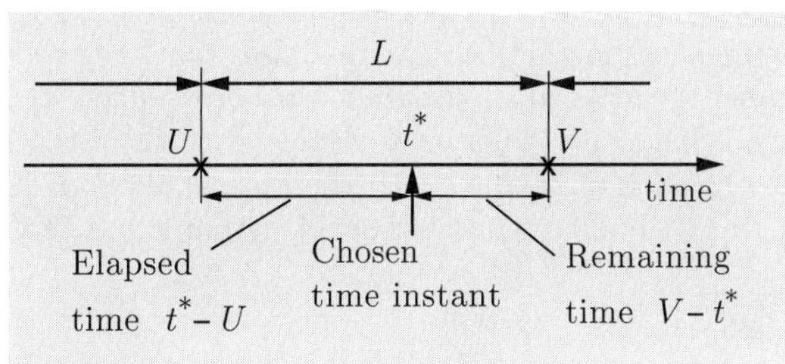

Figure 6.7: Illustration of the random incidence phenomenon. For a fixed time instant t^*, the corresponding interarrival interval $[U, V]$ consists of the elapsed time $t^* - U$ and the remaining time $V - t^*$. These two times are independent and are exponentially distributed with parameter λ, so the PDF of their sum is Erlang of order two.

Random incidence phenomena are often the source of misconceptions and errors, but these can be avoided with careful probabilistic modeling. The key issue is that an observer who arrives at an arbitrary time is more likely to fall in a large rather than a small interarrival interval. As a consequence the expected length seen by the observer is higher, $2/\lambda$ compared with the $1/\lambda$ mean of the exponential PDF. A similar situation arises in the example that follows.

Example 6.17. Random Incidence in a Non-Poisson Arrival Process. Buses arrive at a station deterministically, on the hour, and five minutes after the hour. Thus, the interarrival times alternate between 5 and 55 minutes. The average interarrival time is 30 minutes. A person shows up at the bus station at a "random" time. We interpret "random" to mean a time that is uniformly distributed within a particular hour. Such a person falls into an interarrival interval of length 5 with probability $1/12$, and an interarrival interval of length 55 with probability $11/12$. The expected length of the chosen interarrival interval is

$$5 \cdot \frac{1}{12} + 55 \cdot \frac{11}{12} = 50.83,$$

which is considerably larger than 30, the average interarrival time.

As the preceding example indicates, random incidence is a subtle phenomenon that introduces a bias in favor of larger interarrival intervals, and can manifest itself in contexts other than the Poisson process. In general, whenever different calculations give contradictory results, the reason is that they refer to different probabilistic mechanisms. For instance, considering a fixed **nonrandom** k and the associated random value of the kth interarrival interval is a different experiment from fixing a time t and considering the **random** K such that the Kth interarrival interval contains t.

For a last example with the same flavor, consider a survey of the utilization of town buses. One approach is to select a few buses at random and calculate the average number of riders in the selected buses. An alternative approach is to select a few bus riders at random, look at the buses that they rode and calculate the average number of riders in the latter set of buses. The estimates produced by the two methods have very different statistics, with the second method being biased upwards. The reason is that with the second method, it is much more likely to select a bus with a large number of riders than a bus that is near-empty.

6.3 SUMMARY AND DISCUSSION

In this chapter, we introduced and analyzed two memoryless arrival processes. The Bernoulli process evolves in discrete time, and during each discrete time step, there is a constant probability p of an arrival. The Poisson process evolves in continuous time, and during each small interval of length $\delta > 0$, there is a probability of an arrival approximately equal to $\lambda\delta$. In both cases, the numbers of arrivals in disjoint time intervals are assumed independent. The Poisson process can be viewed as a limiting case of the Bernoulli process, in which the duration of each discrete time slot is taken to be a very small number δ. This fact can be used to draw parallels between the major properties of the two processes, and to transfer insights gained from one process to the other.

Using the memorylessness property of the Bernoulli and Poisson processes, we derived the following.

(a) The PMF of the number of arrivals during a time interval of given length is binomial and Poisson, respectively.

(b) The distribution of the time between successive arrivals is geometric and exponential, respectively.

(c) The distribution of the time until the kth arrival, is Pascal of order k and Erlang of order k, respectively.

Furthermore, we saw that one can start with two independent Bernoulli (respectively, Poisson) processes and "merge" them to form a new Bernoulli (respectively, Poisson) process. Conversely, if one "accepts" each arrival by tossing a coin with success probability p ("splitting"), the process of accepted arrivals is a Bernoulli or Poisson process whose arrival rate is p times the original rate.

We finally considered the "random incidence" phenomenon where an external observer arrives at some given time and measures the interarrival interval within which he arrives. The probabilistic properties of the measured interval are not "typical" because the arriving observer is more likely to fall in a larger interarrival interval. This phenomenon indicates that when talking about a "typical" interval, one must carefully describe the mechanism by which it is selected. Different mechanisms will in general result in different probabilistic properties.

PROBLEMS

SECTION 7.1. The Bernoulli Process

Problem 1. Each of n packages is loaded independently onto either a red truck (with probability p) or onto a green truck (with probability $1-p$). Let R be the total number of items selected for the red truck and let G be the total number of items selected for the green truck.

(a) Determine the PMF, expected value, and variance of the random variable R.

(b) Evaluate the probability that the first item to be loaded ends up being the only one on its truck.

(c) Evaluate the probability that at least one truck ends up with a total of exactly one package.

(d) Evaluate the expected value and the variance of the difference $R - G$.

(e) Assume that $n \geq 2$. Given that both of the first two packages to be loaded go onto the red truck, find the conditional expectation, variance, and PMF of the random variable R.

Problem 2. Dave fails quizzes with probability $1/4$, independent of other quizzes.

(a) What is the probability that Dave fails exactly two of the next six quizzes?

(b) What is the expected number of quizzes that Dave will pass before he has failed three times?

(c) What is the probability that the second and third time Dave fails a quiz will occur when he takes his eighth and ninth quizzes, respectively?

(d) What is the probability that Dave fails two quizzes in a row before he passes two quizzes in a row?

Problem 3. A computer system carries out tasks submitted by two users. Time is divided into slots. A slot can be idle, with probability $p_I = 1/6$, and busy with probability $p_B = 5/6$. During a busy slot, there is probability $p_{1|B} = 2/5$ (respectively, $p_{2|B} = 3/5$) that a task from user 1 (respectively, 2) is executed. We assume that events related to different slots are independent.

(a) Find the probability that a task from user 1 is executed for the first time during the 4th slot.

(b) Given that exactly 5 out of the first 10 slots were idle, find the probability that the 6th idle slot is slot 12.

(c) Find the expected number of slots up to and including the 5th task from user 1.

(d) Find the expected number of busy slots up to and including the 5th task from user 1.

(e) Find the PMF, mean, and variance of the number of tasks from user 2 until the time of the 5th task from user 1.

Problem 4.* Consider a Bernoulli process with probability of success in each trial equal to p.

(a) Relate the number of failures before the rth success (sometimes called a **negative binomial** random variable) to a Pascal random variable and derive its PMF.

(b) Find the expected value and variance of the number of failures before the rth success.

(c) Obtain an expression for the probability that the ith failure occurs before the rth success.

Solution. (a) Let Y be the number of trials until the rth success, which is a Pascal random variable of order r. Let X be the number of failures before the rth success, so that $X = Y - r$. Therefore, $p_X(k) = p_Y(k + r)$, and

$$p_X(k) = \binom{k+r-1}{r-1} p^r (1-p)^k, \qquad k = 0, 1, \ldots.$$

(b) Using the notation of part (a), we have

$$\mathbf{E}[X] = \mathbf{E}[Y] - r = \frac{r}{p} - r = \frac{(1-p)r}{p}.$$

Furthermore,

$$\text{var}(X) = \text{var}(Y) = \frac{(1-p)r}{p^2}.$$

(c) Let again X be the number of failures before the rth success. The ith failure occurs before the rth success if and only if $X \geq i$. Therefore, the desired probability is equal to

$$\sum_{k=i}^{\infty} p_X(k) = \sum_{k=i}^{\infty} \binom{k+r-1}{r-1} p^r (1-p)^k, \qquad i = 1, 2, \ldots.$$

An alternative formula is derived as follows. Consider the first $r + i - 1$ trials. The number of failures in these trials is at least i if and only if the number of successes is less than r. But this is equivalent to the ith failure occurring before the rth success. Hence, the desired probability is the probability that the number of successes in $r+i-1$ trials is less than r, which is

$$\sum_{k=0}^{r-1} \binom{r+i-1}{k} p^k (1-p)^{r+i-1-k}, \qquad i = 1, 2, \ldots.$$

Problem 5.* **Random incidence in the Bernoulli process.** Your cousin has been playing the same video game from time immemorial. Assume that he wins each game with probability p, independent of the outcomes of other games. At midnight, you enter his room and witness his losing the current game. What is the PMF of the number of lost games between his most recent win and his first future win?

Solution. Let t be the number of the game when you enter the room. Let M be the number of the most recent past game that he won, and let N be the number of the first game to be won in the future. The random variable $X = N - t$ is geometrically distributed with parameter p. By symmetry and independence of the games, the random variable $Y = t - M$ is also geometrically distributed with parameter p. The games he lost between his most recent win and his first future win are all the games between M and N. Their number L is given by

$$L = N - M - 1 = X + Y - 1.$$

Thus, $L + 1$ has a Pascal PMF of order two, and

$$\mathbf{P}(L + 1 = k) = \binom{k-1}{1} p^2 (1-p)^{k-2} = (k-1)p^2(1-p)^{k-2}, \qquad k = 2, 3, \ldots.$$

Hence,

$$p_L(i) = \mathbf{P}(L + 1 = i + 1) = i\,p^2(1-p)^{i-1}, \qquad i = 1, 2, \ldots.$$

Problem 6.* Sum of a geometric number of independent geometric random variables. Let $Y = X_1 + \cdots + X_N$, where the random variables X_i are geometric with parameter p, and N is geometric with parameter q. Assume that the random variables $N, X_1, X_2, \ldots$ are independent. Show, without using transforms, that Y is geometric with parameter pq. *Hint:* Interpret the various random variables in terms of a split Bernoulli process.

Solution. We derived this result in Chapter 4, using transforms, but we develop a more intuitive derivation here. We interpret the random variables X_i and N as follows. We view the times X_1, $X_1 + X_2$, etc. as the arrival times in a Bernoulli process with parameter p. Each arrival is rejected with probability $1 - q$ and is accepted with probability q. We interpret N as the number of arrivals until the first acceptance. The process of accepted arrivals is obtained by splitting a Bernoulli process and is therefore itself Bernoulli with parameter pq. The random variable $Y = X_1 + \cdots + X_N$ is the time of the first accepted arrival and is therefore geometric, with parameter pq.

Problem 7.* The bits in a uniform random variable form a Bernoulli process. Let $X_1, X_2, \ldots$ be a sequence of binary random variables taking values in the set $\{0, 1\}$. Let Y be a continuous random variable that takes values in the set $[0, 1]$. We relate X and Y by assuming that Y is the real number whose binary representation is $0.X_1X_2X_3\ldots$. More concretely, we have

$$Y = \sum_{k=1}^{\infty} 2^{-k} X_k.$$

(a) Suppose that the X_i form a Bernoulli process with parameter $p = 1/2$. Show that Y is uniformly distributed. *Hint:* Consider the probability of the event $(i-1)/2^k < Y < i/2^k$, where i and k are positive integers.

(b) Suppose that Y is uniformly distributed. Show that the X_i form a Bernoulli process with parameter $p = 1/2$.

Solution. (a) We have

$$\mathbf{P}\big(Y \in [0, 1/2]\big) = \mathbf{P}(X_1 = 0) = \frac{1}{2} = \mathbf{P}\big(Y \in [1/2, 1]\big).$$

Furthermore,

$$\mathbf{P}\big(Y \in [0, 1/4]\big) = \mathbf{P}(X_1 = 0,\ X_2 = 0) = \frac{1}{4}.$$

Arguing similarly, we consider an interval of the form $\big[(i-1)/2^k, i/2^k\big]$, where i and k are positive integers and $i \leq 2^k$. For Y to fall in the interior of this interval, we need $X_1, \ldots, X_k$ to take on a particular sequence of values (namely, the digits in the binary expansion of $i - 1$). Hence,

$$\mathbf{P}\left((i-1)/2^k < Y < i/2^k\right) = \frac{1}{2^k}.$$

Note also that for any $y \in [0, 1]$, we have $\mathbf{P}(Y = y) = 0$, because the event $\{Y = y\}$ can only occur if infinitely many X_is take on particular values, a zero probability event. Therefore, the CDF of Y is continuous and satisfies

$$\mathbf{P}(Y \leq i/2^k) = i/2^k.$$

Since every number y in $[0, 1]$ can be closely approximated by a number of the form $i/2^k$, we have $\mathbf{P}(Y \leq y) = y$, for every $y \in [0, 1]$, which establishes that Y is uniform.

(b) As in part (a), we observe that every possible zero-one pattern for $X_1, \ldots, X_k$ is associated to one particular interval of the form $\big[(i-1)/2^k, i/2^k\big]$ for Y. These intervals have equal length, and therefore have the same probability $1/2^k$, since Y is uniform. This particular joint PMF for $X_1, \ldots, X_k$, corresponds to independent Bernoulli random variables with parameter $p = 1/2$.

SECTION 7.2. The Poisson Process

Problem 8. During rush hour, from 8 a.m. to 9 a.m., traffic accidents occur according to a Poisson process with a rate of 5 accidents per hour. Between 9 a.m. and 11 a.m., they occur as an independent Poisson process with a rate of 3 accidents per hour. What is the PMF of the total number of accidents between 8 a.m. and 11 a.m.?

Problem 9. An athletic facility has 5 tennis courts. Pairs of players arrive at the courts and use a court for an exponentially distributed time with mean 40 minutes. Suppose a pair of players arrives and finds all courts busy and k other pairs waiting in queue. What is the expected waiting time to get a court?

Problem 10. A fisherman catches fish according to a Poisson process with rate $\lambda = 0.6$ per hour. The fisherman will keep fishing for two hours. If he has caught at least one fish, he quits. Otherwise, he continues until he catches at least one fish.

(a) Find the probability that he stays for more than two hours.

(b) Find the probability that the total time he spends fishing is between two and five hours.

(c) Find the probability that he catches at least two fish.

(d) Find the expected number of fish that he catches.

(e) Find the expected total fishing time, given that he has been fishing for four hours.

Problem 11. Customers depart from a bookstore according to a Poisson process with rate λ per hour. Each customer buys a book with probability p, independent of everything else.

(a) Find the distribution of the time until the first sale of a book.

(b) Find the probability that no books are sold during a particular hour.

(c) Find the expected number of customers who buy a book during a particular hour.

Problem 12. A pizza parlor serves n different types of pizza, and is visited by a number K of customers in a given period of time, where K is a Poisson random variable with mean λ. Each customer orders a single pizza, with all types of pizza being equally likely, independent of the number of other customers and the types of pizza they order. Find the expected number of different types of pizzas ordered.

Problem 13. Transmitters A and B independently send messages to a single receiver in a Poisson manner, with rates of λ_A and λ_B, respectively. All messages are so brief that we may assume that they occupy single points in time. The number of words in a message, regardless of the source that is transmitting it, is a random variable with PMF

$$p_W(w) = \begin{cases} 2/6, & \text{if } w = 1, \\ 3/6, & \text{if } w = 2, \\ 1/6, & \text{if } w = 3, \\ 0, & \text{otherwise,} \end{cases}$$

and is independent of everything else.

(a) What is the probability that during an interval of duration t, a total of exactly nine messages will be received?

(b) Let N be the total number of words received during an interval of duration t. Determine the expected value of N.

(c) Determine the PDF of the time from $t = 0$ until the receiver has received exactly eight three-word messages from transmitter A.

(d) What is the probability that exactly eight of the next twelve messages received will be from transmitter A?

Problem 14. Beginning at time $t = 0$, we start using bulbs, one at a time, to illuminate a room. Bulbs are replaced immediately upon failure. Each new bulb is selected independently by an equally likely choice between a type-A bulb and a type-B bulb. The lifetime, X, of any particular bulb of a particular type is a random variable, independent of everything else, with the following PDF:

$$\text{for type-A Bulbs: } f_X(x) = \begin{cases} e^{-x}, & \text{if } x \geq 0, \\ 0, & \text{otherwise;} \end{cases}$$

$$\text{for type-B Bulbs: } f_X(x) = \begin{cases} 3e^{-3x}, & \text{if } x \geq 0, \\ 0, & \text{otherwise.} \end{cases}$$

(a) Find the expected time until the first failure.

(b) Find the probability that there are no bulb failures before time t.

(c) Given that there are no failures until time t, determine the conditional probability that the first bulb used is a type-A bulb.

(d) Find the variance of the time until the first bulb failure.

(e) Find the probability that the 12th bulb failure is also the 4th type-A bulb failure.

(f) Up to and including the 12th bulb failure, what is the probability that a total of exactly 4 type-A bulbs have failed?

(g) Determine either the PDF or the transform associated with the time until the 12th bulb failure.

(h) Determine the probability that the total period of illumination provided by the first two type-B bulbs is longer than that provided by the first type-A bulb.

(i) Suppose the process terminates as soon as a total of exactly 12 bulb failures have occurred. Determine the expected value and variance of the total period of illumination provided by type-B bulbs while the process is in operation.

(j) Given that there are no failures until time t, find the expected value of the time until the first failure.

Problem 15. A service station handles jobs of two types, A and B. (Multiple jobs can be processed simultaneously.) Arrivals of the two job types are independent Poisson processes with parameters $\lambda_A = 3$ and $\lambda_B = 4$ per minute, respectively. Type A jobs stay in the service station for exactly one minute. Each type B job stays in the service station for a random but integer amount of time which is geometrically distributed, with mean equal to 2, and independent of everything else. The service station started operating at some time in the remote past.

(a) What is the mean, variance, and PMF of the total number of jobs that arrive within a given three-minute interval?

(b) We are told that during a 10-minute interval, exactly 10 new jobs arrived. What is the probability that exactly 3 of them are of type A?

(c) At time 0, no job is present in the service station. What is the PMF of the number of type B jobs that arrive in the future, but before the first type A arrival?

(d) At time $t = 0$, there were exactly two type A jobs in the service station. What is the PDF of the time of the last (before time 0) type A arrival?

(e) At time 1, there was exactly one type B job in the service station. Find the distribution of the time until this type B job departs.

Problem 16. Each morning, as you pull out of your driveway, you would like to make a U-turn rather than drive around the block. Unfortunately, U-turns are illegal in your neighborhood, and police cars drive by according to a Poisson process with rate λ. You decide to make a U-turn once you see that the road has been clear of police cars for τ time units. Let N be the number of police cars you see before you make the U-turn.

(a) Find $\mathbf{E}[N]$.

(b) Find the conditional expectation of the time elapsed between police cars $n-1$ and n, given that $N \geq n$.

(c) Find the expected time that you wait until you make the U-turn. *Hint:* Condition on N.

Problem 17. A wombat in the San Diego zoo spends the day walking from a burrow to a food tray, eating, walking back to the burrow, resting, and repeating the cycle. The amount of time to walk from the burrow to the tray (and also from the tray to the burrow) is 20 secs. The amounts of time spent at the tray and resting are exponentially distributed with mean 30 secs. The wombat, with probability 1/3, will momentarily stand still (for a negligibly small time) during a walk to or from the tray, with all times being equally likely (and independent of what happened in the past). A photographer arrives at a random time and will take a picture at the first time the wombat will stand still. What is the expected value of the length of time the photographer has to wait to snap the wombat's picture?

Problem 18.* Consider a Poisson process. Given that a single arrival occurred in a given interval $[0, t]$, show that the PDF of the arrival time is uniform over $[0, t]$.

Solution. Consider an interval $[a, b] \subset [0, t]$ of length $l = b - a$. Let T be the time of the first arrival, and let A be the event that a single arrival occurred during $[0, t]$. We have

$$\mathbf{P}\big(T \in [a, b] \,|\, A\big) = \frac{\mathbf{P}\big(T \in [a, b] \text{ and } A\big)}{\mathbf{P}(A)}.$$

The numerator is equal to the probability $P(1, l)$ that the Poisson process has exactly one arrival during the length l interval $[a, b]$, times the probability $P(0, t - l)$ that the process has zero arrivals during the set $[0, a) \cup (b, t]$, of total length $t - l$. Hence

$$\mathbf{P}\big(T \in [a, b] \,|\, A\big) = \frac{P(1, l)P(0, t - l)}{P(1, t)} = \frac{(\lambda l)e^{-\lambda l}e^{-\lambda(t-l)}}{(\lambda t)e^{-\lambda t}} = \frac{l}{t},$$

which establishes that T is uniformly distributed.

Problem 19.*

(a) Let X_1 and X_2 be independent and exponentially distributed, with parameters λ_1 and λ_2, respectively. Find the expected value of $\max\{X_1, X_2\}$.

(b) Let Y be exponentially distributed with parameter λ_1. Let Z be Erlang of order 2 with parameter λ_2. Assume that Y and Z are independent. Find the expected value of $\max\{Y, Z\}$.

Solution. A direct but tedious approach would be to find the PDF of the random variable of interest and then evaluate an integral to find its expectation. A much simpler solution is obtained by interpreting the random variables of interest in terms of underlying Poisson processes.

(a) Consider two independent Poisson processes with rates λ_1 and λ_2, respectively. We interpret X_1 as the first arrival time in the first process, and X_2 as the first arrival time in the second process. Let $T = \min\{X_1, X_2\}$ be the first time when one of the processes registers an arrival. Let $S = \max\{X_1, X_2\} - T$ be the additional time until both have registered an arrival. Since the merged process is Poisson with rate $\lambda_1 + \lambda_2$, we have

$$\mathbf{E}[T] = \frac{1}{\lambda_1 + \lambda_2}.$$

Concerning S, there are two cases to consider.

(i) The first arrival comes from the first process; this happens with probability $\lambda_1/(\lambda_1 + \lambda_2)$. We then have to wait for an arrival from the second process, which takes $1/\lambda_2$ time on the average.

(ii) The first arrival comes from the second process; this happens with probability $\lambda_2/(\lambda_1 + \lambda_2)$. We then have to wait for an arrival from the first process, which takes $1/\lambda_1$ time on the average. Putting everything together, we obtain

$$\begin{aligned}\mathbf{E}\big[\max\{X_1, X_2\}\big] &= \frac{1}{\lambda_1 + \lambda_2} + \frac{\lambda_1}{\lambda_1 + \lambda_2} \cdot \frac{1}{\lambda_2} + \frac{\lambda_2}{\lambda_1 + \lambda_2} \cdot \frac{1}{\lambda_1} \\ &= \frac{1}{\lambda_1 + \lambda_2}\left(1 + \frac{\lambda_1}{\lambda_2} + \frac{\lambda_2}{\lambda_1}\right).\end{aligned}$$

(b) Consider two independent Poisson processes with rates λ_1 and λ_2, respectively. We interpret Y as the first arrival time in the first process, and Z as the second arrival time in the second process. Let T be the first time when one of the processes registers an arrival. Since the merged process is Poisson with rate $\lambda_1 + \lambda_2$, we have $\mathbf{E}[T] = 1/(\lambda_1 + \lambda_2)$. There are two cases to consider.

(i) The arrival at time T comes from the first process; this happens with probability $\lambda_1/(\lambda_1 + \lambda_2)$. In this case, we have to wait an additional time until the second process registers two arrivals. This additional time is Erlang of order 2, with parameter λ_2, and its expected value is $2/\lambda_2$.

(ii) The arrival at time T comes from the second process; this happens with probability $\lambda_2/(\lambda_1 + \lambda_2)$. In this case, the additional time S we have to wait is the time until each of the two processes registers an arrival. This is the maximum of two independent exponential random variables and, according to the result of part (a), we have

$$\mathbf{E}[S] = \frac{1}{\lambda_1 + \lambda_2}\left(1 + \frac{\lambda_1}{\lambda_2} + \frac{\lambda_2}{\lambda_1}\right).$$

Putting everything together, we have

$$\mathbf{E}\big[\max\{Y, Z\}\big] = \frac{1}{\lambda_1 + \lambda_2} + \frac{\lambda_1}{\lambda_1 + \lambda_2} \cdot \frac{2}{\lambda_2} + \frac{\lambda_2}{\lambda_1 + \lambda_2} \cdot \mathbf{E}[S],$$

where $\mathbf{E}[S]$ is given by the previous formula.

Problem 20.* Let Y_k be the time of the kth arrival in a Poisson process with rate λ. Show that for all $y > 0$,

$$\sum_{k=1}^{\infty} f_{Y_k}(y) = \lambda.$$

Solution. We have

$$\begin{aligned}
\sum_{k=1}^{\infty} f_{Y_k}(y) &= \sum_{k=1}^{\infty} \frac{\lambda^k y^{k-1} e^{-\lambda y}}{(k-1)!} \\
&= \lambda \sum_{k=1}^{\infty} \frac{\lambda^{k-1} y^{k-1} e^{-\lambda y}}{(k-1)!} \qquad \text{(let } m = k-1\text{)} \\
&= \lambda \sum_{m=0}^{\infty} \frac{\lambda^m y^m e^{-\lambda y}}{m!} \\
&= \lambda.
\end{aligned}$$

The last equality holds because the $\lambda^m y^m e^{-\lambda y}/m!$ terms are the values of a Poisson PMF with parameter λy and must therefore sum to 1.

For a more intuitive derivation, let δ be a small positive number and consider the following events:

A_k: the kth arrival occurs between y and $y+\delta$; the probability of this event is $\mathbf{P}(A_k) \approx f_{Y_k}(y)\delta$;

A: an arrival occurs between y and $y + \delta$; the probability of this event is $\mathbf{P}(A) \approx f_Y(y)\delta$.

Suppose that δ is taken small enough so that the possibility of two or more arrivals during an interval of length δ can be ignored. With this approximation, the events $A_1, A_2, \ldots$ become disjoint, and their union is A. Therefore,

$$\begin{aligned}
\sum_{k=1}^{\infty} f_{Y_k}(y) \cdot \delta &\approx \sum_{k=1}^{\infty} \mathbf{P}(A_k) \\
&\approx \mathbf{P}(A) \\
&\approx \lambda\delta,
\end{aligned}$$

and the desired result follows by canceling δ from both sides.

Problem 21.* Consider an experiment involving two independent Poisson processes with rates λ_1 and λ_2. Let $X_1(k)$ and $X_2(k)$ be the times of the kth arrival in the 1st and the 2nd process, respectively. Show that

$$\mathbf{P}\big(X_1(n) < X_2(m)\big) = \sum_{k=n}^{n+m-1} \binom{n+m-1}{k} \left(\frac{\lambda_1}{\lambda_1+\lambda_2}\right)^k \left(\frac{\lambda_2}{\lambda_1+\lambda_2}\right)^{n+m-1-k}.$$

Solution. Consider the merged Poisson process, which has rate $\lambda_1 + \lambda_2$. Each time there is an arrival in the merged process, this arrival comes from the first process ("success") with probability $\lambda_1/(\lambda_1+\lambda_2)$, and from the second process ("failure") with probability $\lambda_2/(\lambda_1+\lambda_2)$. Consider the situation after $n+m-1$ arrivals. The number of arrivals from the first process is at least n if and only if the number of arrivals from the second process is less than m, which happens if and only if the nth success occurs before the mth failure. Thus, the event $\{X_1(n) < X_2(m)\}$ is the same as the event of having at least n successes in the first $n+m-1$ trials. Therefore, using the binomial PMF for the number of successes in a given number of trials, we have

$$\mathbf{P}\big(X_1(n) < X_2(m)\big) = \sum_{k=n}^{n+m-1} \binom{n+m-1}{k} \left(\frac{\lambda_1}{\lambda_1+\lambda_2}\right)^k \left(\frac{\lambda_2}{\lambda_1+\lambda_2}\right)^{n+m-1-k}.$$

Problem 22.* Sum of a random number of independent Bernoulli random variables. Let $N, X_1, X_2, \ldots$ be independent random variables, where N takes nonnegative integer values, and $X_1, X_2, \ldots$ are Bernoulli with parameter p. Let $Y = X_1 + \cdots + X_N$ for positive values of N and let $Y = 0$ when $N = 0$.

(a) Show that if N is binomial with parameters m and q, then Y is binomial with parameters m and pq.

(b) Show that if N is Poisson with parameter λ, then Y is Poisson with parameter λp.

Solution. (a) Consider splitting the Bernoulli process $X_1, X_2, \ldots$ by keeping successes with probability q and discarding them with probability $1-q$. Then Y represents the number of successes in the split process during the first m trials. Since the split process is Bernoulli with parameter pq, it follows that Y is binomial with parameters m and pq.

(b) Consider splitting a Poisson process with parameter λ by keeping arrivals with probability q and discarding them with probability $1-q$. Then Y represents the number of arrivals in the split process during a unit interval. Since the split process is Poisson with parameter λp, it follows that Y is Poisson with parameter λp.

Problem 23.* Sum of a geometric number of independent exponential random variables. Let $Y = X_1 + \cdots + X_N$, where the random variables X_i are exponential with parameter λ, and N is geometric with parameter p. Assume that the random variables $N, X_1, X_2, \ldots$ are independent. Show, without using transforms, that Y is exponential with parameter λp. *Hint:* Interpret the various random variables in terms of a split Poisson process.

Solution. We derived this result in Chapter 4, using transforms, but we develop a more intuitive derivation here. We interpret the random variables X_i and N as follows. We view the times X_1, $X_1 + X_2$, etc. as the arrival times in a Poisson process with parameter λ. Each arrival is rejected with probability $1-p$ and is accepted with probability p. We interpret N as the number of arrivals until the first acceptance. The process of accepted arrivals is obtained by splitting a Poisson process and is therefore itself Poisson with parameter λp. Note that $Y = X_1 + \cdots + X_N$ is the time of the first accepted arrival and is therefore exponential with parameter λp.

Problem 24.* The number of Poisson arrivals during an exponentially distributed interval. Consider a Poisson process with parameter λ, and an independent random variable T, which is exponential with parameter ν. Find the PMF of the number of Poisson arrivals during the time interval $[0, T]$.

Solution. Let us view T as the first arrival time in a new, independent, Poisson process with parameter ν, and merge this process with the original Poisson process. Each arrival in the merged process comes from the original Poisson process with probability $\lambda/(\lambda+\nu)$, independent of other arrivals. If we view each arrival in the merged process as a trial, and an arrival from the new process as a success, we note that the number K of trials/arrivals until the first success has a geometric PMF, of the form

$$p_K(k) = \left(\frac{\nu}{\lambda+\nu}\right)\left(\frac{\lambda}{\lambda+\nu}\right)^{k-1}, \qquad k = 1, 2, \ldots.$$

Now the number L of arrivals from the original Poisson process until the first "success" is equal to $K - 1$ and its PMF is

$$p_L(l) = p_K(l+1) = \left(\frac{\nu}{\lambda+\nu}\right)\left(\frac{\lambda}{\lambda+\nu}\right)^{l}, \qquad l = 0, 1, \ldots.$$

Problem 25.* An infinite server queue. We consider a queueing system with an infinite number of servers, in which customers arrive according to a Poisson process with rate λ. The ith customer stays in the system for a random amount of time, denoted by X_i. We assume that the random variables X_i are independent identically distributed, and also independent from the arrival process. We also assume, for simplicity, that the X_i take integer values in the range $1, 2, \ldots, n$, with given probabilities. Find the PMF of N_t, the number of customers in the system at time t.

Solution. Let us refer to those customers i whose service time X_i is equal to k as "type-k" customers. We view the overall arrival process as the merging of n Poisson subprocesses; the kth subprocess corresponds to arrivals of type-k customers, is independent of the other arrival subprocesses, and has rate λp_k, where $p_k = \mathbf{P}(X_i = k)$. Let N_t^k be the number of type-k customers in the system at time t. Thus,

$$N_t = \sum_{k=1}^{n} N_t^k,$$

and the random variables N_t^k are independent.

We now determine the PMF of N_t^k. A type-k customer is in the system at time t if and only if that customer arrived between times $t-k$ and t. Thus, N_t^k has a Poisson PMF with mean $\lambda k p_k$. Since the sum of independent Poisson random variables is Poisson, it follows that N_t has a Poisson PMF with parameter

$$\mathbf{E}[N_t] = \lambda \sum_{k=1}^{n} k p_k = \lambda \mathbf{E}[X_i].$$

Problem 26.* Independence of Poisson processes obtained by splitting. Consider a Poisson process whose arrivals are split, with each arrival assigned to one of two subprocesses by tossing an independent coin with success probability p. In

Example 6.13, it was established that each of the subprocesses is a Poisson process. Show that the two subprocesses are independent.

Solution. Let us start with two independent Poisson processes $\mathcal{P}_1$ and $\mathcal{P}_2$, with rates $p\lambda$ and $(1-p)\lambda$, respectively. We merge the two processes and obtain a Poisson process $\mathcal{P}$ with rate λ. We now split the process $\mathcal{P}$ into two new subprocesses $\mathcal{P}_1'$ and $\mathcal{P}_2'$, according to the following rule: an arrival is assigned to subprocess $\mathcal{P}_1'$ (respectively, $\mathcal{P}_2'$) if and only if that arrival corresponds to an arrival from subprocess $\mathcal{P}_1$ (respectively, $\mathcal{P}_2$). Clearly, the two new subprocesses $\mathcal{P}_1'$ and $\mathcal{P}_2'$ are independent, since they are identical to the original subprocesses $\mathcal{P}_1$ and $\mathcal{P}_2$. However, $\mathcal{P}_1'$ and $\mathcal{P}_2'$ were generated by a splitting mechanism that looks different than the one in the problem statement. We will now verify that the new splitting mechanism considered here is statistically identical to the one in the problem statement. It will then follow that the subprocesses constructed in the problem statement have the same statistical properties as $\mathcal{P}_1'$ and $\mathcal{P}_2'$, and are also independent.

So, let us consider the above described splitting mechanism. Given that $\mathcal{P}$ had an arrival at time t, this was due to either an arrival in $\mathcal{P}_1$ (with probability p), or to an arrival in $\mathcal{P}_2$ (probability $1-p$). Therefore, the arrival to $\mathcal{P}$ is assigned to $\mathcal{P}_1'$ or $\mathcal{P}_2'$ with probabilities p and $1-p$, respectively, exactly as in the splitting procedure described in the problem statement. Consider now the kth arrival in $\mathcal{P}$ and let L_k be the event that this arrival originated from subprocess $\mathcal{P}_1$; this is the same as the event that the kth arrival is assigned to subprocess $\mathcal{P}_1'$. As explained in the context of Example 6.14, the events L_k are independent. Thus, the assignments of arrivals to the subprocesses $\mathcal{P}_1'$ and $\mathcal{P}_2'$ are independent for different arrivals, which is the other requirement of the splitting mechanism described in the problem statement.

Problem 27.* Random incidence in an Erlang arrival process. Consider an arrival process in which the interarrival times are independent Erlang random variables of order 2, with mean $2/\lambda$. Assume that the arrival process has been ongoing for a very long time. An external observer arrives at a given time t. Find the PDF of the length of the interarrival interval that contains t.

Solution. We view the Erlang arrival process in the problem statement as part of a Poisson process with rate λ. In particular, the Erlang arrival process registers an arrival once every two arrivals of the Poisson process. For concreteness, let us say that the Erlang process arrivals correspond to even-numbered arrivals in the Poisson process. Let Y_k be the time of the kth arrival in the Poisson process.

Let K be such that $Y_K \le t < Y_{K+1}$. By the discussion of random incidence in Poisson processes in the text, we have that $Y_{K+1} - Y_K$ is Erlang of order 2. The interarrival interval for the Erlang process considered in this problem is of the form $[Y_K, Y_{K+2}]$ or $[Y_{K-1}, Y_{K+1}]$, depending on whether K is even or odd, respectively. In the first case, the interarrival interval in the Erlang process is of the form $(Y_{K+1} - Y_K) + (Y_{K+2} - Y_{K+1})$. We claim that $Y_{K+2} - Y_{K+1}$ is exponential with parameter λ and independent of $Y_{K+1} - Y_K$. Indeed, an observer who arrives at time t and notices that K is even, must first wait until the time Y_{K+1} of the next Poisson arrival. At that time, the Poisson process starts afresh, and the time $Y_{K+2} - Y_{K+1}$ until the next Poisson arrival is independent of the past (hence, independent of $Y_{K+1} - Y_K$) and has an exponential distribution with parameter λ, as claimed. This establishes that, conditioned on K being even, the interarrival interval length $Y_{K+2} - Y_K$ of the Erlang process is Erlang of order 3 (since it is the sum of an exponential random variable and a random variable which is Erlang of order 2). By a symmetrical argument, if we condition on K being odd, the conditional PDF of the interarrival interval length $Y_{K+1} - Y_{K-1}$ of the Erlang process is again the same. Since the conditional PDF of the length of the interarrival interval that contains t is Erlang of order 3, for every conditioning event, it follows that the unconditional PDF is also Erlang of order 3.

7

Markov Chains

Contents

The Bernoulli and Poisson processes studied in the preceding chapter are memoryless, in the sense that the future does not depend on the past: the occurrences of new "successes" or "arrivals" do not depend on the past history of the process. In this chapter, we consider processes where the future depends on and can be predicted to some extent by what has happened in the past.

We emphasize models where the effect of the past on the future is summarized by a **state**, which changes over time according to given probabilities. We restrict ourselves to models in which the state can only take a finite number values and can change according to probabilities that do not depend on the time of the change. We want to analyze the probabilistic properties of the sequence of state values.

The range of applications of the type of models described in this chapter is truly vast. It includes just about any dynamical system whose evolution over time involves uncertainty, provided the state of the system is suitably defined. Such systems arise in a broad variety of fields, such as, for example, communications, automatic control, signal processing, manufacturing, economics, and operations research.

7.1 DISCRETE-TIME MARKOV CHAINS

We will first consider **discrete-time Markov chains**, in which the state changes at certain discrete time instants, indexed by an integer variable n. At each time step n, the state of the chain is denoted by X_n, and belongs to a **finite** set $\mathcal{S}$ of possible states, called the **state space**. Without loss of generality, and unless there is a statement to the contrary, we will assume that $\mathcal{S} = \{1, \ldots, m\}$, for some positive integer m. The Markov chain is described in terms of its **transition probabilities** p_{ij}: whenever the state happens to be i, there is probability p_{ij} that the next state is equal to j. Mathematically,

$$p_{ij} = \mathbf{P}(X_{n+1} = j \mid X_n = i), \qquad i,\, j \in \mathcal{S}.$$

The key assumption underlying Markov chains is that the transition probabilities p_{ij} apply whenever state i is visited, no matter what happened in the past, and no matter how state i was reached. Mathematically, we assume the **Markov property**, which requires that

$$\begin{aligned}\mathbf{P}(X_{n+1} = j \mid X_n = i, X_{n-1} = i_{n-1}, \ldots, X_0 = i_0) &= \mathbf{P}(X_{n+1} = j \mid X_n = i)\\ &= p_{ij},\end{aligned}$$

for all times n, all states $i, j \in \mathcal{S}$, and all possible sequences $i_0, \ldots, i_{n-1}$ of earlier states. Thus, the probability law of the next state X_{n+1} depends on the past only through the value of the present state X_n.

The transition probabilities p_{ij} must be of course nonnegative, and sum to one:

$$\sum_{j=1}^{m} p_{ij} = 1, \qquad \text{for all } i.$$

We will generally allow the probabilities p_{ii} to be positive, in which case it is possible for the next state to be the same as the current one. Even though the state does not change, we still view this as a state transition of a special type (a "self-transition").

Specification of Markov Models

- A Markov chain model is specified by identifying:
  (a) the set of states $\mathcal{S} = \{1, \ldots, m\}$,
  (b) the set of possible transitions, namely, those pairs (i, j) for which $p_{ij} > 0$, and,
  (c) the numerical values of those p_{ij} that are positive.
- The Markov chain specified by this model is a sequence of random variables $X_0, X_1, X_2, \ldots$, that take values in $\mathcal{S}$, and which satisfy

$$\mathbf{P}(X_{n+1} = j \mid X_n = i, X_{n-1} = i_{n-1}, \ldots, X_0 = i_0) = p_{ij},$$

  for all times n, all states $i, j \in \mathcal{S}$, and all possible sequences $i_0, \ldots, i_{n-1}$ of earlier states.

All of the elements of a Markov chain model can be encoded in a **transition probability matrix**, which is simply a two-dimensional array whose element at the ith row and jth column is p_{ij}:

$$\begin{bmatrix} p_{11} & p_{12} & \cdots & p_{1m} \\ p_{21} & p_{22} & \cdots & p_{2m} \\ \vdots & \vdots & \vdots & \vdots \\ p_{m1} & p_{m2} & \cdots & p_{mm} \end{bmatrix}.$$

It is also helpful to lay out the model in the so-called **transition probability graph**, whose nodes are the states and whose arcs are the possible transitions. By recording the numerical values of p_{ij} near the corresponding arcs, one can visualize the entire model in a way that can make some of its major properties readily apparent.

Example 7.1. Alice is taking a probability class and in each week, she can be either up-to-date or she may have fallen behind. If she is up-to-date in a given week, the probability that she will be up-to-date (or behind) in the next week is 0.8 (or 0.2, respectively). If she is behind in the given week, the probability that she will be up-to-date (or behind) in the next week is 0.6 (or 0.4, respectively). We assume that these probabilities do not depend on whether she was up-to-date or behind in previous weeks, so the problem has the typical Markov chain character (the future depends on the past only through the present).

Let us introduce states 1 and 2, and identify them with being up-to-date and behind, respectively. Then, the transition probabilities are

$$p_{11} = 0.8, \qquad p_{12} = 0.2, \qquad p_{21} = 0.6, \qquad p_{22} = 0.4,$$

and the transition probability matrix is

$$\begin{bmatrix} 0.8 & 0.2 \\ 0.6 & 0.4 \end{bmatrix}.$$

The transition probability graph is shown in Fig. 7.1.

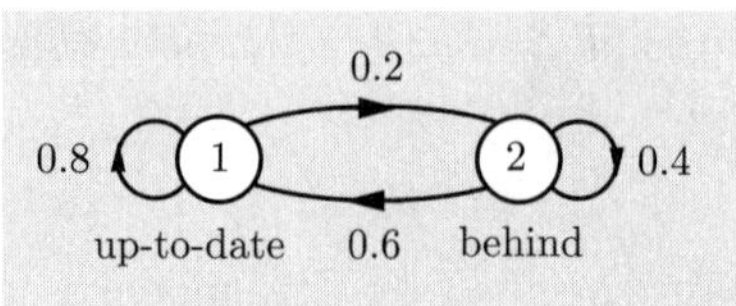

Figure 7.1: The transition probability graph in Example 7.1.

Example 7.2. Spiders and Fly. A fly moves along a straight line in unit increments. At each time period, it moves one unit to the left with probability 0.3, one unit to the right with probability 0.3, and stays in place with probability 0.4, independent of the past history of movements. Two spiders are lurking at positions 1 and m: if the fly lands there, it is captured by a spider, and the process terminates. We want to construct a Markov chain model, assuming that the fly starts in a position between 1 and m.

Let us introduce states $1, 2, \ldots, m$, and identify them with the corresponding positions of the fly. The nonzero transition probabilities are

$$p_{11} = 1, \qquad p_{mm} = 1,$$

$$p_{ij} = \begin{cases} 0.3, & \text{if } j = i-1 \text{ or } j = i+1, \\ 0.4, & \text{if } j = i, \end{cases} \qquad \text{for } i = 2, \ldots, m-1.$$

The transition probability graph and matrix are shown in Fig. 7.2.

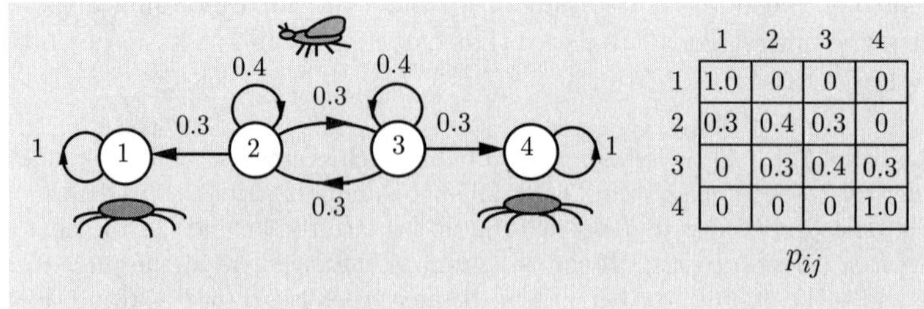

	1	2	3	4
1	1.0	0	0	0
2	0.3	0.4	0.3	0
3	0	0.3	0.4	0.3
4	0	0	0	1.0

Figure 7.2: The transition probability graph and the transition probability matrix in Example 7.2, for the case where $m = 4$.

Example 7.3. Machine Failure, Repair, and Replacement. A machine can be either working or broken down on a given day. If it is working, it will break down in the next day with probability b, and will continue working with probability $1 - b$. If it breaks down on a given day, it will be repaired and be working in the next day with probability r, and will continue to be broken down with probability $1 - r$.

We model the machine by a Markov chain with the following two states:

State 1: Machine is working, State 2: Machine is broken down.

The transition probability graph of the chain is given in Fig. 7.3. The transition probability matrix is

$$\begin{bmatrix} 1-b & b \\ r & 1-r \end{bmatrix}.$$

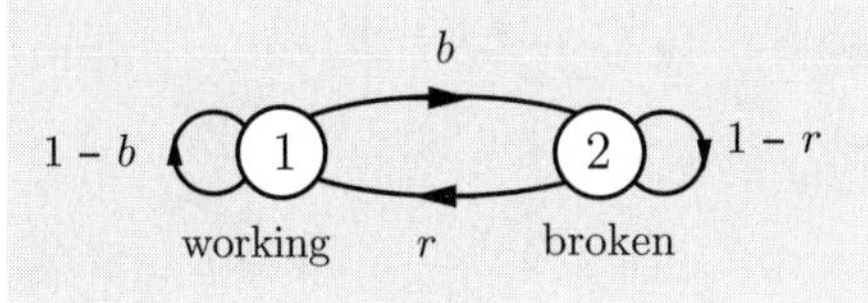

Figure 7.3: Transition probability graph for Example 7.3.

The situation considered here evidently has the Markov property: the state of the machine at the next day depends explicitly only on its state at the present day. However, it is possible to use a Markov chain model even if there is a dependence on the states at several past days. The general idea is to introduce some additional states which encode relevant information from preceding periods, as in the variation that we consider next.

Suppose that whenever the machine remains broken for a given number of ℓ days, despite the repair efforts, it is replaced by a new working machine. To model this as a Markov chain, we replace the single state 2, corresponding to a broken down machine, with several states that indicate the number of days that the machine is broken. These states are

State $(2, i)$: The machine has been broken for i days, $i = 1, 2, \ldots, \ell$.

The transition probability graph is given in Fig. 7.4 for the case where $\ell = 4$.

The second half of the preceding example illustrates that in order to construct a Markov model, there is often a need to introduce new states that capture the dependence of the future on the model's past history. We note that there is some freedom in selecting these additional states, but their number should be generally kept small, for reasons of analytical or computational tractability.

The Probability of a Path

Given a Markov chain model, we can compute the probability of any particular sequence of future states. This is analogous to the use of the multiplication rule

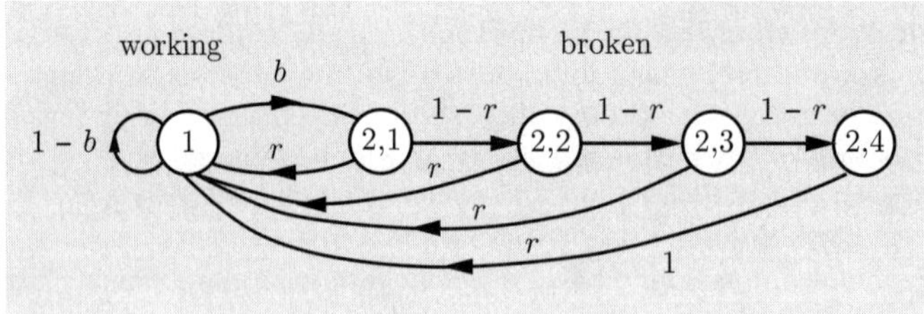

Figure 7.4: Transition probability graph for the second part of Example 7.3. A machine that has remained broken for $\ell = 4$ days is replaced by a new, working machine.

in sequential (tree) probability models. In particular, we have

$$\mathbf{P}(X_0 = i_0, X_1 = i_1, \ldots, X_n = i_n) = \mathbf{P}(X_0 = i_0) p_{i_0 i_1} p_{i_1 i_2} \cdots p_{i_{n-1} i_n}.$$

To verify this property, note that

$$\begin{aligned} \mathbf{P}(X_0 = i_0, X_1 = i_1, \ldots, X_n = i_n) \\ &= \mathbf{P}(X_n = i_n \mid X_0 = i_0, \ldots, X_{n-1} = i_{n-1}) \mathbf{P}(X_0 = i_0, \ldots, X_{n-1} = i_{n-1}) \\ &= p_{i_{n-1} i_n} \mathbf{P}(X_0 = i_0, \ldots, X_{n-1} = i_{n-1}), \end{aligned}$$

where the last equality made use of the Markov property. We then apply the same argument to the term $\mathbf{P}(X_0 = i_0, \ldots, X_{n-1} = i_{n-1})$ and continue similarly, until we eventually obtain the desired expression. If the initial state X_0 is given and is known to be equal to some i_0, a similar argument yields

$$\mathbf{P}(X_1 = i_1, \ldots, X_n = i_n \mid X_0 = i_0) = p_{i_0 i_1} p_{i_1 i_2} \cdots p_{i_{n-1} i_n}.$$

Graphically, a state sequence can be identified with a sequence of arcs in the transition probability graph, and the probability of such a path (given the initial state) is given by the product of the probabilities associated with the arcs traversed by the path.

Example 7.4. For the spider and fly example (Example 7.2), we have

$$\mathbf{P}(X_1 = 2, X_2 = 2, X_3 = 3, X_4 = 4 \mid X_0 = 2) = p_{22} p_{22} p_{23} p_{34} = (0.4)^2 (0.3)^2.$$

We also have

$$\begin{aligned} \mathbf{P}(X_0 = 2, X_1 = 2, X_2 = 2, X_3 = 3, X_4 = 4) &= \mathbf{P}(X_0 = 2) p_{22} p_{22} p_{23} p_{34} \\ &= \mathbf{P}(X_0 = 2)(0.4)^2 (0.3)^2. \end{aligned}$$

Note that in order to calculate a probability of this form, in which there is no conditioning on a fixed initial state, we need to specify a probability law for the initial state X_0.

n-Step Transition Probabilities

Many Markov chain problems require the calculation of the probability law of the state at some future time, conditioned on the current state. This probability law is captured by the **n-step transition probabilities**, defined by

$$r_{ij}(n) = \mathbf{P}(X_n = j \mid X_0 = i).$$

In words, $r_{ij}(n)$ is the probability that the state after n time periods will be j, given that the current state is i. It can be calculated using the following basic recursion, known as the **Chapman-Kolmogorov equation**.

> **Chapman-Kolmogorov Equation for the n-Step Transition Probabilities**
>
> The n-step transition probabilities can be generated by the recursive formula
>
> $$r_{ij}(n) = \sum_{k=1}^{m} r_{ik}(n-1)p_{kj}, \qquad \text{for } n > 1, \text{ and all } i,\ j,$$
>
> starting with
>
> $$r_{ij}(1) = p_{ij}.$$

To verify the formula, we apply the total probability theorem as follows:

$$\begin{aligned}\mathbf{P}(X_n = j \mid X_0 = i) &= \sum_{k=1}^{m} \mathbf{P}(X_{n-1} = k \mid X_0 = i)\,\mathbf{P}(X_n = j \mid X_{n-1} = k,\ X_0 = i)\\ &= \sum_{k=1}^{m} r_{ik}(n-1)p_{kj};\end{aligned}$$

see Fig. 7.5 for an illustration. We have used here the Markov property: once we condition on $X_{n-1} = k$, the conditioning on $X_0 = i$ does not affect the probability p_{kj} of reaching j at the next step.

We can view $r_{ij}(n)$ as the element at the ith row and jth column of a two-dimensional array, called the **n-step transition probability matrix.**† Figures 7.6 and 7.7 give the n-step transition probabilities $r_{ij}(n)$ for the cases of

† Those readers familiar with matrix multiplication, may recognize that the Chapman-Kolmogorov equation can be expressed as follows: the matrix of n-step transition probabilities $r_{ij}(n)$ is obtained by multiplying the matrix of $(n-1)$-step transition probabilities $r_{ik}(n-1)$, with the one-step transition probability matrix. Thus, the n-step transition probability matrix is the nth power of the transition probability matrix.

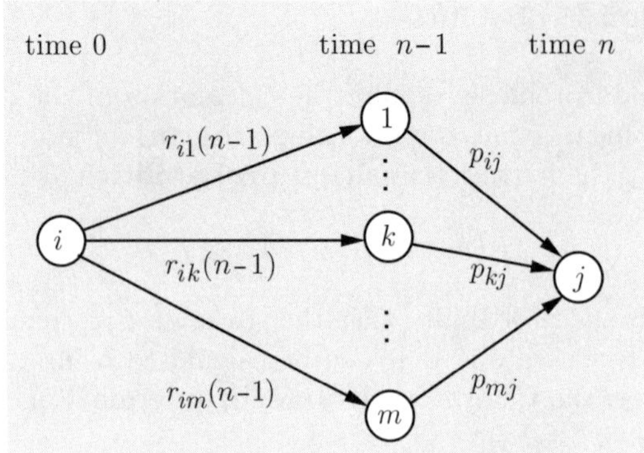

Figure 7.5: Derivation of the Chapman-Kolmogorov equation. The probability of being at state j at time n is the sum of the probabilities $r_{ik}(n-1)p_{kj}$ of the different ways of reaching j.

Examples 7.1 and 7.2, respectively. There are some interesting observations about the limiting behavior of $r_{ij}(n)$ in these two examples. In Fig. 7.6, we see that each $r_{ij}(n)$ converges to a limit, as $n \to \infty$, and this limit does not depend on the initial state i. Thus, each state has a positive "steady-state" probability of being occupied at times far into the future. Furthermore, the probability $r_{ij}(n)$ depends on the initial state i when n is small, but over time this dependence diminishes. Many (but by no means all) probabilistic models that evolve over time have such a character: after a sufficiently long time, the effect of their initial condition becomes negligible.

In Fig. 7.7, we see a qualitatively different behavior: $r_{ij}(n)$ again converges, but the limit depends on the initial state, and can be zero for selected states. Here, we have two states that are "absorbing," in the sense that they are infinitely repeated, once reached. These are the states 1 and 4 that correspond to the capture of the fly by one of the two spiders. Given enough time, it is certain that some absorbing state will be reached. Accordingly, the probability of being at the non-absorbing states 2 and 3 diminishes to zero as time increases. Furthermore, the probability that a particular absorbing state will be reached depends on how "close" we start to that state.

These examples illustrate that there is a variety of types of states and asymptotic occupancy behavior in Markov chains. We are thus motivated to classify and analyze the various possibilities, and this is the subject of the next three sections.

7.2 CLASSIFICATION OF STATES

In the preceding section, we saw some examples indicating that the various states of a Markov chain can have qualitatively different characteristics. In particular,

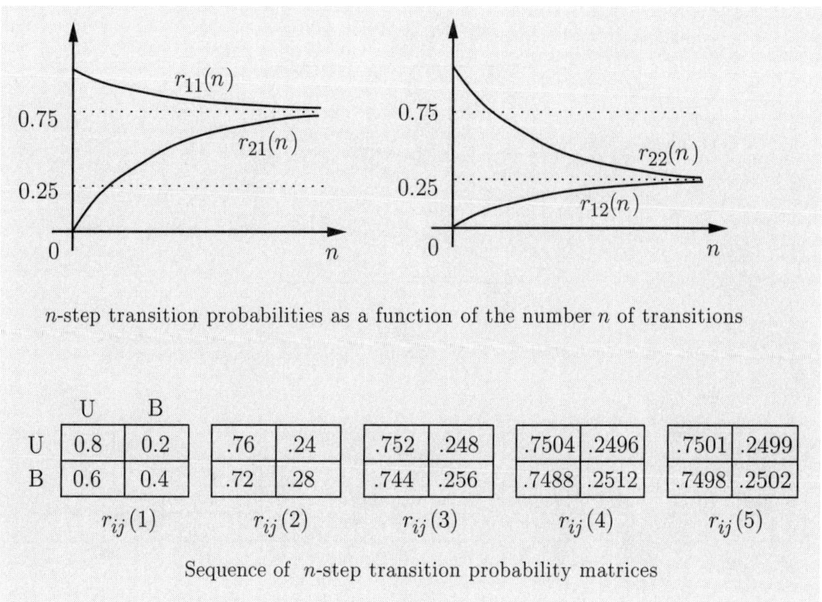

Figure 7.6: n-step transition probabilities for the "up-to-date/behind" Example 7.1. Observe that as n increases, $r_{ij}(n)$ converges to a limit that does not depend on the initial state i.

some states, after being visited once, are certain to be visited again, while for some other states this may not be the case. In this section, we focus on the mechanism by which this occurs. In particular, we wish to classify the states of a Markov chain with a focus on the long-term frequency with which they are visited.

As a first step, we make the notion of revisiting a state precise. Let us say that a state j is **accessible** from a state i if for some n, the n-step transition probability $r_{ij}(n)$ is positive, i.e., if there is positive probability of reaching j, starting from i, after some number of time periods. An equivalent definition is that there is a possible state sequence $i, i_1, \ldots, i_{n-1}, j$, that starts at i and ends at j, in which the transitions $(i, i_1), (i_1, i_2), \ldots, (i_{n-2}, i_{n-1}), (i_{n-1}, j)$ all have positive probability. Let $A(i)$ be the set of states that are accessible from i. We say that i is **recurrent** if for every j that is accessible from i, i is also accessible from j; that is, for all j that belong to $A(i)$ we have that i belongs to $A(j)$.

When we start at a recurrent state i, we can only visit states $j \in A(i)$ from which i is accessible. Thus, from any future state, there is always some probability of returning to i and, given enough time, this is certain to happen. By repeating this argument, if a recurrent state is visited once, it is certain to be revisited an infinite number of times. (See the end-of-chapter problems for a more rigorous version of this argument.)

A state is called **transient** if it is not recurrent. Thus, a state i is transient

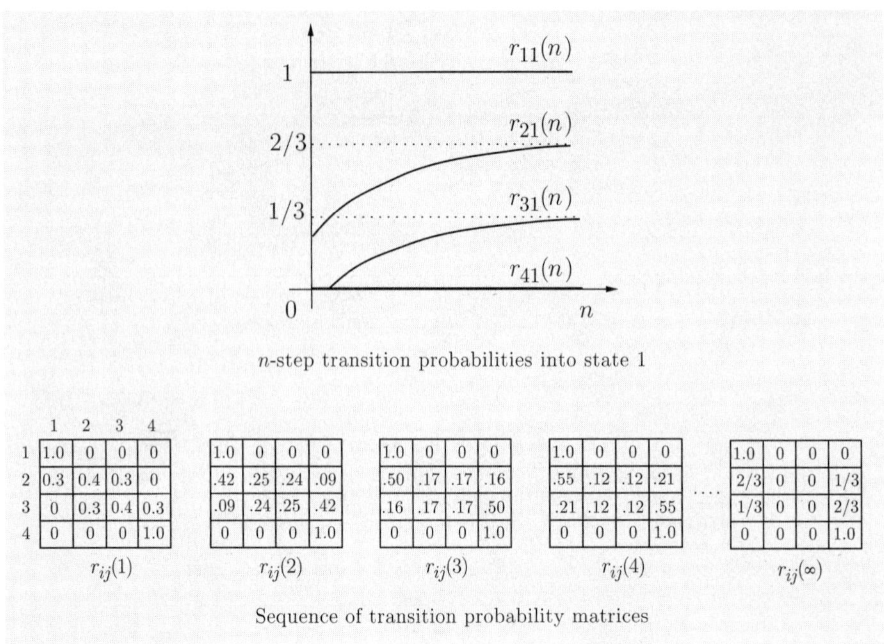

Figure 7.7: The top part of the figure shows the n-step transition probabilities $r_{i1}(n)$ for the "spiders-and-fly" Example 7.2. These are the probabilities of reaching state 1 by time n, starting from state i. We observe that these probabilities converge to a limit, but the limit depends on the starting state. In this example, note that the probabilities $r_{i2}(n)$ and $r_{i3}(n)$ of being at the non-absorbing states 2 or 3, go to zero, as n increases.

if there is a state $j \in A(i)$ such that i is not accessible from j. After each visit to state i, there is positive probability that the state enters such a j. Given enough time, this will happen, and state i cannot be visited after that. Thus, a transient state will only be visited a finite number of times; see again the end-of-chapter problems.

Note that transience or recurrence is determined by the arcs of the transition probability graph [those pairs (i, j) for which $p_{ij} > 0$] and not by the numerical values of the p_{ij}. Figure 7.8 provides an example of a transition probability graph, and the corresponding recurrent and transient states.

If i is a recurrent state, the set of states $A(i)$ that are accessible from i form a **recurrent class** (or simply **class**), meaning that states in $A(i)$ are all accessible from each other, and no state outside $A(i)$ is accessible from them. Mathematically, for a recurrent state i, we have $A(i) = A(j)$ for all j that belong to $A(i)$, as can be seen from the definition of recurrence. For example, in the graph of Fig. 7.8, states 3 and 4 form a class, and state 1 by itself also forms a class.

It can be seen that at least one recurrent state must be accessible from any

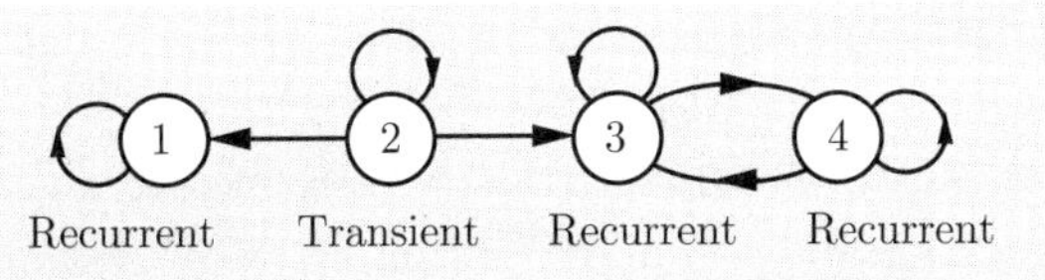

Figure 7.8: Classification of states given the transition probability graph. Starting from state 1, the only accessible state is itself, and so 1 is a recurrent state. States 1, 3, and 4 are accessible from 2, but 2 is not accessible from any of them, so state 2 is transient. States 3 and 4 are accessible from each other, and they are both recurrent.

given transient state. This is intuitively evident and is left as an end-of-chapter problem. It follows that there must exist at least one recurrent state, and hence at least one class. Thus, we reach the following conclusion.

Markov Chain Decomposition

- A Markov chain can be decomposed into one or more recurrent classes, plus possibly some transient states.
- A recurrent state is accessible from all states in its class, but is not accessible from recurrent states in other classes.
- A transient state is not accessible from any recurrent state.
- At least one, possibly more, recurrent states are accessible from a given transient state.

Figure 7.9 provides examples of Markov chain decompositions. Decomposition provides a powerful conceptual tool for reasoning about Markov chains and visualizing the evolution of their state. In particular, we see that:

(a) once the state enters (or starts in) a class of recurrent states, it stays within that class; since all states in the class are accessible from each other, all states in the class will be visited an infinite number of times;

(b) if the initial state is transient, then the state trajectory contains an initial portion consisting of transient states and a final portion consisting of recurrent states from the same class.

For the purpose of understanding long-term behavior of Markov chains, it is important to analyze chains that consist of a single recurrent class. For the purpose of understanding short-term behavior, it is also important to analyze the mechanism by which any particular class of recurrent states is entered starting from a given transient state. These two issues, long-term and short-term behavior, are the focus of Sections 7.3 and 7.4, respectively.

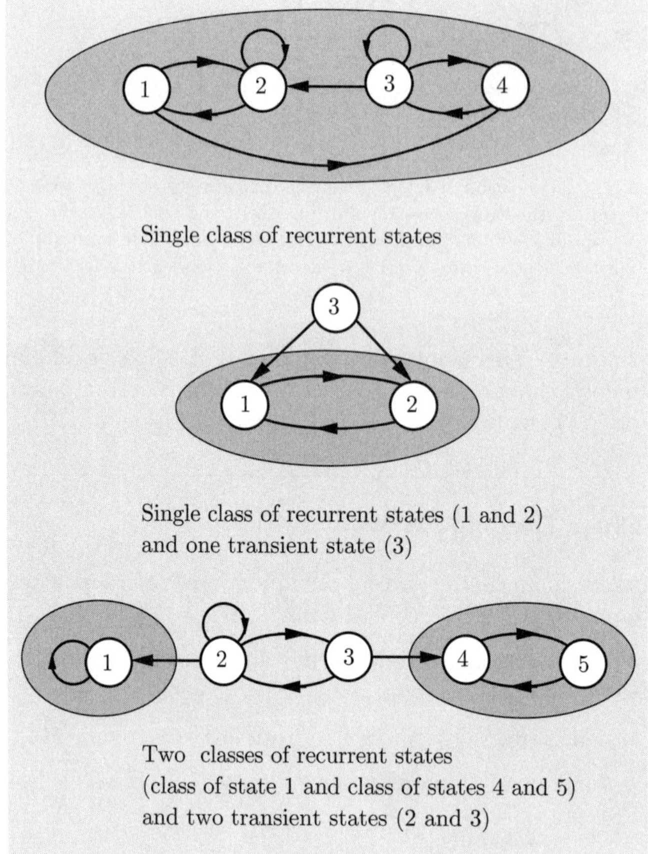

1
2
3
4
Single class of recurrent states
3
1
2
Single class of recurrent states (1 and 2)
and one transient state (3)
1
2
3
4
5
Two classes of recurrent states
(class of state 1 and class of states 4 and 5)
and two transient states (2 and 3)

Figure 7.9: Examples of Markov chain decompositions into recurrent classes and transient states.

Periodicity

There is another important characterization of a recurrent class, which relates to the presence or absence of a certain periodic pattern in the times that a state can be visited. In particular, a recurrent class is said to be **periodic** if its states can be grouped in $d > 1$ disjoint subsets $S_1, \ldots, S_d$ so that all transitions from one subset lead to the next subset; see Fig. 7.10. More precisely,

$$\text{if} \quad i \in S_k \text{ and } p_{ij} > 0, \quad \text{then} \begin{cases} j \in S_{k+1}, & \text{if } k = 1, \ldots, d-1, \\ j \in S_1, & \text{if } k = d. \end{cases}$$

A recurrent class that is not periodic, is said to be **aperiodic**.

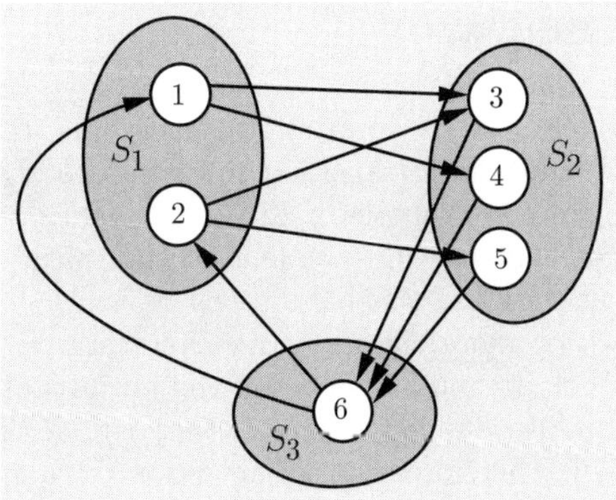

Figure 7.10: Structure of a periodic recurrent class. In this example, $d = 3$.

Thus, in a periodic recurrent class, we move through the sequence of subsets in order, and after d steps, we end up in the same subset. As an example, the recurrent class in the second chain of Fig. 7.9 (states 1 and 2) is periodic, and the same is true of the class consisting of states 4 and 5 in the third chain of Fig. 7.9. All other recurrent classes in the chains of this figure are aperiodic.

Note that given a periodic recurrent class, a positive time n, and a state i in the class, there must exist one or more states j for which $r_{ij}(n) = 0$. The reason is that starting from i, only one of the sets S_k is possible at time n. Thus, a way to verify aperiodicity of a given recurrent class R, is to check whether there is a special time $n \geq 1$ and a special state $i \in R$ from which all states in R can be reached in n steps, i.e., $r_{ij}(n) > 0$ for all $j \in R$. As an example, consider the first chain in Fig. 7.9. Starting from state 1, every state is possible at time $n = 3$, so the unique recurrent class of that chain is aperiodic.

A converse statement, which we do not prove, also turns out to be true: if a recurrent class R is aperiodic, then there exists a time n such that $r_{ij}(n) > 0$ for every i and j in R; see the end-of-chapter problems.

Periodicity

Consider a recurrent class R.

- The class is called **periodic** if its states can be grouped in $d > 1$ disjoint subsets $S_1, \ldots, S_d$, so that all transitions from S_k lead to S_{k+1} (or to S_1 if $k = d$).
- The class is **aperiodic** (not periodic) if and only if there exists a time n such that $r_{ij}(n) > 0$, for all $i, j \in R$.

7.3 STEADY-STATE BEHAVIOR

In Markov chain models, we are often interested in long-term state occupancy behavior, that is, in the n-step transition probabilities $r_{ij}(n)$ when n is very large. We have seen in the example of Fig. 7.6 that the $r_{ij}(n)$ may converge to steady-state values that are independent of the initial state. We wish to understand the extent to which this behavior is typical.

If there are two or more classes of recurrent states, it is clear that the limiting values of the $r_{ij}(n)$ must depend on the initial state (the possibility of visiting j far into the future depends on whether j is in the same class as the initial state i). We will, therefore, restrict attention to chains involving a single recurrent class, plus possibly some transient states. This is not as restrictive as it may seem, since we know that once the state enters a particular recurrent class, it will stay within that class. Thus, the asymptotic behavior of a multiclass chain can be understood in terms of the asymptotic behavior of a single-class chain.

Even for chains with a single recurrent class, the $r_{ij}(n)$ may fail to converge. To see this, consider a recurrent class with two states, 1 and 2, such that from state 1 we can only go to 2, and from 2 we can only go to 1 ($p_{12} = p_{21} = 1$). Then, starting at some state, we will be in that same state after any even number of transitions, and in the other state after any odd number of transitions. Formally,

$$r_{ii}(n) = \begin{cases} 1, & n \text{ even}, \\ 0, & n \text{ odd}. \end{cases}$$

What is happening here is that the recurrent class is periodic, and for such a class, it can be seen that the $r_{ij}(n)$ generically oscillate.

We now assert that for every state j, the probability $r_{ij}(n)$ of being at state j approaches a limiting value that is independent of the initial state i, provided we exclude the two situations discussed above (multiple recurrent classes and/or a periodic class). This limiting value, denoted by π_j, has the interpretation

$$\pi_j \approx \mathbf{P}(X_n = j), \qquad \text{when } n \text{ is large},$$

and is called the **steady-state probability of** j. The following is an important theorem. Its proof is quite complicated and is outlined together with several other proofs in the end-of-chapter problems.

Steady-State Convergence Theorem

Consider a Markov chain with a single recurrent class, which is aperiodic. Then, the states j are associated with steady-state probabilities π_j that have the following properties.

(a) For each j, we have

$$\lim_{n\to\infty} r_{ij}(n) = \pi_j, \qquad \text{for all } i.$$

(b) The π_j are the unique solution to the system of equations below:

$$\pi_j = \sum_{k=1}^{m} \pi_k p_{kj}, \qquad j = 1, \ldots, m,$$

$$1 = \sum_{k=1}^{m} \pi_k.$$

(c) We have

$$\pi_j = 0, \qquad \text{for all transient states } j,$$

$$\pi_j > 0, \qquad \text{for all recurrent states } j.$$

The steady-state probabilities π_j sum to 1 and form a probability distribution on the state space, called the **stationary distribution** of the chain. The reason for the qualification "stationary" is that if the initial state is chosen according to this distribution, i.e., if

$$\mathbf{P}(X_0 = j) = \pi_j, \qquad j = 1, \ldots, m,$$

then, using the total probability theorem, we have

$$\mathbf{P}(X_1 = j) = \sum_{k=1}^{m} \mathbf{P}(X_0 = k) p_{kj} = \sum_{k=1}^{m} \pi_k p_{kj} = \pi_j,$$

where the last equality follows from part (b) of the steady-state convergence theorem. Similarly, we obtain $\mathbf{P}(X_n = j) = \pi_j$, for all n and j. Thus, if the initial state is chosen according to the stationary distribution, the state at any future time will have the same distribution.

The equations

$$\pi_j = \sum_{k=1}^{m} \pi_k p_{kj}, \qquad j = 1, \ldots, m,$$

are called the **balance equations**. They are a simple consequence of part (a) of the theorem and the Chapman-Kolmogorov equation. Indeed, once the convergence of $r_{ij}(n)$ to some π_j is taken for granted, we can consider the equation,

$$r_{ij}(n) = \sum_{k=1}^{m} r_{ik}(n-1) p_{kj},$$

take the limit of both sides as $n \to \infty$, and recover the balance equations.† Together with the **normalization equation**

$$\sum_{k=1}^{m} \pi_k = 1,$$

the balance equations can be solved to obtain the π_j. The following examples illustrate the solution process.

Example 7.5. Consider a two-state Markov chain with transition probabilities

$$p_{11} = 0.8, \qquad p_{12} = 0.2,$$
$$p_{21} = 0.6, \qquad p_{22} = 0.4.$$

(This is the same as the chain of Example 7.1 and Fig. 7.1.) The balance equations take the form

$$\pi_1 = \pi_1 p_{11} + \pi_2 p_{21}, \qquad \pi_2 = \pi_1 p_{12} + \pi_2 p_{22},$$

or

$$\pi_1 = 0.8 \cdot \pi_1 + 0.6 \cdot \pi_2, \qquad \pi_2 = 0.2 \cdot \pi_1 + 0.4 \cdot \pi_2.$$

Note that the above two equations are dependent, since they are both equivalent to

$$\pi_1 = 3\pi_2.$$

This is a generic property, and in fact it can be shown that any one of the balance equations can always be derived from the remaining equations. However, we also know that the π_j satisfy the normalization equation

$$\pi_1 + \pi_2 = 1,$$

which supplements the balance equations and suffices to determine the π_j uniquely. Indeed, by substituting the equation $\pi_1 = 3\pi_2$ into the equation $\pi_1 + \pi_2 = 1$, we obtain $3\pi_2 + \pi_2 = 1$, or

$$\pi_2 = 0.25,$$

which using the equation $\pi_1 + \pi_2 = 1$, yields

$$\pi_1 = 0.75.$$

This is consistent with what we found earlier by iterating the Chapman-Kolmogorov equation (cf. Fig. 7.6).

---

† According to a famous and important theorem from linear algebra (called the Perron-Frobenius theorem), the balance equations always have a nonnegative solution, for any Markov chain. What is special about a chain that has a single recurrent class, which is aperiodic, is that given also the normalization equation, the solution is unique and is equal to the limit of the n-step transition probabilities $r_{ij}(n)$.

Example 7.6. An absent-minded professor has two umbrellas that she uses when commuting from home to office and back. If it rains and an umbrella is available in her location, she takes it. If it is not raining, she always forgets to take an umbrella. Suppose that it rains with probability p each time she commutes, independent of other times. What is the steady-state probability that she gets wet during a commute?

We model this problem using a Markov chain with the following states:

State i: i umbrellas are available in her current location, $\quad i = 0, 1, 2.$

The transition probability graph is given in Fig. 7.11, and the transition probability matrix is

$$\begin{bmatrix} 0 & 0 & 1 \\ 0 & 1-p & p \\ 1-p & p & 0 \end{bmatrix}.$$

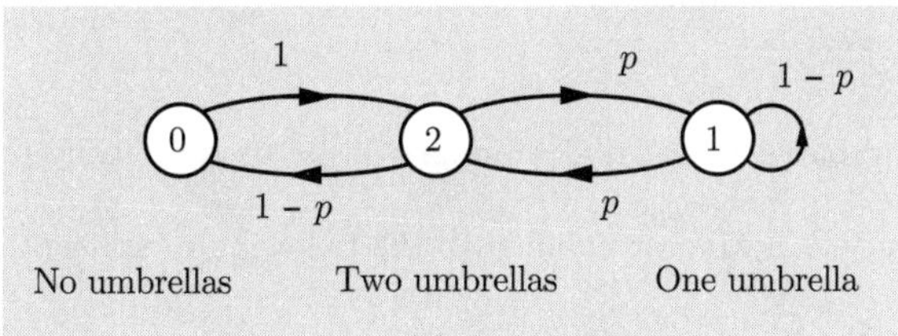

Figure 7.11: Transition probability graph for Example 7.6.

The chain has a single recurrent class that is aperiodic (assuming $0 < p < 1$), so the steady-state convergence theorem applies. The balance equations are

$$\pi_0 = (1-p)\pi_2, \quad \pi_1 = (1-p)\pi_1 + p\pi_2, \quad \pi_2 = \pi_0 + p\pi_1.$$

From the second equation, we obtain $\pi_1 = \pi_2$, which together with the first equation $\pi_0 = (1-p)\pi_2$ and the normalization equation $\pi_0 + \pi_1 + \pi_2 = 1$, yields

$$\pi_0 = \frac{1-p}{3-p}, \quad \pi_1 = \frac{1}{3-p}, \quad \pi_2 = \frac{1}{3-p}.$$

According to the steady-state convergence theorem, the steady-state probability that the professor finds herself in a place without an umbrella is π_0. The steady-state probability that she gets wet is π_0 times the probability of rain p.

Example 7.7. A superstitious professor works in a circular building with m doors, where m is odd, and never uses the same door twice in a row. Instead he uses with probability p (or probability $1-p$) the door that is adjacent in the clockwise direction (or the counterclockwise direction, respectively) to the door he used last.

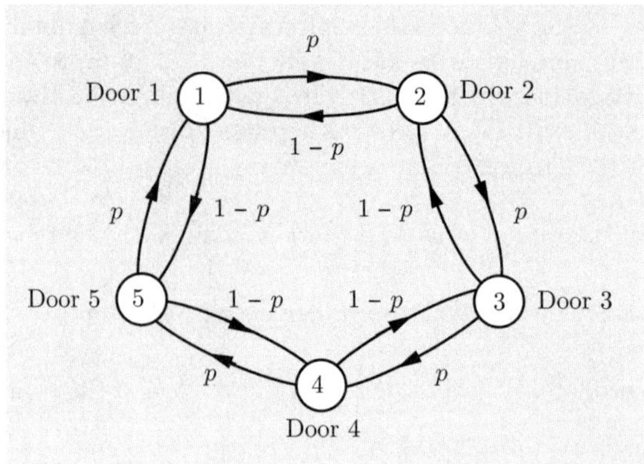

Figure 7.12: Transition probability graph in Example 7.7, for the case of $m = 5$ doors. Assuming that $0 < p < 1$, it is not hard to see that given an initial state, every state j can be reached in exactly 5 steps, and therefore the chain is aperiodic.

What is the probability that a given door will be used on some particular day far into the future?

We introduce a Markov chain with the following m states:

$$\text{State } i\text{: last door used is door } i, \qquad i = 1, \ldots, m.$$

The transition probability graph of the chain is given in Fig. 7.12, for the case $m = 5$. The transition probability matrix is

$$\begin{bmatrix} 0 & p & 0 & 0 & \ldots & 0 & 1-p \\ 1-p & 0 & p & 0 & \ldots & 0 & 0 \\ 0 & 1-p & 0 & p & \ldots & 0 & 0 \\ \vdots & \vdots & \vdots & \vdots & \vdots & \vdots & \vdots \\ p & 0 & 0 & 0 & \ldots & 1-p & 0 \end{bmatrix}.$$

Assuming that $0 < p < 1$, the chain has a single recurrent class that is aperiodic. (To verify aperiodicity, we leave it to the reader to verify that given an initial state, every state j can be reached in exactly m steps, and the criterion for aperiodicity given at the end of the preceding section is satisfied.) The balance equations are

$$\begin{aligned} \pi_1 &= (1-p)\pi_2 + p\pi_m, \\ \pi_i &= p\pi_{i-1} + (1-p)\pi_{i+1}, \qquad i = 2, \ldots, m-1, \\ \pi_m &= (1-p)\pi_1 + p\pi_{m-1}. \end{aligned}$$

These equations are easily solved once we observe that by symmetry, all doors should have the same steady-state probability. This suggests the solution

$$\pi_j = \frac{1}{m}, \qquad j = 1, \ldots, m.$$

Indeed, we see that these π_j satisfy the balance equations as well as the normalization equation, so they must be the desired steady-state probabilities (by the uniqueness part of the steady-state convergence theorem).

Note that if either $p = 0$ or $p = 1$, the chain still has a single recurrent class but is periodic. In this case, the n-step transition probabilities $r_{ij}(n)$ do not converge to a limit, because the doors are used in a cyclic order. Similarly, if m is even, the recurrent class of the chain is periodic, since the states can be grouped into two subsets, the even and the odd numbered states, such that from each subset one can only go to the other subset.

Long-Term Frequency Interpretations

Probabilities are often interpreted as relative frequencies in an infinitely long string of independent trials. The steady-state probabilities of a Markov chain admit a similar interpretation, despite the absence of independence.

Consider, for example, a Markov chain involving a machine, which at the end of any day can be in one of two states, working or broken down. Each time it breaks down, it is immediately repaired at a cost of \$1. How are we to model the long-term expected cost of repair per day? One possibility is to view it as the expected value of the repair cost on a randomly chosen day far into the future; this is just the steady-state probability of the broken down state. Alternatively, we can calculate the total expected repair cost in n days, where n is very large, and divide it by n. Intuition suggests that these two methods of calculation should give the same result. Theory supports this intuition, and in general we have the following interpretation of steady-state probabilities (a justification is given in the end-of-chapter problems).

Steady-State Probabilities as Expected State Frequencies

For a Markov chain with a single class which is aperiodic, the steady-state probabilities π_j satisfy

$$\pi_j = \lim_{n \to \infty} \frac{v_{ij}(n)}{n},$$

where $v_{ij}(n)$ is the expected value of the number of visits to state j within the first n transitions, starting from state i.

Based on this interpretation, π_j is the long-term expected fraction of time that the state is equal to j. Each time that state j is visited, there is probability p_{jk} that the next transition takes us to state k. We conclude that $\pi_j p_{jk}$ can be viewed as the long-term expected fraction of transitions that move the state from j to k.†

Expected Frequency of a Particular Transition

Consider n transitions of a Markov chain with a single class which is aperiodic, starting from a given initial state. Let $q_{jk}(n)$ be the expected number of such transitions that take the state from j to k. Then, regardless of the initial state, we have

$$\lim_{n\to\infty} \frac{q_{jk}(n)}{n} = \pi_j p_{jk}.$$

Given the frequency interpretation of π_j and $\pi_k p_{kj}$, the balance equation

$$\pi_j = \sum_{k=1}^{m} \pi_k p_{kj}$$

has an intuitive meaning. It expresses the fact that the expected frequency π_j of visits to j is equal to the sum of the expected frequencies $\pi_k p_{kj}$ of transitions that lead to j; see Fig. 7.13.

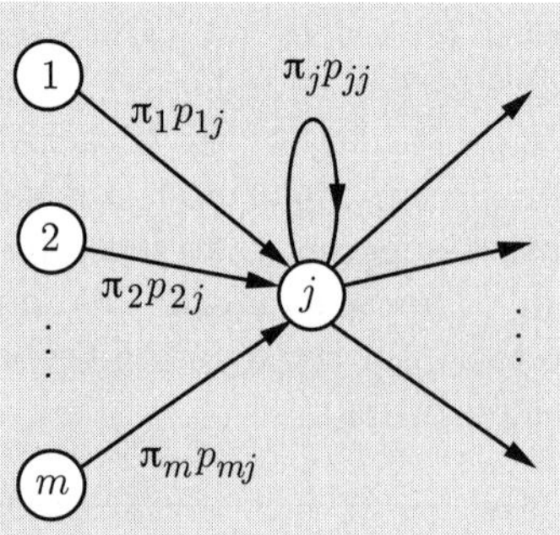

Figure 7.13: Interpretation of the balance equations in terms of frequencies. In a very large number of transitions, we expect a fraction $\pi_k p_{kj}$ that bring the state from k to j. (This also applies to transitions from j to itself, which occur with frequency $\pi_j p_{jj}$.) The sum of the expected frequencies of such transitions is the expected frequency π_j of being at state j.

† In fact, some stronger statements are also true, such as the following. Whenever we carry out a probabilistic experiment and generate a trajectory of the Markov chain over an infinite time horizon, the observed long-term frequency with which state j is visited will be exactly equal to π_j, and the observed long-term frequency of transitions from j to k will be exactly equal to $\pi_j p_{jk}$. Even though the trajectory is random, these equalities hold with essential certainty, that is, with probability 1.

Birth-Death Processes

A **birth-death** process is a Markov chain in which the states are linearly arranged and transitions can only occur to a neighboring state, or else leave the state unchanged. They arise in many contexts, especially in queueing theory. Figure 7.14 shows the general structure of a birth-death process and also introduces some generic notation for the transition probabilities. In particular,

$$b_i = \mathbf{P}(X_{n+1} = i+1 \,|\, X_n = i), \qquad \text{("birth" probability at state } i\text{)},$$
$$d_i = \mathbf{P}(X_{n+1} = i-1 \,|\, X_n = i), \qquad \text{("death" probability at state } i\text{)}.$$

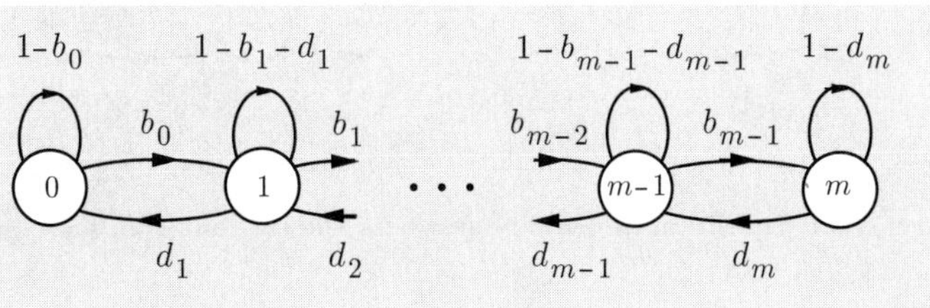

Figure 7.14: Transition probability graph for a birth-death process.

For a birth-death process, the balance equations can be substantially simplified. Let us focus on two neighboring states, say, i and $i+1$. In any trajectory of the Markov chain, a transition from i to $i+1$ has to be followed by a transition from $i+1$ to i, before another transition from i to $i+1$ can occur. Therefore, the expected frequency of transitions from i to $i+1$, which is $\pi_i b_i$, must be equal to the expected frequency of transitions from $i+1$ to i, which is $\pi_{i+1} d_{i+1}$. This leads to the **local balance** equations†

$$\pi_i b_i = \pi_{i+1} d_{i+1}, \qquad i = 0, 1, \ldots, m-1.$$

Using the local balance equations, we obtain

$$\pi_i = \pi_0 \frac{b_0 b_1 \cdots b_{i-1}}{d_1 d_2 \cdots d_i}, \qquad i = 1, \ldots, m,$$

from which, using also the normalization equation $\sum_i \pi_i = 1$, the steady-state probabilities π_i are easily computed.

---

† A more formal derivation that does not rely on the frequency interpretation proceeds as follows. The balance equation at state 0 is $\pi_0(1-b_0) + \pi_1 d_1 = \pi_0$, which yields the first local balance equation $\pi_0 b_0 = \pi_1 d_1$.

The balance equation at state 1 is $\pi_0 b_0 + \pi_1(1 - b_1 - d_1) + \pi_2 d_2 = \pi_1$. Using the local balance equation $\pi_0 b_0 = \pi_1 d_1$ at the previous state, this is rewritten as $\pi_1 d_1 + \pi_1(1 - b_1 - d_1) + \pi_2 d_2 = \pi_1$, which simplifies to $\pi_1 b_1 = \pi_2 d_2$. We can then continue similarly to obtain the local balance equations at all other states.

Example 7.8. Random Walk with Reflecting Barriers. A person walks along a straight line and, at each time period, takes a step to the right with probability b, and a step to the left with probability $1-b$. The person starts in one of the positions $1, 2, \ldots, m$, but if he reaches position 0 (or position $m+1$), his step is instantly reflected back to position 1 (or position m, respectively). Equivalently, we may assume that when the person is in positions 1 or m, he will stay in that position with corresponding probability $1-b$ and b, respectively. We introduce a Markov chain model whose states are the positions $1, \ldots, m$. The transition probability graph of the chain is given in Fig. 7.15.

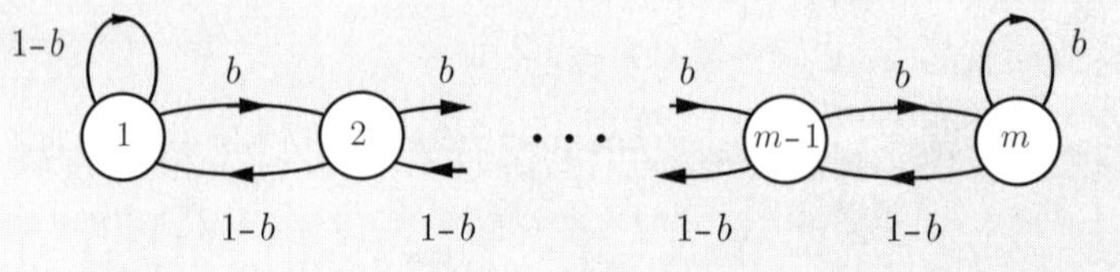

Figure 7.15: Transition probability graph for the random walk Example 7.8.

The local balance equations are

$$\pi_i b = \pi_{i+1}(1-b), \qquad i = 1, \ldots, m-1.$$

Thus, $\pi_{i+1} = \rho\pi_i$, where

$$\rho = \frac{b}{1-b},$$

and we can express all the π_j in terms of π_1, as

$$\pi_i = \rho^{i-1}\pi_1, \qquad i = 1, \ldots, m.$$

Using the normalization equation $1 = \pi_1 + \cdots + \pi_m$, we obtain

$$1 = \pi_1(1 + \rho + \cdots + \rho^{m-1})$$

which leads to

$$\pi_i = \frac{\rho^{i-1}}{1 + \rho + \cdots + \rho^{m-1}}, \qquad i = 1, \ldots, m.$$

Note that if $\rho = 1$ (left and right steps are equally likely), then $\pi_i = 1/m$ for all i.

Example 7.9. Queueing. Packets arrive at a node of a communication network, where they are stored in a buffer and then transmitted. The storage capacity of the buffer is m: if m packets are already present, any newly arriving packets are discarded. We discretize time in very small periods, and we assume that in each period, at most one event can happen that can change the number of packets stored in the node (an arrival of a new packet or a completion of the transmission of an existing packet). In particular, we assume that at each period, exactly one of the following occurs:

(a) one new packet arrives; this happens with a given probability $b > 0$;

(b) one existing packet completes transmission; this happens with a given probability $d > 0$ if there is at least one packet in the node, and with probability 0 otherwise;

(c) no new packet arrives and no existing packet completes transmission; this happens with probability $1 - b - d$ if there is at least one packet in the node, and with probability $1 - b$ otherwise.

We introduce a Markov chain with states $0, 1, \ldots, m$, corresponding to the number of packets in the buffer. The transition probability graph is given in Fig. 7.16.

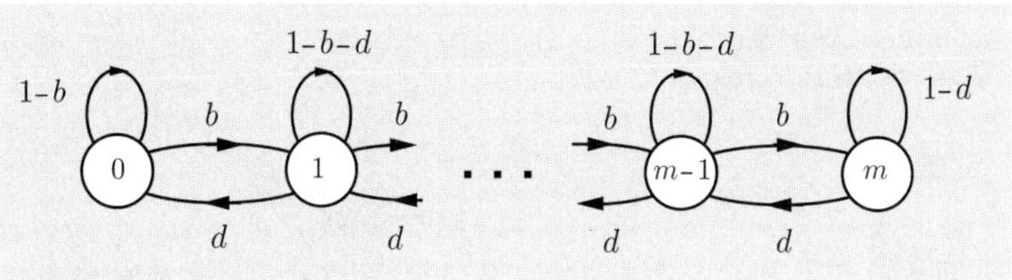

Figure 7.16: Transition probability graph in Example 7.9.

The local balance equations are

$$\pi_i b = \pi_{i+1} d, \qquad i = 0, 1, \ldots, m - 1.$$

We define

$$\rho = \frac{b}{d},$$

and obtain $\pi_{i+1} = \rho \pi_i$, which leads to

$$\pi_i = \rho^i \pi_0, \qquad i = 0, 1, \ldots, m.$$

By using the normalization equation $1 = \pi_0 + \pi_1 + \cdots + \pi_m$, we obtain

$$1 = \pi_0(1 + \rho + \cdots + \rho^m),$$

and

$$\pi_0 = \begin{cases} \dfrac{1 - \rho}{1 - \rho^{m+1}}, & \text{if } \rho \neq 1, \\[2ex] \dfrac{1}{m + 1}, & \text{if } \rho = 1. \end{cases}$$

Using the equation $\pi_i = \rho^i \pi_0$, the steady-state probabilities are

$$\pi_i = \begin{cases} \dfrac{1 - \rho}{1 - \rho^{m+1}} \rho^i, & \text{if } \rho \neq 1, \\[2ex] \dfrac{1}{m + 1}, & \text{if } \rho = 1, \end{cases} \qquad i = 0, 1, \ldots, m.$$

It is interesting to consider what happens when the buffer size m is so large that it can be considered as practically infinite. We distinguish two cases.

(a) Suppose that $b < d$, or $\rho < 1$. In this case, arrivals of new packets are less likely than departures of existing packets. This prevents the number of packets in the buffer from growing, and the steady-state probabilities π_i decrease with i, as in a (truncated) geometric PMF. We observe that as $m \to \infty$, we have $1 - \rho^{m+1} \to 1$, and

$$\pi_i \to \rho^i(1-\rho), \qquad \text{for all } i.$$

We can view these as the steady-state probabilities in a system with an infinite buffer. [As a check, note that we have $\sum_{i=0}^{\infty} \rho^i(1-\rho) = 1$.]

(b) Suppose that $b > d$, or $\rho > 1$. In this case, arrivals of new packets are more likely than departures of existing packets. The number of packets in the buffer tends to increase, and the steady-state probabilities π_i increase with i. As we consider larger and larger buffer sizes m, the steady-state probability of any fixed state i decreases to zero:

$$\pi_i \to 0, \qquad \text{for all } i.$$

Were we to consider a system with an infinite buffer, we would have a Markov chain with a countably infinite number of states. Although we do not have the machinery to study such chains, the preceding calculation suggests that every state will have zero steady-state probability and will be "transient." The number of packets in queue will generally grow to infinity, and any particular state will be visited only a finite number of times.

The preceding analysis provides a glimpse into the character of Markov chains with an infinite number of states. In such chains, even if there is a single and aperiodic recurrent class, the chain may never reach steady-state and a steady-state distribution may not exist.

7.4 ABSORPTION PROBABILITIES AND EXPECTED TIME TO ABSORPTION

In this section, we study the short-term behavior of Markov chains. We first consider the case where the Markov chain starts at a transient state. We are interested in the first recurrent state to be entered, as well as in the time until this happens.

When addressing such questions, the subsequent behavior of the Markov chain (after a recurrent state is encountered) is immaterial. We can therefore focus on the case where every recurrent state k is **absorbing**, i.e.,

$$p_{kk} = 1, \qquad p_{kj} = 0 \text{ for all } j \neq k.$$

If there is a unique absorbing state k, its steady-state probability is 1 (because all other states are transient and have zero steady-state probability), and will be reached with probability 1, starting from any initial state. If there are multiple absorbing states, the probability that one of them will be eventually reached is still 1, but the identity of the absorbing state to be entered is random and the associated probabilities may depend on the starting state. In the sequel, we fix a particular absorbing state, denoted by s, and consider the absorption probability a_i that s is eventually reached, starting from i:

$$a_i = \mathbf{P}(X_n \text{ eventually becomes equal to the absorbing state } s \,|\, X_0 = i).$$

Absorption probabilities can be obtained by solving a system of linear equations, as indicated below.

Absorption Probability Equations

Consider a Markov chain where each state is either transient or absorbing, and fix a particular absorbing state s. Then, the probabilities a_i of eventually reaching state s, starting from i, are the unique solution to the equations

$$a_s = 1,$$

$$a_i = 0, \qquad \text{for all absorbing } i \neq s,$$

$$a_i = \sum_{j=1}^{m} p_{ij} a_j, \qquad \text{for all transient } i.$$

The equations $a_s = 1$, and $a_i = 0$, for all absorbing $i \neq s$, are evident from the definitions. To verify the remaining equations, we argue as follows. Let us consider a transient state i and let A be the event that state s is eventually reached. We have

$$\begin{aligned} a_i &= \mathbf{P}(A \,|\, X_0 = i) \\ &= \sum_{j=1}^{m} \mathbf{P}(A \,|\, X_0 = i, X_1 = j)\mathbf{P}(X_1 = j \,|\, X_0 = i) \qquad \text{(total probability thm.)} \\ &= \sum_{j=1}^{m} \mathbf{P}(A \,|\, X_1 = j) p_{ij} \qquad \text{(Markov property)} \\ &= \sum_{j=1}^{m} a_j p_{ij}. \end{aligned}$$

The uniqueness property of the solution to the absorption probability equations requires a separate argument, which is given in the end-of-chapter problems.

The next example illustrates how we can use the preceding method to calculate the probability of entering a given recurrent class (rather than a given absorbing state).

Example 7.10. Consider the Markov chain shown in Fig. 7.17(a). Note that there are two recurrent classes, namely $\{1\}$ and $\{4, 5\}$. We would like to calculate the probability that the state eventually enters the recurrent class $\{4, 5\}$ starting from one of the transient states. For the purposes of this problem, the possible transitions within the recurrent class $\{4, 5\}$ are immaterial. We can therefore lump the states in this recurrent class and treat them as a single absorbing state (call it state 6), as in Fig. 7.17(b). It then suffices to compute the probability of eventually entering state 6 in this new chain.

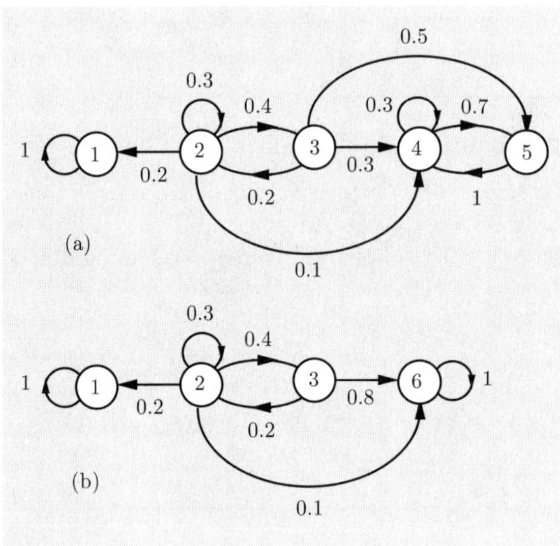

Figure 7.17: (a) Transition probability graph in Example 7.10. (b) A new graph in which states 4 and 5 have been lumped into the absorbing state 6.

The probabilities of eventually reaching state 6, starting from the transient states 2 and 3, satisfy the following equations:

$$a_2 = 0.2a_1 + 0.3a_2 + 0.4a_3 + 0.1a_6,$$
$$a_3 = 0.2a_2 + 0.8a_6.$$

Using the facts $a_1 = 0$ and $a_6 = 1$, we obtain

$$a_2 = 0.3a_2 + 0.4a_3 + 0.1,$$
$$a_3 = 0.2a_2 + 0.8.$$

This is a system of two equations in the two unknowns a_2 and a_3, which can be readily solved to yield $a_2 = 21/31$ and $a_3 = 29/31$.

Example 7.11. Gambler's Ruin. A gambler wins \$1 at each round, with probability p, and loses \$1, with probability $1 - p$. Different rounds are assumed independent. The gambler plays continuously until he either accumulates a target amount of \$$m$, or loses all his money. What is the probability of eventually accumulating the target amount (winning) or of losing his fortune?

We introduce the Markov chain shown in Fig. 7.18 whose state i represents the gambler's wealth at the beginning of a round. The states $i = 0$ and $i = m$ correspond to losing and winning, respectively.

All states are transient, except for the winning and losing states which are absorbing. Thus, the problem amounts to finding the probabilities of absorption at each one of these two absorbing states. Of course, these absorption probabilities depend on the initial state i.

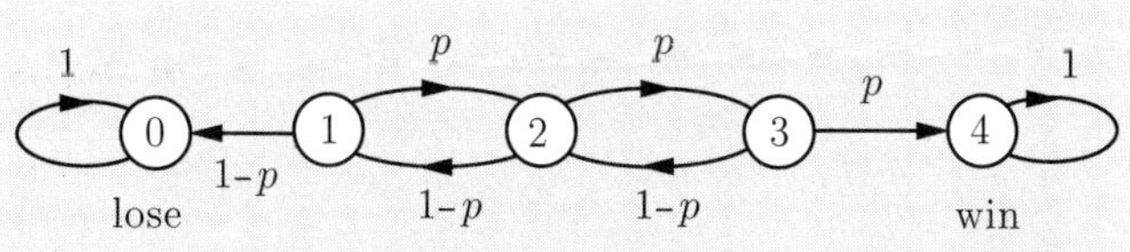

Figure 7.18: Transition probability graph for the gambler's ruin problem (Example 7.11). Here $m = 4$.

Let us set $s = m$ in which case the absorption probability a_i is the probability of winning, starting from state i. These probabilities satisfy

$$\begin{aligned} a_0 &= 0, \\ a_i &= (1-p)a_{i-1} + pa_{i+1}, \qquad i = 1, \ldots, m-1, \\ a_m &= 1. \end{aligned}$$

These equations can be solved in a variety of ways. It turns out there is an elegant method that leads to a nice closed form solution.

Let us write the equations for the a_i as

$$(1-p)(a_i - a_{i-1}) = p(a_{i+1} - a_i), \qquad i = 1, \ldots, m-1.$$

Then, by denoting

$$\delta_i = a_{i+1} - a_i, \qquad i = 0, \ldots, m-1,$$

and

$$\rho = \frac{1-p}{p},$$

the equations are written as

$$\delta_i = \rho\delta_{i-1}, \qquad i = 1, \ldots, m-1,$$

from which we obtain

$$\delta_i = \rho^i\delta_0, \qquad i = 1, \ldots, m-1.$$

This, together with the equation $\delta_0 + \delta_1 + \cdots + \delta_{m-1} = a_m - a_0 = 1$, implies that

$$(1 + \rho + \cdots + \rho^{m-1})\delta_0 = 1,$$

and

$$\delta_0 = \frac{1}{1 + \rho + \cdots + \rho^{m-1}}.$$

Since $a_0 = 0$ and $a_{i+1} = a_i + \delta_i$, the probability a_i of winning starting from a fortune i is equal to

$$\begin{aligned} a_i &= \delta_0 + \delta_1 + \cdots + \delta_{i-1} \\ &= (1 + \rho + \cdots + \rho^{i-1})\delta_0 \\ &= \frac{1 + \rho + \cdots + \rho^{i-1}}{1 + \rho + \cdots + \rho^{m-1}}, \end{aligned}$$

which simplifies to

$$a_i = \begin{cases} \dfrac{1 - \rho^i}{1 - \rho^m}, & \text{if } \rho \neq 1, \\ \dfrac{i}{m}, & \text{if } \rho = 1. \end{cases}$$

The solution reveals that if $\rho > 1$, which corresponds to $p < 1/2$ and unfavorable odds for the gambler, the probability of winning approaches 0 as $m \to \infty$, for any fixed initial fortune. This suggests that if you aim for a large profit under unfavorable odds, financial ruin is almost certain.

Expected Time to Absorption

We now turn our attention to the expected number of steps until a recurrent state is entered (an event that we refer to as "absorption"), starting from a particular transient state. For any state i, we denote

$$\begin{aligned} \mu_i &= \mathbf{E}\big[\text{number of transitions until absorption, starting from } i\big] \\ &= \mathbf{E}\big[\min\{n \geq 0 \,|\, X_n \text{ is recurrent}\} \;\big|\; X_0 = i\big]. \end{aligned}$$

Note that if i is recurrent, then $\mu_i = 0$ according to this definition.

We can derive equations for the μ_i by using the total expectation theorem. We argue that the time to absorption starting from a transient state i is equal to 1 plus the expected time to absorption starting from the next state, which is j with probability p_{ij}. We then obtain a system of linear equations, stated below, which is known to have a unique solution (see Problem 33 for the main idea).

Equations for the Expected Time to Absorption

The expected times to absorption, $\mu_1, \ldots, \mu_m$, are the unique solution to the equations

$$\mu_i = 0, \qquad \text{for all recurrent states } i,$$

$$\mu_i = 1 + \sum_{j=1}^{m} p_{ij}\mu_j, \qquad \text{for all transient states } i.$$

Example 7.12. Spiders and Fly. Consider the spiders-and-fly model of Example 7.2. This corresponds to the Markov chain shown in Fig. 7.19. The states correspond to possible fly positions, and the absorbing states 1 and m correspond to capture by a spider.

Let us calculate the expected number of steps until the fly is captured. We have

$$\mu_1 = \mu_m = 0,$$

and

$$\mu_i = 1 + 0.3\mu_{i-1} + 0.4\mu_i + 0.3\mu_{i+1}, \qquad \text{for } i = 2, \ldots, m-1.$$

We can solve these equations in a variety of ways, such as for example by successive substitution. As an illustration, let $m = 4$, in which case, the equations reduce to

$$\mu_2 = 1 + 0.4\mu_2 + 0.3\mu_3, \qquad \mu_3 = 1 + 0.3\mu_2 + 0.4\mu_3.$$

The first equation yields $\mu_2 = (1/0.6) + (1/2)\mu_3$, which we can substitute in the second equation and solve for μ_3. We obtain $\mu_3 = 10/3$ and by substitution again, $\mu_2 = 10/3$.

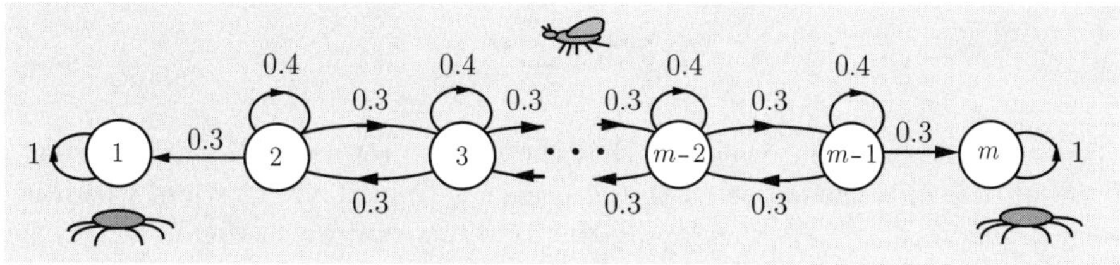

Figure 7.19: Transition probability graph in Example 7.12.

Mean First Passage and Recurrence Times

The idea used to calculate the expected time to absorption can also be used to calculate the expected time to reach a particular recurrent state, starting from any other state. For simplicity, we consider a Markov chain with a single recurrent class. We focus on a special recurrent state s, and we denote by t_i the **mean first passage time from state i to state s**, defined by

$$\begin{aligned} t_i &= \mathbf{E}\big[\text{number of transitions to reach } s \text{ for the first time, starting from } i\big] \\ &= \mathbf{E}\big[\min\{n \geq 0 \,|\, X_n = s\} \;\big|\; X_0 = i\big]. \end{aligned}$$

The transitions out of state s are irrelevant to the calculation of the mean first passage times. We may thus consider a new Markov chain which is identical to the original, except that the special state s is converted into an absorbing state (by setting $p_{ss} = 1$, and $p_{sj} = 0$ for all $j \neq s$). With this transformation, all states other than s become transient. We then compute t_i as the expected number of steps to absorption starting from i, using the formulas given earlier in this section. We have

$$t_i = 1 + \sum_{j=1}^{m} p_{ij} t_j, \qquad \text{for all } i \neq s,$$

$$t_s = 0.$$

This system of linear equations can be solved for the unknowns t_i, and has a unique solution (see the end-of-chapter problems).

The above equations give the expected time to reach the special state s starting from any other state. We may also want to calculate the **mean recurrence time** of the special state s, which is defined as

$$\begin{aligned} t_s^* &= \mathbf{E}[\text{number of transitions up to the first return to } s, \text{ starting from } s] \\ &= \mathbf{E}\big[\min\{n \geq 1 \,|\, X_n = s\} \;\big|\; X_0 = s\big]. \end{aligned}$$

We can obtain t_s^*, once we have the first passage times t_i, by using the equation

$$t_s^* = 1 + \sum_{j=1}^{m} p_{sj} t_j.$$

To justify this equation, we argue that the time to return to s, starting from s, is equal to 1 plus the expected time to reach s from the next state, which is j with probability p_{sj}. We then apply the total expectation theorem.

Example 7.13. Consider the "up-to-date"–"behind" model of Example 7.1. States 1 and 2 correspond to being up-to-date and being behind, respectively, and the transition probabilities are

$$p_{11} = 0.8, \qquad p_{12} = 0.2,$$

$$p_{21} = 0.6, \qquad p_{22} = 0.4.$$

Let us focus on state $s = 1$ and calculate the mean first passage time to state 1, starting from state 2. We have $t_1 = 0$ and

$$t_2 = 1 + p_{21}t_1 + p_{22}t_2 = 1 + 0.4t_2,$$

from which

$$t_2 = \frac{1}{0.6} = \frac{5}{3}.$$

The mean recurrence time to state 1 is given by

$$t_1^* = 1 + p_{11}t_1 + p_{12}t_2 = 1 + 0 + 0.2 \cdot \frac{5}{3} = \frac{4}{3}.$$

Equations for Mean First Passage and Recurrence Times

Consider a Markov chain with a single recurrent class, and let s be a particular recurrent state.

- The mean first passage times t_i to reach state s starting from i, are the unique solution to the system of equations

$$t_s = 0, \qquad t_i = 1 + \sum_{j=1}^{m} p_{ij}t_j, \qquad \text{for all } i \neq s.$$

- The mean recurrence time t_s^* of state s is given by

$$t_s^* = 1 + \sum_{j=1}^{m} p_{sj}t_j.$$

7.5 CONTINUOUS-TIME MARKOV CHAINS

In the Markov chain models that we have considered so far, we have assumed that the transitions between states take unit time. In this section, we consider a related class of models that evolve in continuous time and can be used to study systems involving continuous-time arrival processes. Examples are distribution centers or nodes in communication networks where some events of interest (e.g., arrivals of orders or of new calls) are described in terms of Poisson processes.

As before, we will consider a process that involves transitions from one state to the next, according to given transition probabilities, but we will model the times spent between transitions as continuous random variables. We will still assume that the number of states is finite and, in the absence of a statement to the contrary, we will let the state space be the set $\mathcal{S} = \{1, \ldots, m\}$.

To describe the process, we introduce certain random variables of interest:

X_n : the state right after the nth transition;

Y_n : the time of the nth transition;

T_n : the time elapsed between the $(n-1)$st and the nth transition.

For completeness, we denote by X_0 the initial state, and we let $Y_0 = 0$. We also introduce some assumptions.

Continuous-Time Markov Chain Assumptions

- If the current state is i, the time until the next transition is exponentially distributed with a given parameter ν_i, independent of the past history of the process and of the next state.
- If the current state is i, the next state will be j with a given probability p_{ij}, independent of the past history of the process and of the time until the next transition.

The above assumptions are a complete description of the process and provide an unambiguous method for simulating it: given that we just entered state i, we remain at state i for a time that is exponentially distributed with parameter ν_i, and then move to a next state j according to the transition probabilities p_{ij}. As an immediate consequence, the sequence of states X_n obtained after successive transitions is a discrete-time Markov chain, with transition probabilities p_{ij}, called the **embedded** Markov chain. In mathematical terms, our assumptions can be formulated as follows. Let

$$A = \{T_1 = t_1, \ldots, T_n = t_n,\ X_0 = i_0, \ldots, X_{n-1} = i_{n-1},\ X_n = i\}$$

be an event that captures the history of the process until the nth transition. We then have

$$\begin{aligned}
\mathbf{P}(X_{n+1} = j,\ T_{n+1} \geq t \mid A) &= \mathbf{P}(X_{n+1} = j,\ T_{n+1} \geq t \mid X_n = i) \\
&= \mathbf{P}(X_{n+1} = j \mid X_n = i)\,\mathbf{P}(T_{n+1} \geq t \mid X_n = i) \\
&= p_{ij} e^{-\nu_i t}, \qquad \text{for all } t \geq 0.
\end{aligned}$$

The expected time to the next transition is

$$\mathbf{E}[T_{n+1} \mid X_n = i] = \int_0^\infty \tau \nu_i e^{-\nu_i \tau}\, d\tau = \frac{1}{\nu_i},$$

so we can interpret ν_i as the average number of transitions out of state i, per unit time spent at state i. Consequently, the parameter ν_i is called the **transition**

rate out of state i. Since only a fraction p_{ij} of the transitions out of state i will lead to state j, we may also view

$$q_{ij} = \nu_i p_{ij}$$

as the average number of transitions from i to j, per unit time spent at i. Accordingly, we call q_{ij} the **transition rate from** i **to** j. Note that given the transition rates q_{ij}, one can obtain the transition rates ν_i using the formula

$$\nu_i = \sum_{j=1}^{m} q_{ij},$$

and the transition probabilities using the formula

$$p_{ij} = \frac{q_{ij}}{\nu_i}.$$

Note that the model allows for self-transitions, from a state back to itself, which can indeed happen if a self-transition probability p_{ii} is nonzero. However, such self-transitions have no observable effects: because of the memorylessness of the exponential distribution, the remaining time until the next transition is the same, irrespective of whether a self-transition just occurred or not. For this reason, we can ignore self-transitions and we will henceforth assume that

$$p_{ii} = q_{ii} = 0, \qquad \text{for all } i.$$

Example 7.14. A machine, once in production mode, operates continuously until an alarm signal is generated. The time up to the alarm signal is an exponential random variable with parameter 1. Subsequent to the alarm signal, the machine is tested for an exponentially distributed amount of time with parameter 5. The test results are positive, with probability 1/2, in which case the machine returns to production mode, or negative, with probability 1/2, in which case the machine is taken for repair. The duration of the repair is exponentially distributed with parameter 3. We assume that the above mentioned random variables are all independent and also independent of the test results.

Let states 1, 2, and 3, correspond to production mode, testing, and repair, respectively. The transition rates are $\nu_1 = 1$, $\nu_2 = 5$, and $\nu_3 = 3$. The transition probabilities and the transition rates are given by the following two matrices:

$$P = \begin{bmatrix} 0 & 1 & 0 \\ 1/2 & 0 & 1/2 \\ 1 & 0 & 0 \end{bmatrix}, \qquad Q = \begin{bmatrix} 0 & 1 & 0 \\ 5/2 & 0 & 5/2 \\ 3 & 0 & 0 \end{bmatrix}.$$

See Fig. 7.20 for an illustration.

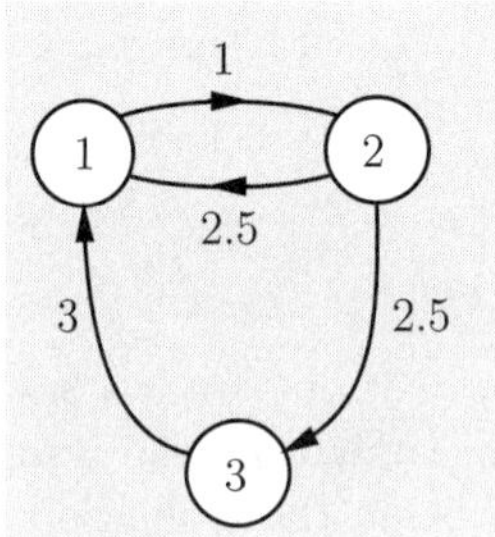

Figure 7.20: Illustration of the Markov chain in Example 7.14. The quantities indicated next to each arc are the transition rates q_{ij}.

We finally note that the continuous-time process we have described has a Markov property similar to its discrete-time counterpart: the future is independent of the past, given the present. To see this, denote by $X(t)$ the state of a continuous-time Markov chain at time $t \geq 0$, and note that it stays constant between transitions.† Let us recall the memorylessness property of the exponential distribution, which in our context implies that for any time t between the nth and $(n+1)$st transition times Y_n and Y_{n+1}, the additional time $Y_{n+1} - t$ until the next transition is independent of the time $t - Y_n$ that the system has been in the current state. It follows that for any time t, and given the present state $X(t)$, the future of the process [the random variables $X(\tau)$ for $\tau > t$], is independent of the past [the random variables $X(\tau)$ for $\tau < t$].

Approximation by a Discrete-Time Markov Chain

We now elaborate on the relation between a continuous-time Markov chain and a corresponding discrete-time version. This relation will lead to an alternative description of a continuous-time Markov chain, and to a set of balance equations characterizing the steady-state behavior.

Let us fix a small positive number δ and consider the discrete-time Markov chain Z_n that is obtained by observing $X(t)$ every δ time units:

$$Z_n = X(n\delta), \qquad n = 0, 1, \ldots.$$

The fact that Z_n is a Markov chain (the future is independent from the past, given the present) follows from the Markov property of $X(t)$. We will use the notation $\overline{p}_{ij}$ to describe the transition probabilities of Z_n.

Given that $Z_n = i$, there is a probability approximately equal to $\nu_i \delta$ that there is a transition between times $n\delta$ and $(n+1)\delta$, and in that case there is a

† If a transition takes place at time t, the notation $X(t)$ is ambiguous. A common convention is to let $X(t)$ refer to the state right after the transition, so that $X(Y_n)$ is the same as X_n.

further probability p_{ij} that the next state is j. Therefore,

$$\overline{p}_{ij} = \mathbf{P}(Z_{n+1} = j \mid Z_n = i) = \nu_i p_{ij}\delta + o(\delta) = q_{ij}\delta + o(\delta), \qquad \text{if } j \neq i,$$

where $o(\delta)$ is a term that is negligible compared to δ, as δ gets smaller. The probability of remaining at i [i.e., no transition occurs between times $n\delta$ and $(n+1)\delta$] is

$$\overline{p}_{ii} = \mathbf{P}(Z_{n+1} = i \mid Z_n = i) = 1 - \sum_{j \neq i} \overline{p}_{ij}.$$

This gives rise to the following alternative description.†

Alternative Description of a Continuous-Time Markov Chain

Given the current state i of a continuous-time Markov chain, and for any $j \neq i$, the state δ time units later is equal to j with probability

$$q_{ij}\delta + o(\delta),$$

independent of the past history of the process.

Example 7.14 (continued). Neglecting $o(\delta)$ terms, the transition probability matrix for the corresponding discrete-time Markov chain Z_n is

$$\begin{bmatrix} 1-\delta & \delta & 0 \\ 5\delta/2 & 1-5\delta & 5\delta/2 \\ 3\delta & 0 & 1-3\delta \end{bmatrix}.$$

Example 7.15. Queueing. Packets arrive at a node of a communication network according to a Poisson process with rate λ. The packets are stored at a buffer with room for up to m packets, and are then transmitted one at a time. However, if a packet finds a full buffer upon arrival, it is discarded. The time required to transmit a packet is exponentially distributed with parameter μ. The transmission times of different packets are independent and are also independent from all the interarrival times.

We will model this system using a continuous-time Markov chain with state $X(t)$ equal to the number of packets in the system at time t [if $X(t) > 0$, then

† Our argument so far shows that a continuous-time Markov chain satisfies this alternative description. Conversely, it can be shown that if we start with this alternative description, the time until a transition out of state i is an exponential random variable with parameter $\nu_i = \sum_j q_{ij}$. Furthermore, given that such a transition has just occurred, the next state is j with probability $q_{ij}/\nu_i = p_{ij}$. This establishes that the alternative description is equivalent to the original one.

$X(t) - 1$ packets are waiting in the queue and one packet is under transmission]. The state increases by one when a new packet arrives and decreases by one when an existing packet departs. To show that $X(t)$ is indeed a Markov chain, we verify that we have the property specified in the above alternative description, and at the same time identify the transition rates q_{ij}.

Consider first the case where the system is empty, i.e., the state $X(t)$ is equal to 0. A transition out of state 0 can only occur if there is a new arrival, in which case the state becomes equal to 1. Since arrivals are Poisson, we have

$$\mathbf{P}\big(X(t+\delta) = 1 \,|\, X(t) = 0\big) = \lambda\delta + o(\delta),$$

and

$$q_{0j} = \begin{cases} \lambda, & \text{if } j = 1, \\ 0, & \text{otherwise.} \end{cases}$$

Consider next the case where the system is full, i.e., the state $X(t)$ is equal to m. A transition out of state m will occur upon the completion of the current packet transmission, at which point the state will become $m-1$. Since the duration of a transmission is exponential (and in particular, memoryless), we have

$$\mathbf{P}\big(X(t+\delta) = m-1 \,|\, X(t) = m\big) = \mu\delta + o(\delta),$$

and

$$q_{mj} = \begin{cases} \mu, & \text{if } j = m-1, \\ 0, & \text{otherwise.} \end{cases}$$

Consider finally the case where $X(t)$ is equal to some intermediate state i, with $0 < i < m$. During the next δ time units, there is a probability $\lambda\delta + o(\delta)$ of a new packet arrival, which will bring the state to $i+1$, and a probability $\mu\delta + o(\delta)$ that a packet transmission is completed, which will bring the state to $i-1$. [The probability of both an arrival and a departure within an interval of length δ is of the order of δ^2 and can be neglected, as is the case with other $o(\delta)$ terms.] Hence,

$$\mathbf{P}\big(X(t+\delta) = i-1 \,|\, X(t) = i\big) = \mu\delta + o(\delta),$$
$$\mathbf{P}\big(X(t+\delta) = i+1 \,|\, X(t) = i\big) = \lambda\delta + o(\delta),$$

and

$$q_{ij} = \begin{cases} \lambda, & \text{if } j = i+1, \\ \mu, & \text{if } j = i-1, \\ 0 & \text{otherwise,} \end{cases} \qquad \text{for } i = 1, 2, \ldots, m-1;$$

see Fig. 7.21.

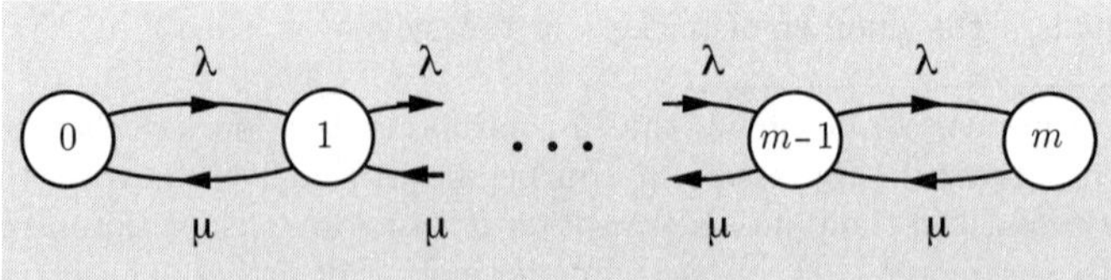

Figure 7.21: Transition graph in Example 7.15.

Steady-State Behavior

We now turn our attention to the long-term behavior of a continuous-time Markov chain and focus on the state occupancy probabilities $\mathbf{P}(X(t) = i)$, in the limit as t gets large. We approach this problem by studying the steady-state probabilities of the corresponding discrete-time chain Z_n.

Since $Z_n = X(n\delta)$, it is clear that the limit π_j of $\mathbf{P}(Z_n = j \,|\, Z_0 = i)$, if it exists, is the same as the limit of $\mathbf{P}(X(t) = j \,|\, X(0) = i)$. It therefore suffices to consider the steady-state probabilities associated with Z_n. Reasoning as in the discrete-time case, we see that for the steady-state probabilities to be independent of the initial state, we need the chain Z_n to have a single recurrent class, which we will henceforth assume. We also note that the Markov chain Z_n is automatically aperiodic. This is because the self-transition probabilities are of the form

$$\overline{p}_{ii} = 1 - \delta \sum_{j \neq i} q_{ij} + o(\delta),$$

which is positive when δ is small, and because chains with nonzero self-transition probabilities are always aperiodic.

The balance equations for the chain Z_n are of the form

$$\pi_j = \sum_{k=1}^{m} \pi_k \, \overline{p}_{kj} \qquad \text{for all } j,$$

or

$$\begin{aligned} \pi_j &= \pi_j \overline{p}_{jj} + \sum_{k \neq j} \pi_k \overline{p}_{kj} \\ &= \pi_j \left(1 - \delta \sum_{k \neq j} q_{jk} + o(\delta) \right) + \sum_{k \neq j} \pi_k \big(q_{kj} \delta + o(\delta) \big). \end{aligned}$$

We cancel out the term π_j that appears on both sides of the equation, divide by δ, and take the limit as δ decreases to zero, to obtain the **balance equations**

$$\pi_j \sum_{k \neq j} q_{jk} = \sum_{k \neq j} \pi_k q_{kj}.$$

We can now invoke the Steady-State Convergence Theorem for the chain Z_n to obtain the following.

Steady-State Convergence Theorem

Consider a continuous-time Markov chain with a single recurrent class. Then, the states j are associated with steady-state probabilities π_j that have the following properties.

(a) For each j, we have

$$\lim_{t \to \infty} \mathbf{P}\big(X(t) = j \,|\, X(0) = i\big) = \pi_j, \qquad \text{for all } i.$$

(b) The π_j are the unique solution to the system of equations below:

$$\pi_j \sum_{k \neq j} q_{jk} = \sum_{k \neq j} \pi_k q_{kj}, \qquad j = 1, \ldots, m,$$

$$1 = \sum_{k=1}^{m} \pi_k.$$

(c) We have

$$\pi_j = 0, \qquad \text{for all transient states } j,$$

$$\pi_j > 0, \qquad \text{for all recurrent states } j.$$

To interpret the balance equations, we view π_j as the expected long-term fraction of time the process spends in state j. It follows that $\pi_k q_{kj}$ can be viewed as the expected frequency of transitions from k to j (expected number of transitions from k to j per unit time). It is seen therefore that the balance equations express the intuitive fact that the frequency of transitions out of state j (the left-hand side term $\pi_j \sum_{k \neq j} q_{jk}$) is equal to the frequency of transitions into state j (the right-hand side term $\sum_{k \neq j} \pi_k q_{kj}$).

Example 7.14 (continued). The balance and normalization equations for this example are

$$\pi_1 = \frac{5}{2}\pi_2 + 3\pi_3, \qquad 5\pi_2 = \pi_1, \qquad 3\pi_3 = \frac{5}{2}\pi_2,$$

$$1 = \pi_1 + \pi_2 + \pi_3.$$

As in the discrete-time case, one of these equations is redundant, e.g., the third equation can be obtained from the first two. Still, there is a unique solution:

$$\pi_1 = \frac{30}{41}, \qquad \pi_2 = \frac{6}{41}, \qquad \pi_3 = \frac{5}{41}.$$

Thus, for example, if we let the process run for a long time, $X(t)$ will be at state 1 with probability $30/41$, independent of the initial state.

The steady-state probabilities π_j are to be distinguished from the steady-state probabilities $\overline{\pi}_j$ of the embedded Markov chain X_n. Indeed, the balance and normalization equations for the embedded Markov chain are

$$\overline{\pi}_1 = \frac{1}{2}\overline{\pi}_2 + \overline{\pi}_3, \qquad \overline{\pi}_2 = \overline{\pi}_1, \qquad \overline{\pi}_3 = \frac{1}{2}\overline{\pi}_2,$$

$$1 = \overline{\pi}_1 + \overline{\pi}_2 + \overline{\pi}_3,$$

yielding the solution

$$\overline{\pi}_1 = \frac{2}{5}, \qquad \overline{\pi}_2 = \frac{2}{5}, \qquad \overline{\pi}_3 = \frac{1}{5}.$$

To interpret the probabilities $\overline{\pi}_j$, we can say, for example, that if we let the process run for a long time, the expected fraction of transitions that lead to state 1 is equal to 2/5.

Note that even though $\overline{\pi}_1 = \overline{\pi}_2$ (that is, there are about as many transitions into state 1 as there are transitions into state 2), we have $\pi_1 > \pi_2$. The reason is that the process tends to spend more time during a typical visit to state 1 than during a typical visit to state 2. Hence, at a given time t, the process $X(t)$ is more likely to be found at state 1. This situation is typical, and the two sets of steady-state probabilities (π_j and $\overline{\pi}_j$) are generically different. The main exception arises in the special case where the transition rates ν_i are the same for all i; see the end-of-chapter problems.

Birth-Death Processes

As in the discrete-time case, the states in a **birth-death process** are linearly arranged and transitions can only occur to a neighboring state, or else leave the state unchanged; formally, we have

$$q_{ij} = 0, \qquad \text{for } |i - j| > 1.$$

In a birth-death process, the long-term expected frequencies of transitions from i to j and of transitions from j to i must be the same, leading to the **local balance equations**

$$\pi_j q_{ji} = \pi_i q_{ij}, \qquad \text{for all } i, j.$$

The local balance equations have the same structure as in the discrete-time case, leading to closed-form formulas for the steady-state probabilities.

Example 7.15 (continued). The local balance equations take the form

$$\pi_i \lambda = \pi_{i+1} \mu, \qquad i = 0, 1, \ldots, m-1,$$

and we obtain $\pi_{i+1} = \rho \pi_i$, where $\rho = \lambda/\mu$. Thus, we have $\pi_i = \rho^i \pi_0$ for all i. The normalization equation $1 = \sum_{i=0}^{m} \pi_i$ yields

$$1 = \pi_0 \sum_{i=0}^{m} \rho^i,$$

and the steady-state probabilities are

$$\pi_i = \frac{\rho^i}{1 + \rho + \cdots + \rho^m}, \qquad i = 0, 1, \ldots, m.$$

7.6 SUMMARY AND DISCUSSION

In this chapter, we have introduced Markov chain models with a finite number of states. In a discrete-time Markov chain, transitions occur at integer times according to given transition probabilities p_{ij}. The crucial property that distinguishes Markov chains from general random processes is that the transition probabilities p_{ij} apply each time that the state is equal to i, independent of the previous values of the state. Thus, given the present, the future of the process is independent of the past.

Coming up with a suitable Markov chain model of a given physical situation is to some extent an art. In general, we need to introduce a rich enough set of states so that the current state summarizes whatever information from the history of the process is relevant to its future evolution. Subject to this requirement, we usually aim at a model that does not involve more states than necessary.

Given a Markov chain model, there are several questions of interest.

(a) Questions referring to the statistics of the process over a finite time horizon. We have seen that we can calculate the probability that the process follows a particular path by multiplying the transition probabilities along the path. The probability of a more general event can be obtained by adding the probabilities of the various paths that lead to the occurrence of the event. In some cases, we can exploit the Markov property to avoid listing each and every path that corresponds to a particular event. A prominent example is the recursive calculation of the n-step transition probabilities, using the Chapman-Kolmogorov equations.

(b) Questions referring to the steady-state behavior of the Markov chain. To address such questions, we classified the states of a Markov chain as transient and recurrent. We discussed how the recurrent states can be divided into disjoint recurrent classes, so that each state in a recurrent class is accessible from every other state in the same class. We also distinguished between periodic and aperiodic recurrent classes. The central result of Markov chain theory is that if a chain consists of a single aperiodic recurrent class, plus possibly some transient states, the probability $r_{ij}(n)$ that the state is equal to some j converges, as time goes to infinity, to a steady-state probability π_j, which does not depend on the initial state i. In other words, the identity of the initial state has no bearing on the statistics of X_n when n is very large. The steady-state probabilities can be found by solving a system of linear equations, consisting of the balance equations and the normalization equation $\sum_j \pi_j = 1$.

(c) Questions referring to the transient behavior of a Markov chain. We discussed the absorption probabilities (the probability that the state eventually enters a given recurrent class, given that it starts at a given transient state), and the mean first passage times (the expected time until a particular recurrent state is entered, assuming that the chain has a single recurrent class). In both cases, we showed that the quantities of interest can be found by considering the unique solution to a system of linear equations.

We finally considered continuous-time Markov chains. In such models, given the current state, the next state is determined by the same mechanism as in discrete-time Markov chains. However, the time until the next transition is an exponentially distributed random variable, whose parameter depends only the current state. Continuous-time Markov chains are in many ways similar to their discrete-time counterparts. They have the same Markov property (the future is independent from the past, given the present). In fact, we can visualize a continuous-time Markov chain in terms of a related discrete-time Markov chain obtained by a fine discretization of the time axis. Because of this correspondence, the steady-state behaviors of continuous-time and discrete-time Markov chains are similar: assuming that there is a single recurrent class, the occupancy probability of any particular state converges to a steady-state probability that does not depend on the initial state. These steady-state probabilities can be found by solving a suitable set of balance and normalization equations.

PROBLEMS

SECTION 7.1. Discrete-Time Markov Chains

Problem 1. The times between successive customer arrivals at a facility are independent and identically distributed random variables with the following PMF:

$$p(k) = \begin{cases} 0.2, & k{=}1, \\ 0.3, & k{=}3, \\ 0.5, & k{=}4, \\ 0, & \text{otherwise.} \end{cases}$$

Construct a four-state Markov chain model that describes the arrival process. In this model, one of the states should correspond to the times when an arrival occurs.

Problem 2. A mouse moves along a tiled corridor with $2m$ tiles, where $m > 1$. From each tile $i \neq 1, 2m$, it moves to either tile $i - 1$ or $i + 1$ with equal probability. From tile 1 or tile $2m$, it moves to tile 2 or $2m - 1$, respectively, with probability 1. Each time the mouse moves to a tile $i \leq m$ or $i > m$, an electronic device outputs a signal L or R, respectively. Can the generated sequence of signals L and R be described as a Markov chain with states L and R?

Problem 3. Consider the Markov chain in Example 7.2, for the case where $m = 4$, as in Fig. 7.2, and assume that the process starts at any of the four states, with equal probability. Let $Y_n = 1$ whenever the Markov chain is at state 1 or 2, and $Y_n = 2$ whenever the Markov chain is at state 3 or 4. Is the process Y_n a Markov chain?

SECTION 7.2. Classification of States

Problem 4. A spider and a fly move along a straight line in unit increments. The spider always moves towards the fly by one unit. The fly moves towards the spider by one unit with probability 0.3, moves away from the spider by one unit with probability 0.3, and stays in place with probability 0.4. The initial distance between the spider and the fly is integer. When the spider and the fly land in the same position, the spider captures the fly.

(a) Construct a Markov chain that describes the relative location of the spider and fly.

(b) Identify the transient and recurrent states.

Problem 5. Consider a Markov chain with states $1, 2, \ldots, 9$, and the following transition probabilities: $p_{12} = p_{17} = 1/2$, $p_{i(i+1)} = 1$ for $i \neq 1, 6, 9$, and $p_{61} = p_{91} = 1$. Is the recurrent class of the chain periodic or not?

Problem 6.* Existence of a recurrent state. Show that in a Markov chain at least one recurrent state must be accessible from any given state, i.e., for any i, there is at least one recurrent j in the set $A(i)$ of accessible states from i.

Solution. Fix a state i. If i is recurrent, then every $j \in A(i)$ is also recurrent so we are done. Assume that i is transient. Then, there exists a state $i_1 \in A(i)$ such that $i \notin A(i_1)$. If i_1 is recurrent, then we have found a recurrent state that is accessible from i, and we are done. Suppose now that i_1 is transient. Then, $i_1 \neq i$ because otherwise the assumptions $i_1 \in A(i)$ and $i \notin A(i_1)$ would yield $i \in A(i)$ and $i \notin A(i)$, which is a contradiction. Since i_1 is transient, there exists some i_2 such that $i_2 \in A(i_1)$ and $i_1 \notin A(i_2)$. In particular, $i_2 \in A(i)$. If i_2 is recurrent, we are done. So, suppose that i_2 is transient. The same argument as before shows that $i_2 \neq i_1$. Furthermore, we must also have $i_2 \neq i$. This is because if we had $i_2 = i$, we would have $i_1 \in A(i) = A(i_2)$, contradicting the assumption $i_1 \notin A(i_2)$. Continuing this process, at the kth step, we will either obtain a recurrent state i_k which is accessible from i, or else we will obtain a transient state i_k which is different than all the preceding states $i, i_1, \ldots, i_{k-1}$. Since there is only a finite number of states, a recurrent state must ultimately be obtained.

Problem 7.* Consider a Markov chain with some transient and some recurrent states.

(a) Show that for some numbers c and γ, with $c > 0$ and $0 < \gamma < 1$, we have

$$\mathbf{P}(X_n \text{ is transient} \mid X_0 = i) \leq c\gamma^n, \qquad \text{for all } i \text{ and } n \geq 1.$$

(b) Let T be the first time n at which X_n is recurrent. Show that such a time is certain to exist (i.e., the probability of the event that there exists a time n at which X_n is recurrent is equal to 1) and that $\mathbf{E}[T] < \infty$.

Solution. (a) For notational convenience, let

$$q_i(n) = \mathbf{P}(X_n \text{ transient} \mid X_0 = i).$$

A recurrent state that is accessible from state i can be reached in at most m steps, where m is the number of states. Therefore, $q_i(m) < 1$. Let

$$\beta = \max_{i=1,\ldots,m} q_i(m)$$

and note that for all i, we have $q_i(m) \leq \beta < 1$. If a recurrent state has not been reached by time m, which happens with probability at most β, the conditional probability that a recurrent state is not reached in the next m steps is at most β as well, which suggests that $q_i(2m) \leq \beta^2$. Indeed, conditioning on the possible values of X_m, we obtain

$$\begin{aligned}
q_i(2m) &= \mathbf{P}(X_{2m} \text{ transient} \mid X_0 = i) \\
&= \sum_{j \text{ transient}} \mathbf{P}(X_{2m} \text{ transient} \mid X_m = j, X_0 = i)\, \mathbf{P}(X_m = j \mid X_0 = i) \\
&= \sum_{j \text{ transient}} \mathbf{P}(X_{2m} \text{ transient} \mid X_m = j)\, \mathbf{P}(X_m = j \mid X_0 = i) \\
&= \sum_{j \text{ transient}} \mathbf{P}(X_m \text{ transient} \mid X_0 = j)\, \mathbf{P}(X_m = j \mid X_0 = i)
\end{aligned}$$

$$\leq \beta \sum_{j \text{ transient}} \mathbf{P}(X_m = j \,|\, X_0 = i)$$
$$= \beta \mathbf{P}(X_m \text{ transient} \,|\, X_0 = i)$$
$$\leq \beta^2.$$

Continuing similarly, we obtain

$$q_i(km) \leq \beta^k, \qquad \text{for all } i \text{ and } k \geq 1.$$

Let n be any positive integer, and let k be the integer such that $km \leq n < (k+1)m$. Then, we have

$$q_i(n) \leq q_i(km) \leq \beta^k = \beta^{-1}\left(\beta^{1/m}\right)^{(k+1)m} \leq \beta^{-1}\left(\beta^{1/m}\right)^n.$$

Thus, the desired relation holds with $c = \beta^{-1}$ and $\gamma = \beta^{1/m}$.

(b) Let A be the event that the state never enters the set of recurrent states. Using the result from part (a), we have

$$\mathbf{P}(A) \leq \mathbf{P}(X_n \text{ transient}) \leq c\gamma^n.$$

Since this is true for every n and since $\gamma < 1$, we must have $\mathbf{P}(A) = 0$. This establishes that there is certainty (probability equal to 1) that there is a finite time T that a recurrent state is first entered. We then have

$$\begin{aligned}
\mathbf{E}[T] &= \sum_{n=1}^{\infty} n\mathbf{P}(X_{n-1} \text{ transient},\ X_n \text{ recurrent}) \\
&\leq \sum_{n=1}^{\infty} n\mathbf{P}(X_{n-1} \text{ transient}) \\
&\leq \sum_{n=1}^{\infty} nc\gamma^{n-1} \\
&= \frac{c}{1-\gamma}\sum_{n=1}^{\infty} n(1-\gamma)\gamma^{n-1} \\
&= \frac{c}{(1-\gamma)^2},
\end{aligned}$$

where the last equality is obtained using the expression for the mean of the geometric distribution.

Problem 8.* Recurrent states. Show that if a recurrent state is visited once, the probability that it will be visited again in the future is equal to 1 (and, therefore, the probability that it will be visited an infinite number of times is equal to 1). *Hint:* Modify the chain to make the recurrent state of interest the only recurrent state, and use the result of Problem 7(b).

Solution. Let s be a recurrent state, and suppose that s has been visited once. From then on, the only possible states are those in the same recurrence class as s. Therefore, without loss of generality, we can assume that there is a single recurrent class. Suppose that the current state is some $i \neq s$. We want to show that s is guaranteed to be visited some time in the future.

Consider a new Markov chain in which the transitions out of state s are disabled, so that $p_{ss} = 1$. The transitions out of states i, for $i \neq s$ are unaffected. Clearly, s is recurrent in the new chain. Furthermore, for any state $i \neq s$, there is a positive probability path from i to s in the original chain (since s is recurrent in the original chain), and the same holds true in the new chain. Since i is not accessible from s in the new chain, it follows that every $i \neq s$ in the new chain is transient. By the result of Problem 7(b), state s will be eventually visited by the new chain (with probability 1). But the original chain is identical to the new one until the time that s is first visited. Hence, state s is guaranteed to be eventually visited by the original chain s. By repeating this argument, we see that s is guaranteed to be visited an infinite number of times (with probability 1).

Problem 9.* Periodic classes. Consider a recurrent class R. Show that exactly one of the following two alternatives must hold:

(i) The states in R can be grouped in $d > 1$ disjoint subsets $S_1, \ldots, S_d$, so that all transitions from S_k lead to S_{k+1}, or to S_1 if $k = d$. (In this case, R is periodic.)

(ii) There exists a time n such that $r_{ij}(n) > 0$ for all $i, j \in R$. (In this case R is aperiodic.)

Hint: Fix a state i and let d be the greatest common divisor of the elements of the set $Q = \{n \mid r_{ii}(n) > 0\}$. If $d = 1$, use the following fact from elementary number theory: if the positive integers $\alpha_1, \alpha_2, \ldots$ have no common divisor other than 1, then every positive integer n outside a finite set can be expressed in the form $n = k_1\alpha_1 + k_2\alpha_2 + \cdots + k_t\alpha_t$ for some nonnegative integers $k_1, \ldots, k_t$, and some $t \geq 1$.

Solution. Fix a state i and consider the set $Q = \{n \mid r_{ii}(n) > 0\}$. Let d be the greatest common divisor of the elements of Q. Consider first the case where $d \neq 1$. For $k = 1, \ldots, d$, let S_k be the set of all states that are reachable from i in $ld + k$ steps for some nonnegative integer l. Suppose that $s \in S_k$ and $p_{ss'} > 0$. Since $s \in S_k$, we can reach s from i in $ld + k$ steps for some l, which implies that we can reach s' from i in $ld + k + 1$ steps. This shows that $s' \in S_{k+1}$ if $k < d$, and that $s' \in S_1$ if $k = d$. It only remains to show that the sets $S_1, \ldots, S_d$ are disjoint. Suppose, to derive a contradiction, that $s \in S_k$ and $s \in S_{k'}$ for some $k \neq k'$. Let q be the length of some positive probability path from s to i. Starting from i, we can get to s in $ld + k$ steps, and then return to i in q steps. Hence $ld + k + q$ belongs to Q, which implies that d divides $k + q$. By the same argument, d must also divide $k' + q$. Hence d divides $k - k'$, which is a contradiction because $1 \leq |k - k'| \leq d - 1$.

Consider next the case where $d = 1$. Let $Q = \{\alpha_1, \alpha_2, \ldots\}$. Since these are the possible lengths of positive probability paths that start and end at i, it follows that any integer n of the form $n = k_1\alpha_1 + k_2\alpha_2 + \cdots + k_t\alpha_t$ is also in Q. (To see this, use k_1 times a path of length α_1, followed by using k_2 times a path of length α_2, etc.) By the number-theoretic fact given in the hint, the set Q contains all but finitely many positive integers. Let n_i be such that

$$r_{ii}(n) > 0, \qquad \text{for all } n > n_i.$$

Fix some $j \neq i$ and let q be the length of a shortest positive probability path from i to j, so that $q < m$, where m is the number of states. Consider some n that satisfies

$n > n_i + m$, and note that $n - q > n_i + m - q > n_i$. Thus, we can go from i to itself in $n - q$ steps, and then from i to j in q steps. Therefore, there is a positive probability path from i to j, of length n, so that $r_{ij}(n) > 0$.

We have so far established that at least one of the alternatives given in the problem statement must hold. To establish that they cannot hold simultaneously, note that the first alternative implies that $r_{ii}(n) = 0$ whenever n is not an integer multiple of d, which is incompatible with the second alternative.

For completeness, we now provide a proof of the number-theoretic fact that was used in this problem. We start with the set of positive integers $\alpha_1, \alpha_2, \ldots$, and assume that they have no common divisor other than 1. We define M as the set of all positive integers the form $\sum_{i=1}^{t} k_i \alpha_i$, where the k_i are nonnegative integers. Note that this set is closed under addition (the sum of two elements of M is of the same form and must also belong to M). Let g be the smallest difference between two distinct elements of M. Then, $g \geq 1$ and $g \leq \alpha_i$ for all i, since α_i and $2\alpha_i$ both belong to M.

Suppose that $g > 1$. Since the greatest common divisor of the α_i is 1, there exists some α_{i^*} which is not divisible by g. We then have

$$\alpha_{i^*} = \ell g + r,$$

for some positive integer ℓ, where the remainder r satisfies $0 < r < g$. Furthermore, in view of the definition of g, there exist nonnegative integers $k_1, k_1', \ldots, k_t, k_t'$ such that

$$\sum_{i=1}^{t} k_i \alpha_i = \sum_{i=1}^{t} k_i' \alpha_i + g.$$

Multiplying this equation by ℓ and using the equation $\alpha_{i^*} = \ell g + r$, we obtain

$$\sum_{i=1}^{t} (\ell k_i) \alpha_i = \sum_{i=1}^{t} (\ell k_i') \alpha_i + \ell g = \sum_{i=1}^{t} (\ell k_i') \alpha_i + \alpha_{i^*} - r.$$

This shows that there exist two numbers in the set M, whose difference is equal to r. Since $0 < r < g$, this contradicts our definition of g as the smallest possible difference. This contradiction establishes that g must be equal to 1.

Since $g = 1$, there exists some positive integer x such that $x \in M$ and $x + 1 \in M$. We will now show that every integer n larger than $\alpha_1 x$ belongs to M. Indeed, by dividing n by α_1, we obtain $n = k\alpha_1 + r$, where $k \geq x$ and where the remainder r satisfies $0 \leq r < \alpha_1$. We rewrite n in the form

$$n = x(\alpha_1 - r) + (x + 1)r + (k - x)\alpha_1.$$

Since x, $x + 1$, and α_1 all belong to M, this shows that n is the sum of elements of M and must also belong to M, as desired.

SECTION 7.3. Steady-State Behavior

Problem 10. Consider the two models of machine failure and repair in Example 7.3. Find conditions on b and r for the chain to have a single recurrent class which is aperiodic and, under those conditions, find closed form expressions for the steady-state probabilities.

Problem 11. A professor gives tests that are hard, medium, or easy. If she gives a hard test, her next test will be either medium or easy, with equal probability. However, if she gives a medium or easy test, there is a 0.5 probability that her next test will be of the same difficulty, and a 0.25 probability for each of the other two levels of difficulty. Construct an appropriate Markov chain and find the steady-state probabilities.

Problem 12. Alvin likes to sail each Saturday to his cottage on a nearby island off the coast. Alvin is an avid fisherman, and enjoys fishing off his boat on the way to and from the island, as long as the weather is good. Unfortunately, the weather is good on the way to or from the island with probability p, independent of what the weather was on any past trip (so the weather could be nice on the way to the island, but poor on the way back). Now, if the weather is nice, Alvin will take one of his n fishing rods for the trip, but if the weather is bad, he will not bring a fishing rod with him. We want to find the probability that on a given leg of the trip to or from the island the weather will be nice, but Alvin will not fish because all his fishing rods are at his other home.

(a) Formulate an appropriate Markov chain model with $n+1$ states and find the steady-state probabilities.

(b) What is the steady-state probability that on a given trip, Alvin sails with nice weather but without a fishing rod?

Problem 13. Consider the Markov chain in Fig. 7.22. Let us refer to a transition that results in a state with a higher (respectively, lower) index as a birth (respectively, death). Calculate the following quantities, assuming that when we start observing the chain, it is already in steady-state.

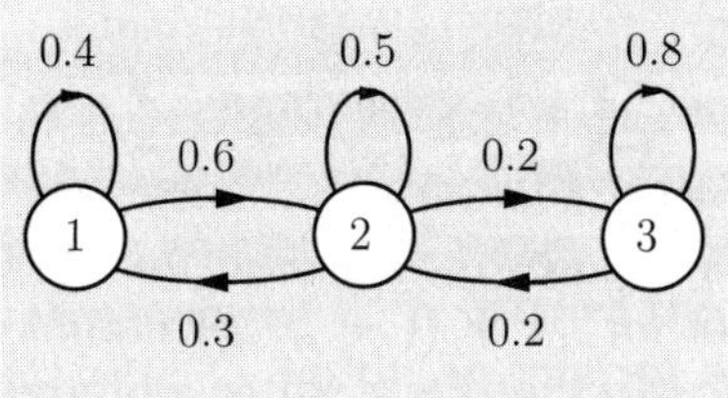

Figure 7.22: Transition probability graph for Problem 11.

(a) For each state i, the probability that the current state is i.

(b) The probability that the first transition we observe is a birth.

(c) The probability that the first change of state we observe is a birth.

(d) The conditional probability that the process was in state 2 before the first transition that we observe, given that this transition was a birth.

(e) The conditional probability that the process was in state 2 before the first change of state that we observe, given that this change of state was a birth.

(f) The conditional probability that the first observed transition is a birth given that it resulted in a change of state.

(g) The conditional probability that the first observed transition leads to state 2, given that it resulted in a change of state.

Problem 14. Consider a Markov chain with given transition probabilities and with a single recurrent class that is aperiodic. Assume that for $n \geq 500$, the n-step transition probabilities are very close to the steady-state probabilities.

(a) Find an approximate formula for $\mathbf{P}(X_{1000} = j, X_{1001} = k, X_{2000} = l \mid X_0 = i)$.

(b) Find an approximate formula for $\mathbf{P}(X_{1000} = i \mid X_{1001} = j)$.

Problem 15. Ehrenfest model of diffusion. We have a total of n balls, some of them black, some white. At each time step, we either do nothing, which happens with probability ϵ, where $0 < \epsilon < 1$, or we select a ball at random, so that each ball has probability $(1 - \epsilon)/n > 0$ of being selected. In the latter case, we change the color of the selected ball (if white it becomes black, and vice versa), and the process is repeated indefinitely. What is the steady-state distribution of the number of white balls?

Problem 16. Bernoulli-Laplace model of diffusion. Each of two urns contains m balls. Out of the total of the $2m$ balls, m are white and m are black. A ball is simultaneously selected from each urn and moved to the other urn, and the process is indefinitely repeated. What is the steady-state distribution of the number of white balls in each urn?

Problem 17. Consider a Markov chain with two states denoted 1 and 2, and transition probabilities

$$p_{11} = 1 - \alpha, \qquad p_{12} = \alpha,$$

$$p_{21} = \beta, \qquad p_{22} = 1 - \beta,$$

where α and β are such that $0 < \alpha < 1$ and $0 < \beta < 1$.

(a) Show that the two states of the chain form a recurrent and aperiodic class.

(b) Use induction to show that for all n, we have

$$r_{11}(n) = \frac{\beta}{\alpha + \beta} + \frac{\alpha(1 - \alpha - \beta)^n}{\alpha + \beta}, \qquad r_{12}(n) = \frac{\alpha}{\alpha + \beta} - \frac{\alpha(1 - \alpha - \beta)^n}{\alpha + \beta},$$

$$r_{21}(n) = \frac{\beta}{\alpha + \beta} - \frac{\beta(1 - \alpha - \beta)^n}{\alpha + \beta}, \qquad r_{22}(n) = \frac{\alpha}{\alpha + \beta} + \frac{\beta(1 - \alpha - \beta)^n}{\alpha + \beta}.$$

(c) What are the steady-state probabilities π_1 and π_2?

Problem 18. The parking garage at MIT has installed a card-operated gate, which, unfortunately, is vulnerable to absent-minded faculty and staff. In particular, in each day, a car crashes the gate with probability p, in which case a new gate must be installed. Also a gate that has survived for m days must be replaced as a matter of periodic maintenance. What is the long-term expected frequency of gate replacements?

Problem 19.* Steady-state convergence. Consider a Markov chain with a single recurrent class, and assume that there exists a time $\overline{n}$ such that

$$r_{ij}(\overline{n}) > 0,$$

for all i and all recurrent j. (This is equivalent to assuming that the class is aperiodic.) We wish to show that for any i and j, the limit

$$\lim_{n\to\infty} r_{ij}(n)$$

exists and does not depend on i. To derive this result, we need to show that the choice of the initial state has no long-term effect. To quantify this effect, we consider two different initial states i and k, and consider two independent Markov chains, X_n and Y_n, with the same transition probabilities and with $X_0 = i$, $Y_0 = k$. Let $T = \min\{n \,|\, X_n = Y_n\}$ be the first time that the two chains enter the same state.

(a) Show that there exist positive constants c and $\gamma < 1$ such that

$$\mathbf{P}(T \geq n) \leq c\gamma^n.$$

(b) Show that if the states of the two chains became equal by time n, their occupancy probabilities at time n are the same, that is,

$$\mathbf{P}(X_n = j \,|\, T \leq n) = \mathbf{P}(Y_n = j \,|\, T \leq n).$$

(c) Show that $|r_{ij}(n) - r_{kj}(n)| \leq c\gamma^n$, for all i, j, k, and n. *Hint:* Condition on the two events $\{T > n\}$ and $\{T \leq n\}$.

(d) Let $q_j^+(n) = \max_i r_{ij}(n)$ and $q_j^-(n) = \min_i r_{ij}(n)$. Show that

$$q_j^-(n) \leq q_j^-(n+1) \leq q_j^+(n+1) \leq q_j^+(n), \qquad \text{for all } n.$$

(e) Show that the sequence $r_{ij}(n)$ converges to a limit that does not depend on i. *Hint:* Combine the results of parts (c) and (d) to show that the two sequences $q_j^-(n)$ and $q_j^+(n)$ converge and have the same limit.

Solution. (a) The argument is similar to the one used to bound the PMF of the time until a recurrent state is entered (Problem 7). Let l be some recurrent state and let $\beta = \min_i r_{il}(\overline{n}) > 0$. No matter what is the current state of X_n and Y_n, there is probability of at least β^2 that both chains are at state l after $\overline{n}$ time steps. Thus,

$$\mathbf{P}(T > \overline{n}) \leq 1 - \beta^2.$$

Similarly,

$$\mathbf{P}(T > 2\overline{n}) = \mathbf{P}(T > \overline{n})\,\mathbf{P}(T > 2\overline{n} \,|\, T > \overline{n}) \leq (1 - \beta^2)^2,$$

and

$$\mathbf{P}(T > k\overline{n}) \leq (1 - \beta^2)^k.$$

This implies that

$$\mathbf{P}(T \geq n) \leq c\gamma^n,$$

where $\gamma = (1-\beta^2)^{1/\overline{n}}$, and $c = 1/(1-\beta^2)^{\overline{n}}$.

(b) We condition on the possible values of T and on the common state l of the two chains at time T, and use the total probability theorem. We have

$$\begin{aligned}
\mathbf{P}(X_n = j \,|\, T \leq n) &= \sum_{t=0}^{n}\sum_{l=1}^{m} \mathbf{P}(X_n = j \,|\, T = t, X_t = l)\, \mathbf{P}(T = t, X_t = l \,|\, T \leq n) \\
&= \sum_{t=0}^{n}\sum_{l=1}^{m} \mathbf{P}(X_n = j \,|\, X_t = l)\, \mathbf{P}(T = t, X_t = l \,|\, T \leq n) \\
&= \sum_{t=0}^{n}\sum_{l=1}^{m} r_{lj}(n-t)\, \mathbf{P}(T = t, X_t = l \,|\, T \leq n).
\end{aligned}$$

Similarly,

$$\mathbf{P}(Y_n = j \,|\, T \leq n) = \sum_{t=0}^{n}\sum_{l=1}^{m} r_{lj}(n-t)\, \mathbf{P}(T = t, Y_t = l \,|\, T \leq n).$$

But the events $\{T = t, X_t = l\}$ and $\{T = t, Y_t = l\}$ are identical, and therefore have the same probability, which implies that $\mathbf{P}(X_n = j \,|\, T \leq n) = \mathbf{P}(Y_n = j \,|\, T \leq n)$.

(c) We have

$$r_{ij}(n) = \mathbf{P}(X_n = j) = \mathbf{P}(X_n = j \,|\, T \leq n)\, \mathbf{P}(T \leq n) + \mathbf{P}(X_n = j \,|\, T > n)\, \mathbf{P}(T > n)$$

and

$$r_{kj}(n) = \mathbf{P}(Y_n = j) = \mathbf{P}(Y_n = j \,|\, T \leq n)\, \mathbf{P}(T \leq n) + \mathbf{P}(Y_n = j \,|\, T > n)\, \mathbf{P}(T > n).$$

By subtracting these two equations, using the result of part (b) to eliminate the first terms in their right-hand sides, and by taking the absolute value of both sides, we obtain

$$\begin{aligned}
|r_{ij}(n) - r_{kj}(n)| &\leq \big|\mathbf{P}(X_n = j \,|\, T > n)\, \mathbf{P}(T > n) - \mathbf{P}(Y_n = j \,|\, T > n)\, \mathbf{P}(T > n)\big| \\
&\leq \mathbf{P}(T > n) \\
&\leq c\gamma^n.
\end{aligned}$$

(d) By conditioning on the state after the first transition, and using the total probability theorem, we have the following variant of the Chapman-Kolmogorov equation:

$$r_{ij}(n+1) = \sum_{k=1}^{m} p_{ik} r_{kj}(n).$$

Using this equation, we obtain

$$q_j^+(n+1) = \max_i r_{ij}(n+1) = \max_i \sum_{k=1}^m p_{ik} r_{kj}(n) \le \max_i \sum_{k=1}^m p_{ik} q_j^+(n) = q_j^+(n).$$

The inequality $q_j^-(n) \le q_j^-(n+1)$ is established by a symmetrical argument. The inequality $q_j^-(n+1) \le q_j^+(n+1)$ is a consequence of the definitions.

(e) The sequences $q_j^-(n)$ and $q_j^+(n)$ converge because they are monotonic. The inequality $|r_{ij}(n) - r_{kj}(n)| \le c\gamma^n$, for all i and k, implies that $q_j^+(n) - q_j^-(n) \le c\gamma^n$. Taking the limit as $n \to \infty$, we obtain that the limits of $q_j^+(n)$ and $q_j^-(n)$ are the same. Let π_j denote this common limit. Since $q_j^-(n) \le r_{ij}(n) \le q_j^+(n)$, it follows that $r_{ij}(n)$ also converges to π_j, and the limit is independent of i.

Problem 20.* Uniqueness of solutions to the balance equations. Consider a Markov chain with a single recurrent class, plus possibly some transient states.

(a) Assuming that the recurrent class is aperiodic, show that the balance equations together with the normalization equation have a unique nonnegative solution. *Hint:* Given a solution different from the steady-state probabilities, let it be the PMF of X_0 and consider what happens as time goes to infinity.

(b) Show that the uniqueness result of part (a) is also true when the recurrent class is periodic. *Hint:* Introduce self-transitions in the Markov chain, in a manner that results in an equivalent set of balance equations, and use the result of part (a).

Solution. (a) Let $\pi_1, \ldots, \pi_m$ be the steady-state probabilities, that is, the limits of the $r_{ij}(n)$. These satisfy the balance and normalization equations. Suppose that there is another nonnegative solution $\overline{\pi}_1, \ldots, \overline{\pi}_m$. Let us initialize the Markov chain according to these probabilities, so that $\mathbf{P}(X_0 = j) = \overline{\pi}_j$ for all j. Using the argument given in the text, we obtain $\mathbf{P}(X_n = j) = \overline{\pi}_j$, for all times. Thus,

$$\begin{aligned} \overline{\pi}_j &= \lim_{n\to\infty} \mathbf{P}(X_n = j) \\ &= \lim_{n\to\infty} \sum_{k=1}^m \overline{\pi}_k r_{kj}(n) \\ &= \sum_{k=1}^m \overline{\pi}_k \pi_j \\ &= \pi_j. \end{aligned}$$

(b) Consider a new Markov chain, whose transition probabilities $\overline{p}_{ij}$ are given by

$$\overline{p}_{ii} = (1-\alpha)p_{ii} + \alpha, \qquad \overline{p}_{ij} = (1-\alpha)p_{ij}, \quad j \ne i.$$

Here, α is a number satisfying $0 < \alpha < 1$. The balance equations for the new Markov chain take the form

$$\pi_j = \pi_j\big((1-\alpha)p_{jj} + \alpha\big) + \sum_{i\ne j} \pi_i (1-\alpha) p_{ij},$$

or

$$(1-\alpha)\pi_j = (1-\alpha)\sum_{i=1}^{m}\pi_i p_{ij}.$$

These equations are equivalent to the balance equations for the original chain. Notice that the new chain is aperiodic, because self-transitions have positive probability. This establishes uniqueness of solutions for the new chain, and implies the same for the original chain.

Problem 21.* Expected long-term frequency interpretation. Consider a Markov chain with a single recurrent class which is aperiodic. Show that

$$\pi_j = \lim_{n\to\infty}\frac{v_{ij}(n)}{n}, \qquad \text{for all } i,\, j = 1,\ldots,m,$$

where the π_j are the steady-state probabilities, and $v_{ij}(n)$ is the expected value of the number of visits to state j within the first n transitions, starting from state i. *Hint:* Use the following fact from analysis. If a sequence a_n converges to a number a, the sequence b_n defined by $b_n = (1/n)\sum_{k=1}^{n} a_k$ also converges to a.

Solution. We first assert that for all n, i, and j, we have

$$v_{ij}(n) = \sum_{k=1}^{n} r_{ij}(k).$$

To see this, note that

$$v_{ij}(n) = \mathbf{E}\left[\sum_{k=1}^{n} I_k \;\middle|\; X_0 = i\right],$$

where I_k is the random variable that takes the value 1 if $X_k = j$, and the value 0 otherwise, so that

$$\mathbf{E}\big[I_k \,|\, X_0 = i\big] = r_{ij}(k).$$

Since

$$\frac{v_{ij}(n)}{n} = \frac{1}{n}\sum_{k=1}^{n} r_{ij}(k),$$

and $r_{ij}(k)$ converges to π_j, it follows that $v_{ij}(n)/n$ also converges to π_j, which is the desired result.

For completeness, we also provide the proof of the fact given in the hint, and which was used in the last step of the above argument. Consider a sequence a_n that converges to some a, and let $b_n = (1/n)\sum_{k=1}^{n} a_k$. Fix some $\epsilon > 0$. Since a_n converges to a, there exists some n_0 such that $a_k \le a + (\epsilon/2)$, for all $k > n_0$. Let also $c = \max_k a_k$. We then have

$$b_n = \frac{1}{n}\sum_{k=1}^{n_0} a_k + \frac{1}{n}\sum_{k=n_0+1}^{n} a_k \le \frac{n_0}{n}c + \frac{n-n_0}{n}\left(a + \frac{\epsilon}{2}\right).$$

The limit of the right-hand side, as n tends to infinity, is $a + (\epsilon/2)$. Therefore, there exists some n_1 such that $b_n \le a + \epsilon$, for every $n \ge n_1$. By a symmetrical argument, there exists some n_2 such that $b_n \geq a - \epsilon$, for every $n \geq n_2$. We have shown that for every $\epsilon > 0$, there exists some n_3 (namely, $n_3 = \max\{n_1, n_2\}$) such that $|b_n - a| \leq \epsilon$, for all $n \geq n_3$. This means that b_n converges to a.

Problem 22.* Doubly stochastic matrices. Consider a Markov chain with a single recurrent class which is aperiodic, and whose transition probability matrix is **doubly stochastic**, i.e., it has the property that the entries in any column (as well as in any row) add to unity, so that

$$\sum_{i=1}^{m} p_{ij} = 1, \qquad j = 1, \ldots, m.$$

(a) Show that the transition probability matrix of the chain in Example 7.7 is doubly stochastic.

(b) Show that the steady-state probabilities are

$$\pi_j = \frac{1}{m}, \qquad j = 1, \ldots, m.$$

(c) Suppose that the recurrent class of the chain is instead periodic. Show that $\pi_1 = \cdots = \pi_m = 1/m$ is the unique solution to the balance and normalization equations. Discuss your answer in the context of Example 7.7 for the case where m is even.

Solution. (a) Indeed the rows and the columns of the transition probability matrix in this example all add to 1.

(b) We have

$$\sum_{i=1}^{m} \frac{1}{m} p_{ij} = \frac{1}{m}.$$

Thus, the given probabilities $\pi_j = 1/m$ satisfy the balance equations and must therefore be the steady-state probabilities.

(c) Let $(\pi_1, \ldots, \pi_m)$ be a solution to the balance and normalization equations. Consider a particular j such that $\pi_j \geq \pi_i$ for all i, and let $q = \pi_j$. The balance equation for state j yields

$$q = \pi_j = \sum_{i=1}^{m} \pi_i p_{ij} \leq q \sum_{i=1}^{m} p_{ij} = q,$$

where the last step follows because the transition probability matrix is doubly stochastic. It follows that the above inequality is actually an equality and

$$\sum_{i=1}^{m} \pi_i p_{ij} = \sum_{i=1}^{m} q p_{ij}.$$

Since $\pi_i \leq q$ for all i, we must have $\pi_i p_{ij} = q p_{ij}$ for every i, Thus, $\pi_i = q$ for every state i from which a transition to j is possible. By repeating this argument, we see that $\pi_i = q$ for every state i such that there is a positive probability path from i to j. Since all states are recurrent and belong to the same class, all states i have this property, and therefore π_i is the same for all i. Since the π_i add to 1, we obtain $\pi_1 = 1/m$ for all i.

If m is even in Example 7.7, the chain is periodic with period 2. Despite this fact, the result we have just established shows that $\pi_j = 1/m$ is the unique solution to the balance and normalization equations.

Problem 23.* **Queueing.** Consider the queueing Example 7.9, but assume that the probabilities of a packet arrival and a packet transmission depend on the state of the queue. In particular, in each period where there are i packets in the node, one of the following occurs:

(i) one new packet arrives; this happens with a given probability b_i. We assume that $b_i > 0$ for $i < m$, and $b_m = 0$.

(ii) one existing packet completes transmission; this happens with a given probability $d_i > 0$ if $i \geq 1$, and with probability 0 otherwise;

(iii) no new packet arrives and no existing packet completes transmission; this happens with probability $1 - b_i - d_i$ if $i \geq 1$, and with probability $1 - b_i$ if $i = 0$.

Calculate the steady-state probabilities of the corresponding Markov chain.

Solution. We introduce a Markov chain where the states are $0, 1, \ldots, m$, and correspond to the number of packets currently stored at the node. The transition probability graph is given in Fig. 7.23.

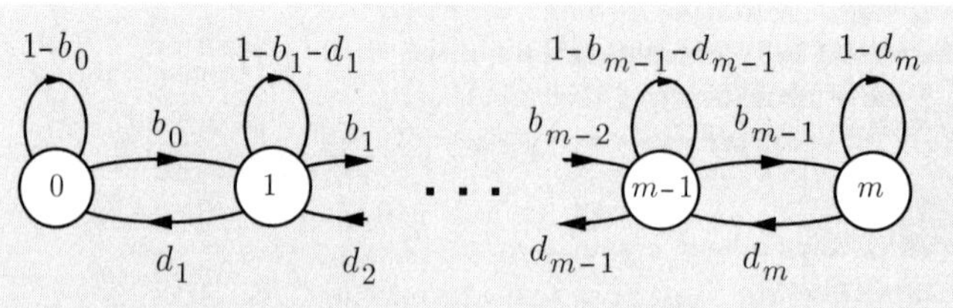

Figure 7.23: Transition probability graph for Problem 23.

Similar to Example 7.9, we write down the local balance equations, which take the form

$$\pi_i b_i = \pi_{i+1} d_{i+1}, \qquad i = 0, 1, \ldots, m-1.$$

Thus we have $\pi_{i+1} = \rho_i \pi_i$, where

$$\rho_i = \frac{b_i}{d_{i+1}}.$$

Hence $\pi_i = (\rho_0 \cdots \rho_{i-1})\pi_0$ for $i = 1, \ldots, m$. By using the normalization equation $1 = \pi_0 + \pi_1 + \cdots + \pi_m$, we obtain

$$1 = \pi_0(1 + \rho_0 + \rho_0\rho_1 + \cdots + \rho_0 \cdots \rho_{m-1}),$$

from which

$$\pi_0 = \frac{1}{1 + \rho_0 + \rho_0\rho_1 + \cdots + \rho_0 \cdots \rho_{m-1}}.$$

The remaining steady-state probabilities are

$$\pi_i = \frac{\rho_0 \cdots \rho_{i-1}}{1 + \rho_0 + \rho_0\rho_1 + \cdots + \rho_0 \cdots \rho_{m-1}}, \qquad i = 1, \ldots, m.$$

Problem 24.* Dependence of the balance equations. Show that if we add the first $m-1$ balance equations $\pi_j = \sum_{k=1}^m \pi_k p_{kj}$, for $j = 1, \ldots, m-1$, we obtain the last equation $\pi_m = \sum_{k=1}^m \pi_k p_{km}$.

Solution. By adding the first $m-1$ balance equations, we obtain

$$\begin{aligned}\sum_{j=1}^{m-1} \pi_j &= \sum_{j=1}^{m-1} \sum_{k=1}^{m} \pi_k p_{kj} \\ &= \sum_{k=1}^{m} \pi_k \sum_{j=1}^{m-1} p_{kj} \\ &= \sum_{k=1}^{m} \pi_k (1 - p_{km}) \\ &= \pi_m + \sum_{k=1}^{m-1} \pi_k - \sum_{k=1}^{m} \pi_k p_{km}.\end{aligned}$$

This equation is equivalent to the last balance equation $\pi_m = \sum_{k=1}^m \pi_k p_{km}$.

Problem 25.* Local balance equations. We are given a Markov chain that has a single recurrent class which is aperiodic. Suppose that we have found a solution $\pi_1, \ldots, \pi_m$ to the following system of local balance and normalization equations:

$$\pi_i p_{ij} = \pi_j p_{ji}, \qquad i, j = 1, \ldots, m,$$

$$\sum_{i=1}^{m} \pi_i = 1, \qquad i = 1, \ldots, m.$$

(a) Show that the π_j are the steady-state probabilities.

(b) What is the interpretation of the equations $\pi_i p_{ij} = \pi_j p_{ji}$ in terms of expected long-term frequencies of transitions between i and j?

(c) Construct an example where the local balance equations are not satisfied by the steady-state probabilities.

Solution. (a) By adding the local balance equations $\pi_i p_{ij} = \pi_j p_{ji}$ over i, we obtain

$$\sum_{i=1}^{m} \pi_i p_{ij} = \sum_{i=1}^{m} \pi_j p_{ji} = \pi_j,$$

so the π_j also satisfy the balance equations. Therefore, they are equal to the steady-state probabilities.

(b) We know that $\pi_i p_{ij}$ can be interpreted as the expected long-term frequency of transitions from i to j, so the local balance equations imply that the expected long-term frequency of any transition is equal to the expected long-term frequency of the reverse transition. (This property is also known as *time reversibility* of the chain.)

(c) We need a minimum of three states for such an example. Let the states be $1, 2, 3$, and let $p_{12} > 0$, $p_{13} > 0$, $p_{21} > 0$, $p_{32} > 0$, with all other transition probabilities being 0. The chain has a single recurrent aperiodic class. The local balance equations do not hold because the expected frequency of transitions from 1 to 3 is positive, but the expected frequency of reverse transitions is 0.

Problem 26.* Sampled Markov chains. Consider a Markov chain X_n with transition probabilities p_{ij}, and let $r_{ij}(n)$ be the n-step transition probabilities.

(a) Show that for all $n \geq 1$ and $l \geq 1$, we have

$$r_{ij}(n+l) = \sum_{k=1}^{m} r_{ik}(n) r_{kj}(l).$$

(b) Suppose that there is a single recurrent class, which is aperiodic. We sample the Markov chain every l transitions, thus generating a process Y_n, where $Y_n = X_{ln}$. Show that the sampled process can be modeled by a Markov chain with a single aperiodic recurrent class and transition probabilities $r_{ij}(l)$.

(c) Show that the Markov chain of part (b) has the same steady-state probabilities as the original process.

Solution. (a) We condition on X_n and use the total probability theorem. We have

$$\begin{aligned}
r_{ij}(n+l) &= \mathbf{P}(X_{n+l} = j \,|\, X_0 = i) \\
&= \sum_{k=1}^{m} \mathbf{P}(X_n = k \,|\, X_0 = i)\, \mathbf{P}(X_{n+l} = j \,|\, X_n = k,\, X_0 = i) \\
&= \sum_{k=1}^{m} \mathbf{P}(X_n = k \,|\, X_0 = i)\, \mathbf{P}(X_{n+l} = j \,|\, X_n = k) \\
&= \sum_{k=1}^{m} r_{ik}(n) r_{kj}(l),
\end{aligned}$$

where in the third equality we used the Markov property.

(b) Since X_n is Markov, once we condition on X_{ln}, the past of the process (the states X_k for $k < ln$) becomes independent of the future (the states X_k for $k > ln$). This implies that given Y_n, the past (the states Y_k for $k < n$) is independent of the future (the states Y_k for $k > n$). Thus, Y_n has the Markov property. Because of our assumptions on X_n, there is a time $\overline{n}$ such that

$$\mathbf{P}(X_n = j \,|\, X_0 = i) > 0,$$

for every $n \geq \overline{n}$, every state i, and every state j in the single recurrent class R of the process X_n. This implies that

$$\mathbf{P}(Y_n = j \,|\, Y_0 = i) > 0,$$

for every $n \geq \overline{n}$, every i, and every $j \in R$. Therefore, the process Y_n has a single recurrent class, which is aperiodic.

(c) The n-step transition probabilities $r_{ij}(n)$ of the process X_n converge to the steady-state probabilities π_j. The n-step transition probabilities of the process Y_n are of the form $r_{ij}(ln)$, and therefore also converge to the same limits π_j. This establishes that the π_j are the steady-state probabilities of the process Y_n.

Problem 27.* Given a Markov chain X_n with a single recurrent class which is aperiodic, consider the Markov chain whose state at time n is (X_{n-1}, X_n). Thus, the state in the new chain can be associated with the last transition in the original chain.

(a) Show that the steady-state probabilities of the new chain are

$$\eta_{ij} = \pi_i p_{ij},$$

where the π_i are the steady-state probabilities of the original chain.

(b) Generalize part (a) to the case of the Markov chain $(X_{n-k}, X_{n-k+1}, \ldots, X_n)$, whose state can be associated with the last k transitions of the original chain.

Solution. (a) For every state (i, j) of the new Markov chain, we have

$$\mathbf{P}\big((X_{n-1}, X_n) = (i, j)\big) = \mathbf{P}(X_{n-1} = i)\,\mathbf{P}(X_n = j \,|\, X_{n-1} = i) = \mathbf{P}(X_{n-1} = i)\,p_{ij}.$$

Since the Markov chain X_n has a single recurrent class which is aperiodic, $\mathbf{P}(X_{n-1} = i)$ converges to the steady-state probability π_i, for every i. It follows that $\mathbf{P}\big((X_{n-1}, X_n) = (i, j)\big)$ converges to $\pi_i p_{ij}$, which is therefore the steady-state probability of (i, j).

(b) Using the multiplication rule, we have

$$\mathbf{P}\big((X_{n-k}, \ldots, X_n) = (i_0, \ldots, i_k)\big) = \mathbf{P}(X_{n-k} = i_0)\,p_{i_0 i_1} \cdots p_{i_{k-1} i_k}.$$

Therefore, by an argument similar to the one in part (a), the steady-state probability of state $(i_0, \ldots, i_k)$ is equal to $\pi_{i_0} p_{i_0 i_1} \cdots p_{i_{k-1} i_k}$.

SECTION 7.4. Absorption Probabilities and Expected Time to Absorption

Problem 28. There are m classes offered by a particular department, and each year, the students rank each class from 1 to m, in order of difficulty, with rank m being the highest. Unfortunately, the ranking is completely arbitrary. In fact, any given class is equally likely to receive any given rank on a given year (two classes may not receive the same rank). A certain professor chooses to remember only the highest ranking his class has ever gotten.

(a) Find the transition probabilities of the Markov chain that models the ranking that the professor remembers.

(b) Find the recurrent and the transient states.

(c) Find the expected number of years for the professor to achieve the highest ranking given that in the first year he achieved the ith ranking.

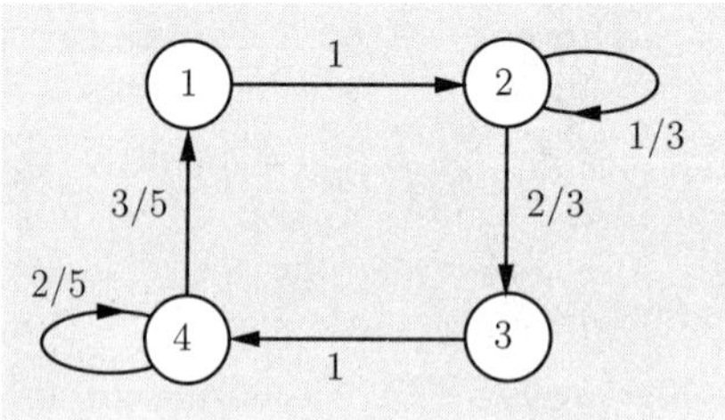

Figure 7.24: Transition probability graph for Problem 29.

Problem 29. Consider the Markov chain specified in Fig. 7.24. The steady-state probabilities are known to be:

$$\pi_1 = \frac{6}{31}, \qquad \pi_2 = \frac{9}{31}, \qquad \pi_3 = \frac{6}{31}, \qquad \pi_4 = \frac{10}{31}.$$

Assume that the process is in state 1 just before the first transition.

(a) What is the probability that the process will be in state 1 just after the sixth transition?

(b) Determine the expected value and variance of the number of transitions up to and including the next transition during which the process returns to state 1.

(c) What is (approximately) the probability that the state of the system resulting from transition 1000 is neither the same as the state resulting from transition 999 nor the same as the state resulting from transition 1001?

Problem 30. Consider the Markov chain specified in Fig. 7.25.

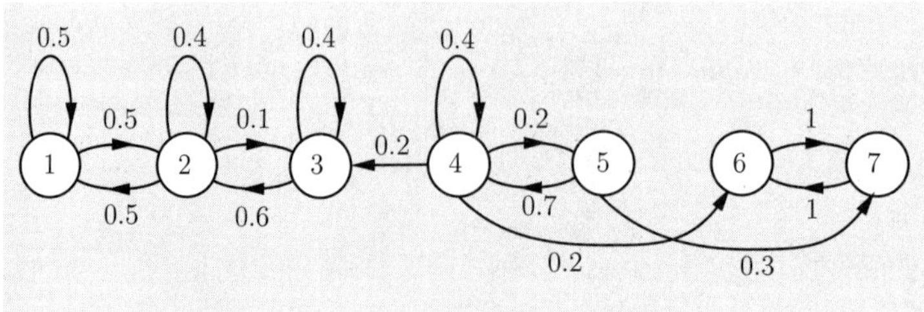

Figure 7.25: Transition probability graph for Problem 30.

(a) Identify the transient and recurrent states. Also, determine the recurrent classes and indicate which ones, if any are periodic.

(b) Do there exist steady-state probabilities given that the process starts in state 1? If so, what are they?

(c) Do there exist steady-state probabilities given that the process starts in state 6? If so, what are they?

(d) Assume that the process starts in state 1 but we begin observing it after it reaches steady-state.

(i) Find the probability that the state increases by one during the first transition we observe.

(ii) Find the conditional probability that the process was in state 2 when we started observing it, given that the state increased by one during the first transition that we observed.

(iii) Find the probability that the state increased by one during the first change of state that we observed.

(e) Assume that the process starts in state 4.

(i) For each recurrent class, determine the probability that we eventually reach that class.

(ii) What is the expected number of transitions up to and including the transition at which we reach a recurrent state for the first time?

Problem 31.* Absorption probabilities. Consider a Markov chain where each state is either transient or absorbing. Fix an absorbing state s. Show that the probabilities a_i of eventually reaching s starting from a state i are the unique solution to the equations

$$\begin{aligned} a_s &= 1, \\ a_i &= 0, \qquad \text{for all absorbing } i \neq s, \\ a_i &= \sum_{j=1}^{m} p_{ij} a_j, \qquad \text{for all transient } i. \end{aligned}$$

Hint: If there are two solutions, find a system of equations that is satisfied by their difference, and look for its solutions.

Solution. The fact that the a_i satisfy these equations was established in the text, using the total probability theorem. To show uniqueness, let $\overline{a}_i$ be another solution, and let $\delta_i = \overline{a}_i - a_i$. Denoting by A the set of absorbing states and using the fact $\delta_j = 0$ for all $j \in A$, we have

$$\delta_i = \sum_{j=1}^{m} p_{ij}\delta_j = \sum_{j \notin A} p_{ij}\delta_j, \qquad \text{for all transient } i.$$

Applying this relation m successive times, we obtain

$$\delta_i = \sum_{j_1 \notin A} p_{ij_1} \sum_{j_2 \notin A} p_{j_1 j_2} \cdots \sum_{j_m \notin A} p_{j_{m-1} j_m} \cdot \delta_{j_m}.$$

Hence

$$\begin{aligned} |\delta_i| &\leq \sum_{j_1 \notin A} p_{ij_1} \sum_{j_2 \notin A} p_{j_1 j_2} \cdots \sum_{j_m \notin A} p_{j_{m-1} j_m} \cdot |\delta_{j_m}| \\ &= \mathbf{P}(X_1 \notin A, \ldots, X_m \notin A \,|\, X_0 = i) \cdot |\delta_{j_m}| \\ &\leq \mathbf{P}(X_1 \notin A, \ldots, X_m \notin A \,|\, X_0 = i) \cdot \max_{j \notin A} |\delta_j|. \end{aligned}$$

The above relation holds for all transient i, so we obtain

$$\max_{j \notin A} |\delta_j| \leq \beta \cdot \max_{j \notin A} |\delta_j|,$$

where

$$\beta = \mathbf{P}(X_1 \notin A, \ldots, X_m \notin A \,|\, X_0 = i).$$

Note that $\beta < 1$, because there is positive probability that X_m is absorbing, regardless of the initial state. It follows that $\max_{j \notin A} |\delta_j| = 0$, or $a_i = \overline{a}_i$ for all i that are not absorbing. We also have $a_j = \overline{a}_j$ for all absorbing j, so $a_i = \overline{a}_i$ for all i.

Problem 32.* Multiple recurrent classes. Consider a Markov chain that has more that one recurrent class, as well as some transient states. Assume that all the recurrent classes are aperiodic.

(a) For any transient state i, let $a_i(k)$ be the probability that starting from i we will reach a state in the kth recurrent class. Derive a system of equations whose solution are the $a_i(k)$.

(b) Show that each of the n-step transition probabilities $r_{ij}(n)$ converges to a limit, and discuss how these limits can be calculated.

Solution. (a) We introduce a new Markov chain that has only transient and absorbing states. The transient states correspond to the transient states of the original, while the absorbing states correspond to the recurrent classes of the original. The transition probabilities $\hat{p}_{ij}$ of the new chain are as follows: if i and j are transient, $\hat{p}_{ij} = p_{ij}$; if i is a transient state and k corresponds to a recurrent class, $\hat{p}_{ik}$ is the sum of the transition probabilities from i to states in the recurrent class in the original Markov chain.

The desired probabilities $a_i(k)$ are the absorption probabilities in the new Markov chain and are given by the corresponding formulas:

$$a_i(k) = \hat{p}_{ik} + \sum_{j:\ \text{transient}} \hat{p}_{ij} a_j(k), \qquad \text{for all transient } i.$$

(b) If i and j are recurrent but belong to different classes, $r_{ij}(n)$ is always 0. If i and j are recurrent but belong to the same class, $r_{ij}(n)$ converges to the steady-state probability of j in a chain consisting of only this particular recurrent class. If j is transient, $r_{ij}(n)$ converges to 0. Finally, if i is transient and j is recurrent, then $r_{ij}(n)$ converges to the product of two probabilities: (1) the probability that starting from i we will reach a state in the recurrent class of j, and (2) the steady-state probability of j conditioned on the initial state being in the class of j.

Problem 33.* Mean first passage times. Consider a Markov chain with a single recurrent class, and let s be a fixed recurrent state. Show that the system of equations

$$t_s = 0, \qquad t_i = 1 + \sum_{j=1}^{m} p_{ij} t_j, \qquad \text{for all } i \neq s,$$

satisfied by the mean first passage times, has a unique solution. *Hint:* If there are two solutions, find a system of equations that is satisfied by their difference, and look for its solutions.

Solution. Let t_i be the mean first passage times. These satisfy the given system of equations. To show uniqueness, let $\overline{t}_i$ be another solution. Then we have for all $i \neq s$

$$t_i = 1 + \sum_{j \neq s} p_{ij} t_j, \qquad \overline{t}_i = 1 + \sum_{j \neq s} p_{ij} \overline{t}_j,$$

and by subtraction, we obtain

$$\delta_i = \sum_{j \neq s} p_{ij} \delta_j,$$

where $\delta_i = \overline{t}_i - t_i$. By applying m successive times this relation, if follows that

$$\delta_i = \sum_{j_1 \neq s} p_{ij_1} \sum_{j_2 \neq s} p_{j_1 j_2} \cdots \sum_{j_m \neq s} p_{j_{m-1} j_m} \cdot \delta_{j_m}.$$

Hence, we have for all $i \neq s$

$$\begin{aligned} |\delta_i| &\leq \sum_{j_1 \neq s} p_{ij_1} \sum_{j_2 \neq s} p_{j_1 j_2} \cdots \sum_{j_m \neq s} p_{j_{m-1} j_m} \cdot \max_j |\delta_j| \\ &= \mathbf{P}(X_1 \neq s, \ldots, X_m \neq s \,|\, X_0 = i) \cdot \max_j |\delta_j|. \end{aligned}$$

On the other hand, we have $\mathbf{P}(X_1 \neq s, \ldots, X_m \neq s \,|\, X_0 = i) < 1$. This is because starting from any state there is positive probability that s is reached in m steps. It follows that all the δ_i must be equal to zero.

Problem 34.* Balance equations and mean recurrence times. Consider a Markov chain with a single recurrent class, and let s be a fixed recurrent state. For any state i, let

$$\rho_i = \mathbf{E}\big[\text{Number of visits to } i \text{ between two successive visits to } s\big],$$

where by convention, $\rho_s = 1$.

(a) Show that for all i, we have

$$\rho_i = \sum_{k=1}^{m} \rho_k p_{ki}.$$

(b) Show that the numbers

$$\pi_i = \frac{\rho_i}{t_s^*}, \qquad i = 1, \ldots, m,$$

sum to 1 and satisfy the balance equations, where t_s^* is the mean recurrence time of s (the expected number of transitions up to the first return to s, starting from s).

(c) Show that if $\pi_1, \ldots, \pi_m$ are nonnegative, satisfy the balance equations, and sum to 1, then

$$\pi_i = \begin{cases} \dfrac{1}{t_i^*}, & \text{if } i \text{ is recurrent,} \\ 0, & \text{if } i \text{ is transient,} \end{cases}$$

where t_i^* is the mean recurrence time of i.

(d) Show that the distribution of part (b) is the unique probability distribution that satisfies the balance equations.

Note: This problem not only provides an alternative proof of the existence and uniqueness of probability distributions that satisfy the balance equations, but also makes an intuitive connection between steady-state probabilities and mean recurrence times. The main idea is to break the process into "cycles," with a new cycle starting each time that the recurrent state s is visited. The steady-state probability of s can be interpreted as the long-term expected frequency of visits to state s, which is inversely proportional to the average time between consecutive visits (the mean recurrence time); cf. part (c). Furthermore, if a state i is expected to be visited, say, twice as often as some other state j during a typical cycle, it is plausible that the long-term expected frequency π_i of state i will be twice as large as π_j. Thus, the steady-state probabilities π_i should be proportional to the expected number of visits ρ_i during a cycle; cf. part (b).

Solution. (a) Consider the Markov chain X_n, initialized with $X_0 = s$. We claim that for all i

$$\rho_i = \sum_{n=1}^{\infty} \mathbf{P}(X_1 \neq s, \ldots, X_{n-1} \neq s, X_n = i).$$

To see this, we first consider the case $i \neq s$, and let I_n be the random variable that takes the value 1 if $X_1 \neq s, \ldots, X_{n-1} \neq s$, and $X_n = i$, and the value 0 otherwise. Then, the number of visits to state i before the next visit to state s is equal to $\sum_{n=1}^{\infty} I_n$. Thus,†

$$\rho_i = \mathbf{E}\left[\sum_{n=1}^{\infty} I_n\right] = \sum_{n=1}^{\infty} \mathbf{E}[I_n] = \sum_{n=1}^{\infty} \mathbf{P}(X_1 \neq s, \ldots, X_{n-1} \neq s, X_n = i).$$

---

† The interchange of the infinite summation and the expectation in the subsequent calculation can be justified by the following argument. We have for any $k > 0$,

$$\mathbf{E}\left[\sum_{n=1}^{\infty} I_n\right] = \mathbf{E}\left[\sum_{n=1}^{k} I_n\right] + \mathbf{E}\left[\sum_{n=k+1}^{\infty} I_n\right] = \sum_{n=1}^{k} \mathbf{E}[I_n] + \mathbf{E}\left[\sum_{n=k+1}^{\infty} I_n\right].$$

Let T be the first positive time that state s is visited. Then,

$$\mathbf{E}\left[\sum_{n=k+1}^{\infty} I_n\right] = \sum_{t=k+2}^{\infty} \mathbf{P}(T = t)\mathbf{E}\left[\sum_{n=k+1}^{\infty} I_n \;\middle|\; T = t\right] \leq \sum_{t=k+2}^{\infty} t\,\mathbf{P}(T = t).$$

Since the mean recurrence time $\sum_{t=1}^{\infty} t\,\mathbf{P}(T = t)$ is finite, the limit, as $k \to \infty$ of $\sum_{t=k+2}^{\infty} t\,\mathbf{P}(T = t)$ is equal to zero. We take the limit of both sides of the earlier equation, as $k \to \infty$, to obtain the desired relation

$$\mathbf{E}\left[\sum_{n=1}^{\infty} I_n\right] = \sum_{n=1}^{\infty} \mathbf{E}[I_n].$$

When $i = s$, the events

$$\{X_1 \neq s, \ldots, X_{n-1} \neq s, X_n = s\},$$

for the different values of n, form a partition of the sample space, because they correspond to the different possibilities for the time of the next visit to state s. Thus,

$$\sum_{n=1}^{\infty} \mathbf{P}(X_1 \neq s, \ldots, X_{n-1} \neq s, X_n = s) = 1 = \rho_s,$$

which completes the verification of our assertion.

We next use the total probability theorem to write for $n \geq 2$,

$$\mathbf{P}(X_1 \neq s, \ldots, X_{n-1} \neq s, X_n = i) = \sum_{k \neq s} \mathbf{P}(X_1 \neq s, \ldots, X_{n-2} \neq s, X_{n-1} = k) p_{ki}.$$

We thus obtain

$$\begin{aligned}
\rho_i &= \sum_{n=1}^{\infty} \mathbf{P}(X_1 \neq s, \ldots, X_{n-1} \neq s, X_n = i) \\
&= p_{si} + \sum_{n=2}^{\infty} \mathbf{P}(X_1 \neq s, \ldots, X_{n-1} \neq s, X_n = i) \\
&= p_{si} + \sum_{n=2}^{\infty} \sum_{k \neq s} \mathbf{P}(X_1 \neq s, \ldots, X_{n-2} \neq s, X_{n-1} = k) p_{ki} \\
&= p_{si} + \sum_{k \neq s} p_{ki} \sum_{n=2}^{\infty} \mathbf{P}(X_1 \neq s, \ldots, X_{n-2} \neq s, X_{n-1} = k) \\
&= \rho_s p_{si} + \sum_{k \neq s} p_{ki} \rho_k \\
&= \sum_{k=1}^{m} \rho_k p_{ki}.
\end{aligned}$$

(b) Dividing both sides of the relation established in part (a) by t_s^*, we obtain

$$\pi_i = \sum_{k=1}^{m} \pi_k p_{ki},$$

where $\pi_i = \rho_i / t_s^*$. Thus, the π_i solve the balance equations. Furthermore, the π_i are nonnegative, and we clearly have $\sum_{i=1}^{m} \rho_i = t_s^*$ or $\sum_{i=1}^{m} \pi_i = 1$. Hence, $(\pi_1, \ldots, \pi_m)$ is a probability distribution.

(c) Consider a probability distribution $(\pi_1, \ldots, \pi_m)$ that satisfies the balance equations. Fix a recurrent state s, let t_s^* be the mean recurrence time of s, and let t_i be the mean first passage time from a state $i \neq s$ to state s. We will show that $\pi_s t_s^* = 1$. Indeed, we have

$$t_s^* = 1 + \sum_{j \neq s} p_{sj} t_j,$$

$$t_i = 1 + \sum_{j \neq s} p_{ij} t_j, \qquad \text{for all } i \neq s.$$

Multiplying these equations with π_s and π_i, respectively, and adding, we obtain

$$\pi_s t_s^* + \sum_{i \neq s} \pi_i t_i = 1 + \sum_{i=1}^{m} \pi_i \sum_{j \neq s} p_{ij} t_j.$$

By using the balance equations, the right-hand side is equal to

$$1 + \sum_{i=1}^{m} \pi_i \sum_{j \neq s} p_{ij} t_j = 1 + \sum_{j \neq s} t_j \sum_{i=1}^{m} \pi_i p_{ij} = 1 + \sum_{j \neq s} t_j \pi_j.$$

By combining the last two equations, we obtain $\pi_s t_s^* = 1$.

Since the probability distribution $(\pi_1, \ldots, \pi_m)$ satisfies the balance equations, if the initial state X_0 is chosen according to this distribution, all subsequent states X_n have the same distribution. If we start at a transient state i, the probability of being at that state at time n diminishes to 0 as $n \to \infty$. It follows that we must have $\pi_i = 0$.

(d) Part (b) shows that there exists at least one probability distribution that satisfies the balance equations. Part (c) shows that there can be only one such probability distribution.

Problem 35.* The strong law of large numbers for Markov chains. Consider a finite-state Markov chain in which all states belong to a single recurrent class which is aperiodic. For a fixed state s, let Y_k be the time of the kth visit to state s. Let also V_n be the number of visits to state s during the first n transitions.

(a) Show that Y_k/k converges with probability 1 to the mean recurrence time t_s^* of state s.

(b) Show that V_n/n converges with probability 1 to $1/t_s^*$.

(c) Can you relate the limit of V_n/n to the steady-state probability of state s?

Solution. (a) Let us fix an initial state i, not necessarily the same as s. Thus, the random variables $Y_{k+1} - Y_k$, for $k \geq 1$, correspond to the time between successive visits to state s. Because of the Markov property (the past is independent of the future, given the present), the process "starts fresh" at each revisit to state s and, therefore, the random variables $Y_{k+1} - Y_k$ are independent and identically distributed, with mean equal to the mean recurrence time t_s^*. Using the strong law of large numbers, we obtain

$$\lim_{k \to \infty} \frac{Y_k}{k} = \lim_{k \to \infty} \frac{Y_1}{k} + \lim_{k \to \infty} \frac{(Y_2 - Y_1) + \cdots + (Y_k - Y_{k-1})}{k} = 0 + t_s^*,$$

with probability 1.

(b) Let us fix an element of the sample space (a trajectory of the Markov chain). Let y_k and v_n be the values of the random variables Y_k and V_n, respectively. Furthermore, let us assume that the sequence y_k/k converges to t_s^*; according to the result of part (a), the set of trajectories with this property has probability 1. Let us consider some n between the time of the kth visit to state s and the time just before the next visit to that state:

$$y_k \leq n < y_{k+1}.$$

For every n in this range, we have $v_n = k$, and also

$$\frac{1}{y_{k+1}} < \frac{1}{n} \leq \frac{1}{y_k},$$

from which we obtain

$$\frac{k}{y_{k+1}} \leq \frac{v_n}{n} \leq \frac{k}{y_k}.$$

Note that

$$\lim_{k\to\infty} \frac{k}{y_{k+1}} = \lim_{k\to\infty} \frac{k+1}{y_{k+1}} \cdot \lim_{k\to\infty} \frac{k}{k+1} = \lim_{k\to\infty} \frac{k}{y_k} = \frac{1}{t_s^*}.$$

If we now let n go to infinity, the corresponding values of k, chosen to satisfy $y_k \leq n < y_{k+1}$ also go to infinity. Therefore, the sequence v_n/n is between two sequences both of which converge to $1/t_s^*$, which implies that the sequence v_n/n converges to $1/t_s^*$ as well. Since this happens for every trajectory in a set of trajectories that has probability equal to 1, we conclude that V_n/n converges to $1/t_s^*$, with probability 1.

(c) We have $1/t_s^* = \pi_s$, as established in Problem 35. This implies the intuitive result that V_n/n converges to π_s, with probability 1. *Note*: It is tempting to try to establish the convergence of V_n/n to π_s by combining the facts that V_n/n converges [part (b)] together with the fact that $\mathbf{E}[V_n]/n$ converges to π_s (cf. the long-term expected frequency interpretation of steady-state probabilities in Section 7.3). However, this line of reasoning is not valid. This is because it is generally possible for a sequence Y_n of random variables to converge with probability 1 to a constant, while the expected values converge to a different constant. An example is the following. Let X be uniformly distributed in the unit interval $[0, 1]$. let

$$Y_n = \begin{cases} 0, & \text{if } X \geq 1/n, \\ n, & \text{if } X < 1/n. \end{cases}$$

As long as X is nonzero (which happens with probability 1), the sequence Y_n converges to zero. On the other hand, it can be seen that

$$\mathbf{E}[Y_n] = \mathbf{P}(X < 1/n)\,\mathbf{E}[Y_n \,|\, X < 1/n] = \frac{1}{n} \cdot \frac{n}{2} = \frac{1}{2}, \qquad \text{for all } n.$$

SECTION 7.5. Continuous-Time Markov Chains

Problem 36. A facility of m identical machines is sharing a single repairperson. The time to repair a failed machine is exponentially distributed with mean $1/\lambda$. A machine once operational, fails after a time that is exponentially distributed with mean $1/\mu$. All failure and repair times are independent.

(a) Find the steady-state probability that there is no operational machine.

(b) Find the expected number of operational machines, in steady-state.

Problem 37. Empty taxis pass by a street corner at a Poisson rate of two per minute and pick up a passenger if one is waiting there. Passengers arrive at the street corner at a Poisson rate of one per minute and wait for a taxi only if there are less than four persons waiting; otherwise they leave and never return. Penelope arrives at the street corner at a given time. Find her expected waiting time, given that she joins the queue. Assume that the process is in steady-state.

Problem 38. There are m users who share a computer system. Each user alternates between "thinking" intervals whose durations are independent exponentially distributed with parameter λ, and an "active" mode that starts by submitting a service request. The server can only serve one request at a time, and will serve a request completely before serving other requests. The service times of different requests are independent exponentially distributed random variables with parameter μ, and also independent of the thinking times of the users. Construct a Markov chain model and derive the steady-state distribution of the number of pending requests, including the one presently served, if any.

Problem 39.* Consider a continuous-time Markov chain in which the transition rates ν_i are the same for all i. Assume that the process has a single recurrent class.

(a) Explain why the sequence Y_n of transition times form a Poisson process.

(b) Show that the steady-state probabilities of the Markov chain $X(t)$ are the same as the steady-state probabilities of the embedded Markov chain X_n.

Solution. (a) Denote by ν the common value of the transition rates ν_i. The sequence $\{Y_n\}$ is a sequence of independent exponentially distributed time intervals with parameter ν. Therefore they can be associated with the arrival times of a Poisson process with rate ν.

(b) The balance and normalization equations for the continuous-time chain are

$$\pi_j \sum_{k \neq j} q_{jk} = \sum_{k \neq j} \pi_k q_{kj}, \qquad j = 1, \ldots, m,$$

$$1 = \sum_{k=1}^{m} \pi_k.$$

By using the relation $q_{jk} = \nu p_{jk}$, and by canceling the common factor ν, these equations are written as

$$\pi_j \sum_{k \neq j} p_{jk} = \sum_{k \neq j} \pi_k p_{kj}, \qquad j = 1, \ldots, m,$$

$$1 = \sum_{k=1}^{m} \pi_k.$$

We have $\sum_{k \neq j} p_{jk} = 1 - p_{jj}$, so the first of these two equations is written as

$$\pi_j(1 - p_{jj}) = \sum_{k \neq j} \pi_k p_{kj},$$

or

$$\pi_j = \sum_{k=1}^{m} \pi_k p_{kj}, \qquad j = 1, \ldots, m.$$

These are the balance equations for the embedded Markov chain, which have a unique solution since the embedded Markov chain has a single recurrent class, which is aperiodic. Hence the π_j are the steady-state probabilities for the embedded Markov chain.

8

Bayesian Statistical Inference

Contents

Statistical inference is the process of extracting information about an unknown variable or an unknown model from available data. In this chapter and the next one we aim to:

(a) Develop an appreciation of the main two approaches (Bayesian and classical), their differences, and similarities.

(b) Present the main categories of inference problems (parameter estimation, hypothesis testing, and significance testing).

(c) Discuss the most important methodologies (maximum a posteriori probability rule, least mean squares estimation, maximum likelihood, regression, likelihood ratio tests, etc.).

(d) Illustrate the theory with some concrete examples.

Probability versus Statistics

Statistical inference differs from probability theory in some fundamental ways. Probability is a self-contained mathematical subject, based on the axioms introduced in Chapter 1. In probabilistic reasoning, we *assume* a fully specified probabilistic model that obeys these axioms. We then use mathematical methods to quantify the consequences of this model or answer various questions of interest. In particular, every unambiguous question has a unique correct answer, even if this answer is sometimes hard to find. The model is taken for granted and, in principle, it need not bear any resemblance to reality (although for the model to be useful, this would better be the case).

Statistics is different, and it involves an element of art. For any particular problem, there may be several reasonable methods, yielding different answers. In general, there is no principled way for selecting the "best" method, unless one makes several assumptions and imposes additional constraints on the inference problem. For instance, given the history of stock market returns over the last five years, there is no single "best" method for estimating next year's returns.

We can narrow down the search for the "right" method by requiring certain desirable properties (e.g., that the method make a correct inference when the number of available data is unlimited). The choice of one method over another usually hinges on several factors: performance guarantees, past experience, common sense, as well as the consensus of the statistics community on the applicability of a particular method on a particular problem type. We will aim to introduce the reader to some of the most popular methods/choices, and the main approaches for their analysis and comparison.

Bayesian versus Classical Statistics

Within the field of statistics there are two prominent schools of thought, with opposing views: the **Bayesian** and the **classical** (also called **frequentist**). Their fundamental difference relates to the nature of the unknown models or variables. In the Bayesian view, they are treated as random variables with known distributions. In the classical view, they are treated as deterministic quantities that happen to be unknown.

The Bayesian approach essentially tries to move the field of statistics back to the realm of probability theory, where every question has a unique answer. In particular, when trying to infer the nature of an unknown model, it views the model as chosen randomly from a given model class. This is done by introducing a random variable Θ that characterizes the model, and by postulating a **prior** probability distribution $p_\Theta(\theta)$. Given the observed data x, one can, in principle, use Bayes' rule to derive a **posterior** probability distribution $p_{\Theta|X}(\theta\,|\,x)$. This captures all the information that x can provide about θ.

By contrast, the classical approach views the unknown quantity θ as a constant that happens to be unknown. It then strives to develop an estimate of θ that has some performance guarantees. This introduces an important conceptual difference from other methods in this book: we are not dealing with a single probabilistic model, but rather with multiple candidate probabilistic models, one for each possible value of θ.

The debate between the two schools has been ongoing for about a century, often with philosophical overtones. Furthermore, each school has constructed examples to show that the methods of the competing school can sometimes produce unreasonable or unappealing answers. We briefly review some of the arguments in this debate.

Suppose that we are trying to measure a physical constant, say the mass of the electron, by means of noisy experiments. The classical statistician will argue that the mass of the electron, while unknown, is just a constant, and that there is no justification for modeling it as a random variable. The Bayesian statistician will counter that a prior distribution simply reflects our state of knowledge. For example, if we already know from past experiments a rough range for this quantity, we can express this knowledge by postulating a prior distribution which is concentrated over that range.

A classical statistician will often object to the arbitrariness of picking a particular prior. A Bayesian statistician will counter that every statistical procedure contains some hidden choices. Furthermore, in some cases, classical methods turn out to be equivalent to Bayesian ones, for a particular choice of a prior. By locating all of the assumptions in one place, in the form of a prior, the Bayesian statistician contends that these assumptions are brought to the surface and are amenable to scrutiny.

Finally, there are practical considerations. In many cases, Bayesian methods are computationally intractable, e.g., when they require the evaluation of multidimensional integrals. On the other hand, with the availability of faster computation, much of the recent research in the Bayesian community focuses on making Bayesian methods practical.

Model versus Variable Inference

Applications of statistical inference tend to be of two different types. In **model inference**, the object of study is a real phenomenon or process for which we wish to construct or validate a model on the basis of available data (e.g., do planets follow elliptical trajectories?). Such a model can then be used to make predictions about the future, or to infer some hidden underlying causes. In **variable inference**, we wish to estimate the value of one or more unknown variables by using some related, possibly noisy information (e.g., what is my current position, given a few GPS readings?).

The distinction between model and variable inference is not sharp; for example, by describing a model in terms of a set of variables, we can cast a model inference problem as a variable inference problem. In any case, we will not emphasize this distinction in the sequel, because the same methodological principles apply to both types of inference.

In some applications, both model and variable inference issues may arise. For example, we may collect some initial data, use them to build a model, and then use the model to make inferences about the values of certain variables.

Example 8.1. A Noisy Channel. A transmitter sends a sequence of binary messages $s_i \in \{0, 1\}$, and a receiver observes

$$X_i = as_i + W_i, \qquad i = 1, \ldots, n,$$

where the W_i are zero mean normal random variables that model channel noise, and a is a scalar that represents the channel attenuation. In a model inference setting, a is unknown. The transmitter sends a pilot signal consisting of a sequence of messages $s_1, \ldots, s_n$, whose values are known by the receiver. On the basis of the observations $X_1, \ldots, X_n$, the receiver wishes to estimate the value of a, that is, build a model of the channel.

Alternatively, in a variable inference setting, a is assumed to be known (possibly because it has already been inferred using a pilot signal, as above). The receiver observes $X_1, \ldots, X_n$, and wishes to infer the values of $s_1, \ldots, s_n$.

A Rough Classification of Statistical Inference Problems

We describe here a few different types of inference problems. In an **estimation** problem, a model is fully specified, except for an unknown, possibly multidimensional, parameter θ, which we wish to estimate. This parameter can be viewed as either a random variable (Bayesian approach) or as an unknown constant (classical approach). The usual objective is to arrive at an estimate of θ that is close to the true value in some sense. For example:

(a) In the noisy transmission problem of Example 8.1, use the knowledge of the pilot sequence and the observations to estimate a.

(b) Using polling data, estimate the fraction of a voter population that prefers candidate A over candidate B.

(c) On the basis of historical stock market data, estimate the mean and variance of the daily movement in the price of a particular stock.

In a **binary hypothesis testing** problem, we start with two hypotheses and use the available data to decide which of the two is true. For example:

(a) In the noisy transmission problem of Example 8.1, use the knowledge of a and X_i to decide whether s_i was 0 or 1.

(b) Given a noisy picture, decide whether there is a person in the picture or not.

(c) Given a set of trials with two alternative medical treatments, decide which treatment is most effective.

More generally, in an m-**ary hypothesis testing** problem, there is a finite number m of competing hypotheses. The performance of a particular method is typically judged by the probability that it makes an erroneous decision. Again, both Bayesian and classical approaches are possible.

In this chapter, we focus primarily on problems of Bayesian estimation, but also discuss hypothesis testing. In the next chapter, in addition to estimation, we discuss a broader range of hypothesis testing problems. Our treatment is introductory and far from exhausts the range of statistical inference problems encountered in practice. As an illustration of a different type of problem, consider the construction of a model of the form $Y = g(X) + W$ that relates two random variables X and Y. Here W is zero mean noise, and g is an unknown function to be estimated. Problems of this type, where the uncertain object (the function g in this case) cannot be described by a fixed number of parameters, are called **nonparametric** and are beyond our scope.

Major Terms, Problems, and Methods in this Chapter

- **Bayesian statistics** treats unknown parameters as random variables with known prior distributions.
- In **parameter estimation**, we want to generate estimates that are close to the true values of the parameters in some probabilistic sense.
- In **hypothesis testing**, the unknown parameter takes one of a finite number of values, corresponding to competing hypotheses; we want to choose one of the hypotheses, aiming to achieve a small probability of error.
- Principal Bayesian inference methods:

  (a) **Maximum a posteriori probability** (MAP) rule: Out of the possible parameter values/hypotheses, select one with maximum conditional/posterior probability given the data (Section 8.2).

  (b) **Least mean squares** (LMS) estimation: Select an estimator/function of the data that minimizes the mean squared error between the parameter and its estimate (Section 8.3).

(c) **Linear least mean squares** estimation: Select an estimator which is a linear function of the data and minimizes the mean squared error between the parameter and its estimate (Section 8.4). This may result in higher mean squared error, but requires simple calculations, based only on the means, variances, and covariances of the random variables involved.

8.1 BAYESIAN INFERENCE AND THE POSTERIOR DISTRIBUTION

In Bayesian inference, the unknown quantity of interest, which we denote by Θ, is modeled as a random variable or as a finite collection of random variables. Here, Θ may represent physical quantities, such as the position and velocity of a vehicle, or a set of unknown parameters of a probabilistic model. For simplicity, unless the contrary is explicitly stated, we view Θ as a single random variable.

We aim to extract information about Θ, based on observing a collection $X = (X_1, \ldots, X_n)$ of related random variables, called **observations**, **measurements**, or an **observation vector**. For this, we assume that we know the joint distribution of Θ and X. Equivalently, we assume that we know:

(a) A prior distribution p_Θ or f_Θ, depending on whether Θ is discrete or continuous.

(b) A conditional distribution $p_{X|\Theta}$ or $f_{X|\Theta}$, depending on whether X is discrete or continuous.

Once a particular value x of X has been observed, *a complete answer to the Bayesian inference problem is provided by the posterior distribution* $p_{\Theta|X}(\theta \,|\, x)$ *or* $f_{\Theta|X}(\theta \,|\, x)$ *of* Θ; see Fig. 8.1. This distribution is determined by the appropriate form of Bayes' rule. It encapsulates everything there is to know about Θ, given the available information, and it is the starting point for further analysis.

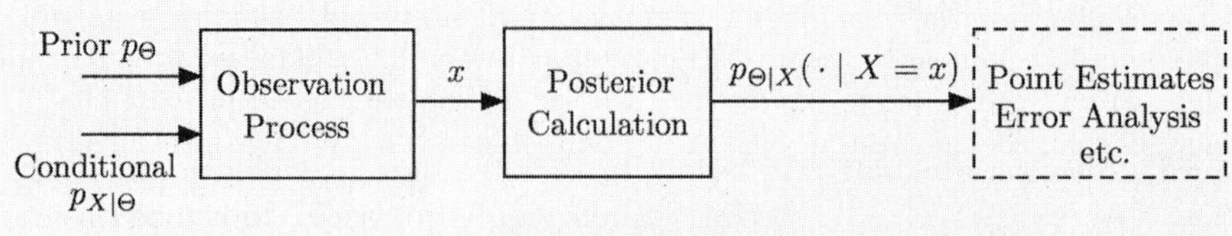

Figure 8.1. A summary of a Bayesian inference model. The starting point is the joint distribution of the parameter Θ and observation X, or equivalently the prior and conditional PMF/PDF. Given the value x of the observation X, the posterior PMF/PDF is formed using Bayes' rule. The posterior can be used to answer additional inference questions; for example the calculation of estimates of Θ, and associated probabilities or error variances.

Summary of Bayesian Inference

- We start with a prior distribution p_Θ or f_Θ for the unknown random variable Θ.
- We have a model $p_{X|\Theta}$ or $f_{X|\Theta}$ of the observation vector X.
- After observing the value x of X, we form the posterior distribution of Θ, using the appropriate version of Bayes' rule.

Note that there are four different versions of Bayes' rule, which we repeat here for easy reference. They correspond to the different combinations of discrete or continuous Θ and X. Yet, all four versions are syntactically similar: starting from the simplest version (all variables are discrete), we only need to replace a PMF with a PDF and a sum with an integral when a continuous random variable is involved. Furthermore, when Θ is multidimensional, the corresponding sums or integrals are to be understood as multidimensional.

The Four Versions of Bayes' Rule

- Θ discrete, X discrete:

$$p_{\Theta|X}(\theta\,|\,x) = \frac{p_\Theta(\theta)p_{X|\Theta}(x\,|\,\theta)}{\sum_{\theta'} p_\Theta(\theta')p_{X|\Theta}(x\,|\,\theta')}.$$

- Θ discrete, X continuous:

$$p_{\Theta|X}(\theta\,|\,x) = \frac{p_\Theta(\theta)f_{X|\Theta}(x\,|\,\theta)}{\sum_{\theta'} p_\Theta(\theta')f_{X|\Theta}(x\,|\,\theta')}.$$

- Θ continuous, X discrete:

$$f_{\Theta|X}(\theta\,|\,x) = \frac{f_\Theta(\theta)p_{X|\Theta}(x\,|\,\theta)}{\int f_\Theta(\theta')p_{X|\Theta}(x\,|\,\theta')\,d\theta'}.$$

- Θ continuous, X continuous:

$$f_{\Theta|X}(\theta\,|\,x) = \frac{f_\Theta(\theta)f_{X|\Theta}(x\,|\,\theta)}{\int f_\Theta(\theta')f_{X|\Theta}(x\,|\,\theta')\,d\theta'}.$$

Let us illustrate the calculation of the posterior with some examples.

Example 8.2. Romeo and Juliet start dating, but Juliet will be late on any date by a random amount X, uniformly distributed over the interval $[0, \theta]$. The parameter θ is unknown and is modeled as the value of a random variable Θ, uniformly distributed between zero and one hour. Assuming that Juliet was late by an amount x on their first date, how should Romeo use this information to update the distribution of Θ?

Here the prior PDF is

$$f_\Theta(\theta) = \begin{cases} 1, & \text{if } 0 \le \theta \le 1, \\ 0, & \text{otherwise,} \end{cases}$$

and the conditional PDF of the observation is

$$f_{X|\Theta}(x \,|\, \theta) = \begin{cases} 1/\theta, & \text{if } 0 \le x \le \theta, \\ 0, & \text{otherwise.} \end{cases}$$

Using Bayes' rule, and taking into account that $f_\Theta(\theta)\, f_{X|\Theta}(x \,|\, \theta)$ is nonzero only if $0 \le x \le \theta \le 1$, we find that for any $x \in [0, 1]$, the posterior PDF is

$$f_{\Theta|X}(\theta \,|\, x) = \frac{f_\Theta(\theta)\, f_{X|\Theta}(x \,|\, \theta)}{\displaystyle\int_0^1 f_\Theta(\theta')\, f_{X|\Theta}(x \,|\, \theta')\, d\theta'} = \frac{1/\theta}{\displaystyle\int_x^1 \frac{1}{\theta'}\, d\theta'} = \frac{1}{\theta \cdot |\log x|}, \qquad \text{if } x \le \theta \le 1,$$

and $f_{\Theta|X}(\theta \,|\, x) = 0$ if $\theta < x$ or $\theta > 1$.

Consider now a variation involving the first n dates. Assume that Juliet is late by random amounts $X_1, \ldots, X_n$, which given $\Theta = \theta$, are uniformly distributed in the interval $[0, \theta]$, and conditionally independent. Let $X = (X_1, \ldots, X_n)$ and $x = (x_1, \ldots, x_n)$. Similar to the case where $n = 1$, we have

$$f_{X|\Theta}(x \,|\, \theta) = \begin{cases} 1/\theta^n, & \text{if } \overline{x} \le \theta \le 1, \\ 0, & \text{otherwise,} \end{cases}$$

where

$$\overline{x} = \max\{x_1, \ldots, x_n\}.$$

The posterior PDF is

$$f_{\Theta|X}(\theta \,|\, x) = \begin{cases} \dfrac{c(\overline{x})}{\theta^n}, & \text{if } \overline{x} \le \theta \le 1, \\ 0, & \text{otherwise,} \end{cases}$$

where $c(\overline{x})$ is a normalizing constant that depends only on $\overline{x}$:

$$c(\overline{x}) = \frac{1}{\displaystyle\int_{\overline{x}}^1 \frac{1}{(\theta')^n}\, d\theta'}.$$

Example 8.3. Inference of a Common Mean of Normal Random Variables. We observe a collection $X = (X_1, \ldots, X_n)$ of random variables, with an unknown common mean whose value we wish to infer. We assume that given the value of the common mean, the X_i are normal and independent, with known variances $\sigma_1^2, \ldots, \sigma_n^2$. In a Bayesian approach to this problem, we model the common mean as a random variable Θ, with a given prior. For concreteness, we assume a normal prior, with known mean x_0 and known variance σ_0^2.

Let us note, for future reference, that our model is equivalent to one of the form

$$X_i = \Theta + W_i, \qquad i = 1, \ldots, n,$$

where the random variables $\Theta, W_1, \ldots, W_n$ are independent and normal, with known means and variances. In particular, for any value θ,

$$\mathbf{E}[W_i] = \mathbf{E}[W_i \,|\, \Theta = \theta] = 0, \qquad \text{var}(W_i) = \text{var}(X_i \,|\, \Theta = \theta) = \sigma_i^2.$$

A model of this type is common in many engineering applications involving several independent measurements of an unknown quantity.

According to our assumptions, we have

$$f_\Theta(\theta) = c_1 \cdot \exp\left\{ -\frac{(\theta - x_0)^2}{2\sigma_0^2} \right\},$$

and

$$f_{X|\Theta}(x \,|\, \theta) = c_2 \cdot \exp\left\{ -\frac{(x_1 - \theta)^2}{2\sigma_1^2} \right\} \cdots \exp\left\{ -\frac{(x_n - \theta)^2}{2\sigma_n^2} \right\},$$

where c_1 and c_2 are normalizing constants that do not depend on θ. We invoke Bayes' rule,

$$f_{\Theta|X}(\theta \,|\, x) = \frac{f_\Theta(\theta) f_{X|\Theta}(x \,|\, \theta)}{\displaystyle\int f_\Theta(\theta') f_{X|\Theta}(x \,|\, \theta')\, d\theta'},$$

and note that the numerator term, $f_\Theta(\theta) f_{X|\Theta}(x \,|\, \theta)$, is of the form

$$c_1 c_2 \cdot \exp\left\{ -\sum_{i=0}^{n} \frac{(x_i - \theta)^2}{2\sigma_i^2} \right\}.$$

After some algebra, which involves completing the square inside the exponent, we find that the numerator is of the form

$$d \cdot \exp\left\{ -\frac{(\theta - m)^2}{2v} \right\},$$

where

$$m = \frac{\displaystyle\sum_{i=0}^{n} x_i/\sigma_i^2}{\displaystyle\sum_{i=0}^{n} 1/\sigma_i^2}, \qquad v = \frac{1}{\displaystyle\sum_{i=0}^{n} 1/\sigma_i^2},$$

and d is a constant, which depends on the x_i but does not depend on θ. Since the denominator term in Bayes' rule does not depend on θ either, we conclude that the posterior PDF has the form

$$f_{\Theta|X}(\theta\,|\,x) = a \cdot \exp\left\{-\frac{(\theta - m)^2}{2v}\right\},$$

for some normalizing constant a, which depends on the x_i, but not on θ. We recognize this as a normal PDF, and we conclude that the posterior PDF is normal with mean m and variance v.

As a special case suppose that $\sigma_0^2, \sigma_1^2, \ldots, \sigma_n^2$ have a common value σ^2. Then, the posterior PDF of Θ is normal with mean and variance

$$m = \frac{x_0 + \cdots + x_n}{n+1}, \qquad v = \frac{\sigma^2}{n+1},$$

respectively. In this case, the prior mean x_0 acts just as another observation, and contributes equally to determine the posterior mean m of Θ. Notice also that the standard deviation of the posterior PDF of Θ tends to zero, at the rough rate of $1/\sqrt{n}$, as the number of observations increases.

If the variances σ_i^2 are different, the formula for the posterior mean m is still a weighted average of the x_i, but with a larger weight on the observations with smaller variance.

The preceding example has the remarkable property that the posterior distribution of Θ is in the same family as the prior distribution, namely, the family of normal distributions. This is appealing for two reasons:

(a) The posterior can be characterized in terms of only two numbers, the mean and the variance.

(b) The form of the solution opens up the possibility of efficient **recursive inference**. Suppose that after $X_1, \ldots, X_n$ are observed, an additional observation X_{n+1} is obtained. Instead of solving the inference problem from scratch, we can view $f_{\Theta|X_1,\ldots,X_n}$ as our prior, and use the new observation to obtain the new posterior $f_{\Theta|X_1,\ldots,X_n,X_{n+1}}$. We may then apply the solution of Example 8.3 to this new problem. It is then plausible (and can be formally established) that the new posterior of Θ will have mean

$$\frac{(m/v) + (x_{n+1}/\sigma_{n+1}^2)}{(1/v) + (1/\sigma_{n+1}^2)}$$

and variance

$$\frac{1}{(1/v) + (1/\sigma_{n+1}^2)},$$

where m and v are the mean and variance of the old posterior $f_{\Theta|X_1,\ldots,X_n}$.

This situation where the posterior is in the same family of distributions as the prior is not very common. Besides the normal family, another prominent example involves coin tosses/Bernoulli trials and binomial distributions.

Example 8.4. Beta Priors on the Bias of a Coin. We wish to estimate the probability of heads, denoted by θ, of a biased coin. We model θ as the value of a random variable Θ with known prior PDF f_Θ. We consider n independent tosses and let X be the number of heads observed. From Bayes' rule, the posterior PDF of Θ has the form, for $\theta \in [0, 1]$,

$$\begin{aligned} f_{\Theta|X}(\theta \,|\, k) &= c f_\Theta(\theta) p_{X|\Theta}(k \,|\, \theta) \\ &= d\, f_\Theta(\theta)\theta^k(1-\theta)^{n-k}, \end{aligned}$$

where c is a normalizing constant (independent of θ), and $d = c\binom{n}{k}$.

Suppose now that the prior is a beta density with integer parameters $\alpha > 0$ and $\beta > 0$, of the form

$$f_\Theta(\theta) = \begin{cases} \dfrac{1}{\mathrm{B}(\alpha,\beta)}\,\theta^{\alpha-1}(1-\theta)^{\beta-1}, & \text{if } 0 < \theta < 1, \\ 0, & \text{otherwise,} \end{cases}$$

where $\mathrm{B}(\alpha, \beta)$ is a normalizing constant, known as the Beta function, given by

$$\mathrm{B}(\alpha,\beta) = \int_0^1 \theta^{\alpha-1}(1-\theta)^{\beta-1}\,d\theta = \frac{(\alpha-1)!\,(\beta-1)!}{(\alpha+\beta-1)!};$$

the last equality can be obtained from integration by parts, or through a probabilistic argument (see Problem 30 in Chapter 3). Then, the posterior PDF of Θ is of the form

$$f_{\Theta|X}(\theta \,|\, k) = \frac{d}{\mathrm{B}(\alpha,\beta)}\theta^{k+\alpha-1}(1-\theta)^{n-k+\beta-1}, \qquad 0 \le \theta \le 1,$$

and hence is a beta density with parameters

$$\alpha' = k + \alpha, \qquad \beta' = n - k + \beta.$$

In the special case where $\alpha = \beta = 1$, the prior f_Θ is the uniform density over $[0, 1]$. In this case, the posterior density is beta with parameters $k + 1$ and $n - k + 1$.

The beta density arises often in inference applications and has interesting properties. In particular, if Θ has a beta density with integer parameters $\alpha > 0$ and $\beta > 0$, its mth moment is given by

$$\begin{aligned} \mathbf{E}[\Theta^m] &= \frac{1}{\mathrm{B}(\alpha,\beta)}\int_0^1 \theta^{m+\alpha-1}(1-\theta)^{\beta-1}\,d\theta \\ &= \frac{\mathrm{B}(\alpha+m,\beta)}{\mathrm{B}(\alpha,\beta)} \\ &= \frac{\alpha(\alpha+1)\cdots(\alpha+m-1)}{(\alpha+\beta)(\alpha+\beta+1)\cdots(\alpha+\beta+m-1)}. \end{aligned}$$

The preceding examples involved a continuous random variable Θ, and were typical of parameter estimation problems. The following is a discrete example, and is typical of those arising in binary hypothesis testing problems.

Example 8.5. Spam Filtering. An email message may be "spam" or "legitimate." We introduce a parameter Θ, taking values 1 and 2, corresponding to spam and legitimate, respectively, with given probabilities $p_\Theta(1)$ and $p_\Theta(2)$. Let $\{w_1, \ldots, w_n\}$ be a collection of special words (or combinations of words) whose appearance suggests a spam message. For each i, let X_i be the Bernoulli random variable that models the appearance of w_i in the message ($X_i = 1$ if w_i appears and $X_i = 0$ if it does not). We assume that the conditional probabilities $p_{X_i|\Theta}(x_i \,|\, 1)$ and $p_{X_i|\Theta}(x_i \,|\, 2)$, $x_i = 0, 1$, are known. For simplicity we also assume that conditioned on Θ, the random variables $X_1, \ldots, X_n$ are independent.

We calculate the posterior probabilities of spam and legitimate messages using Bayes' rule. We have

$$\mathbf{P}(\Theta = m \,|\, X_1 = x_1, \ldots, X_n = x_n) = \frac{p_\Theta(m) \displaystyle\prod_{i=1}^{n} p_{X_i|\Theta}(x_i \,|\, m)}{\displaystyle\sum_{j=1}^{2} p_\Theta(j) \prod_{i=1}^{n} p_{X_i|\Theta}(x_i \,|\, j)}, \qquad m = 1, 2.$$

These posterior probabilities may be used to classify the message as spam or legitimate, by using methods to be discussed later.

Multiparameter Problems

Our discussion has so far focused on the case of a single unknown parameter. The case of multiple unknown parameters is entirely similar. Our next example involves a two-dimensional parameter.

Example 8.6. Localization Using a Sensor Network. There are n acoustic sensors, spread out over a geographic region of interest. The coordinates of the ith sensor are (a_i, b_i). A vehicle with known acoustic signature is located in this region, at unknown Cartesian coordinates $\Theta = (\Theta_1, \Theta_2)$. Every sensor has a distance-dependent probability of detecting the vehicle (i.e., "picking up" the vehicle's signature). Based on which sensors detected the vehicle, and which ones did not, we would like to infer as much as possible about its location.

The prior PDF f_Θ is meant to capture our beliefs on the location of the vehicle, possibly based on previous observations. Let us assume, for simplicity, that Θ_1 and Θ_2 are independent normal random variables with zero mean and unit variance, so that

$$f_\Theta(\theta_1, \theta_2) = \frac{1}{2\pi} e^{-(\theta_1^2 + \theta_2^2)/2}.$$

Let X_i be equal to 1 (respectively, 0) if sensor i has detected the vehicle. To model signal attenuation, we assume that the probability of detection decreases exponentially with the distance between the vehicle and the sensor, which we denote by $d_i(\theta_1, \theta_2)$. More precisely, we use the model

$$\mathbf{P}\big(X_i = 1 \,|\, \Theta = (\theta_1, \theta_2)\big) = p_{X_i|\Theta}(1 \,|\, \theta_1, \theta_2) = e^{-d_i(\theta_1, \theta_2)},$$

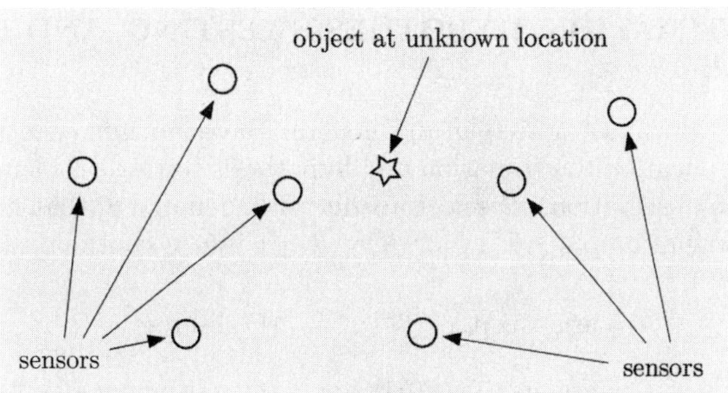

Figure 8.2. Localization using a sensor network.

where $d_i^2(\theta_1, \theta_2) = (a_i - \theta_1)^2 + (b_i - \theta_2)^2$. Furthermore, we assume that conditioned on the vehicle's location Θ, the X_i are independent.

To calculate the posterior PDF, let S be the set of sensors for which $X_i = 1$. The numerator term in Bayes' formula for $f_{\Theta|X}(\theta \,|\, x)$ is

$$f_\Theta(\theta) p_{X|\Theta}(x \,|\, \theta) = \frac{1}{2\pi} e^{-(\theta_1^2 + \theta_2^2)/2} \prod_{i \in S} e^{-d_i(\theta_1,\theta_2)} \prod_{i \notin S} \left(1 - e^{-d_i(\theta_1,\theta_2)}\right),$$

where x is the vector with components $x_i = 1$ for $i \in S$, and $x_i = 0$ for $i \notin S$. The denominator is obtained by integrating the numerator over θ_1 and θ_2.

As Example 8.6 illustrates, the principles for calculating the posterior PDF $f_{\Theta|X}(\theta \,|\, x)$ are essentially the same, regardless of whether Θ consists of one or of multiple components. However, while the posterior PDF can be obtained in principle using Bayes' rule, a closed form solution should not be expected in general. In practice, some numerical computation may be required. Often, calculating the normalizing constant in the denominator of Bayes' formula can be challenging. In Example 8.6, the denominator is a double integral, over the variables θ_1 and θ_2, and its numerical evaluation is manageable. If, however, Θ is high-dimensional, numerical integration becomes formidable. There exist sophisticated methods that can often carry out this integration approximately, based on random sampling, but they are beyond our scope.

When $\Theta = (\Theta_1, \ldots, \Theta_m)$ is multidimensional, we are sometimes interested in just a single component of Θ, say Θ_1. We may then focus on $f_{\Theta_1|X}(\theta_1 \,|\, x)$, the marginal posterior PDF of Θ_1, which can be obtained from the formula

$$f_{\Theta_1|X}(\theta_1 \,|\, x) = \int \cdots \int f_{\Theta|X}(\theta_1, \theta_2, \ldots, \theta_m \,|\, x) \, d\theta_2 \cdots d\theta_m.$$

Note, however, that when Θ is high-dimensional, evaluating this multiple integral can be difficult.

8.2 POINT ESTIMATION, HYPOTHESIS TESTING, AND THE MAP RULE

We will now introduce a simple and general Bayesian inference method, and discuss its application to estimation and hypothesis testing problems. Given the value x of the observation, we select a value of θ, denoted $\hat{\theta}$, that maximizes the posterior distribution $p_{\Theta|X}(\theta \,|\, x)$ [or $f_{\Theta|X}(\theta \,|\, x)$, if Θ is continuous]:

$$\hat{\theta} = \arg\max_{\theta} p_{\Theta|X}(\theta \,|\, x), \qquad (\Theta \text{ discrete}),$$

$$\hat{\theta} = \arg\max_{\theta} f_{\Theta|X}(\theta \,|\, x), \qquad (\Theta \text{ continuous}).$$

This is called the **Maximum a Posteriori probability (MAP)** rule (see Fig. 8.3).

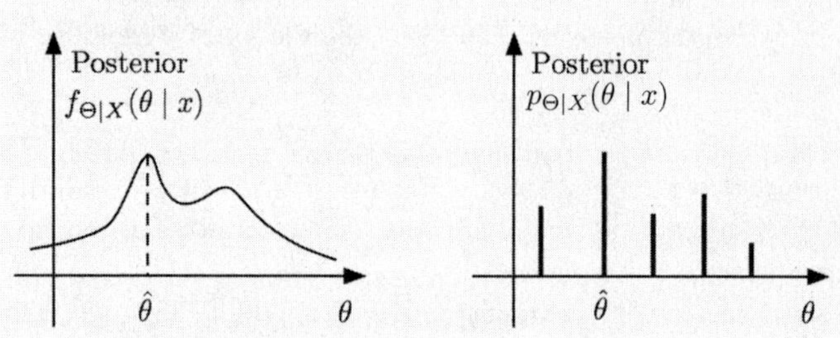

Figure 8.3. Illustration of the MAP rule for inference of a continuous parameter (left figure) and a discrete parameter (right figure).

When Θ is discrete, the MAP rule has an important optimality property: since it chooses $\hat{\theta}$ to be the most likely value of Θ, it maximizes the probability of correct decision for any given value x. This implies that it also maximizes (over all decision rules) the overall (averaged over all possible values x) probability of correct decision. Equivalently, the MAP rule minimizes the probability of an incorrect decision [for each observation value x, as well as the overall probability of error (averaged over x)].†

† To state this more precisely, let us consider a general decision rule, which upon observing the value x, selects a value of θ denoted by $g(x)$. Denote also by $g_{\text{MAP}}(\cdot)$ the MAP rule. Let I and I_{MAP} be Bernoulli random variables that are equal to 1 whenever the general decision rule (respectively, the MAP rule) makes a correct decision; thus, the event $\{I = 1\}$ is the same as the event $\{g(X) = \Theta\}$, and similarly for g_{MAP}. By the definition of the MAP rule,

$$\mathbf{E}[I \,|\, X] = \mathbf{P}\big(g(X) = \Theta \,|\, X\big) \leq \mathbf{P}\big(g_{\text{MAP}}(X) = \Theta \,|\, X\big) = \mathbf{E}[I_{\text{MAP}} \,|\, X],$$

for any possible realization of X. Using the law of iterated expectations, we obtain

The form of the posterior distribution, as given by Bayes' rule, allows an important computational shortcut: the denominator is the same for all θ and depends only on the value x of the observation. Thus, to maximize the posterior, we only need to choose a value of θ that maximizes the numerator $p_\Theta(\theta)p_{X|\Theta}(x\,|\,\theta)$ if Θ and X are discrete, or similar expressions if Θ and/or X are continuous. Calculation of the denominator is unnecessary.

The Maximum a Posteriori Probability (MAP) Rule

- Given the observation value x, the MAP rule selects a value $\hat{\theta}$ that maximizes over θ the posterior distribution $p_{\Theta|X}(\theta\,|\,x)$ (if Θ is discrete) or $f_{\Theta|X}(\theta\,|\,x)$ (if Θ is continuous).
- Equivalently, it selects $\hat{\theta}$ that maximizes over θ:

$$\begin{aligned} &p_\Theta(\theta)p_{X|\Theta}(x\,|\,\theta) && \text{(if } \Theta \text{ and } X \text{ are discrete)},\\ &p_\Theta(\theta)f_{X|\Theta}(x\,|\,\theta) && \text{(if } \Theta \text{ is discrete and } X \text{ is continuous)},\\ &f_\Theta(\theta)p_{X|\Theta}(x\,|\,\theta) && \text{(if } \Theta \text{ is continuous and } X \text{ is discrete)},\\ &f_\Theta(\theta)f_{X|\Theta}(x\,|\,\theta) && \text{(if } \Theta \text{ and } X \text{ are continuous)}. \end{aligned}$$

- If Θ takes only a finite number of values, the MAP rule minimizes (over all decision rules) the probability of selecting an incorrect hypothesis. This is true for both the unconditional probability of error and the conditional one, given any observation value x.

Let us illustrate the MAP rule by revisiting some of the examples in the preceding section.

Example 8.3 (continued). Here, Θ is a normal random variable, with mean x_0 and variance σ_0^2. We observe a collection $X = (X_1, \ldots, X_n)$ of random variables which conditioned on the value θ of Θ, are independent normal random variables with mean θ and variances $\sigma_1^2, \ldots, \sigma_n^2$. We found that the posterior PDF is normal

---

$\mathbf{E}[I] \le \mathbf{E}[I_{\text{MAP}}]$, or

$$\mathbf{P}\big(g(X) = \Theta\big) \le \mathbf{P}\big(g_{\text{MAP}}(X) = \Theta\big).$$

Thus, the MAP rule maximizes the overall probability of a correct decision over all decision rules g. Note that this argument is mostly relevant when Θ is discrete. If Θ, when conditioned on $X = x$, is a continuous random variable, the probability of a correct decision is 0 under any rule.

with mean m and variance v, given by

$$m = \mathbf{E}[\Theta \,|\, X = x] = \frac{\sum_{i=0}^{n} x_i/\sigma_i^2}{\sum_{i=0}^{n} 1/\sigma_i^2}, \qquad v = \text{var}(\Theta \,|\, X = x) = \frac{1}{\sum_{i=0}^{n} 1/\sigma_i^2}.$$

Since the normal PDF is maximized at its mean, the MAP estimate is $\hat{\theta} = m$.

Example 8.5 (continued). Here the parameter Θ takes values 1 and 2, corresponding to spam and legitimate messages, respectively, with given probabilities $p_\Theta(1)$ and $p_\Theta(2)$, and X_i is the Bernoulli random variable that models the appearance of w_i in the message ($X_i = 1$ if w_i appears and $X_i = 0$ if it does not). We have calculated the posterior probabilities of spam and legitimate messages as

$$\mathbf{P}(\Theta = m \,|\, X_1 = x_1, \ldots, X_n = x_n) = \frac{p_\Theta(m) \prod_{i=1}^{n} p_{X_i|\Theta}(x_i \,|\, m)}{\sum_{j=1}^{2} p_\Theta(j) \prod_{i=1}^{n} p_{X_i|\Theta}(x_i \,|\, j)}, \qquad m = 1, 2.$$

Suppose we want to classify a message as spam or legitimate based on the corresponding vector $(x_1, \ldots, x_n)$. Then, the MAP rule decides that the message is spam if

$$\mathbf{P}(\Theta = 1 \,|\, X_1 = x_1, \ldots, X_n = x_n) > \mathbf{P}(\Theta = 2 \,|\, X_1 = x_1, \ldots, X_n = x_n),$$

or equivalently, if

$$p_\Theta(1) \prod_{i=1}^{n} p_{X_i|\Theta}(x_i \,|\, 1) > p_\Theta(2) \prod_{i=1}^{n} p_{X_i|\Theta}(x_i \,|\, 2).$$

Point Estimation

In an estimation problem, given the observed value x of X, the posterior distribution captures all the relevant information provided by x. On the other hand, we may be interested in certain quantities that summarize properties of the posterior. For example, we may select a **point estimate**, which is a single numerical value that represents our best guess of the value of Θ.

Let us introduce some concepts and terminology relating to estimation. For simplicity, we assume that Θ is one-dimensional, but the methodology extends to other cases. We use the term **estimate** to refer to the numerical value $\hat{\theta}$ that we choose to report on the basis of the actual observation x. The value of $\hat{\theta}$ is to be determined by applying some function g to the observation x, resulting in $\hat{\theta} = g(x)$. The random variable $\hat{\Theta} = g(X)$ is called an **estimator**, and its realized value equals $g(x)$ whenever the random variable X takes the value x.

The reason that $\hat{\Theta}$ is a random variable is that the outcome of the estimation procedure depends on the random value of the observation.

We can use different functions g to form different estimators; some will be better than others. For an extreme example, consider the function that satisfies $g(x) = 0$ for all x. The resulting estimator, $\hat{\Theta} = 0$, makes no use of the data, and is therefore not a good choice. We have already seen two of the most popular estimators:

(a) The **Maximum a Posteriori Probability (MAP)** estimator. Here, having observed x, we choose an estimate $\hat{\theta}$ that maximizes the posterior distribution over all θ, breaking ties arbitrarily.

(b) The **Conditional Expectation** estimator, introduced in Section 4.3. Here, we choose the estimate $\hat{\theta} = \mathbf{E}[\Theta \mid X = x]$.

The conditional expectation estimator will be discussed in detail in the next section. It will be called there the least mean squares (LMS) estimator because it has an important property: it minimizes the mean squared error over all estimators, as we show later. Regarding the MAP estimator, we have a few remarks.

(a) If the posterior distribution of Θ is symmetric around its (conditional) mean and unimodal (i.e., has a single maximum), the maximum occurs at the mean. Then, the MAP estimator coincides with the conditional expectation estimator. This is the case, for example, if the posterior distribution is guaranteed to be normal, as in Example 8.3.

(b) If Θ is continuous, the actual evaluation of the MAP estimate $\hat{\theta}$ can sometimes be carried out analytically; for example, if there are no constraints on θ, by setting to zero the derivative of $f_{\Theta|X}(\theta \mid x)$, or of $\log f_{\Theta|X}(\theta \mid x)$, and solving for θ. In other cases, however, a numerical search may be required.

Point Estimates

- An **estimator** is a random variable of the form $\hat{\Theta} = g(X)$, for some function g. Different choices of g correspond to different estimators.
- An **estimate** is the value $\hat{\theta}$ of an estimator, as determined by the realized value x of the observation X.
- Once the value x of X is observed, the **Maximum a Posteriori Probability (MAP)** estimator, sets the estimate $\hat{\theta}$ to a value that maximizes the posterior distribution over all possible values of θ.
- Once the value x of X is observed, the **Conditional Expectation (LMS)** estimator sets the estimate $\hat{\theta}$ to $\mathbf{E}[\Theta \mid X = x]$.

Example 8.7. Consider Example 8.2, in which Juliet is late on the first date by a random amount X. The distribution of X is uniform over the interval $[0, \Theta]$, and Θ is an unknown random variable with a uniform prior PDF f_Θ over the interval $[0, 1]$. In that example, we saw that for $x \in [0, 1]$, the posterior PDF is

$$f_{\Theta|X}(\theta \,|\, x) = \begin{cases} \dfrac{1}{\theta \cdot |\log x|}, & \text{if } x \le \theta \le 1, \\ 0, & \text{otherwise.} \end{cases}$$

For a given x, $f_{\Theta|X}(\theta \,|\, x)$ is decreasing in θ, over the range $[x, 1]$ of possible values of Θ. Thus, the MAP estimate is equal to x. Note that this is an "optimistic" estimate. If Juliet is late by a small amount on the first date ($x \approx 0$), the estimate of future lateness is also small.

The conditional expectation estimate turns out to be less "optimistic." In particular, we have

$$\mathbf{E}[\Theta \,|\, X = x] = \int_x^1 \theta \, \frac{1}{\theta \cdot |\log x|} \, d\theta = \frac{1 - x}{|\log x|}.$$

The two estimates are plotted as functions of x in Fig. 8.4. It can be seen that for any observed lateness x, $\mathbf{E}[\Theta \,|\, X = x]$ is larger than the MAP estimate of Θ.

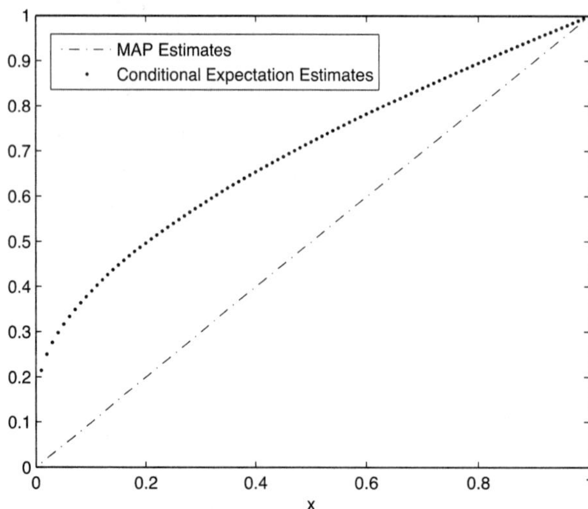

Figure 8.4. MAP and conditional expectation estimates as functions of the observation x in Example 8.7.

Example 8.8. We consider the model in Example 8.4, where we observe the number X of heads in n independent tosses of a biased coin. We assume that the prior distribution of Θ, the probability of heads, is uniform over $[0, 1]$. We will derive the MAP and conditional expectation estimators of Θ.

As shown in Example 8.4, when $X = k$, the posterior PDF of Θ is a beta density with parameters $\alpha = k + 1$ and $\beta = n - k + 1$:

$$f_{\Theta|X}(\theta \,|\, k) = \begin{cases} \dfrac{1}{\mathrm{B}(k+1, n-k+1)}\, \theta^k \,(1-\theta)^{n-k}, & \text{if } \theta \in [0,1], \\ 0, & \text{otherwise.} \end{cases}$$

This posterior PDF turns out to have a single peak. To find its location, we differentiate the expression $\theta^k \,(1-\theta)^{n-k}$ with respect to θ, and set the derivative to zero, to obtain

$$k\theta^{k-1}(1-\theta)^{n-k} - (n-k)\theta^k(1-\theta)^{n-k-1} = 0,$$

which yields

$$\hat{\theta} = \frac{k}{n}.$$

This is the MAP estimate.

To obtain the conditional expectation estimate, we use the formula for the mean of the beta density (cf. Example 8.4):

$$\mathbf{E}[\Theta \,|\, X = k] = \frac{k+1}{n+2}.$$

Note that for large values of n, the MAP and conditional expectation estimates nearly coincide.

In the absence of additional assumptions, a point estimate carries no guarantees on its accuracy. For example, the MAP estimate may lie quite far from the bulk of the posterior distribution. Thus, it is usually desirable to also report some additional information, such as the conditional mean squared error $\mathbf{E}\big[(\hat{\Theta} - \Theta)^2 \,|\, X = x\big]$. In the next section, we will discuss this issue further. In particular, we will revisit the two preceding examples and we will calculate the conditional mean squared error for the MAP and conditional expectation estimates.

Hypothesis Testing

In a hypothesis testing problem, Θ takes one of m values, $\theta_1, \ldots, \theta_m$, where m is usually a small integer; often $m = 2$, in which case we are dealing with a binary hypothesis testing problem. We refer to the event $\{\Theta = \theta_i\}$ as the ith hypothesis, and denote it by H_i.

Once the value x of X is observed, we may use Bayes' rule to calculate the posterior probabilities $\mathbf{P}(\Theta = \theta_i \,|\, X = x) = p_{\Theta|X}(\theta_i \,|\, x)$, for each i. We may then select the hypothesis whose posterior probability is largest, according to the MAP rule. (If there is a tie, with several hypotheses attaining the largest posterior probability, we can choose among them arbitrarily.) As mentioned earlier, the MAP rule is optimal in the sense that it maximizes the probability of correct decision over all decision rules.

The MAP Rule for Hypothesis Testing

- Given the observation value x, the MAP rule selects a hypothesis H_i for which the value of the posterior probability $\mathbf{P}(\Theta = \theta_i \,|\, X = x)$ is largest.
- Equivalently, it selects a hypothesis H_i for which $p_\Theta(\theta_i) p_{X|\Theta}(x \,|\, \theta_i)$ (if X is discrete) or $p_\Theta(\theta_i) f_{X|\Theta}(x \,|\, \theta_i)$ (if X is continuous) is largest.
- The MAP rule minimizes the probability of selecting an incorrect hypothesis for any observation value x, as well as the probability of error over all decision rules.

Once we have derived the MAP rule, we may also compute the corresponding probability of a correct decision (or error), as a function of the observation value x. In particular, if $g_{\rm MAP}(x)$ is the hypothesis selected by the MAP rule when $X = x$, the probability of correct decision is

$$\mathbf{P}\big(\Theta = g_{\rm MAP}(x) \,|\, X = x\big).$$

Furthermore, if S_i is the set of all x such that the MAP rule selects hypothesis H_i, the overall probability of correct decision is

$$\mathbf{P}\big(\Theta = g_{\rm MAP}(X)\big) = \sum_i \mathbf{P}(\Theta = \theta_i, X \in S_i),$$

and the corresponding probability of error is

$$\sum_i \mathbf{P}(\Theta \neq \theta_i, X \in S_i).$$

The following is a typical example of MAP rule calculations for the case of two hypotheses.

Example 8.9. We have two biased coins, referred to as coins 1 and 2, with probabilities of heads equal to p_1 and p_2, respectively. We choose a coin at random (either coin is equally likely to be chosen) and we want to infer its identity, based on the outcome of a single toss. Let $\Theta = 1$ and $\Theta = 2$ be the hypotheses that coin 1 or 2, respectively, was chosen. Let X be equal to 1 or 0, depending on whether the outcome of the toss was a head or a tail, respectively.

Using the MAP rule, we compare $p_\Theta(1) p_{X|\Theta}(x \,|\, 1)$ and $p_\Theta(2) p_{X|\Theta}(x \,|\, 2)$, and decide in favor of the coin for which the corresponding expression is largest. Since $p_\Theta(1) = p_\Theta(2) = 1/2$, we only need to compare $p_{X|\Theta}(x \,|\, 1)$ and $p_{X|\Theta}(x \,|\, 2)$, and select the hypothesis under which the observed toss outcome is most likely. Thus, for example, if $p_1 = 0.46$, $p_2 = 0.52$, and the outcome was a tail we notice that

$$\mathbf{P}(\text{tail} \,|\, \Theta = 1) = 1 - 0.46 > 1 - 0.52 = \mathbf{P}(\text{tail} \,|\, \Theta = 2),$$

and decide in favor of coin 1.

Suppose now that we toss the selected coin n times and let X be the number of heads obtained. Then, the preceding argument is still valid and shows that according to the MAP rule, we should select the hypothesis under which the observed outcome is most likely [this critically depends on the assumption $p_\Theta(1) = p_\Theta(2) = 1/2$]. Thus, if $X = k$, we should decide that $\Theta = 1$ if

$$p_1^k(1-p_1)^{n-k} > p_2^k(1-p_2)^{n-k},$$

and decide that $\Theta = 2$ otherwise. Figure 8.5 illustrates the MAP rule.

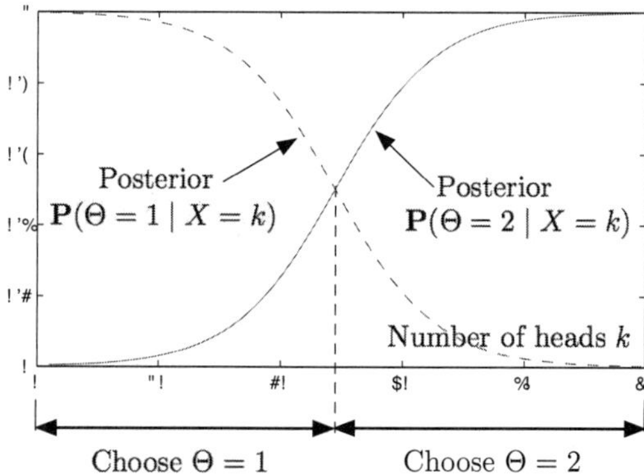

Figure 8.5. The MAP rule for Example 8.9, in the case where $n = 50$, $p_1 = 0.46$, and $p_2 = 0.52$. It calculates the posterior probabilities

$$\begin{aligned}\mathbf{P}(\Theta = i \,|\, X = k) &= c(k)\, p_\Theta(i)\, \mathbf{P}(X = k \,|\, \Theta = i)\\ &= c(k)\, p_\Theta(i)\, p_i^k(1-p_i)^{n-k}, \qquad i = 1, 2,\end{aligned}$$

where $c(k)$ is a positive normalizing constant, and chooses the hypothesis $\Theta = i$ that has the largest posterior probability. Because $p_\Theta(1) = p_\Theta(2) = 1/2$ in this example, the MAP rule chooses the hypothesis $\Theta = i$ for which $p_i^k(1-p_i)^{n-k}$ is largest. The rule is to accept $\Theta = 1$ if $k \le k^*$, where $k^* = 24$, and to accept $\Theta = 2$ otherwise.

The character of the MAP rule, as illustrated in Fig. 8.5, is typical of decision rules in binary hypothesis testing problems: it is specified by a partition of the observation space into the two disjoint sets in which each of the two hypotheses is chosen. In this example, the MAP rule is specified by a single threshold k^*: accept $\Theta = 1$ if $k \le k^*$ and accept $\Theta = 2$ otherwise. The overall probability of error is obtained by using the total probability theorem:

$$\begin{aligned}\mathbf{P}(\text{error}) &= \mathbf{P}(\Theta = 1, X > k^*) + \mathbf{P}(\Theta = 2, X \le k^*)\\ &= p_\Theta(1) \sum_{k=k^*+1}^{n} c(k) p_1^k(1-p_1)^{n-k} + p_\Theta(2) \sum_{k=1}^{k^*} c(k) p_2^k (1-p_2)^{n-k}\end{aligned}$$

$$= \frac{1}{2}\left(\sum_{k=k^*+1}^{n} c(k) p_1^k (1-p_1)^{n-k} + \sum_{k=1}^{k^*} c(k) p_2^k (1-p_2)^{n-k} \right),$$

where $c(k)$ is a positive normalizing constant. Figure 8.6 gives the probability of error for a threshold-type of decision rule, as a function of the threshold k^*. The MAP rule, which in the current example corresponds to $k^* = 24$, maximizes the probability of a correct decision, and hence gives the minimal probability of error.

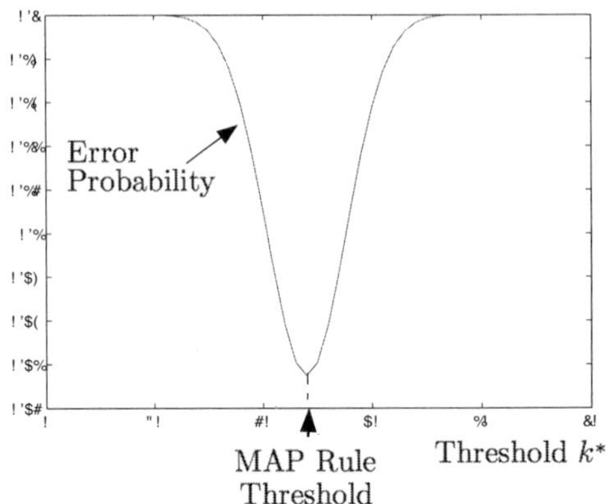

Figure 8.6. A plot of the probability of error for a threshold-type of decision rule that accepts $\Theta = 1$ if $k \leq k^*$ and accepts $\Theta = 2$ otherwise, as a function of the threshold k^* (cf. Example 8.9). The problem data here are $n = 50$, $p_1 = 0.46$, and $p_2 = 0.52$, the same as in Fig. 8.5. The threshold of the MAP rule is $k^* = 24$, and minimizes the probability of error.

The following is a classical example from communication engineering.

Example 8.10. Signal Detection and the Matched Filter. A transmitter sends one of two possible messages. Let $\Theta = 1$ or $\Theta = 2$, depending on whether the first or the second message is transmitted. We assume that the two messages are equally likely, i.e., the prior probabilities are $p_\Theta(1) = p_\Theta(2) = 1/2$.

In order to enhance the resiliency of the transmission with respect to noise, the transmitter uses a signal that extends the transmitted message over time. In particular, the transmitter sends a signal $S = (S_1, \ldots, S_n)$, where each S_i is a real number. If $\Theta = 1$ (respectively, $\Theta = 2$), then S is a fixed sequence $(a_1, \ldots, a_n)$ [respectively, $(b_1, \ldots, b_n)$]. We assume that the two candidate signals have the same "energy," i.e., $a_1^2 + \cdots + a_n^2 = b_1^2 + \cdots + b_n^2$. The receiver observes the transmitted signal, but corrupted by additive noise. More specifically, it obtains the observations

$$X_i = S_i + W_i, \qquad i = 1, \ldots, n,$$

where we assume that the W_i are standard normal random variables, independent of each other and independent of the signal.

Under the hypothesis $\Theta = 1$, the X_i are independent normal random variables, with mean a_i and unit variance. Thus,

$$f_{X|\Theta}(x\,|\,1) = \frac{1}{(\sqrt{2\pi})^n} e^{-\big((x_1 - a_1)^2 + \cdots + (x_n - a_n)^2\big)/2}.$$

Similarly,

$$f_{X|\Theta}(x\,|\,2) = \frac{1}{(\sqrt{2\pi})^n} e^{-\big((x_1 - b_1)^2 + \cdots + (x_n - b_n)^2\big)/2}.$$

From Bayes' rule, the probability that the first message was transmitted is

$$\frac{\exp\Big\{-\big((x_1 - a_1)^2 + \cdots + (x_n - a_n)^2\big)/2\Big\}}{\exp\Big\{-\big((x_1 - a_1)^2 + \cdots + (x_n - a_n)^2\big)/2\Big\} + \exp\Big\{-\big((x_1 - b_1)^2 + \cdots + (x_n - b_n)^2\big)/2\Big\}}.$$

After expanding the squared terms and using the assumption $a_1^2 + \cdots + a_n^2 = b_1^2 + \cdots + b_n^2$, this expression simplifies to

$$\mathbf{P}(\Theta = 1\,|\,X = x) = p_{\Theta|X}(1\,|\,x) = \frac{e^{(a_1x_1 + \cdots + a_nx_n)}}{e^{(a_1x_1 + \cdots + a_nx_n)} + e^{(b_1x_1 + \cdots + b_nx_n)}}.$$

The formula for $\mathbf{P}(\Theta = 2\,|\,X = x)$ is similar, with the a_i in the numerator replaced by b_i.

According to the MAP rule, we should choose the hypothesis with maximum posterior probability, which yields:

$$\text{select } \Theta = 1 \text{ if } \quad \sum_{i=1}^n a_ix_i > \sum_{i=1}^n b_ix_i,$$

$$\text{select } \Theta = 2 \text{ if } \quad \sum_{i=1}^n a_ix_i < \sum_{i=1}^n b_ix_i.$$

(If the inner products above are equal, either hypothesis may be selected.) This particular structure for deciding which signal was transmitted is called a **matched filter**: we "match" the received signal $(x_1, \ldots, x_n)$ with each of the two candidate signals by forming the inner products $\sum_{i=1}^n a_ix_i$ and $\sum_{i=1}^n b_ix_i$; we then select the hypothesis that yields the higher value (the "best match").

This example can be generalized to the case of $m > 2$ equally likely messages. We assume that for message k, the transmitter sends a fixed signal $(a_1^k, \ldots, a_n^k)$, where $(a_1^k)^2 + \cdots + (a_n^k)^2$ is the same for all k. Then, under the same noise model, a similar calculation shows that the MAP rule decodes a received signal $(x_1, \ldots, x_n)$ as the message k for which $\sum_{i=1}^n a_i^k x_i$ is largest.

8.3 BAYESIAN LEAST MEAN SQUARES ESTIMATION

In this section, we discuss in more detail the conditional expectation estimator. In particular, we show that it results in the least possible mean squared error (hence the abbreviation LMS for least mean squares), and we explore some of its other properties.

We start by considering the simpler problem of estimating Θ with a constant $\hat{\theta}$, in the absence of an observation X. The estimation error $\hat{\theta}-\Theta$ is random (because Θ is random), but the mean squared error $\mathbf{E}\big[(\Theta-\hat{\theta})^2\big]$ is a number that depends on $\hat{\theta}$, and can be minimized over $\hat{\theta}$. With this criterion, it turns out that the best possible estimate is to set $\hat{\theta}$ equal to $\mathbf{E}[\Theta]$, as we proceed to verify.

For any estimate $\hat{\theta}$, we have

$$\mathbf{E}\big[(\Theta-\hat{\theta})^2\big] = \text{var}(\Theta-\hat{\theta}) + \big(\mathbf{E}[\Theta-\hat{\theta}]\big)^2 = \text{var}(\Theta) + \big(\mathbf{E}[\Theta]-\hat{\theta}\big)^2;$$

the first equality uses the formula $\mathbf{E}[Z^2] = \text{var}(Z) + \big(\mathbf{E}[Z]\big)^2$, and the second holds because when the constant $\hat{\theta}$ is subtracted from the random variable Θ, the variance is unaffected while the mean is reduced by $\hat{\theta}$. We now note that the term $\text{var}(\Theta)$ does not depend on $\hat{\theta}$. Therefore, we should choose $\hat{\theta}$ to minimize the term $\big(\mathbf{E}[\Theta]-\hat{\theta}\big)^2$, which leads to $\hat{\theta} = \mathbf{E}[\Theta]$ (see Fig. 8.7).

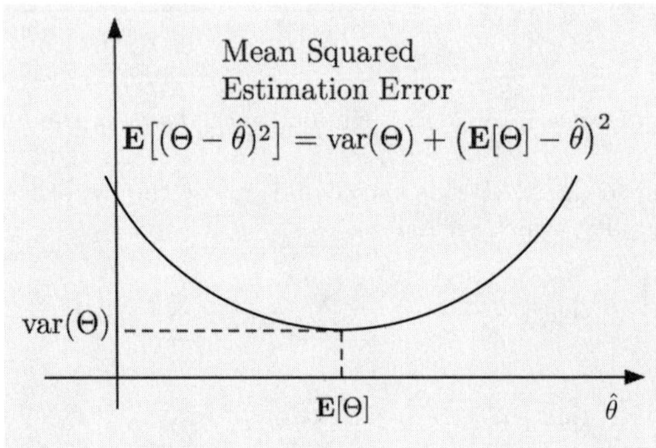

Figure 8.7: The mean squared error $\mathbf{E}\big[(\Theta-\hat{\theta})^2\big]$, as a function of the estimate $\hat{\theta}$, is a quadratic in $\hat{\theta}$, and is minimized when $\hat{\theta} = \mathbf{E}[\Theta]$. The minimum value of the mean squared error is $\text{var}(\Theta)$.

Suppose now that we use an observation X to estimate Θ, so as to minimize the mean squared error. Once we know the value x of X, the situation is identical to the one considered earlier, except that we are now in a new "universe," where everything is conditioned on $X = x$. We can therefore adapt our earlier conclusion and assert that the conditional expectation $\mathbf{E}[\Theta \,|\, X = x]$ minimizes the conditional mean squared error $\mathbf{E}\big[(\Theta-\hat{\theta})^2 \,|\, X = x\big]$ over all constants $\hat{\theta}$.

Generally, the (unconditional) mean squared estimation error associated with an estimator $g(X)$ is defined as

$$\mathbf{E}\Big[\big(\Theta - g(X)\big)^2\Big].$$

If we view $\mathbf{E}[\Theta \,|\, X]$ as an estimator/function of X, the preceding analysis shows that out of all possible estimators, the mean squared estimation error is minimized when $g(X) = \mathbf{E}[\Theta \,|\, X]$.†

Key Facts About Least Mean Squares Estimation

- In the absence of any observations, $\mathbf{E}\big[(\Theta - \hat{\theta})^2\big]$ is minimized when $\hat{\theta} = \mathbf{E}[\Theta]$:

$$\mathbf{E}\Big[\big(\Theta - \mathbf{E}[\Theta]\big)^2\Big] \le \mathbf{E}\big[(\Theta - \hat{\theta})^2\big], \qquad \text{for all } \hat{\theta}.$$

- For any given value x of X, $\mathbf{E}\big[(\Theta - \hat{\theta})^2 \,|\, X = x\big]$ is minimized when $\hat{\theta} = \mathbf{E}[\Theta \,|\, X = x]$:

$$\mathbf{E}\Big[\big(\Theta - \mathbf{E}[\Theta \,|\, X = x]\big)^2 \;\Big|\; X = x\Big] \le \mathbf{E}\big[(\Theta - \hat{\theta})^2 \,|\, X = x\big], \quad \text{for all } \hat{\theta}.$$

- Out of all estimators $g(X)$ of Θ based on X, the mean squared estimation error $\mathbf{E}\Big[\big(\Theta - g(X)\big)^2\Big]$ is minimized when $g(X) = \mathbf{E}[\Theta \,|\, X]$:

$$\mathbf{E}\Big[\big(\Theta - \mathbf{E}[\Theta \,|\, X]\big)^2\Big] \le \mathbf{E}\Big[\big(\Theta - g(X)\big)^2\Big], \qquad \text{for all estimators } g(X).$$

---

† For any given value x of X, $g(x)$ is a number, and therefore,

$$\mathbf{E}\Big[\big(\Theta - \mathbf{E}[\Theta \,|\, X = x]\big)^2 \;\Big|\; X = x\Big] \le \mathbf{E}\Big[\big(\Theta - g(x)\big)^2 \;\Big|\; X = x\Big].$$

Thus,

$$\mathbf{E}\Big[\big(\Theta - \mathbf{E}[\Theta \,|\, X]\big)^2 \;\Big|\; X\Big] \le \mathbf{E}\Big[\big(\Theta - g(X)\big)^2 \;\Big|\; X\Big],$$

which is now an inequality between random variables (functions of X). We take expectations of both sides, and use the law of iterated expectations, to conclude that

$$\mathbf{E}\Big[\big(\Theta - \mathbf{E}[\Theta \,|\, X]\big)^2\Big] \le \mathbf{E}\Big[\big(\Theta - g(X)\big)^2\Big],$$

for all estimators $g(X)$.

Example 8.11. Let Θ be uniformly distributed over the interval $[4, 10]$ and suppose that we observe Θ with some random error W. In particular, we observe the value of the random variable

$$X = \Theta + W,$$

where we assume that W is uniformly distributed over the interval $[-1, 1]$ and independent of Θ.

To calculate $\mathbf{E}[\Theta \mid X = x]$, we note that $f_\Theta(\theta) = 1/6$, if $4 \leq \theta \leq 10$, and $f_\Theta(\theta) = 0$, otherwise. Conditioned on Θ being equal to some θ, X is the same as $\theta + W$, and is uniformly distributed over the interval $[\theta - 1, \theta + 1]$. Thus, the joint PDF is given by

$$f_{\Theta,X}(\theta, x) = f_\Theta(\theta) f_{X|\Theta}(x \mid \theta) = \frac{1}{6} \cdot \frac{1}{2} = \frac{1}{12},$$

if $4 \leq \theta \leq 10$ and $\theta - 1 \leq x \leq \theta + 1$, and is zero for all other values of (θ, x). The parallelogram in the right-hand side of Fig. 8.8 is the set of pairs (θ, x) for which $f_{\Theta,X}(\theta, x)$ is nonzero.

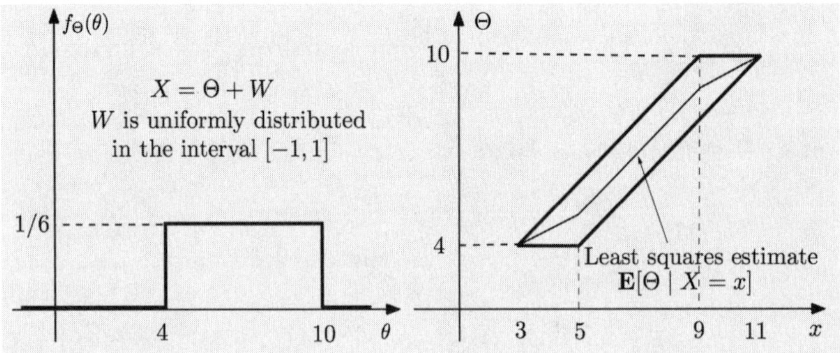

Figure 8.8: The PDFs in Example 8.11. The joint PDF of Θ and X is uniform over the parallelogram shown on the right. The LMS estimate of Θ, given the value x of the random variable $X = \Theta + W$, depends on x and is represented by the piecewise linear function shown on the right.

Given that $X = x$, the posterior PDF $f_{\Theta|X}$ is uniform on the corresponding vertical section of the parallelogram. Thus $\mathbf{E}[\Theta \mid X = x]$ is the midpoint of that section, which in this example happens to be a piecewise linear function of x. Conditioned on a particular value x of X, the mean squared error, defined as $\mathbf{E}\big[\big(\Theta - \mathbf{E}[\Theta \mid X]\big)^2 \mid X = x\big]$, is the conditional variance of Θ. It is a function of x, illustrated in Fig. 8.9.

Example 8.12. Consider Example 8.7, in which Juliet is late on the first date by a random amount X that is uniformly distributed over the interval $[0, \Theta]$. Here, Θ is an unknown random variable with a uniform prior PDF f_Θ over the interval $[0, 1]$. In that example, we saw that the MAP estimate is equal to x and that the

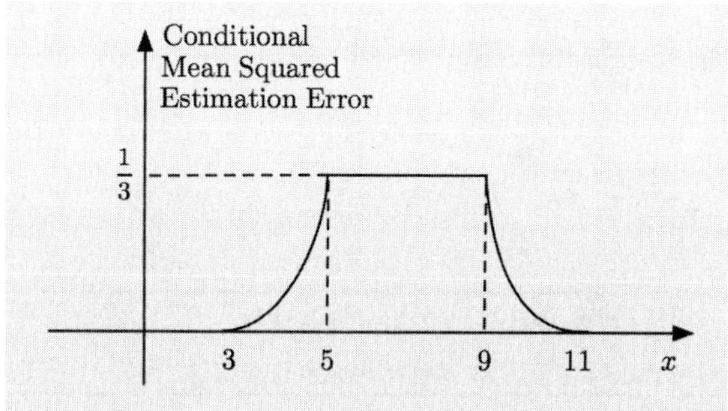

Figure 8.9: The conditional mean squared error in Example 8.11, as a function of the observed value x of X. Note that certain values of the observation are more favorable than others. For example, if $X = 3$, then we are certain that $\Theta = 4$, and the conditional mean squared error is zero.

LMS estimate is

$$\mathbf{E}[\Theta \mid X = x] = \int_x^1 \theta \frac{1}{\theta \cdot |\log x|}\, d\theta = \frac{1 - x}{|\log x|}.$$

Let us calculate the conditional mean squared error for the MAP and the LMS estimates. Given $X = x$, for any estimate $\hat{\theta}$, we have

$$\begin{aligned}\mathbf{E}\left[(\hat{\theta} - \Theta)^2 \mid X = x\right] &= \int_x^1 (\hat{\theta} - \theta)^2 \cdot \frac{1}{\theta |\log x|}\, d\theta \\ &= \int_x^1 (\hat{\theta}^2 - 2\hat{\theta}\theta + \theta^2) \cdot \frac{1}{\theta |\log x|}\, d\theta \\ &= \hat{\theta}^2 - \hat{\theta}\frac{2(1 - x)}{|\log x|} + \frac{1 - x^2}{2|\log x|}.\end{aligned}$$

For the MAP estimate, $\hat{\theta} = x$, the conditional mean squared error is

$$\mathbf{E}\left[(\hat{\theta} - \Theta)^2 \mid X = x\right] = x^2 + \frac{3x^2 - 4x + 1}{2|\log x|}.$$

For the LMS estimate, $\hat{\theta} = (1 - x)/|\log x|$, the conditional mean squared error is

$$\mathbf{E}\left[(\hat{\theta} - \Theta)^2 \mid X = x\right] = \frac{1 - x^2}{2|\log x|} - \left(\frac{1 - x}{\log x}\right)^2.$$

The conditional mean squared errors of the two estimates are plotted in Fig. 8.10, as functions of x, and it can be seen that the LMS estimator has uniformly smaller mean squared error. This is a manifestation of the general optimality property of the LMS estimator that we have established.

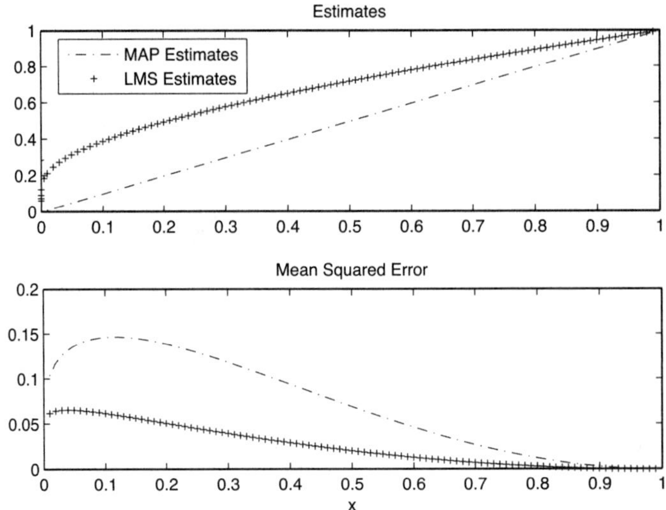

Figure 8.10. MAP and LMS estimates, and their conditional mean squared errors as functions of the observation x in Example 8.12.

Example 8.13. We consider the model in Example 8.8, where we observe the number X of heads in n independent tosses of a biased coin. We assume that the prior distribution of Θ, the probability of heads, is uniform over $[0, 1]$. In that example, we saw that when $X = k$, the posterior density is beta with parameters $\alpha = k+1$ and $\beta = n-k+1$, and that the MAP estimate is equal to k/n. By using the formula for the moments of the beta density (cf. Example 8.4), we have

$$\mathbf{E}[\Theta^m \,|\, X = k] = \frac{(k+1)(k+2)\cdots(k+m)}{(n+2)(n+3)\cdots(n+m+1)},$$

and in particular, the LMS estimate is

$$\mathbf{E}[\Theta \,|\, X = k] = \frac{k+1}{n+2}.$$

Given $X = k$, the conditional mean squared error for any estimate $\hat{\theta}$ is

$$\begin{aligned}\mathbf{E}\left[(\hat{\theta} - \Theta)^2 \,|\, X = k\right] &= \hat{\theta}^2 - 2\,\hat{\theta}\,\mathbf{E}\left[\Theta \,|\, X = k\right] + \mathbf{E}\left[\Theta^2 \,|\, X = k\right] \\ &= \hat{\theta}^2 - 2\,\hat{\theta}\,\frac{k+1}{n+2} + \frac{(k+1)(k+2)}{(n+2)(n+3)}.\end{aligned}$$

The conditional mean squared error of the MAP estimate is

$$\begin{aligned}\mathbf{E}\left[(\hat{\theta} - \Theta)^2 \;\middle|\; X = k\right] &= \mathbf{E}\left[\left(\frac{k}{n} - \Theta\right)^2 \;\middle|\; X = k\right] \\ &= \frac{k}{n}^2 - 2\,\frac{k}{n} \cdot \frac{k+1}{n+2} + \frac{(k+1)(k+2)}{(n+2)(n+3)}.\end{aligned}$$

The conditional mean squared error of the LMS estimate is

$$\begin{aligned}\mathbf{E}\left[(\hat{\theta}-\Theta)^2 \mid X=k\right] &= \mathbf{E}\left[\Theta^2 \mid X=k\right] - \left(\mathbf{E}[\Theta \mid X=k]\right)^2 \\ &= \frac{(k+1)(k+2)}{(n+2)(n+3)} - \left(\frac{k+1}{n+2}\right)^2.\end{aligned}$$

The results are plotted in Fig. 8.11 for the case of $n = 15$ tosses. Note that, as in the preceding example, the LMS estimate has a uniformly smaller conditional mean squared error.

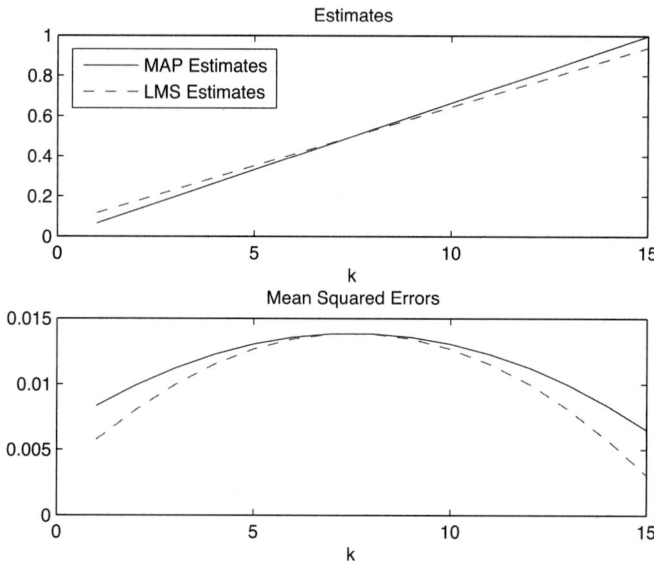

Figure 8.11. MAP and LMS estimates, and corresponding conditional mean squared errors as functions of the observed number of heads k in $n = 15$ tosses (cf. Example 8.13).

Some Properties of the Estimation Error

Let us use the notation

$$\hat{\Theta} = \mathbf{E}[\Theta \mid X], \qquad \tilde{\Theta} = \hat{\Theta} - \Theta,$$

for the LMS estimator and the associated estimation error, respectively. The random variables $\hat{\Theta}$ and $\tilde{\Theta}$ have a number of useful properties, which were derived in Section 4.3 and for easy reference are repeated below. (Note the change in notation: while in Section 4.3, the observation was denoted by Y and the estimated parameter was denoted by X, here they are denoted by X and Θ, respectively.)

Properties of the Estimation Error

- The estimation error $\tilde{\Theta}$ is **unbiased**, i.e., it has zero unconditional and conditional mean:

$$\mathbf{E}[\tilde{\Theta}] = 0, \qquad \mathbf{E}[\tilde{\Theta} \mid X = x] = 0, \quad \text{for all } x.$$

- The estimation error $\tilde{\Theta}$ is uncorrelated with the estimate $\hat{\Theta}$:

$$\text{cov}(\hat{\Theta}, \tilde{\Theta}) = 0.$$

- The variance of Θ can be decomposed as

$$\text{var}(\Theta) = \text{var}(\hat{\Theta}) + \text{var}(\tilde{\Theta}).$$

Example 8.14. Let us say that the observation X is *uninformative* if the mean squared estimation error $\mathbf{E}[\tilde{\Theta}^2] = \text{var}(\tilde{\Theta})$ is the same as $\text{var}(\Theta)$, the unconditional variance of Θ. When is this the case?

Using the formula

$$\text{var}(\Theta) = \text{var}(\hat{\Theta}) + \text{var}(\tilde{\Theta}),$$

we see that X is uninformative if and only if $\text{var}\big(\hat{\Theta}\big) = 0$. The variance of a random variable is zero if and only if that random variable is a constant, equal to its mean. We conclude that X is uninformative if and only if the estimate $\hat{\Theta} = \mathbf{E}[\Theta \mid X]$ is equal to $\mathbf{E}[\Theta]$, for every value of X.

If Θ and X are independent, we have $\mathbf{E}[\Theta \mid X = x] = \mathbf{E}[\Theta]$ for all x, and X is indeed uninformative, which is quite intuitive. The converse, however, is not true: it is possible for $\mathbf{E}[\Theta \mid X = x]$ to be always equal to the constant $\mathbf{E}[\Theta]$, without Θ and X being independent. (Can you construct an example?)

The Case of Multiple Observations and Multiple Parameters

The preceding discussion was phrased as if X were a single random variable. However, the entire argument and its conclusions apply even if X is a vector of random variables, $X = (X_1, \ldots, X_n)$. Thus, the mean squared estimation error is minimized if we use $\mathbf{E}[\Theta \mid X_1, \ldots, X_n]$ as our estimator, i.e.,

$$\mathbf{E}\Big[\big(\Theta - \mathbf{E}[\Theta \mid X_1, \ldots, X_n]\big)^2\Big] \leq \mathbf{E}\Big[\big(\Theta - g(X_1, \ldots, X_n)\big)^2\Big],$$

for all estimators $g(X_1, \ldots, X_n)$.

This provides a complete solution to the general problem of LMS estimation, but is often difficult to implement, for the following reasons:

(a) In order to compute the conditional expectation $\mathbf{E}[\Theta \mid X_1, \ldots, X_n]$, we need a complete probabilistic model, that is, the joint PDF $f_{\Theta, X_1, \ldots, X_n}$.

(b) Even if this joint PDF is available, $\mathbf{E}[\Theta \mid X_1, \ldots, X_n]$ can be a very complicated function of $X_1, \ldots, X_n$.

As a consequence, practitioners often resort to approximations of the conditional expectation or focus on estimators that are not optimal but are simple and easy to implement. The most common approach, discussed in the next section, involves a restriction to linear estimators.

Finally, let us consider the case where we want to estimate multiple parameters $\Theta_1, \ldots, \Theta_m$. It is then natural to consider the criterion

$$\mathbf{E}\big[(\Theta_1 - \hat{\Theta}_1)^2\big] + \cdots + \mathbf{E}\big[(\Theta_m - \hat{\Theta}_m)^2\big],$$

and minimize it over all estimators $\hat{\Theta}_1, \ldots, \hat{\Theta}_m$. But this is equivalent to finding, for each i, an estimator $\hat{\Theta}_i$ that minimizes $\mathbf{E}\big[(\Theta_i - \hat{\Theta}_i)^2\big]$, so that we are essentially dealing with m decoupled estimation problems, one for each unknown parameter Θ_i, yielding $\hat{\Theta}_i = \mathbf{E}[\Theta_i \mid X_1, \ldots, X_n]$, for all i.

8.4 BAYESIAN LINEAR LEAST MEAN SQUARES ESTIMATION

In this section, we derive an estimator that minimizes the mean squared error within a restricted class of estimators: those that are linear functions of the observations. While this estimator may result in higher mean squared error, it has a significant practical advantage: it requires simple calculations, involving only means, variances, and covariances of the parameters and observations. It is thus a useful alternative to the conditional expectation/LMS estimator in cases where the latter is hard to compute.

A linear estimator of a random variable Θ, based on observations $X_1, \ldots, X_n$, has the form

$$\hat{\Theta} = a_1 X_1 + \cdots + a_n X_n + b.$$

Given a particular choice of the scalars $a_1, \ldots, a_n, b$, the corresponding mean squared error is

$$\mathbf{E}\big[(\Theta - a_1 X_1 - \cdots - a_n X_n - b)^2\big].$$

The linear LMS estimator chooses $a_1, \ldots, a_n, b$ to minimize the above expression. We first develop the solution for the case where $n = 1$, and then generalize.

Linear Least Mean Squares Estimation Based on a Single Observation

We are interested in finding a and b that minimize the mean squared estimation error $\mathbf{E}\big[(\Theta - aX - b)^2\big]$ associated with a linear estimator $aX + b$ of Θ. Suppose that a has already been chosen. How should we choose b? This is the same as choosing a constant b to estimate the random variable $\Theta - aX$. By the discussion in the beginning of Section 8.3, the best choice is

$$b = \mathbf{E}[\Theta - aX] = \mathbf{E}[\Theta] - a\mathbf{E}[X].$$

With this choice of b, it remains to minimize, with respect to a, the expression

$$\mathbf{E}\Big[\big(\Theta - aX - \mathbf{E}[\Theta] + a\mathbf{E}[X]\big)^2\Big].$$

We write this expression as

$$\text{var}(\Theta - aX) = \sigma_\Theta^2 + a^2\sigma_X^2 + 2\text{cov}(\Theta, -aX) = \sigma_\Theta^2 + a^2\sigma_X^2 - 2a \cdot \text{cov}(\Theta, X),$$

where σ_Θ and σ_X are the standard deviations of Θ and X, respectively, and

$$\text{cov}(\Theta, X) = \mathbf{E}\Big[\big(\Theta - \mathbf{E}[\Theta]\big)\big(X - \mathbf{E}[X]\big)\Big]$$

is the covariance of Θ and X. To minimize $\text{var}(\Theta - aX)$ (a quadratic function of a), we set its derivative to zero and solve for a. This yields

$$a = \frac{\text{cov}(\Theta, X)}{\sigma_X^2} = \frac{\rho\sigma_\Theta\sigma_X}{\sigma_X^2} = \rho\frac{\sigma_\Theta}{\sigma_X},$$

where

$$\rho = \frac{\text{cov}(\Theta, X)}{\sigma_\Theta\sigma_X}$$

is the correlation coefficient. With this choice of a, the mean squared estimation error of the resulting linear estimator $\hat{\Theta}$ is given by

$$\begin{aligned}\text{var}(\Theta - \hat{\Theta}) &= \sigma_\Theta^2 + a^2\sigma_X^2 - 2a \cdot \text{cov}(\Theta, X) \\ &= \sigma_\Theta^2 + \rho^2\frac{\sigma_\Theta^2}{\sigma_X^2}\sigma_X^2 - 2\rho\frac{\sigma_\Theta}{\sigma_X}\rho\sigma_\Theta\sigma_X \\ &= (1 - \rho^2)\sigma_\Theta^2.\end{aligned}$$

Linear LMS Estimation Formulas

- The linear LMS estimator $\hat{\Theta}$ of Θ based on X is

$$\hat{\Theta} = \mathbf{E}[\Theta] + \frac{\text{cov}(\Theta, X)}{\text{var}(X)}\big(X - \mathbf{E}[X]\big) = \mathbf{E}[\Theta] + \rho\frac{\sigma_\Theta}{\sigma_X}\big(X - \mathbf{E}[X]\big),$$

  where

$$\rho = \frac{\text{cov}(\Theta, X)}{\sigma_\Theta\sigma_X}$$

  is the correlation coefficient.

- The resulting mean squared estimation error is equal to

$$(1 - \rho^2)\sigma_\Theta^2.$$

The formula for the linear LMS estimator only involves the means, variances, and covariance of Θ and X. Furthermore, it has an intuitive interpretation. Suppose, for concreteness, that the correlation coefficient ρ is positive. The estimator starts with the baseline estimate $\mathbf{E}[\Theta]$ for Θ, which it then adjusts by taking into account the value of $X - \mathbf{E}[X]$. For example, when X is larger than its mean, the positive correlation between X and Θ suggests that Θ is expected to be larger than its mean. Accordingly, the resulting estimate is set to a value larger than $\mathbf{E}[\Theta]$. The value of ρ also affects the quality of the estimate. When $|\rho|$ is close to 1, the two random variables are highly correlated, and knowing X allows us to accurately estimate Θ, resulting in a small mean squared error.

We finally note that the properties of the estimation error presented in Section 8.3 can be shown to hold when $\hat{\Theta}$ is the linear LMS estimator; see the end-of-chapter problems.

Example 8.15. We revisit the model in Examples 8.2, 8.7, and 8.12, in which Juliet is always late by an amount X that is uniformly distributed over the interval $[0, \Theta]$, and Θ is a random variable with a uniform prior PDF $f_\Theta(\theta)$ over the interval $[0, 1]$. Let us derive the linear LMS estimator of Θ based on X.

Using the fact that $\mathbf{E}[X \mid \Theta] = \Theta/2$ and the law of iterated expectations, the expected value of X is

$$\mathbf{E}[X] = \mathbf{E}\big[\mathbf{E}[X \mid \Theta]\big] = \mathbf{E}\left[\frac{\Theta}{2}\right] = \frac{\mathbf{E}[\Theta]}{2} = \frac{1}{4}.$$

Furthermore, using the law of total variance (this is the same calculation as in Example 4.17 of Chapter 4), we have

$$\text{var}(X) = \frac{7}{144}.$$

We now find the covariance of X and Θ, using the formula

$$\text{cov}(\Theta, X) = \mathbf{E}[\Theta X] - \mathbf{E}[\Theta]\,\mathbf{E}[X],$$

and the fact

$$\mathbf{E}[\Theta^2] = \text{var}(\Theta) + \big(\mathbf{E}[\Theta]\big)^2 = \frac{1}{12} + \frac{1}{4} = \frac{1}{3}.$$

We have

$$\mathbf{E}[\Theta X] = \mathbf{E}\big[\mathbf{E}[\Theta X \mid \Theta]\big] = \mathbf{E}\big[\Theta\,\mathbf{E}[X \mid \Theta]\big] = \mathbf{E}\left[\frac{\Theta^2}{2}\right] = \frac{1}{6},$$

where the first equality follows from the law of iterated expectations, and the second equality holds since for all θ,

$$\mathbf{E}[\Theta X \mid \Theta = \theta] = \mathbf{E}[\theta X \mid \Theta = \theta] = \theta\,\mathbf{E}[X \mid \Theta = \theta].$$

Thus,

$$\text{cov}(\Theta, X) = \mathbf{E}[\Theta X] - \mathbf{E}[\Theta]\,\mathbf{E}[X] = \frac{1}{6} - \frac{1}{2}\cdot\frac{1}{4} = \frac{1}{24}.$$

The linear LMS estimator is

$$\hat{\Theta} = \mathbf{E}[\Theta] + \frac{\text{cov}(\Theta, X)}{\text{var}(X)}\big(X - \mathbf{E}[X]\big) = \frac{1}{2} + \frac{1/24}{7/144}\left(X - \frac{1}{4}\right) = \frac{6}{7}X + \frac{2}{7}.$$

The corresponding conditional mean squared error is calculated using the formula derived in Example 8.12,

$$\mathbf{E}\left[(\hat{\theta} - \Theta)^2 \,|\, X = x\right] = \hat{\theta}^2 - \hat{\theta}\frac{2(1-x)}{|\log x|} + \frac{1-x^2}{2|\log x|},$$

and substituting the expression just derived, $\hat{\theta} = (6/7)x + (2/7)$. In Fig. 8.12, we compare the linear LMS estimator with the MAP estimator and the LMS estimator (cf. Examples 8.2, 8.7, and 8.12). Note that the LMS and linear LMS estimators are nearly identical for much of the region of interest, and so are the corresponding conditional mean squared errors. The MAP estimator has significantly larger mean squared error than the other two estimators. For x close to 1, the linear LMS estimator performs worse than the other two estimators, and indeed may give an estimate $\hat{\theta} > 1$, which is outside the range of possible values of Θ.

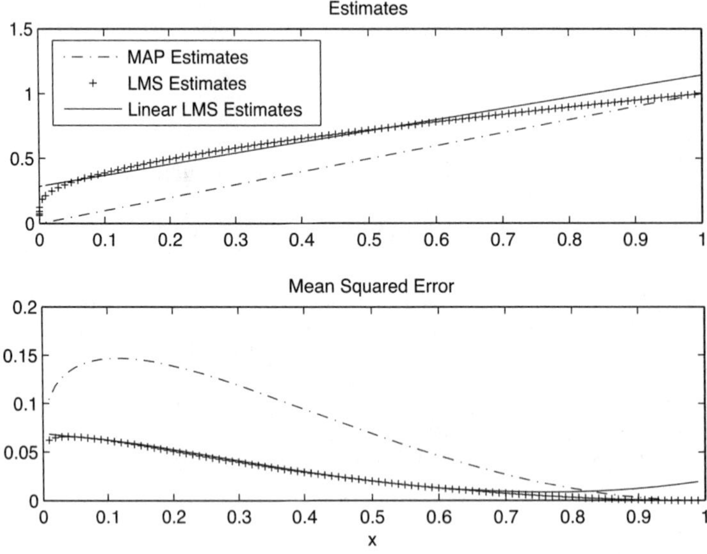

Figure 8.12. Three estimators and their mean squared errors, as functions of the observed value x, for the problem in Example 8.15.

Example 8.16. Linear LMS Estimation of the Bias of a Coin. We revisit the coin tossing problem of Examples 8.4, 8.8, and 8.13, and derive the linear LMS estimator. Here, the probability of heads of the coin is modeled as a random variable Θ whose prior distribution is uniform over the interval $[0, 1]$. The coin is tossed n times, independently, resulting in a random number of heads, denoted by X. Thus, if Θ is equal to θ, the random variable X has a binomial distribution with parameters n and θ.

We calculate the various coefficients that appear in the formula for the linear LMS estimator. We have $\mathbf{E}[\Theta] = 1/2$, and

$$\mathbf{E}[X] = \mathbf{E}\big[\mathbf{E}[X \mid \Theta]\big] = \mathbf{E}[n\Theta] = \frac{n}{2}.$$

The variance of Θ is $1/12$, so that $\sigma_\Theta = 1/\sqrt{12}$. Also, as calculated in the previous example, $\mathbf{E}[\Theta^2] = 1/3$. If Θ takes the value θ, the (conditional) variance of X is $n\theta(1-\theta)$. Using the law of total variance, we obtain

$$\begin{aligned} \text{var}(X) &= \mathbf{E}\big[\text{var}(X \mid \Theta)\big] + \text{var}\big(\mathbf{E}[X \mid \Theta]\big) \\ &= \mathbf{E}\big[n\Theta(1-\Theta)\big] + \text{var}(n\Theta) \\ &= \frac{n}{2} - \frac{n}{3} + \frac{n^2}{12} \\ &= \frac{n(n+2)}{12}. \end{aligned}$$

In order to find the covariance of X and Θ, we use the formula

$$\text{cov}(\Theta, X) = \mathbf{E}[\Theta X] - \mathbf{E}[\Theta]\,\mathbf{E}[X] = \mathbf{E}[\Theta X] - \frac{n}{4}.$$

Similar to Example 8.15, we have

$$\mathbf{E}[\Theta X] = \mathbf{E}\big[\mathbf{E}[\Theta X \mid \Theta]\big] = \mathbf{E}\big[\Theta\,\mathbf{E}[X \mid \Theta]\big] = \mathbf{E}[n\Theta^2] = \frac{n}{3},$$

so that

$$\text{cov}(\Theta, X) = \frac{n}{3} - \frac{n}{4} = \frac{n}{12}.$$

Putting everything together, we conclude that the linear LMS estimator takes the form

$$\hat{\Theta} = \frac{1}{2} + \frac{n/12}{n(n+2)/12}\left(X - \frac{n}{2}\right) = \frac{1}{2} + \frac{1}{n+2}\left(X - \frac{n}{2}\right) = \frac{X+1}{n+2}.$$

Notice that this agrees with the LMS estimator that we derived in Example 8.13. This should not be a surprise: if the LMS estimator turns out to be linear, as was the case in Example 8.13, then that estimator is also optimal within the smaller class of linear estimators.

The Case of Multiple Observations and Multiple Parameters

The linear LMS methodology extends to the case of multiple observations. There is an analogous formula for the linear LMS estimator, derived in a similar manner. It involves only means, variances, and covariances between various pairs of random variables. Also, if there are multiple parameters Θ_i to be estimated, we may consider the criterion

$$\mathbf{E}\big[(\Theta_1 - \hat{\Theta}_1)^2\big] + \cdots + \mathbf{E}\big[(\Theta_m - \hat{\Theta}_m)^2\big],$$

and minimize it over all estimators $\hat{\Theta}_1, \ldots, \hat{\Theta}_m$ that are linear functions of the observations. This is equivalent to finding, for each i, a linear estimator $\hat{\Theta}_i$ that minimizes $\mathbf{E}\big[(\Theta_i - \hat{\Theta}_i)^2\big]$, so that we are essentially dealing with m decoupled linear estimation problems, one for each unknown parameter.

In the case where there are multiple observations with a certain independence property, the formula for the linear LMS estimator simplifies as we will now describe. Let Θ be a random variable with mean μ and variance σ_0^2, and let $X_1, \ldots, X_n$ be observations of the form

$$X_i = \Theta + W_i,$$

where the W_i are random variables with mean 0 and variance σ_i^2, which represent observation errors. Under the assumption that the random variables $\Theta, W_1, \ldots, W_n$ are uncorrelated, the linear LMS estimator of Θ, based on the observations $X_1, \ldots, X_n$, turns out to be

$$\hat{\Theta} = \frac{\mu/\sigma_0^2 + \displaystyle\sum_{i=1}^{n} X_i/\sigma_i^2}{\displaystyle\sum_{i=0}^{n} 1/\sigma_i^2}.$$

The derivation involves forming the function

$$h(a_1, \ldots, a_n, b) = \mathbf{E}\big[(\Theta - a_1X_1 - \cdots - a_nX_n - b)^2\big],$$

and minimizing it by setting to zero its partial derivatives with respect to $a_1, \ldots, a_n, b$. After some calculation (given in the end-of-chapter problems), this results in

$$b = \frac{\mu/\sigma_0^2}{\displaystyle\sum_{i=0}^{n} 1/\sigma_i^2}, \qquad a_j = \frac{1/\sigma_j^2}{\displaystyle\sum_{i=0}^{n} 1/\sigma_i^2}, \qquad j = 1, \ldots, n,$$

from which the formula for the linear LMS estimator given earlier follows.

Linear Estimation and Normal Models

The linear LMS estimator is generally different from and, therefore, inferior to the LMS estimator $\mathbf{E}[\Theta \mid X_1, \ldots, X_n]$. However, if the LMS estimator happens to be linear in the observations $X_1, \ldots, X_n$, then it is also the linear LMS estimator, i.e., the two estimators coincide.

An important example where this occurs is the estimation of a normal random variable Θ on the basis of observations $X_i = \Theta + W_i$, where the W_i are independent zero mean normal noise terms, independent of Θ. This is the same model as in Example 8.3, where we saw that the posterior distribution of Θ is normal and that the conditional mean is a linear function of the observations. Thus, the LMS and the linear LMS estimators coincide. Indeed, the formula for the linear LMS estimator given in this section is consistent with the expression for the posterior mean $\hat{\theta}$ in Example 8.3 (the notation μ here corresponds to x_0 in Example 8.3). This is a manifestation of a property that can be shown to hold more generally: if $\Theta, X_1, \ldots, X_n$ are all linear functions of a collection of independent normal random variables, then the LMS and the linear LMS estimators coincide. They also coincide with the MAP estimator, since the normal distribution is symmetric and unimodal.

The above discussion leads to an interesting interpretation of linear LMS estimation: the estimator is the same as the one that would have been obtained if we were to pretend that the random variables involved were normal, with the given means, variances, and covariances. Thus, there are two alternative perspectives on linear LMS estimation: either as a computational shortcut (avoid the evaluation of a possibly complicated formula for $\mathbf{E}[\Theta \mid X]$), or as a model simplification (replace less tractable distributions by normal ones).

The Choice of Variables in Linear Estimation

Let us point out an important difference between LMS and linear LMS estimation. Consider an unknown random variable Θ, observations $X_1, \ldots, X_n$, and transformed observations $Y_i = h(X_i)$, $i = 1, \ldots, n$, where the function h is one-to-one. The transformed observations Y_i convey the same information as the original observations X_i, and therefore the LMS estimator based on $Y_1, \ldots, Y_n$ is the same as the one based on $X_1, \ldots, X_n$:

$$\mathbf{E}\big[\Theta \mid h(X_1), \ldots, h(X_n)\big] = \mathbf{E}[\Theta \mid X_1, \ldots, X_n].$$

On the other hand, linear LMS estimation is based on the premise that the class of linear functions of the observations $X_1, \ldots, X_n$ contains reasonably good estimators of Θ; this may not always be the case. For example, suppose that Θ is the unknown variance of some distribution and $X_1, \ldots, X_n$ represent independent random variables drawn from that distribution. Then, it would be unreasonable to expect that a good estimator of Θ can be obtained with a linear function of $X_1, \ldots, X_n$. This suggests that it may be helpful to transform the observations so that good estimators of Θ can be found within the class of linear functions of the transformed observations. Suitable transformations may not always be obvious, but intuition into the structure of the given problem, may suggest some good choices; see Problem 17 for a simple example.

8.5 SUMMARY AND DISCUSSION

We have introduced statistical inference methods that aim to extract information about unknown variables or models from probabilistically related observations. We have focused on the case where the unknown is a (possibly multidimensional) parameter θ, and we have discussed hypothesis testing and estimation problems.

We have drawn a distinction between the Bayesian and classical inference approaches. In this chapter, we have discussed Bayesian methods, which treat the parameter as a random variable Θ with known prior distribution. The key object of interest here is the posterior distribution of Θ given the observations. The posterior can in principle be calculated using Bayes' rule, although in practice, this may be difficult.

The MAP rule, which maximizes the posterior over θ, is a general inference method that can address both estimation and hypothesis testing problems. We discussed two other methods for parameter estimation: the LMS (or conditional expectation) estimator and the linear LMS estimator, both of which are based on minimization of the mean squared error between Θ and its estimate. The latter estimator results in higher mean squared error, but requires simple calculations, involving only means, variances, and covariances of the parameters and observations. Under normality assumptions on the parameter and observations, the MAP and the two LMS estimators coincide.

PROBLEMS

SECTION 8.1. Bayesian Inference and the Posterior Distribution

Problem 1. Artemisia moves to a new house and she is "fifty-percent sure" that the phone number is 2537267. To verify this, she uses the house phone to dial 2537267, she obtains a busy signal, and concludes that this is indeed the correct number. Assuming that the probability of a typical seven-digit phone number being busy at any given time is 1%, what is the probability that Artemisia's conclusion was correct?

Problem 2. Nefeli, a student in a probability class, takes a multiple-choice test with 10 questions and 3 choices per question. For each question, there are two equally likely possibilities, independent of other questions: either she knows the answer, in which case she answers the question correctly, or else she guesses the answer with probability of success 1/3.

(a) Given that Nefeli answered correctly the first question, what is the probability that she knew the answer to that question?

(b) Given that Nefeli answered correctly 6 out of the 10 questions, what is the posterior PMF of the number of questions of which she knew the answer?

SECTION 8.2. Point Estimation, Hypothesis Testing, and the MAP Rule

Problem 3. The number of minutes between successive bus arrivals at Alvin's bus stop is exponentially distributed with parameter Θ, and Alvin's prior PDF of Θ is

$$f_\Theta(\theta) = \begin{cases} 10\,\theta, & \text{if } \theta \in [0, 1/5], \\ 0, & \text{otherwise.} \end{cases}$$

(a) Alvin arrives on Monday at the bus stop and has to wait 30 minutes for the bus to arrive. What is the posterior PDF, and the MAP and conditional expectation estimates of Θ?

(b) Following his Monday experience, Alvin decides to estimate Θ more accurately, and records his waiting times for five days. These are 30, 25, 15, 40, and 20 minutes, and Alvin assumes that his observations are independent. What is the posterior PDF, and the MAP and conditional expectation estimates of Θ given the five-day data?

Problem 4. Students in a probability class take a multiple-choice test with 10 questions and 3 choices per question. A student who knows the answer to a question will answer it correctly, while a student that does not will guess the answer with probability of success 1/3. Each student is equally likely to belong to one of three categories

$i = 1, 2, 3$: those who know the answer to each question with corresponding probabilities θ_i, where $\theta_1 = 0.3$, $\theta_2 = 0.7$, and $\theta_3 = 0.95$ (independent of other questions). Suppose that a randomly chosen student answers k questions correctly.

(a) For each possible value of k, derive the MAP estimate of the category that this student belongs to.

(b) Let M be the number of questions that the student knows how to answer. Derive the posterior PMF, and the MAP and LMS estimates of M given that the student answered correctly 5 questions.

Problem 5. Consider a variation of the biased coin problem in Example 8.4, and assume the probability of heads, Θ, is distributed over $[0, 1]$ according to the PDF

$$f_\Theta(\theta) = 2 - 4\left|\frac{1}{2} - \theta\right|, \qquad \theta \in [0, 1].$$

Find the MAP estimate of Θ, assuming that n independent coin tosses resulted in k heads and $n - k$ tails.

Problem 6. Professor May B. Hard, who has a tendency to give difficult problems in probability quizzes, is concerned about one of the problems she has prepared for an upcoming quiz. She therefore asks her TA to solve the problem and record the solution time. May's prior probability that the problem is difficult is 0.3, and she knows from experience that the conditional PDF of her TA's solution time X, in minutes, is

$$f_{T|\Theta}(x \mid \Theta = 1) = \begin{cases} c_1 e^{-0.04x}, & \text{if } 5 \le x \le 60, \\ 0 & \text{otherwise,} \end{cases}$$

if $\Theta = 1$ (problem is difficult), and is

$$f_{T|\Theta}(x \mid \Theta = 2) = \begin{cases} c_2 e^{-0.16x}, & \text{if } 5 \le x \le 60, \\ 0 & \text{otherwise,} \end{cases}$$

if $\Theta = 2$ (problem is not difficult), where c_1 and c_2 are normalizing constants. She uses the MAP rule to decide whether the problem is difficult.

(a) Given that the TA's solution time was 20 minutes, which hypothesis will she accept and what will be the probability of error?

(b) Not satisfied with the reliability of her decision, May asks four more TAs to solve the problem. The TAs' solution times are conditionally independent and identically distributed with the solution time of the first TA. The recorded solution times are 10, 25, 15, and 35 minutes. On the basis of the five observations, which hypothesis will she now accept, and what will be the probability of error?

Problem 7. We have two boxes, each containing three balls: one black and two white in box 1; two black and one white in box 2. We choose one of the boxes at random, where the probability of choosing box 1 is equal to some given p, and then draw a ball.

(a) Describe the MAP rule for deciding the identity of the box based on whether the drawn ball is black or white.

(b) Assuming that $p = 1/2$, find the probability of an incorrect decision and compare it with the probability of error if no ball had been drawn.

Problem 8. The probability of heads of a given coin is known to be either q_0 (hypothesis H_0) or q_1 (hypothesis H_1). We toss the coin repeatedly and independently, and record the number of heads before a tail is observed for the first time. We assume that $0 < q_0 < q_1 < 1$, and that we are given prior probabilities $\mathbf{P}(H_0)$ and $\mathbf{P}(H_1)$. For parts (a) and (b), we also assume that $\mathbf{P}(H_0) = \mathbf{P}(H_1) = 1/2$.

(a) Calculate the probability that hypothesis H_1 is true, given that there were exactly k heads before the first tail.

(b) Consider the decision rule that decides in favor of hypothesis H_1 if $k \geq k^*$, where k^* is some nonnegative integer, and decides in favor of hypothesis H_0 otherwise. Give a formula for the probability of error in terms of k^*, q_0, and q_1. For what value of k^* is the probability of error minimized? Is there another type of decision rule that would lead to an even lower probability of error?

(c) Assume that $q_0 = 0.3$, $q_1 = 0.7$, and $\mathbf{P}(H_1) > 0.7$. How does the optimal choice of k^* (the one that minimizes the probability of error) change as $\mathbf{P}(H_1)$ increases from 0.7 to 1.0?

Problem 9.* Consider a Bayesian hypothesis testing problem involving m hypotheses, and an observation vector $X = (X_1, \ldots, X_n)$. Let $g_n(X_1, \ldots, X_n)$ be the decision resulting from the MAP rule based on $X_1, \ldots, X_n$, and $g_{n-1}(X_1, \ldots, X_{n-1})$ the decision resulting from the MAP rule based on $X_1, \ldots, X_{n-1}$ (i.e., the MAP rule that uses only the first $n - 1$ components of the observation vector). Let $x = (x_1, \ldots, x_n)$ be the realized value of the observation vector, and let

$$e_n(x_1, \ldots, x_n) = \mathbf{P}\big(\Theta \neq g_n(x_1, \ldots, x_n) \bigm| X_1 = x_1, \ldots, X_n = x_n\big),$$
$$e_{n-1}(x_1, \ldots, x_{n-1}) = \mathbf{P}\big(\Theta \neq g_{n-1}(x_1, \ldots, x_{n-1}) \bigm| X_1 = x_1, \ldots, X_{n-1} = x_{n-1}\big),$$

be the corresponding probabilities of error. Show that

$$e_n(x_1, \ldots, x_n) \leq e_{n-1}(x_1, \ldots, x_{n-1}),$$

so making the MAP decision with extra data cannot increase the probability of error.

Solution. We view $g_{n-1}(X_1, \ldots, X_{n-1})$ as a special case of a decision rule based on all components $X_1, \ldots, X_n$ of the observation vector. Since the MAP rule $g_n(X_1, \ldots, X_n)$ minimizes the probability of error over all decision rules based on $X_1, \ldots, X_n$, the result follows.

SECTION 8.3. Bayesian Least Mean Squares Estimation

Problem 10. A police radar always overestimates the speed of incoming cars by an amount that is uniformly distributed between 0 and 5 miles/hour. Assume that car speeds are uniformly distributed between 55 and 75 miles/hour. What is the LMS estimate of a car's speed based on the radar's measurement?

Problem 11. The number Θ of shopping carts in a store is uniformly distributed between 1 and 100. Carts are sequentially numbered between 1 and Θ. You enter the store, observe the number X on the first cart you encounter, assumed uniformly distributed over the range $1, \ldots, \Theta$, and use this information to estimate Θ. Find and plot the MAP estimator and the LMS estimator. *Hint*: Note the resemblance with Example 8.2.

Problem 12. Consider the multiple observation variant of Example 8.2: given that $\Theta = \theta$, the random variables $X_1, \ldots, X_n$ are independent and uniformly distributed on the interval $[0, \theta]$, and the prior distribution of Θ is uniform on the interval $[0, 1]$. Assume that $n > 3$.

(a) Find the LMS estimate of Θ, given the values $x_1, \ldots, x_n$ of $X_1, \ldots, X_n$.

(b) Plot the conditional mean squared error of the MAP and LMS estimators, as functions of $\overline{x} = \max\{x_1, \ldots, x_n\}$, for the case $n = 5$.

(c) If $\overline{x}$ is held fixed at $\overline{x} = 0.5$, how do the MAP and the LMS estimates, and the corresponding conditional mean squared errors behave as $n \to \infty$?

Problem 13.*

(a) Let $Y_1, \ldots, Y_n$ be independent identically distributed random variables and let $Y = Y_1 + \cdots + Y_n$. Show that

$$\mathbf{E}[Y_1 \,|\, Y] = \frac{Y}{n}.$$

(b) Let Θ and W be independent zero-mean normal random variables, with positive integer variances k and m, respectively. Use the result of part (a) to find $\mathbf{E}[\Theta \,|\, \Theta + W]$, and verify that this agrees with the conditional expectation formula in Example 8.3. *Hint:* Think of Θ and W as sums of independent random variables.

(c) Repeat part (b) for the case where Θ and W are independent Poisson random variables with integer means λ and μ, respectively.

Solution. (a) By symmetry, we see that $\mathbf{E}[Y_i \,|\, Y]$ is the same for all i. Furthermore,

$$\mathbf{E}[Y_1 + \cdots + Y_n \,|\, Y] = \mathbf{E}[Y \,|\, Y] = Y.$$

Therefore, $\mathbf{E}[Y_1 \,|\, Y] = Y/n$.

(b) We can think of Θ and W as sums of independent standard normal random variables:

$$\Theta = \Theta_1 + \cdots + \Theta_k, \qquad W = W_1 + \cdots + W_m.$$

We identify Y with $\Theta + W$ and use the result from part (a), to obtain

$$\mathbf{E}[\Theta_i \,|\, \Theta + W] = \frac{\Theta + W}{k + m}.$$

Thus,

$$\mathbf{E}[\Theta \,|\, \Theta + W] = \mathbf{E}[\Theta_1 + \cdots + \Theta_k \,|\, \Theta + W] = \frac{k}{k+m}(\Theta + W).$$

The formula for the conditional mean derived in Example 8.3, specialized to the current context (zero prior mean and a single measurement) shows that the conditional expectation is of the form

$$\frac{(\Theta+W)/\sigma_W^2}{(1/\sigma_\Theta^2)+(1/\sigma_W^2)} = \frac{\sigma_\Theta^2}{\sigma_\Theta^2+\sigma_W^2}(\Theta+W) = \frac{k}{k+m}(\Theta+W),$$

consistent with the answer obtained here.

(c) We recall that the sum of independent Poisson random variables is Poisson. Thus the argument in part (b) goes through, by thinking of Θ and W as sums of λ (respectively, μ) independent Poisson random variables with mean one. We then obtain

$$\mathbf{E}[\Theta \mid \Theta+W] = \frac{\lambda}{\lambda+\mu}(\Theta+W).$$

SECTION 8.4. Bayesian Linear Least Mean Squares Estimation

Problem 14. Consider the random variables Θ and X in Example 8.11. Find the linear LMS estimator of Θ based on X, and the associated mean squared error.

Problem 15. For the model in the shopping cart problem (Problem 11), derive and plot the conditional mean squared error, as a function of the number on the observed cart, for the MAP, LMS, and linear LMS estimators.

Problem 16. The joint PDF of random variables X and Θ is of the form

$$f_{X,\Theta}(x,\theta) = \begin{cases} c, & \text{if } (x,\theta)\in S, \\ 0, & \text{otherwise,} \end{cases}$$

where c is a constant and S is the set

$$S = \big\{(x,\theta) \,|\, 0\le x\le 2,\ 0\le\theta\le 2,\ x-1\le\theta\le x\big\}.$$

We want to estimate Θ based on X.

(a) Find the LMS estimator $g(X)$ of Θ.

(b) Calculate $\mathbf{E}\big[(\Theta-g(X))^2 \,|\, X=x\big]$, $\mathbf{E}\big[g(X)\big]$, and $\mathrm{var}\big(g(X)\big)$.

(c) Calculate the mean squared error $\mathbf{E}\big[(\Theta-g(X))^2\big]$. Is it the same as $\mathbf{E}\big[\mathrm{var}(\Theta \,|\, X)\big]$?

(d) Calculate $\mathrm{var}(\Theta)$ using the law of total variance.

(e) Derive the linear LMS estimator of Θ based on X, and calculate its mean squared error.

Problem 17. Let Θ be a positive random variable, with known mean μ and variance σ^2, to be estimated on the basis of a measurement X of the form $X=\sqrt{\Theta}\,W$. We assume that W is independent of Θ with zero mean, unit variance, and known fourth moment $\mathbf{E}[W^4]$. Thus, the conditional mean and variance of X given Θ are 0 and

Θ, respectively, so we are essentially trying to estimate the variance of X given an observed value. Find the linear LMS estimator of Θ based on X, and the linear LMS estimator of Θ based on X^2.

Problem 18. Swallowed Buffon's needle. A doctor is treating a patient who has accidentally swallowed a needle. The key factor in whether to operate on the patient is the length Θ of the needle, which is unknown, but is assumed to be uniformly distributed between 0 and $l > 0$. We wish to form an estimate of Θ based on X, its projected length in an X-ray. We introduce a two-dimensional coordinate system and write

$$X = \Theta \cos W,$$

where W is the acute angle formed by the needle and one of the axes. We assume that W is uniformly distributed in the interval $[0, \pi/2]$, and is independent from Θ.

(a) Find the LMS estimator $\mathbf{E}[\Theta \,|\, X]$. In particular, derive $F_{X|\Theta}(x \,|\, \theta)$, $f_{X|\Theta}(x \,|\, \theta)$, $f_X(x)$, $f_{\Theta|X}(\theta \,|\, x)$, and then compute $\mathbf{E}[\Theta \,|\, X = x]$. *Hint:* You may find the following integration formulas useful:

$$\int_a^b \frac{1}{\sqrt{\alpha^2 - c^2}}\, d\alpha = \log\left(\alpha + \sqrt{\alpha^2 - c^2}\right)\Big|_a^b, \qquad \int_a^b \frac{\alpha}{\sqrt{\alpha^2 - c^2}}\, d\alpha = \sqrt{\alpha^2 - c^2}\,\Big|_a^b.$$

(b) Find the linear LMS estimate of Θ based on X, and the associated mean squared error.

Problem 19. Consider a photodetector in an optical communications system that counts the number of photons arriving during a certain interval. A user conveys information by switching a photon transmitter on or off. Assume that the probability of the transmitter being on is p. If the transmitter is on, the number of photons transmitted over the interval of interest is a Poisson random variable Θ with mean λ. If the transmitter is off, the number of photons transmitted is zero.

Unfortunately, regardless of whether or not the transmitter is on or off, photons may still be detected due to a phenomenon called "shot noise." The number N of detected shot noise photons is a Poisson random variable with mean μ. Thus, the total number X of detected photons is equal to $\Theta + N$ if the transmitter is on, and is equal to N otherwise. We assume that N and Θ are independent, so that $\Theta + N$ is also Poisson with mean $\lambda + \mu$.

(a) What is the probability that the transmitter was on, given that the photodetector detected k photons?

(c) Describe the MAP rule for deciding whether the transmitter was on.

(d) Find the linear LMS estimator of the number of transmitted photons, based on the number of detected photons.

Problem 20.* Estimation with spherically invariant PDFs. Let Θ and X be continuous random variables with joint PDF of the form

$$f_{\Theta,X}(\theta, x) = h\big(q(\theta, x)\big),$$

where h is a nonnegative scalar function, and $q(\theta, x)$ is a quadratic function of the form

$$q(\theta, x) = a(\theta - \overline{\theta})^2 + b(x - \overline{x})^2 - 2c(\theta - \overline{\theta})(x - \overline{x}).$$

Here $a, b, c, \overline{\theta}, \overline{x}$ are some scalars with $a \neq 0$. Derive the LMS and linear LMS estimates, for any x such that $\mathbf{E}[\Theta \,|\, X = x]$ is well-defined and finite. Assuming that $q(\theta, x) \geq 0$ for all x, θ, and that h is monotonically decreasing, derive the MAP estimate and show that it coincides with the LMS and linear LMS estimates.

Solution. The posterior is given by

$$f_{\Theta|X}(\theta \,|\, x) = \frac{f_{\Theta,X}(\theta, x)}{f_X(x)} = \frac{h\big(q(\theta, x)\big)}{f_X(x)}.$$

To motivate the derivation of the LMS and linear LMS estimates, consider first the MAP estimate, assuming that $q(\theta, x) \geq 0$ for all x, θ, and that h is monotonically decreasing. The MAP estimate maximizes $h\big(q(\theta, x)\big)$ and, since h is a decreasing function, it minimizes $q(\theta, x)$ over θ. By setting to 0 the derivative of $q(\theta, x)$ with respect to θ, we obtain

$$\hat{\theta} = \overline{\theta} + \frac{c}{a}(x - \overline{x}).$$

(We are using here the fact that a nonnegative quadratic function of one variable is minimized at a point where its derivative is equal to 0.)

We will now show that $\hat{\theta}$ is equal to the LMS and linear LMS estimates [without the assumption that $q(\theta, x) \geq 0$ for all x, θ, and that h is monotonically decreasing]. We write

$$\theta - \overline{\theta} = \theta - \hat{\theta} + \frac{c}{a}(x - \overline{x}),$$

and substitute in the formula for $q(\theta, x)$ to obtain after some algebra

$$q(\theta, x) = a(\theta - \hat{\theta})^2 + \left(b - \frac{c^2}{a}\right)(x - \overline{x})^2.$$

Thus, for any given x, the posterior is a function of θ that is symmetric around $\hat{\theta}$. This implies that $\hat{\theta}$ is equal to the conditional mean $\mathbf{E}[\Theta \,|\, X = x]$, whenever $\mathbf{E}[\Theta \,|\, X = x]$ is well-defined and finite. Furthermore, we have

$$\mathbf{E}[\Theta \,|\, X] = \overline{\theta} + \frac{c}{a}(X - \overline{x}).$$

Since $\mathbf{E}[\Theta \,|\, X]$ is linear in X, it is also the linear LMS estimator.

Problem 21.* Linear LMS estimation based on two observations. Consider three random variables Θ, X, and Y, with known variances and covariances. Assume that $\text{var}(X) > 0$, $\text{var}(Y) > 0$, and that $\big|\rho(X, Y)\big| \neq 1$. Give a formula for the linear LMS estimator of Θ based on X and Y, assuming that X and Y are uncorrelated, and also in the general case.

Solution. We consider a linear estimator of the form $\hat{\Theta} = aX + bY + c$ and choose a, b, and c to minimize the mean squared error $\mathbf{E}\big[(\Theta - aX - bY - c)^2\big]$. Suppose that a and b have already been chosen. Then, c must minimize $\mathbf{E}\big[(\Theta - aX - bY - c)^2\big]$, so

$$c = \mathbf{E}[\Theta] - a\mathbf{E}[X] - b\mathbf{E}[Y].$$

It follows that a and b minimize

$$\mathbf{E}\Big[\big((\Theta - \mathbf{E}[\Theta]) - a(X - \mathbf{E}[X]) - b(Y - \mathbf{E}[Y])\big)^2\Big].$$

We may thus assume that Θ, X, and Y are zero mean, and in the final formula subtract the means. Under this assumption, the mean squared error is equal to

$$\mathbf{E}\big[(\Theta - aX - bY)^2\big] = \mathbf{E}[\Theta^2] + a^2\mathbf{E}[X^2] + b^2\mathbf{E}[Y^2] - 2a\mathbf{E}[\Theta X] - 2b\mathbf{E}[\Theta Y] + 2ab\mathbf{E}[XY].$$

Assume that X and Y are uncorrelated, so that $\mathbf{E}[XY] = \mathbf{E}[X]\,\mathbf{E}[Y] = 0$. We differentiate the expression for the mean squared error with respect to a and b, and set the derivatives to zero to obtain

$$a = \frac{\mathbf{E}[\Theta X]}{\mathbf{E}[X^2]} = \frac{\text{cov}(\Theta, X)}{\text{var}(X)}, \qquad b = \frac{\mathbf{E}[\Theta Y]}{\mathbf{E}[Y^2]} = \frac{\text{cov}(\Theta, Y)}{\text{var}(Y)}.$$

Thus, the linear LMS estimator is

$$\hat{\Theta} = \mathbf{E}[\Theta] + \frac{\text{cov}(\Theta, X)}{\text{var}(X)}\big(X - \mathbf{E}[X]\big) + \frac{\text{cov}(\Theta, Y)}{\text{var}(Y)}(Y - \mathbf{E}[Y]).$$

If X and Y are correlated, we similarly set the derivatives of the mean squared error to zero. We obtain and then solve a system of two linear equations in the unknowns a and b, whose solution is

$$a = \frac{\text{var}(Y)\text{cov}(\Theta, X) - \text{cov}(\Theta, Y)\text{cov}(X, Y)}{\text{var}(X)\text{var}(Y) - \text{cov}^2(X, Y)},$$

$$b = \frac{\text{var}(X)\text{cov}(\Theta, Y) - \text{cov}(\Theta, X)\text{cov}(X, Y)}{\text{var}(X)\text{var}(Y) - \text{cov}^2(X, Y)}.$$

Note that the assumption $\big|\rho(X, Y)\big| \neq 1$ guarantees that the denominator in the preceding two equations is nonzero.

Problem 22.* Linear LMS estimation based on multiple observations. Let Θ be a random variable with mean μ and variance σ_0^2, and let $X_1, \ldots, X_n$ be observations of the form

$$X_i = \Theta + W_i,$$

where the observation errors W_i are random variables with mean 0 and variance σ_i^2. We assume that the random variables $\Theta, W_1, \ldots, W_n$ are independent. Verify that the linear LMS estimator of Θ based on $X_1, \ldots, X_n$ is

$$\hat{\Theta} = \frac{\mu/\sigma_0^2 + \displaystyle\sum_{i=1}^{n} X_i/\sigma_i^2}{\displaystyle\sum_{i=0}^{n} 1/\sigma_i^2},$$

by minimizing over $a_1, \ldots, a_n, b$ the function

$$h(a_1, \ldots, a_n, b) = \frac{1}{2}\mathbf{E}\big[(\Theta - a_1X_1 - \cdots - a_nX_n - b)^2\big].$$

Solution. We will show that the minimizing values of $a_1, \ldots, a_n, b$ are

$$b^* = \frac{\mu/\sigma_0^2}{\sum_{i=0}^{n} 1/\sigma_i^2}, \qquad a_j^* = \frac{1/\sigma_j^2}{\sum_{i=0}^{n} 1/\sigma_i^2}, \quad j = 1, \ldots, n.$$

To this end, it is sufficient to show that the partial derivatives of h, with respect to $a_1, \ldots, a_n, b$, are all equal to 0 when evaluated at $a_1^*, \ldots, a_n^*, b^*$. (Because the quadratic function h is nonnegative, it can be shown that any point at which its derivatives are zero must be a minimum.)

By differentiating h, we obtain

$$\left.\frac{\partial h}{\partial b}\right|_{a_i^*, b^*} = \mathbf{E}\left[\left(\sum_{i=1}^{n} a_i^* - 1\right)\Theta + \sum_{i=1}^{n} a_i^* W_i + b^*\right],$$

$$\left.\frac{\partial h}{\partial a_i}\right|_{a_i^*, b^*} = \mathbf{E}\left[X_i\left(\left(\sum_{i=1}^{n} a_i^* - 1\right)\Theta + \sum_{i=1}^{n} a_i^* W_i + b^*\right)\right].$$

From the expressions for b^* and a_i^*, we see that

$$\sum_{i=1}^{n} a_i^* - 1 = -\frac{b^*}{\mu}.$$

Using this equality and the facts

$$\mathbf{E}[\Theta] = \mu, \qquad \mathbf{E}[W_i] = 0,$$

it follows that

$$\left.\frac{\partial h}{\partial b}\right|_{a_i^*, b^*} = \mathbf{E}\left[\left(-\frac{b^*}{\mu}\right)\Theta + \sum_{i=1}^{n} a_i^* W_i + b^*\right] = 0.$$

Using, in addition, the equations

$$\mathbf{E}\big[X_i(\mu - \Theta)\big] = \mathbf{E}\big[(\Theta - \mu + W_i + \mu)(\mu - \Theta)\big] = -\sigma_0^2,$$

$$\mathbf{E}[X_i W_i] = \mathbf{E}\big[(\Theta + W_i)W_i\big] = \sigma_i^2, \qquad \text{for all } i,$$

$$\mathbf{E}[X_j W_i] = \mathbf{E}\big[(\Theta + W_j)W_i\big] = 0, \qquad \text{for all } i \text{ and } j \text{ with } i \neq j,$$

we obtain

$$\begin{aligned}
\left.\frac{\partial h}{\partial a_i}\right|_{a_i^*, b^*} &= \mathbf{E}\left[X_i\left(\left(-\frac{b^*}{\mu}\right)\Theta + \sum_{i=1}^{n} a_i^* W_i + b^*\right)\right] \\
&= \mathbf{E}\left[X_i\left((\mu - \Theta)\frac{b^*}{\mu} + \sum_{i=1}^{n} a_i^* W_i\right)\right] \\
&= -\sigma_0^2 \frac{b^*}{\mu} + a_i^* \sigma_i^2 \\
&= 0,
\end{aligned}$$

where the last equality holds in view of the definitions of a_i^* and b^*.

Problem 23.* Properties of LMS estimation. Let Θ and X be two random variables with positive variances. Let $\hat{\Theta}_L$ be the linear LMS estimator of Θ based on X, and let $\tilde{\Theta}_L = \hat{\Theta}_L - \Theta$ be the associated error. Similarly, let $\hat{\Theta}$ be the LMS estimator $\mathbf{E}[\Theta \,|\, X]$ of Θ based on X, and let $\tilde{\Theta} = \hat{\Theta} - \Theta$ be the associated error.

(a) Show that the estimation error $\tilde{\Theta}_L$ satisfies

$$\mathbf{E}[\tilde{\Theta}_L] = 0.$$

(b) Show that the estimation error $\tilde{\Theta}_L$ is uncorrelated with the observation X.

(c) Show that the variance of Θ can be decomposed as

$$\text{var}(\Theta) = \text{var}(\hat{\Theta}) + \text{var}(\tilde{\Theta}_L).$$

(d) Show that the LMS estimation error $\tilde{\Theta}$ is uncorrelated with any function $h(X)$ of the observation X.

(e) Show that $\tilde{\Theta}$ is not necessarily independent from X.

(f) Show that the linear LMS estimation error $\tilde{\Theta}_L$ is not necessarily uncorrelated with every function $h(X)$ of the observation X, and that $\mathbf{E}[\tilde{\Theta}_L \,|\, X = x]$ need not be equal to zero for all x.

Solution. (a) We have

$$\hat{\Theta}_L = \mathbf{E}[\Theta] + \frac{\text{cov}(\Theta, X)}{\sigma_X^2}\big(X - \mathbf{E}[X]\big).$$

Taking expectations of both sides, we obtain $\mathbf{E}[\hat{\Theta}_L] = \mathbf{E}[\Theta]$, or $\mathbf{E}[\tilde{\Theta}_L] = 0$.

(b) Using the formula for $\hat{\Theta}_L$, we obtain

$$\begin{aligned}
\mathbf{E}\big[(\hat{\Theta}_L - \Theta)X\big] &= \mathbf{E}\left[\left(\mathbf{E}[\Theta] + \frac{\text{cov}(\Theta, X)}{\sigma_X^2}\big(X - \mathbf{E}[X]\big)\right)X - \Theta X\right] \\
&= \mathbf{E}\left[\mathbf{E}[\Theta]\,X + \frac{\text{cov}(\Theta, X)}{\sigma_X^2}\big(X^2 - X\mathbf{E}[X]\big) - \Theta X\right] \\
&= \frac{\text{cov}(\Theta, X)\mathbf{E}[X^2]}{\sigma_X^2} - \frac{\text{cov}(\Theta, X)\big(\mathbf{E}[X]\big)^2}{\sigma_X^2} - \big(\mathbf{E}[\Theta X] - \mathbf{E}[\Theta]\,\mathbf{E}[X]\big) \\
&= \text{cov}(\Theta, X)\left(\frac{\mathbf{E}[X^2]}{\sigma_X^2} - \frac{\big(\mathbf{E}[X]\big)^2}{\sigma_X^2} - 1\right) \\
&= \text{cov}(\Theta, X)\left(\frac{\sigma_X^2}{\sigma_X^2} - 1\right) \\
&= 0.
\end{aligned}$$

The fact $\mathbf{E}[\tilde{\Theta}_L X] = 0$ we just established, together with the fact $\mathbf{E}[\tilde{\Theta}_L] = 0$ from part (a), imply that

$$\text{cov}(\tilde{\Theta}_L, X) = \mathbf{E}[\tilde{\Theta}_L X] - \mathbf{E}[\tilde{\Theta}_L]\,\mathbf{E}[X] = 0.$$

(c) Since $\text{cov}(\tilde{\Theta}_L, X) = 0$ and $\hat{\Theta}_L$ is a linear function of X, we obtain $\text{cov}(\tilde{\Theta}_L, \hat{\Theta}_L) = 0$. Thus,

$$\begin{aligned}\text{var}(\Theta) &= \text{var}(\hat{\Theta}_L - \tilde{\Theta}_L) = \text{var}(\hat{\Theta}_L) + \text{var}(-\tilde{\Theta}_L) + 2\text{cov}(\hat{\Theta}_L, -\tilde{\Theta}_L) \\ &= \text{var}(\hat{\Theta}_L) + \text{var}(\tilde{\Theta}_L) - 2\text{cov}(\hat{\Theta}_L, \tilde{\Theta}_L) = \text{var}(\hat{\Theta}_L) + \text{var}(\tilde{\Theta}_L).\end{aligned}$$

(d) This is because $\mathbf{E}[\tilde{\Theta}] = 0$ and

$$\begin{aligned}\mathbf{E}\big[\tilde{\Theta}h(X)\big] &= \mathbf{E}\big[\big(\mathbf{E}[\Theta \,|\, X] - \Theta\big)h(X)\big] \\ &= \mathbf{E}\big[\mathbf{E}[\Theta \,|\, X]h(X)\big] - \mathbf{E}\big[\Theta h(X)\big] \\ &= \mathbf{E}\big[\mathbf{E}[\Theta h(X) \,|\, X]\big] - \mathbf{E}\big[\Theta h(X)\big] \\ &= \mathbf{E}\big[\Theta h(X)\big] - \mathbf{E}\big[\Theta h(X)\big] \\ &= 0.\end{aligned}$$

(e) Let Θ and X be discrete random variables with joint PMF

$$p_{\Theta,X}(\theta, x) = \begin{cases} 1/3, & \text{for } (\theta, x) = (0,0),\ (1,1),\ (-1,1), \\ 0, & \text{otherwise.} \end{cases}$$

In this example, we have $X = |\Theta|$, so that X and Θ are not independent. Note that $\mathbf{E}[\Theta \,|\, X = x] = 0$ for all possible values x, so $\mathbf{E}[\Theta \,|\, X] = 0$. Thus, we have $\tilde{\Theta} = -\Theta$. Since Θ and X are not independent, $\tilde{\Theta}$ and X are not independent either.

(f) Let Θ and X be discrete random variables with joint PMF

$$p_{\Theta,X}(\theta, x) = \begin{cases} 1/3, & \text{for } (\theta, x) = (0,0),\ (1,1),\ (1,-1), \\ 0, & \text{otherwise.} \end{cases}$$

In this example, we have $\Theta = |X|$. Note that $\mathbf{E}[X] = 0$ and $\mathbf{E}[\Theta X] = 0$, so that X and Θ are uncorrelated. We have $\hat{\Theta}_L = \mathbf{E}[\Theta] = 2/3$, and $\tilde{\Theta}_L = (2/3) - \Theta = (2/3) - |X|$, which is not independent from X. Furthermore, we have $\mathbf{E}[\tilde{\Theta}_L \,|\, X = x] = (2/3) - |x|$, which takes the different values $2/3$ and $-1/3$, depending on whether $x = 0$ or $|x| = 1$.

Problem 24.* Properties of linear LMS estimation based on multiple observations. Let $\Theta, X_1, \ldots, X_n$ be random variables with given variances and covariances. Let $\hat{\Theta}_L$ be the linear LMS estimator of Θ based on $X_1, \ldots, X_n$, and let $\tilde{\Theta}_L = \hat{\Theta}_L - \Theta$ be the associated error. Show that $\mathbf{E}[\tilde{\Theta}_L] = 0$ and that $\tilde{\Theta}_L$ is uncorrelated with X_i for every i.

Solution. We start by showing that $\mathbf{E}[\tilde{\Theta}_L X_i] = 0$, for all i. Consider a new linear estimator of the form $\hat{\Theta}_L + \alpha X_i$, where α is a scalar parameter. Since $\hat{\Theta}_L$ is a linear LMS estimator, its mean squared error $\mathbf{E}\big[(\hat{\Theta}_L - \Theta)^2\big]$ is no larger than the mean squared error $h(\alpha) = \mathbf{E}\big[(\hat{\Theta}_L + \alpha X_i - \Theta)^2\big]$ of the new estimator. Therefore, the function $h(\alpha)$ attains its minimum value when $\alpha = 0$, and $(dh/d\alpha)(0) = 0$. Note that

$$h(\alpha) = \mathbf{E}\big[(\tilde{\Theta}_L + \alpha X_i)^2\big] = \mathbf{E}[\tilde{\Theta}_L^2] + \alpha\mathbf{E}[\tilde{\Theta}_L X_i] + \alpha^2\mathbf{E}[X_i^2].$$

The condition $(dh/d\alpha)(0) = 0$ yields $\mathbf{E}[\tilde{\Theta}_L X_i] = 0$.

Let us now repeat the above argument, but with the constant 1 replacing the random variable X_i. Following the same steps, we obtain $\mathbf{E}[\tilde{\Theta}_L] = 0$. Finally, note that

$$\text{cov}(\tilde{\Theta}_L, X_i) = \mathbf{E}[\tilde{\Theta}_L X_i] - \mathbf{E}[\tilde{\Theta}_L]\,\mathbf{E}[X_i] = 0 - 0 \cdot \mathbf{E}[X_i] = 0,$$

so that $\tilde{\Theta}_L$ and X_i are uncorrelated.

9

Classical Statistical Inference

Contents

In the preceding chapter, we developed the Bayesian approach to inference, where unknown parameters are modeled as random variables. In all cases we worked within a single, fully-specified probabilistic model, and we based most of our derivations and calculations on judicious application of Bayes' rule.

By contrast, in the present chapter we adopt a fundamentally different philosophy: we view the unknown parameter θ as a deterministic (not random) but unknown quantity. The observation X is random and its distribution $p_X(x;\theta)$ [if X is discrete] or $f_X(x;\theta)$ [if X is continuous] depends on the value of θ (see Fig. 9.1). Thus, instead of working within a single probabilistic model, we will be dealing simultaneously with *multiple candidate models*, one model for each possible value of θ. In this context, a "good" hypothesis testing or estimation procedure will be one that possesses certain desirable properties *under every candidate model*, that is, for every possible value of θ. In some cases, this may be considered to be a worst case viewpoint: a procedure is not considered to fulfill our specifications unless it does so against the worst possible value that θ can take.

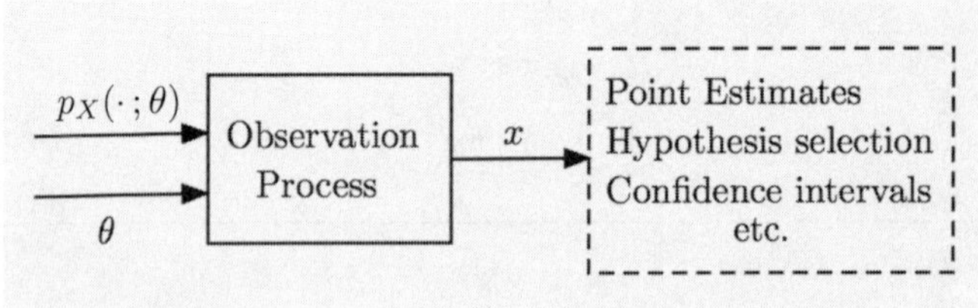

Figure 9.1. Summary of a classical inference model. For each value of θ, we have a distribution $p_X(x;\theta)$. The value x of the observation X is used to compute a point estimate, or select a hypothesis, etc.

Our notation will generally indicate the dependence of probabilities and expected values on θ. For example, we will denote by $\mathbf{E}_\theta\big[h(X)\big]$ the expected value of a random variable $h(X)$ as a function of θ. Similarly, we will use the notation $\mathbf{P}_\theta(A)$ to denote the probability of an event A. Note that this only indicates a functional dependence, not conditioning in the probabilistic sense.

The first two sections focus on parameter estimation, with a special emphasis on the maximum likelihood and the linear regression methods, often involving independent, identically distributed (i.i.d.) observations. The issues here are similar to the ones in Bayesian estimation, discussed in the preceding chapter. We are interested in estimators (functions of the observations) that have some desirable properties. However, the criteria for desirability are somewhat different because they should be satisfied for all possible values of the unknown parameter. For example, we may require that the expected value of the estimation error be zero, or that the estimation error be small with high probability, for all possible values of the unknown parameter.

The third section deals with binary hypothesis testing problems. Here, we develop methods that bear similarity with the (Bayesian) MAP method, discussed in the preceding chapter. In particular, we calculate the "likelihood" of each hypothesis under the observed data, and we choose a hypothesis by comparing the ratio of the two likelihoods with a suitably chosen threshold.

The last section addresses different types of hypothesis testing problems. For an example, suppose a coin is tossed n times, the resulting sequence of heads and tails is observed, and we wish to decide whether the coin is fair or not. The main hypothesis that we wish to test is whether $p = 1/2$, where p denotes the unknown probability of heads. The alternative hypothesis ($p \neq 1/2$) is **composite**, in the sense that it consists of several, possibly infinitely many, subhypotheses (e.g., $p = 0.1$, $p = 0.4999$, etc.). It is clear that no method can reliably distinguish between a coin with $p = 0.5$ and a coin with $p = 0.4999$ on the basis of a moderate number of observations. Such problems are usually approached using the methodology of **significance testing**. Here, one asks the question: "are the observed data compatible with the hypothesis that $p = 0.5$?" Roughly speaking, a hypothesis is rejected if the observed data are unlikely to have been generated "accidentally" or "by chance," under that hypothesis.

Major Terms, Problems, and Methods in this Chapter

- **Classical statistics** treats unknown parameters as constants to be determined. A separate probabilistic model is assumed for each possible value of the unknown parameter.
- In **parameter estimation**, we want to generate estimates that are nearly correct under any possible value of the unknown parameter.
- In **hypothesis testing**, the unknown parameter takes a finite number m of values ($m \geq 2$), corresponding to competing hypotheses; we want to choose one of the hypotheses, aiming to achieve a small probability of error under any of the possible hypotheses.
- In **significance testing**, we want to accept or reject a single hypothesis, while keeping the probability of false rejection suitably small.
- Principal classical inference methods in this chapter:
  - (a) **Maximum likelihood (ML) estimation**: Select the parameter that makes the observed data "most likely," i.e., maximizes the probability of obtaining the data at hand (Section 9.1).
  - (b) **Linear regression**: Find the linear relation that matches best a set of data pairs, in the sense that it minimizes the sum of the squares of the discrepancies between the model and the data (Section 9.2).

(c) **Likelihood ratio test**: Given two hypotheses, select one based on the ratio of their "likelihoods," so that certain error probabilities are suitably small (Section 9.3).

(d) **Significance testing**: Given a hypothesis, reject it if and only if the observed data falls within a certain rejection region. This region is specially designed to keep the probability of false rejection below some threshold (Section 9.4).

9.1 CLASSICAL PARAMETER ESTIMATION

In this section, we focus on parameter estimation, using the classical approach where the parameter θ is not random, but is rather viewed as an unknown constant. We first introduce some definitions and associated properties of estimators. We then discuss the maximum likelihood estimator, which may be viewed as the classical counterpart of the Bayesian MAP estimator. We finally focus on the simple but important example of estimating an unknown mean, and possibly an unknown variance. We also discuss the associated issue of constructing an interval that contains the unknown parameter with high probability (a "confidence interval"). The methods that we develop rely heavily on the laws of large numbers and the central limit theorem (cf. Chapter 5).

Properties of Estimators

Given observations $X = (X_1, \ldots, X_n)$, an **estimator** is a random variable of the form $\hat{\Theta} = g(X)$, for some function g. Note that since the distribution of X depends on θ, the same is true for the distribution of $\hat{\Theta}$. We use the term **estimate** to refer to an actual realized value of $\hat{\Theta}$.

Sometimes, particularly when we are interested in the role of the number of observations n, we use the notation $\hat{\Theta}_n$ for an estimator. It is then also appropriate to view $\hat{\Theta}_n$ as a sequence of estimators (one for each value of n). The mean and variance of $\hat{\Theta}_n$ are denoted $\mathbf{E}_\theta[\hat{\Theta}_n]$ and $\text{var}_\theta(\hat{\Theta}_n)$, respectively, and are defined in the usual way. Both $\mathbf{E}_\theta[\hat{\Theta}_n]$ and $\text{var}_\theta(\hat{\Theta}_n)$ are numerical functions of θ, but for simplicity, when the context is clear we sometimes do not show this dependence.

We introduce some terminology related to various properties of estimators.

Terminology Regarding Estimators

Let $\hat{\Theta}_n$ be an **estimator** of an unknown parameter θ, that is, a function of n observations $X_1, \ldots, X_n$ whose distribution depends on θ.

- The **estimation error**, denoted by $\tilde{\Theta}_n$, is defined by $\tilde{\Theta}_n = \hat{\Theta}_n - \theta$.
- The **bias** of the estimator, denoted by $\mathrm{b}_\theta(\hat{\Theta}_n)$, is the expected value of the estimation error:

$$\mathrm{b}_\theta(\hat{\Theta}_n) = \mathbf{E}_\theta[\hat{\Theta}_n] - \theta.$$

- The expected value, the variance, and the bias of $\hat{\Theta}_n$ depend on θ, while the estimation error depends in addition on the observations $X_1, \ldots, X_n$.
- We call $\hat{\Theta}_n$ **unbiased** if $\mathbf{E}_\theta[\hat{\Theta}_n] = \theta$, for every possible value of θ.
- We call $\hat{\Theta}_n$ **asymptotically unbiased** if $\lim_{n\to\infty} \mathbf{E}_\theta[\hat{\Theta}_n] = \theta$, for every possible value of θ.
- We call $\hat{\Theta}_n$ **consistent** if the sequence $\hat{\Theta}_n$ converges to the true value of the parameter θ, in probability, for every possible value of θ.

An estimator, being a function of the random observations, cannot be expected to be exactly equal to the unknown value θ. Thus, the estimation error will be generically nonzero. On the other hand, if the average estimation error is zero, for every possible value of θ, then we have an unbiased estimator, and this is a desirable property. Asymptotic unbiasedness only requires that the estimator become unbiased as the number n of observations increases, and this is desirable when n is large.

Besides the bias $\mathrm{b}_\theta(\hat{\Theta}_n)$, we are usually interested in the size of the estimation error. This is captured by the mean squared error $\mathbf{E}_\theta[\tilde{\Theta}_n^2]$, which is related to the bias and the variance of $\hat{\Theta}_n$ according to the following formula:†

$$\mathbf{E}_\theta[\tilde{\Theta}_n^2] = \mathrm{b}_\theta^2(\hat{\Theta}_n) + \mathrm{var}_\theta(\hat{\Theta}_n).$$

This formula is important because in many statistical problems, there is a trade-off between the two terms on the right-hand-side. Often a reduction in the variance is accompanied by an increase in the bias. Of course, a good estimator is one that manages to keep both terms small.

We will now discuss some specific estimation approaches, starting with maximum likelihood estimation. This is a general method that bears similarity to MAP estimation, introduced in the context of Bayesian inference. We will subsequently consider the simple but important case of estimating the mean and variance of a random variable. This will bring about a connection with our discussion of the laws of large numbers in Chapter 5.

---

† This is an application of the formula $\mathbf{E}[X^2] = \big(\mathbf{E}[X]\big)^2 + \mathrm{var}(X)$, with $X = \tilde{\Theta}_n$ and where the expectation is taken with respect to the distribution corresponding to θ; we are also using the facts $\mathbf{E}_\theta[\tilde{\Theta}_n] = \mathrm{b}_\theta(\hat{\Theta}_n)$ and $\mathrm{var}_\theta(\tilde{\Theta}_n) = \mathrm{var}_\theta(\hat{\Theta}_n - \theta) = \mathrm{var}_\theta(\hat{\Theta}_n)$.

Maximum Likelihood Estimation

Let the vector of observations $X = (X_1, \ldots, X_n)$ be described by a joint PMF $p_X(x; \theta)$ whose form depends on an unknown (scalar or vector) parameter θ. Suppose we observe a particular value $x = (x_1, \ldots, x_n)$ of X. Then, a **maximum likelihood** (ML) estimate is a value of the parameter that maximizes the numerical function $p_X(x_1, \ldots, x_n; \theta)$ over all θ (see Fig. 9.2):

$$\hat{\theta}_n = \arg\max_{\theta} p_X(x_1, \ldots, x_n; \theta).$$

For the case where X is continuous, the same approach applies with $p_X(x; \theta)$ replaced by the joint PDF $f_X(x; \theta)$, so that

$$\hat{\theta}_n = \arg\max_{\theta} f_X(x_1, \ldots, x_n; \theta).$$

We refer to $p_X(x; \theta)$ [or $f_X(x; \theta)$ if X is continuous] as the **likelihood function**.

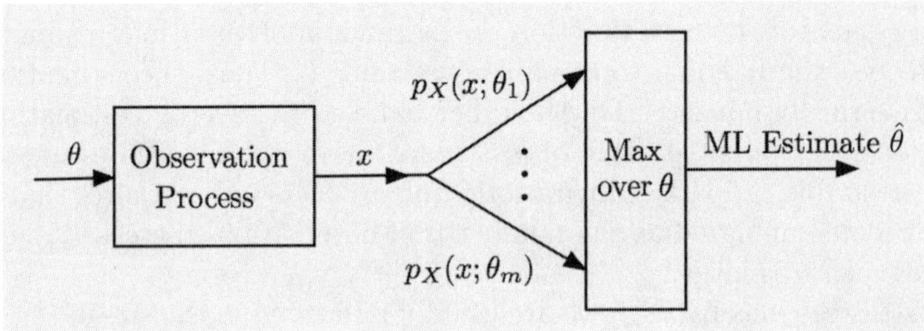

Figure 9.2. Illustration of ML estimation, assuming X is discrete and θ takes one of the m values $\theta_1, \ldots, \theta_m$. Given the value of the observation $X = x$, the values of the likelihood function $p_X(x; \theta_i)$ become available for all i, and a value of θ that maximizes $p_X(x; \theta)$ is selected.

In many applications, the observations X_i are assumed to be independent, in which case, the likelihood function is of the form

$$p_X(x_1, \ldots, x_n; \theta) = \prod_{i=1}^{n} p_{X_i}(x_i; \theta)$$

(for discrete X_i). In this case, it is often analytically or computationally convenient to maximize its logarithm, called the **log-likelihood function**,

$$\log p_X(x_1, \ldots, x_n; \theta) = \log \prod_{i=1}^{n} p_{X_i}(x_i; \theta) = \sum_{i=1}^{n} \log p_{X_i}(x_i; \theta),$$

over θ. When X is continuous, there is a similar possibility, with PMFs replaced by PDFs: we maximize over θ the expression

$$\log f_X(x_1, \ldots, x_n; \theta) = \log \prod_{i=1}^{n} f_{X_i}(x_i; \theta) = \sum_{i=1}^{n} \log f_{X_i}(x_i; \theta).$$

The term "likelihood" needs to be interpreted properly. In particular, having observed the value x of X, $p_X(x; \theta)$ *is not* the probability that the unknown parameter is equal to θ. Instead, it is the probability that the observed value x can arise when the parameter is equal to θ. Thus, in maximizing the likelihood, we are asking the question: "What is the value of θ under which the observations we have seen are most likely to arise?"

Recall that in Bayesian MAP estimation, the estimate is chosen to maximize the expression $p_\Theta(\theta)\, p_{X|\Theta}(x \,|\, \theta)$ over all θ, where $p_\Theta(\theta)$ is the prior PMF of an unknown discrete parameter θ. Thus, if we view $p_X(x; \theta)$ as a conditional PMF, we may interpret ML estimation as MAP estimation with a **flat prior**, i.e., a prior which is the same for all θ, indicating the absence of any useful prior knowledge. Similarly, in the case of continuous θ with a bounded range of possible values, we may interpret ML estimation as MAP estimation with a uniform prior: $f_\Theta(\theta) = c$ for all θ and some constant c.

Example 9.1. Let us revisit Example 8.2, in which Juliet is always late by an amount X that is uniformly distributed over the interval $[0, \theta]$, and θ is an unknown parameter. In that example, we used a random variable Θ with flat prior PDF $f_\Theta(\theta)$ (uniform over the interval $[0, 1]$) to model the parameter, and we showed that the MAP estimate is the value x of X. In the classical context of this section, there is no prior, and θ is treated as a constant, but the ML estimate is also $\hat{\theta} = x$.

Example 9.2. Estimating the Mean of a Bernoulli Random Variable. We want to estimate the probability of heads, θ, of a biased coin, based on the outcomes of n independent tosses $X_1, \ldots, X_n$ ($X_i = 1$ for a head, and $X_i = 0$ for a tail). This is similar to the Bayesian setting of Example 8.8, where we assumed a flat prior. We found there that the peak of the posterior PDF (the MAP estimate) is located at $\theta = k/n$, where k is the number of heads observed. It follows that k/n is also the ML estimate of θ, so that the ML estimator is

$$\hat{\Theta}_n = \frac{X_1 + \cdots + X_n}{n}.$$

This estimator is unbiased. It is also consistent, because $\hat{\Theta}_n$ converges to θ in probability, by the weak law of large numbers.

It is interesting to compare the ML estimator with the LMS estimator, obtained with a Bayesian approach in Example 8.8. We showed there that with a flat prior, the posterior mean is $(k+1)/(n+2)$. Thus, the ML estimate, k/n, is similar but not identical to the LMS estimate obtained with the Bayesian approach. However, as $n \to \infty$, the two estimates asymptotically coincide.

Example 9.3. Estimating the Parameter of an Exponential Random Variable. Customers arrive to a facility, with the ith customer arriving at time Y_i. We assume that the ith interarrival time, $X_i = Y_i - Y_{i-1}$ (with the convention $Y_0 = 0$) is exponentially distributed with unknown parameter θ, and that the random variables $X_1, \ldots, X_n$ are independent. (This is the Poisson arrivals model, studied in Chapter 6.) We wish to estimate the value of θ (interpreted as the arrival rate), on the basis of the observations $X_1, \ldots, X_n$.

The corresponding likelihood function is

$$f_X(x;\theta) = \prod_{i=1}^{n} f_{X_i}(x_i;\theta) = \prod_{i=1}^{n} \theta e^{-\theta x_i},$$

and the log-likelihood function is

$$\log f_X(x;\theta) = n\log\theta - \theta y_n,$$

where

$$y_n = \sum_{i=1}^{n} x_i.$$

The derivative with respect to θ is $(n/\theta) - y_n$, and by setting it to 0, we see that the maximum of $\log f_X(x;\theta)$, over $\theta \geq 0$, is attained at $\hat{\theta}_n = n/y_n$. The resulting estimator is

$$\hat{\Theta}_n = \left(\frac{Y_n}{n}\right)^{-1}.$$

It is the inverse of the sample mean of the interarrival times, and it can be interpreted as an empirical arrival rate.

Note that by the weak law of large numbers, Y_n/n converges in probability to $\mathbf{E}[X_i] = 1/\theta$, as $n \to \infty$. This can be used to show that $\hat{\Theta}_n$ converges to θ in probability, so the estimator is consistent.

Our discussion and examples so far have focused on the case of a single unknown parameter θ. The next example involves a two-dimensional parameter.

Example 9.4. Estimating the Mean and Variance of a Normal. Consider the problem of estimating the mean μ and variance v of a normal distribution using n independent observations $X_1, \ldots, X_n$. The parameter vector here is $\theta = (\mu, v)$. The corresponding likelihood function is

$$f_X(x;\mu,v) = \prod_{i=1}^{n} f_{X_i}(x_i;\mu,v) = \prod_{i=1}^{n} \frac{1}{\sqrt{2\pi v}} e^{-(x_i-\mu)^2/2v}.$$

After some calculation it can be written as†

$$f_X(x;\mu,v) = \frac{1}{(2\pi v)^{n/2}} \cdot \exp\left\{-\frac{ns_n^2}{2v}\right\} \cdot \exp\left\{-\frac{n(m_n-\mu)^2}{2v}\right\},$$

where m_n is the realized value of the random variable

$$M_n = \frac{1}{n}\sum_{i=1}^{n} X_i,$$

and s_n^2 is the realized value of the random variable

$$\overline{S}_n^2 = \frac{1}{n}\sum_{i=1}^{n} (X_i - M_n)^2.$$

The log-likelihood function is

$$\log f_X(x;\mu,v) = -\frac{n}{2}\cdot\log(2\pi) - \frac{n}{2}\cdot\log v - \frac{ns_n^2}{2v} - \frac{n(m_n-\mu)^2}{2v}.$$

Setting to zero the derivatives of this function with respect to μ and v, we obtain the estimate and estimator, respectively,

$$\hat{\theta}_n = (m_n, s_n^2), \qquad \hat{\Theta}_n = (M_n, \overline{S}_n^2).$$

Note that M_n is the sample mean, while $\overline{S}_n^2$ may be viewed as a "sample variance." As will be shown shortly, $\mathbf{E}_\theta[\overline{S}_n^2]$ converges to v as n increases, so that $\overline{S}_n^2$ is asymptotically unbiased. Using also the weak law of large numbers, it can be shown that M_n and $\overline{S}_n^2$ are consistent estimators of μ and v, respectively.

Maximum likelihood estimation has some appealing properties. For example, it obeys the **invariance principle**: if $\hat{\Theta}_n$ is the ML estimate of θ, then for any one-to-one function h of θ, the ML estimate of the parameter $\zeta = h(\theta)$ is $h(\hat{\Theta}_n)$. Also, when the observations are i.i.d., and under some mild additional assumptions, it can be shown that the ML estimator is consistent.

Another interesting property is that when θ is a scalar parameter, then under some mild conditions, the ML estimator has an **asymptotic normality** property. In particular, it can be shown that the distribution of $(\hat{\Theta}_n - \theta)/\sigma(\hat{\Theta}_n)$, where $\sigma^2(\hat{\Theta}_n)$ is the variance of $\hat{\Theta}_n$, approaches a standard normal distribution. Thus, if we are able to also estimate $\sigma(\hat{\Theta}_n)$, we can use it to derive an error variance estimate based on a normal approximation. When θ is a vector parameter, a similar statement applies to each one of its components.

---

† To verify this, write for $i = 1, \ldots, n$,

$$(x_i - \mu)^2 = (x_i - m_n + m_n - \mu)^2 = (x_i - m_n)^2 + (m_n - \mu)^2 + 2(x_i - m_n)(m_n - \mu),$$

sum over i, and note that

$$\sum_{i=1}^{n}(x_i - m_n)(m_n - \mu) = (m_n - \mu)\sum_{i=1}^{n}(x_i - m_n) = 0.$$

Maximum Likelihood Estimation

- We are given the realization $x = (x_1, \ldots, x_n)$ of a random vector $X = (X_1, \ldots, X_n)$, distributed according to a PMF $p_X(x; \theta)$ or PDF $f_X(x; \theta)$.
- The maximum likelihood (ML) estimate is a value of θ that maximizes the likelihood function, $p_X(x; \theta)$ or $f_X(x; \theta)$, over all θ.
- The ML estimate of a one-to-one function $h(\theta)$ of θ is $h(\hat{\theta}_n)$, where $\hat{\theta}_n$ is the ML estimate of θ (the invariance principle).
- When the random variables X_i are i.i.d., and under some mild additional assumptions, each component of the ML estimator is consistent and asymptotically normal.

Estimation of the Mean and Variance of a Random Variable

We now discuss the simple but important problem of estimating the mean and the variance of a probability distribution. This is similar to the preceding Example 9.4, but in contrast with that example, we do not assume that the distribution is normal. In fact, the estimators presented here do not require knowledge of the distributions $p_X(x; \theta)$ [or $f_X(x; \theta)$ if X is continuous].

Suppose that the observations $X_1, \ldots, X_n$ are i.i.d., with an unknown common mean θ. The most natural estimator of θ is the **sample mean**:

$$M_n = \frac{X_1 + \cdots + X_n}{n}.$$

This estimator is unbiased, since $\mathbf{E}_\theta[M_n] = \mathbf{E}_\theta[X] = \theta$. Its mean squared error is equal to its variance, which is v/n, where v is the common variance of the X_i. Note that the mean squared error does not depend on θ. Furthermore, by the weak law of large numbers, this estimator converges to θ in probability, and is therefore consistent.

The sample mean is not necessarily the estimator with the smallest variance. For example, consider the estimator $\hat{\Theta}_n = 0$, which ignores the observations and always yields an estimate of zero. The variance of $\hat{\Theta}_n$ is zero, but its bias is $\mathrm{b}_\theta(\hat{\Theta}_n) = -\theta$. In particular, the mean squared error depends on θ and is equal to θ^2.

The next example compares the sample mean with a Bayesian MAP estimator that we derived in Section 8.2 under certain assumptions.

Example 9.5. Suppose that the observations $X_1, \ldots, X_n$ are normal, i.i.d., with an unknown common mean θ, and known variance v. In Example 8.3, we used a Bayesian approach, and assumed a normal prior distribution on the parameter

θ. For the case where the prior mean of θ was zero, we arrived at the following estimator:

$$\hat{\Theta}_n = \frac{X_1 + \cdots + X_n}{n+1}.$$

This estimator is biased, because $\mathbf{E}_\theta[\hat{\Theta}_n] = n\theta/(n+1)$ and $\mathrm{b}_\theta(\hat{\Theta}_n) = -\theta/(n+1)$. However, $\lim_{n\to\infty} \mathrm{b}_\theta(\hat{\Theta}_n) = 0$, so $\hat{\Theta}_n$ is asymptotically unbiased. Its variance is

$$\mathrm{var}_\theta(\hat{\Theta}_n) = \frac{v\,n}{(n+1)^2},$$

and it is slightly smaller than the variance v/n of the sample mean. Note that in the special case of this example, $\mathrm{var}_\theta(\hat{\Theta}_n)$ is independent of θ. The mean squared error is equal to

$$\mathbf{E}_\theta[\tilde{\Theta}_n^2] = \mathrm{b}_\theta^2(\hat{\Theta}_n) + \mathrm{var}_\theta(\hat{\Theta}_n) = \frac{\theta^2}{(n+1)^2} + \frac{v\,n}{(n+1)^2}.$$

Suppose that in addition to the sample mean/estimator of θ,

$$M_n = \frac{X_1 + \cdots + X_n}{n},$$

we are interested in an estimator of the variance v. A natural one is

$$\overline{S}_n^2 = \frac{1}{n}\sum_{i=1}^{n}(X_i - M_n)^2,$$

which coincides with the ML estimator derived in Example 9.4 under a normality assumption.

Using the facts

$$\mathbf{E}_{(\theta,v)}[M_n] = \theta, \qquad \mathbf{E}_{(\theta,v)}[X_i^2] = \theta^2 + v, \qquad \mathbf{E}_{(\theta,v)}[M_n^2] = \theta^2 + \frac{v}{n},$$

we have

$$\begin{aligned}
\mathbf{E}_{(\theta,v)}\big[\overline{S}_n^2\big] &= \frac{1}{n}\mathbf{E}_{(\theta,v)}\left[\sum_{i=1}^{n} X_i^2 - 2M_n\sum_{i=1}^{n} X_i + nM_n^2\right] \\
&= \mathbf{E}_{(\theta,v)}\left[\frac{1}{n}\sum_{i=1}^{n} X_i^2 - 2M_n^2 + M_n^2\right] \\
&= \mathbf{E}_{(\theta,v)}\left[\frac{1}{n}\sum_{i=1}^{n} X_i^2 - M_n^2\right] \\
&= \theta^2 + v - \left(\theta^2 + \frac{v}{n}\right) \\
&= \frac{n-1}{n}v.
\end{aligned}$$

Thus, $\overline{S}_n^2$ is not an unbiased estimator of v, although it is asymptotically unbiased.

We can obtain an unbiased variance estimator after some suitable scaling. This is the estimator

$$\hat{S}_n^2 = \frac{1}{n-1}\sum_{i=1}^{n}(X_i - M_n)^2 = \frac{n}{n-1}\overline{S}_n^2.$$

The preceding calculation shows that

$$\mathbf{E}_{(\theta,v)}\big[\hat{S}_n^2\big] = v,$$

so $\hat{S}_n^2$ is an unbiased estimator of v, for all n. However, for large n, the estimators $\hat{S}_n^2$ and $\overline{S}_n^2$ are essentially the same.

Estimates of the Mean and Variance of a Random Variable

Let the observations $X_1, \ldots, X_n$ be i.i.d., with mean θ and variance v that are unknown.

- The sample mean
$$M_n = \frac{X_1 + \cdots + X_n}{n}$$
is an unbiased estimator of θ, and its mean squared error is v/n.
- Two variance estimators are
$$\overline{S}_n^2 = \frac{1}{n}\sum_{i=1}^{n}(X_i - M_n)^2, \qquad \hat{S}_n^2 = \frac{1}{n-1}\sum_{i=1}^{n}(X_i - M_n)^2.$$
- The estimator $\overline{S}_n^2$ coincides with the ML estimator if the X_i are normal. It is biased but asymptotically unbiased. The estimator $\hat{S}_n^2$ is unbiased. For large n, the two variance estimators essentially coincide.

Confidence Intervals

Consider an estimator $\hat{\Theta}_n$ of an unknown parameter θ. Besides the numerical value provided by an estimate, we are often interested in constructing a so-called confidence interval. Roughly speaking, this is an interval that contains θ with a certain high probability, for every possible value of θ.

For a precise definition, let us first fix a desired **confidence level**, $1 - \alpha$, where α is typically a small number. We then replace the point estimator $\hat{\Theta}_n$ by a lower estimator $\hat{\Theta}_n^-$ and an upper estimator $\hat{\Theta}_n^+$, designed so that $\hat{\Theta}_n^- \leq \hat{\Theta}_n^+$, and

$$\mathbf{P}_\theta\big(\hat{\Theta}_n^- \leq \theta \leq \hat{\Theta}_n^+\big) \geq 1 - \alpha,$$

for every possible value of θ. Note that, similar to estimators, $\hat{\Theta}_n^-$ and $\hat{\Theta}_n^+$, are functions of the observations, and hence random variables whose distributions depend on θ. We call $[\hat{\Theta}_n^-, \hat{\Theta}_n^+]$ a $1 - \alpha$ **confidence interval**.

Example 9.6. Suppose that the observations X_i are i.i.d. normal, with unknown mean θ and known variance v. Then, the sample mean estimator

$$\hat{\Theta}_n = \frac{X_1 + \cdots + X_n}{n}$$

is normal,† with mean θ and variance v/n. Let $\alpha = 0.05$. Using the CDF $\Phi(z)$ of the standard normal (available in the normal tables), we have $\Phi(1.96) = 0.975 = 1-\alpha/2$ and we obtain

$$\mathbf{P}_\theta\left(\frac{|\hat{\Theta}_n - \theta|}{\sqrt{v/n}} \leq 1.96\right) = 0.95.$$

We can rewrite this statement in the form

$$\mathbf{P}_\theta\left(\hat{\Theta}_n - 1.96\sqrt{\frac{v}{n}} \leq \theta \leq \hat{\Theta}_n + 1.96\sqrt{\frac{v}{n}}\right) = 0.95,$$

which implies that

$$\left[\hat{\Theta}_n - 1.96\sqrt{\frac{v}{n}},\ \hat{\Theta}_n + 1.96\sqrt{\frac{v}{n}}\right]$$

is a 95% confidence interval, where we identify $\hat{\Theta}_n^-$ and $\hat{\Theta}_n^+$ with $\hat{\Theta}_n - 1.96\sqrt{v/n}$ and $\hat{\Theta}_n + 1.96\sqrt{v/n}$, respectively.

In the preceding example, we may be tempted to describe the concept of a 95% confidence interval by a statement such as "the true parameter lies in the confidence interval with probability 0.95." Such statements, however, can be ambiguous. For example, suppose that after the observations are obtained, the confidence interval turns out to be $[-2.3, 4.1]$. We cannot say that θ lies in $[-2.3, 4.1]$ with probability 0.95, because the latter statement does not involve any random variables; after all, in the classical approach, θ is a constant. Instead, the random entity in the phrase "the true parameter lies in the confidence interval" is the confidence interval, not the true parameter.

For a concrete interpretation, suppose that θ is fixed. We construct a confidence interval many times, using the same statistical procedure, i.e., each

† We are using here the important fact that the sum of independent normal random variables is normal (see Chapter 4).

time, we obtain an independent collection of n observations and construct the corresponding 95% confidence interval. We then expect that about 95% of these confidence intervals will include θ. This should be true regardless of what the value of θ is.

Confidence Intervals

- A **confidence interval** for a scalar unknown parameter θ is an interval whose endpoints $\hat{\Theta}_n^-$ and $\hat{\Theta}_n^+$ bracket θ with a given high probability.
- $\hat{\Theta}_n^-$ and $\hat{\Theta}_n^+$ are random variables that depend on the observations $X_1, \ldots, X_n$.
- A $1-\alpha$ confidence interval is one that satisfies

$$\mathbf{P}_\theta\big(\hat{\Theta}_n^- \leq \theta \leq \hat{\Theta}_n^+\big) \geq 1-\alpha,$$

for all possible values of θ.

Confidence intervals are usually constructed by forming an interval around an estimator $\hat{\Theta}_n$ (cf. Example 9.6). Furthermore, out of a variety of possible confidence intervals, one with the smallest possible width is usually desirable. However, this construction is sometimes hard because the distribution of the error $\hat{\Theta}_n - \theta$ is either unknown or depends on θ. Fortunately, for many important models, $\hat{\Theta}_n - \theta$ is asymptotically normal and asymptotically unbiased. By this we mean that the CDF of the random variable

$$\frac{\hat{\Theta}_n - \theta}{\sqrt{\mathrm{var}_\theta(\hat{\Theta}_n)}}$$

approaches the standard normal CDF as n increases, for every value of θ. We may then proceed exactly as in our earlier normal example (Example 9.6), provided that $\mathrm{var}_\theta(\hat{\Theta}_n)$ is known or can be approximated, as we now discuss.

Confidence Intervals Based on Estimator Variance Approximations

Suppose that the observations X_i are i.i.d. with mean θ and variance v that are unknown. We may estimate θ with the sample mean

$$\hat{\Theta}_n = \frac{X_1 + \cdots + X_n}{n},$$

and estimate v with the unbiased estimator

$$\hat{S}_n^2 = \frac{1}{n-1}\sum_{i=1}^n (X_i - \hat{\Theta}_n)^2$$

that we introduced earlier. In particular, we may estimate the variance v/n of the sample mean by $\hat{S}_n^2/n$. Then, for a given α, we may use these estimates and the central limit theorem to construct an (approximate) $1-\alpha$ confidence interval. This is the interval

$$\left[\hat{\Theta}_n - z\,\frac{\hat{S}_n}{\sqrt{n}},\ \hat{\Theta}_n + z\,\frac{\hat{S}_n}{\sqrt{n}}\right],$$

where z is obtained from the relation

$$\Phi(z) = 1 - \frac{\alpha}{2}$$

and the normal tables, and where $\hat{S}_n$ is the positive square root of $\hat{S}_n^2$. For instance, if $\alpha = 0.05$, we use the fact $\Phi(1.96) = 0.975 = 1 - \alpha/2$ (from the normal tables) and obtain an approximate 95% confidence interval of the form

$$\left[\hat{\Theta}_n - 1.96\,\frac{\hat{S}_n}{\sqrt{n}},\ \hat{\Theta}_n + 1.96\,\frac{\hat{S}_n}{\sqrt{n}}\right].$$

Note that in this approach, there are two different approximations in effect. First, we are treating $\hat{\Theta}_n$ as if it were a normal random variable; second, we are replacing the true variance v/n of $\hat{\Theta}_n$ by its estimate $\hat{S}_n^2/n$.

Even in the special case where the X_i are normal random variables, the confidence interval produced by the preceding procedure is still approximate. The reason is that $\hat{S}_n^2$ is only an approximation to the true variance v, and the random variable

$$T_n = \frac{\sqrt{n}\,(\hat{\Theta}_n - \theta)}{\hat{S}_n}$$

is not normal. However, for normal X_i, it can be shown that the PDF of T_n does not depend on θ and v, and can be computed explicitly. It is called the ***t*-distribution with $n-1$ degrees of freedom.**† Like the standard normal PDF, it is symmetric and bell-shaped, but it is a little more spread out and has heavier tails (see Fig. 9.3). The probabilities of various intervals of interest are available in tables, similar to the normal tables. Thus, when the X_i are normal (or nearly normal) and n is relatively small, a more accurate confidence interval is of the form

$$\left[\hat{\Theta}_n - z\,\frac{\hat{S}_n}{\sqrt{n}},\ \hat{\Theta}_n + z\,\frac{\hat{S}_n}{\sqrt{n}}\right],$$

† The t-distribution has interesting properties and can be expressed in closed form, but the precise formula is not important for our purposes. Sometimes it is called the "Student's distribution." It was published in 1908 by William Gosset while he was employed at a brewery in Dublin. He wrote his paper under the pseudonym Student because he was prohibited from publishing under his own name. Gosset was concerned with the selection of the best yielding varieties of barley and had to work with small sample sizes.

where z is obtained from the relation

$$\Psi_{n-1}(z) = 1 - \frac{\alpha}{2},$$

and $\Psi_{n-1}(z)$ is the CDF of the t-distribution with $n-1$ degrees of freedom, available in tables. These tables may be found in many sources. An abbreviated version is given in the opposite page.

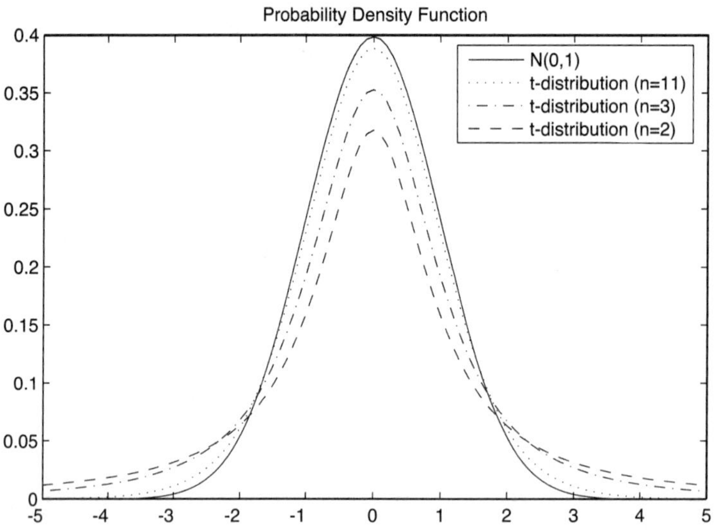

Figure 9.3. The PDF of the t-distribution with $n-1$ degrees of freedom in comparison with the standard normal PDF.

On the other hand, when n is moderately large (e.g., $n \geq 50$), the t-distribution is very close to the normal distribution, and the normal tables can be used.

Example 9.7. The weight of an object is measured eight times using an electronic scale that reports the true weight plus a random error that is normally distributed with zero mean and unknown variance. Assume that the errors in the observations are independent. The following results are obtained:

$$0.5547, \quad 0.5404, \quad 0.6364, \quad 0.6438, \quad 0.4917, \quad 0.5674, \quad 0.5564, \quad 0.6066.$$

We compute a 95% confidence interval ($\alpha = 0.05$) using the t-distribution. The value of the sample mean $\hat{\Theta}_n$ is 0.5747, and the estimated variance of $\hat{\Theta}_n$ is

$$\frac{\hat{S}_n^2}{n} = \frac{1}{n(n-1)} \sum_{i=1}^{n} (X_i - \hat{\Theta}_n)^2 = 3.2952 \cdot 10^{-4},$$

	0.100	0.050	0.025	0.010	0.005	0.001
1	3.078	6.314	12.71	31.82	63.66	318.3
2	1.886	2.920	4.303	6.965	9.925	22.33
3	1.638	2.353	3.182	4.541	5.841	10.21
4	1.533	2.132	2.776	3.747	4.604	7.173
5	1.476	2.015	2.571	3.365	4.032	5.893
6	1.440	1.943	2.447	3.143	3.707	5.208
7	1.415	1.895	2.365	2.998	3.499	4.785
8	1.397	1.860	2.306	2.896	3.355	4.501
9	1.383	1.833	2.262	2.821	3.250	4.297
10	1.372	1.812	2.228	2.764	3.169	4.144
11	1.363	1.796	2.201	2.718	3.106	4.025
12	1.356	1.782	2.179	2.681	3.055	3.930
13	1.350	1.771	2.160	2.650	3.012	3.852
14	1.345	1.761	2.145	2.624	2.977	3.787
15	1.341	1.753	2.131	2.602	2.947	3.733
20	1.325	1.725	2.086	2.528	2.845	3.552
30	1.310	1.697	2.042	2.457	2.750	3.385
60	1.296	1.671	2.000	2.390	2.660	3.232
120	1.289	1.658	1.980	2.358	2.617	3.160
∞	1.282	1.645	1.960	2.326	2.576	3.090

The t-tables for the CDF $\Psi_{n-1}(z)$ of the t-distribution with a given number $n-1$ of degrees of freedom. The entries in this table are:

Left column: Number of degrees of freedom $n-1$.

Top row: A desired tail probability β.

Entries under the top row: A value z such that $\Psi_{n-1}(z) = 1-\beta$.

so that $\hat{S}_n/\sqrt{n} = 0.0182$. From the t-distribution tables, we obtain $1-\Psi_7(2.365) = 0.025 = \alpha/2$, so that

$$\mathbf{P}_\theta\left(\frac{|\hat{\Theta}_n - \theta|}{\hat{S}_n/\sqrt{n}} \leq 2.365\right) = 0.95.$$

Thus,

$$\left[\hat{\Theta}_n - 2.365\,\frac{\hat{S}_n}{\sqrt{n}},\ \hat{\Theta}_n + 2.365\,\frac{\hat{S}_n}{\sqrt{n}}\right] = [0.531, 0.618]$$

is a 95% confidence interval. It is interesting to compare it with the confidence interval

$$\left[\hat{\Theta}_n - 1.96\,\frac{\hat{S}_n}{\sqrt{n}},\ \hat{\Theta}_n + 1.96\,\frac{\hat{S}_n}{\sqrt{n}}\right] = [0.539, 0.610]$$

obtained from the normal tables, which is narrower and therefore more optimistic about the precision of the estimate $\hat{\theta} = 0.5747$.

The approximate confidence intervals constructed so far relied on the particular estimator $\hat{S}_n^2$ for the unknown variance v. However, different estimators or approximations of the variance are possible. For example, suppose that the observations $X_1, \ldots, X_n$ are i.i.d. Bernoulli with unknown mean θ, and variance $v = \theta(1-\theta)$. Then, instead of $\hat{S}_n^2$, the variance could be approximated by $\hat{\Theta}_n(1-\hat{\Theta}_n)$. Indeed, as n increases, $\hat{\Theta}_n$ converges to θ, in probability, from which it can be shown that $\hat{\Theta}_n(1-\hat{\Theta}_n)$ converges to v. Another possibility is to just observe that $\theta(1-\theta) \leq 1/4$ for all $\theta \in [0,1]$, and use $1/4$ as a conservative estimate of the variance. The following example illustrates these alternatives.

Example 9.8. Polling. Consider the polling problem of Section 5.4 (Example 5.11), where we wish to estimate the fraction θ of voters who support a particular candidate for office. We collect n independent sample voter responses $X_1, \ldots, X_n$, where X_i is viewed as a Bernoulli random variable, with $X_i = 1$ if the ith voter supports the candidate. We estimate θ with the sample mean $\hat{\Theta}_n$, and construct a confidence interval based on a normal approximation and different ways of estimating or approximating the unknown variance. For concreteness, suppose that 684 out of a sample of $n = 1200$ voters support the candidate, so that $\hat{\Theta}_n = 684/1200 = 0.57$.

(a) If we use the variance estimate

$$\begin{aligned}\hat{S}_n^2 &= \frac{1}{n-1}\sum_{i=1}^{n}(X_i - \hat{\Theta}_n)^2 \\ &= \frac{1}{1199}\left(684 \cdot \left(1 - \frac{684}{1200}\right)^2 + (1200-684)\cdot\left(0 - \frac{684}{1200}\right)^2\right) \\ &\approx 0.245,\end{aligned}$$

and treat $\hat{\Theta}_n$ as a normal random variable with mean θ and variance 0.245, we obtain the 95% confidence interval

$$\begin{aligned}\left[\hat{\Theta}_n - 1.96\,\frac{\hat{S}_n}{\sqrt{n}},\ \hat{\Theta}_n + 1.96\,\frac{\hat{S}_n}{\sqrt{n}}\right] &= \left[0.57 - \frac{1.96\cdot\sqrt{0.245}}{\sqrt{1200}},\ 0.57 + \frac{1.96\cdot\sqrt{0.245}}{\sqrt{1200}}\right] \\ &= [0.542,\ 0.598].\end{aligned}$$

(b) The variance estimate

$$\hat{\Theta}_n(1-\hat{\Theta}_n) = \frac{684}{1200}\cdot\left(1 - \frac{684}{1200}\right) = 0.245$$

is the same as the previous one (up to three decimal place accuracy), and the resulting 95% confidence interval

$$\left[\hat{\Theta}_n - 1.96\,\frac{\sqrt{\hat{\Theta}_n(1-\hat{\Theta}_n)}}{\sqrt{n}},\ \hat{\Theta}_n + 1.96\,\frac{\sqrt{\hat{\Theta}_n(1-\hat{\Theta}_n)}}{\sqrt{n}}\right]$$

is again $[0.542,\ 0.598]$.

(c) The conservative upper bound of 1/4 for the variance results in the confidence interval

$$\left[\hat{\Theta}_n - 1.96\,\frac{1/2}{\sqrt{n}},\ \hat{\Theta}_n + 1.96\,\frac{1/2}{\sqrt{n}}\right] = \left[0.57 - \frac{1.96\cdot(1/2)}{\sqrt{1200}},\ 0.57 + \frac{1.96\cdot(1/2)}{\sqrt{1200}}\right]$$
$$= [0.542,\ 0.599],$$

which is only slightly wider, but practically the same as before.

Figure 9.4 illustrates the confidence intervals obtained using methods (b) and (c), for a fixed value $\hat{\Theta}_n = 0.57$ and a range of sample sizes from $n = 10$ to $n = 10{,}000$. We see that when n is in the hundreds, as is typically the case in voter polling, the difference is slight. On the other hand, for small values of n, the different approaches result in fairly different confidence intervals, and therefore some care is required.

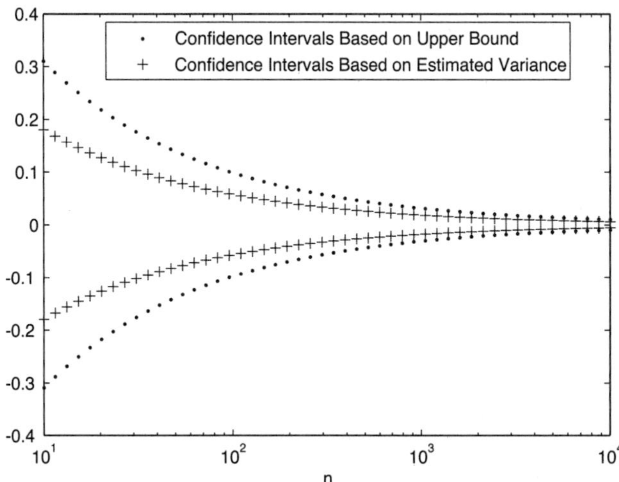

Figure 9.4. The distance of the confidence interval endpoints from $\hat{\Theta}_n$ for methods (b) and (c) of approximating the variance in the polling Example 9.8, when $\hat{\Theta}_n = 0.57$ and for a range of sample sizes from $n = 10$ to $n = 10{,}000$.

9.2 LINEAR REGRESSION

In this section, we develop the linear regression methodology for building a model of the relation between two or more variables of interest on the basis of available data. An interesting feature of this methodology is that it may be explained and developed simply as a least squares approximation procedure, without any probabilistic assumptions. Yet, the linear regression formulas may also be interpreted in the context of various probabilistic frameworks, which provide perspective and a mechanism for quantitative analysis.

We first consider the case of only two variables, and then generalize. We wish to model the relation between two variables of interest, x and y (e.g., years of education and income), based on a collection of data pairs (x_i, y_i), $i = 1, \ldots, n$. For example, x_i could be the years of education and y_i the annual income of the ith person in the sample. Often a two-dimensional plot of these samples indicates a systematic, approximately linear relation between x_i and y_i. Then, it is natural to attempt to build a linear model of the form

$$y \approx \theta_0 + \theta_1 x,$$

where θ_0 and θ_1 are unknown parameters to be estimated.

In particular, given some estimates $\hat{\theta}_0$ and $\hat{\theta}_1$ of the resulting parameters, the value y_i corresponding to x_i, as predicted by the model, is

$$\hat{y}_i = \hat{\theta}_0 + \hat{\theta}_1 x_i.$$

Generally, $\hat{y}_i$ will be different from the given value y_i, and the corresponding difference

$$\tilde{y}_i = y_i - \hat{y}_i,$$

is called the ith **residual**. A choice of estimates that results in small residuals is considered to provide a good fit to the data. With this motivation, the linear regression approach chooses the parameter estimates $\hat{\theta}_0$ and $\hat{\theta}_1$ that minimize the sum of the squared residuals,

$$\sum_{i=1}^{n} (y_i - \hat{y}_i)^2 = \sum_{i=1}^{n} (y_i - \theta_0 - \theta_1 x_i)^2,$$

over all θ_1 and θ_2; see Fig. 9.5 for an illustration.

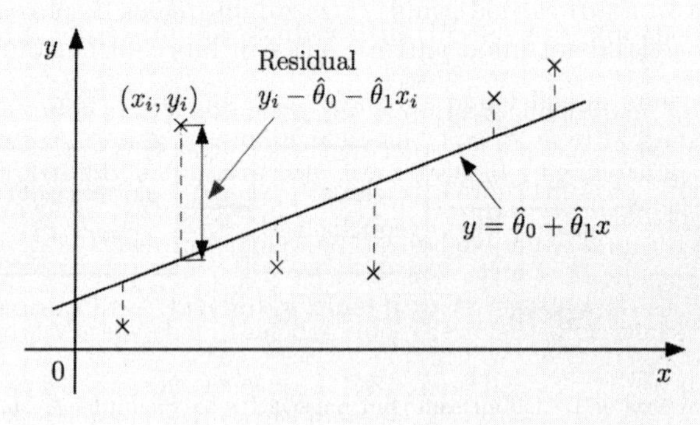

Figure 9.5: Illustration of a set of data pairs (x_i, y_i), and a linear model $y = \hat{\theta}_0 + \hat{\theta}_1 x$, obtained by minimizing over θ_0, θ_1 the sum of the squares of the residuals $y_i - \theta_0 - \theta_1 x_i$.

Note that the postulated linear model may or may not be true. For example, the true relation between the two variables may be nonlinear. The linear least squares approach aims at finding the best possible linear model, and involves an implicit hypothesis that a linear model is valid. In practice, there is often an additional phase where we examine whether the hypothesis of a linear model is supported by the data and try to validate the estimated model.

To derive the formulas for the linear regression estimates $\hat{\theta}_0$ and $\hat{\theta}_1$, we observe that once the data are given, the sum of the squared residuals is a quadratic function of θ_0 and θ_1. To perform the minimization, we set to zero the partial derivatives with respect to θ_0 and θ_1. We obtain two linear equations in θ_0 and θ_1, which can be solved explicitly. After some algebra, we find that the solution has a simple and appealing form, summarized below.

Linear Regression

Given n data pairs (x_i, y_i), the estimates that minimize the sum of the squared residuals are given by

$$\hat{\theta}_1 = \frac{\sum_{i=1}^{n}(x_i - \overline{x})(y_i - \overline{y})}{\sum_{i=1}^{n}(x_i - \overline{x})^2}, \qquad \hat{\theta}_0 = \overline{y} - \hat{\theta}_1 \overline{x},$$

where

$$\overline{x} = \frac{1}{n}\sum_{i=1}^{n} x_i, \qquad \overline{y} = \frac{1}{n}\sum_{i=1}^{n} y_i.$$

Example 9.9. The leaning tower of Pisa continuously tilts over time. Measurements between years 1975 and 1987 of the "lean" of a fixed point on the tower (the distance in meters of the actual position of the point, and its position if the tower were straight) have produced the following table.

Year	1975	1976	1977	1978	1979	1980	1981
Lean	2.9642	2.9644	2.9656	2.9667	2.9673	2.9688	2.9696
Year	1982	1983	1984	1985	1986	1987	
Lean	2.9698	2.9713	2.9717	2.9725	2.9742	2.9757	

Let us use linear regression to estimate the parameters θ_0 and θ_1 in a model of the form $y = \theta_0 + \theta_1 x$, where x is the year and y is the lean. Using the regression formulas, we obtain

$$\hat{\theta}_1 = \frac{\sum_{i=1}^{n}(x_i - \overline{x})(y_i - \overline{y})}{\sum_{i=1}^{n}(x_i - \overline{x})^2} = 0.0009, \qquad \hat{\theta}_0 = \overline{y} - \hat{\theta}_1\overline{x} = 1.1233,$$

where

$$\overline{x} = \frac{1}{n}\sum_{i=1}^{n} x_i = 1981, \qquad \overline{y} = \frac{1}{n}\sum_{i=1}^{n} y_i = 2.9694.$$

The estimated linear model is

$$y = 0.0009x + 1.1233,$$

and is illustrated in Figure 9.6.

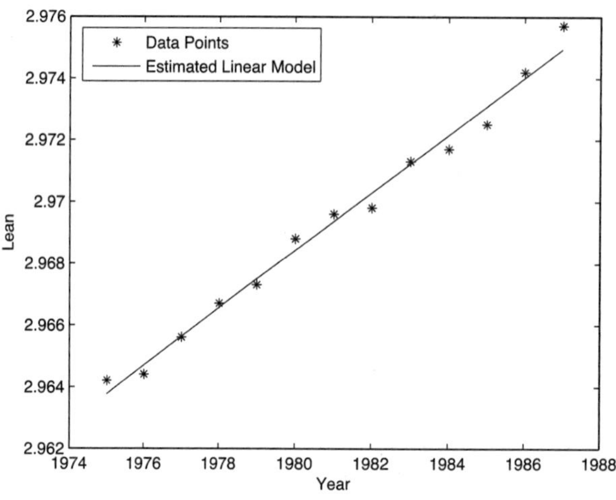

Figure 9.6: The data and the estimated linear model for the lean of the tower of Pisa (Example 9.9).

Justification of the Least Squares Formulation†

The least squares formulation can be justified on the basis of probabilistic considerations in several different ways, based on different sets of assumptions.

(a) **Maximum likelihood (linear model, normal noise).** We assume that the x_i are given numbers (not random variables). We assume that y_i is the

† This subsection can be skipped without loss of continuity.

realization of a random variable Y_i, generated according to a model of the form

$$Y_i = \theta_0 + \theta_1 x_i + W_i, \qquad i = 1, \ldots, n,$$

where the W_i are i.i.d. normal random variables with mean zero and variance σ^2. It follows that the Y_i are independent normal variables, where Y_i has mean $\theta_0 + \theta_1 x_i$ and variance σ^2. The likelihood function takes the form

$$f_Y(y; \theta) = \prod_{i=1}^{n} \frac{1}{\sqrt{2\pi}\sigma} \exp\left\{ -\frac{(y_i - \theta_0 - \theta_1 x_i)^2}{2\sigma^2} \right\}.$$

Maximizing the likelihood function is the same as maximizing the exponent in the above expression, which is equivalent to minimizing the sum of the squared residuals. Thus, the linear regression estimates can be viewed as ML estimates within a suitable linear/normal context. In fact they can be shown to be unbiased estimates in this context. Furthermore, the variances of the estimates can be calculated using convenient formulas (see the end-of-chapter problems), and then used to construct confidence intervals using the methodology of Section 9.1.

(b) **Approximate Bayesian linear LMS estimation (under a possibly nonlinear model).** Suppose now that both x_i and y_i are realizations of random variables X_i and Y_i. The different pairs (X_i, Y_i) are i.i.d., but with unknown joint distribution. Consider an additional independent pair (X_0, Y_0), with the same joint distribution. Suppose we observe X_0 and wish to estimate Y_0 using a linear estimator of the form $\hat{Y}_0 = \theta_0 + \theta_1 X_0$. We know from Section 8.4 that the linear LMS estimator of Y_0, given X_0, is of the form

$$\mathbf{E}[Y_0] + \frac{\text{cov}(X_0, Y_0)}{\text{var}(X_0)}\big(X_0 - \mathbf{E}[X_0]\big),$$

yielding

$$\theta_1 = \frac{\text{cov}(X_0, Y_0)}{\text{var}(X_0)}, \qquad \theta_0 = \mathbf{E}[Y_0] - \theta_1 \mathbf{E}[X_0].$$

Since we do not know the distribution of (X_0, Y_0), we use $\overline{x}$ as an estimate of $\mathbf{E}[X_0]$, $\overline{y}$ as an estimate of $\mathbf{E}[Y_0]$, $\sum_{i=1}^{n}(x_i - \overline{x})(y_i - \overline{y})/n$ as an estimate of $\text{cov}(X_0, Y_0)$, and $\sum_{i=1}^{n}(x_i - \overline{x})^2/n$ as an estimate of $\text{var}(X_0)$. By substituting these estimates into the above formulas for θ_0 and θ_1, we recover the expressions for the linear regression parameter estimates given earlier. Note that this argument does not assume that a linear model is valid.

(c) **Approximate Bayesian LMS estimation (linear model).** Let the pairs (X_i, Y_i) be random and i.i.d. as in (b) above. Let us also make the additional assumption that the pairs satisfy a linear model of the form

$$Y_i = \theta_0 + \theta_1 X_i + W_i,$$

where the W_i are i.i.d., zero mean noise terms, independent of the X_i. From the least mean squares property of conditional expectations, we know that

$\mathbf{E}[Y_0 \mid X_0]$ minimizes the mean squared estimation error $\mathbf{E}\big[\big(Y_0 - g(X_0)\big)^2\big]$, over all functions g. Under our assumptions, $\mathbf{E}[Y_0 \mid X_0] = \theta_0 + \theta_1 X_0$. Thus, the true parameters θ_0 and θ_1 minimize

$$\mathbf{E}\big[(Y_0 - \theta_0' - \theta_1' X_0)^2\big],$$

over all θ_0' and θ_1'. By the weak law of large numbers, this expression is the limit as $n \to \infty$ of

$$\frac{1}{n}\sum_{i=1}^{n}(Y_i - \theta_0' - \theta_1' X_i)^2.$$

This indicates that we will obtain a good approximation of the minimizers of $\mathbf{E}\big[(Y_0 - \theta_0' - \theta_1' X_0)^2\big]$ (the true parameters), by minimizing the above expression (with X_i and Y_i replaced by their observed values x_i and y_i, respectively). But minimizing this expression is the same as minimizing the sum of the squared residuals.

Bayesian Linear Regression†

Linear models and regression are not exclusively tied to classical inference methods. They can also be studied within a Bayesian framework, as we now explain. In particular, we may model $x_1, \ldots, x_n$ as given numbers, and $y_1, \ldots, y_n$ as the observed values of a vector $Y = (Y_1, \ldots, Y_n)$ of random variables that obey a linear relation

$$Y_i = \Theta_0 + \Theta_1 x_i + W_i.$$

Here, $\Theta = (\Theta_0, \Theta_1)$ is the parameter to be estimated, and $W_1, \ldots, W_n$ are i.i.d. random variables with mean zero and known variance σ^2. Consistent with the Bayesian philosophy, we model Θ_0 and Θ_1 as random variables. We assume that $\Theta_0, \Theta_1, W_1, \ldots, W_n$ are independent, and that Θ_0, Θ_1 have mean zero and variances σ_0^2, σ_1^2, respectively.

We may now derive a Bayesian estimator based on the MAP approach and the assumption that Θ_0, Θ_1, and $W_1, \ldots, W_n$ are normal random variables. We maximize over θ_0, θ_1 the posterior PDF $f_{\Theta|Y}(\theta_0, \theta_1 \mid y_1, \ldots, y_n)$. By Bayes' rule, the posterior PDF is‡

$$f_\Theta(\theta_0, \theta_1) f_{Y|\Theta}(y_1, \ldots, y_n \mid \theta_0, \theta_1),$$

divided by a positive normalizing constant that does not depend on (θ_0, θ_1). Under our normality assumptions, this expression can be written as

$$c \cdot \exp\left\{-\frac{\theta_0^2}{2\sigma_0^2}\right\} \cdot \exp\left\{-\frac{\theta_1^2}{2\sigma_1^2}\right\} \cdot \prod_{i=1}^{n} \exp\left\{-\frac{(y_i - \theta_0 - x_i\theta_i)^2}{2\sigma^2}\right\},$$

† This subsection can be skipped without loss of continuity.

‡ Note that in this paragraph, we use conditional probability notation since we are dealing with a Bayesian framework.

where c is a normalizing constant that does not depend on (θ_0, θ_1). Equivalently, we minimize over θ_0 and θ_1 the expression

$$\frac{\theta_0^2}{2\sigma_0^2} + \frac{\theta_1^2}{2\sigma_1^2} + \sum_{i=1}^{n} \frac{(y_i - \theta_0 - x_i\theta_1)^2}{2\sigma^2}.$$

Note the similarity with the expression $\sum_{i=1}^{n}(y_i - \theta_0 - x_i\theta_1)^2$, which is minimized in the earlier classical linear regression formulation. (The two minimizations would be identical if σ_0 and σ_1 were so large that the terms $\theta_0^2/2\sigma_0^2$ and $\theta_1^2/2\sigma_1^2$ could be neglected.) The minimization is carried out by setting to zero the partial derivatives with respect to θ_0 and θ_1. After some algebra, we obtain the following solution.

Bayesian Linear Regression

- **Model:**
  (a) We assume a linear relation $Y_i = \Theta_0 + \Theta_1 x_i + W_i$.
  (b) The x_i are modeled as known constants.
  (c) The random variables $\Theta_0, \Theta_1, W_1, \ldots, W_n$ are normal and independent.
  (d) The random variables Θ_0 and Θ_1 have mean zero and variances σ_0^2, σ_1^2, respectively.
  (e) The random variables W_i have mean zero and variance σ^2.
- **Estimation Formulas:**

  Given the data pairs (x_i, y_i), the MAP estimates of Θ_0 and Θ_1 are

$$\hat{\theta}_1 = \frac{\sigma_1^2}{\sigma^2 + \sigma_1^2 \sum_{i=1}^{n}(x_i - \overline{x})^2} \cdot \sum_{i=1}^{n}(x_i - \overline{x})(y_i - \overline{y}),$$

$$\hat{\theta}_0 = \frac{n\sigma_0^2}{\sigma^2 + n\sigma_0^2}(\overline{y} - \hat{\theta}_1\overline{x}),$$

  where

$$\overline{x} = \frac{1}{n}\sum_{i=1}^{n} x_i, \qquad \overline{y} = \frac{1}{n}\sum_{i=1}^{n} y_i.$$

We make a few remarks:

(a) If σ^2 is very large compared to σ_0^2 and σ_1^2, we obtain $\hat{\theta}_0 \approx 0$ and $\hat{\theta}_1 \approx 0$. What is happening here is that the observations are too noisy and are essentially ignored, so that the estimates become the same as the prior means, which we assumed to be zero.

(b) If we let the prior variances σ_0^2 and σ_1^2 increase to infinity, we are indicating the absence of any useful prior information on Θ_0 and Θ_1. In this case, the MAP estimates become independent of σ^2, and they agree with the classical linear regression formulas that we derived earlier.

(c) Suppose, for simplicity, that $\overline{x} = 0$. When estimating Θ_1, the values y_i of the observations Y_i are weighted in proportion to the associated values x_i. This is intuitive: when x_i is large, the contribution of $\Theta_1 x_i$ to Y_i is relatively large, and therefore Y_i contains useful information on Θ_1. Conversely, if x_i is zero, the observation Y_i is independent of Θ_1 and can be ignored.

(d) The estimates $\hat{\theta}_0$ and $\hat{\theta}_1$ are linear functions of the y_i, but not of the x_i. Recall, however, that the x_i are treated as exogenous, non-random quantities, whereas the y_i are observed values of the random variables Y_i. Thus the MAP estimators $\hat{\Theta}_0$, $\hat{\Theta}_1$ are linear estimators, in the sense defined in Section 8.4. It follows, in view of our normality assumptions, that the estimators are also Bayesian linear LMS estimators as well as LMS estimators (cf. the discussion near the end of Section 8.4).

Multiple Linear Regression

Our discussion of linear regression so far involved a single **explanatory variable**, namely x, a special case known as **simple regression**. The objective was to build a model that explains the observed values y_i on the basis of the values x_i. Many phenomena, however, involve multiple underlying or explanatory variables. (For example, we may consider a model that tries to explain annual income as a function of both age and years of education.) Models of this type are called **multiple regression** models.

For instance, suppose that our data consist of triples of the form (x_i, y_i, z_i) and that we wish to estimate the parameters θ_j of a model of the form

$$y \approx \theta_0 + \theta_1 x + \theta_2 z.$$

As an example, y_i may be the income, x_i the age, and z_i the years of education of the ith person in a random sample. We then seek to minimize the sum of the squared residuals

$$\sum_{i=1}^{n} (y_i - \theta_0 - \theta_1 x_i - \theta_2 z_i)^2,$$

over all θ_0, θ_1, and θ_2. More generally, there is no limit on the number of explanatory variables to be employed. The calculation of the regression estimates

$\hat{\theta}_i$ is conceptually the same as for the case of a single explanatory variable, but of course the formulas are more complicated.

As a special case, suppose that $z_i = x_i^2$, in which case we are dealing with a model of the form

$$y \approx \theta_0 + \theta_1 x + \theta_2 x^2.$$

Such a model would be appropriate if there are good reasons to expect a quadratic dependence of y_i on x_i. (Of course, higher order polynomial models are also possible.) While such a quadratic dependence is nonlinear, the underlying model is still said to be linear, in the sense that the unknown parameters θ_j are linearly related to the observed random variables Y_i. More generally, we may consider a model of the form

$$y \approx \theta_0 + \sum_{j=1}^{m} \theta_j h_j(x),$$

where the h_j are functions that capture the general form of the anticipated dependence of y on x. We may then obtain parameters $\hat{\theta}_0, \hat{\theta}_1, \ldots, \hat{\theta}_m$ by minimizing over $\theta_0, \theta_1, \ldots, \theta_m$ the expression

$$\sum_{i=1}^{n} \left(y_i - \theta_0 - \sum_{j=1}^{m} \theta_j h_j(x_i) \right)^2.$$

This minimization problem is known to admit closed form as well as efficient numerical solutions.

Nonlinear Regression

There are nonlinear extensions of the linear regression methodology to situations where the assumed model structure is nonlinear in the unknown parameters. In particular, we assume that the variables x and y obey a relation of the form

$$y \approx h(x; \theta),$$

where h is a given function and θ is a parameter to be estimated. We are given data pairs (x_i, y_i), $i = 1, \ldots, n$, and we seek a value of θ that minimizes the sum of the squared residuals

$$\sum_{i=1}^{n} \big(y_i - h(x_i; \theta) \big)^2.$$

Unlike linear regression, this minimization problem does not admit, in general, a closed form solution. However, fairly efficient computational methods are available for solving it in practice. Similar to linear regression, nonlinear least squares can be motivated as ML estimation with a model of the form

$$Y_i = h(x_i; \theta) + W_i,$$

where the W_i are i.i.d. normal random variables with zero mean. To see this, note that the likelihood function takes the form

$$f_Y(y;\theta) = \prod_{i=1}^{n} \frac{1}{\sqrt{2\pi}\sigma} \exp\left\{ -\frac{\big(y_i - h(x_i;\theta)\big)^2}{2\sigma^2} \right\},$$

where σ^2 is the variance of W_i. Maximizing the likelihood function is the same as maximizing the exponent in the above expression, which is equivalent to minimizing the sum of the squared residuals.

Practical Considerations

Regression is used widely in many contexts, from engineering to the social sciences. Yet its application often requires caution. We discuss here some important issues that need to be kept in mind, and the main ways in which regression may fail to produce reliable estimates.

(a) **Heteroskedasticity.** The motivation of linear regression as ML estimation in the presence of normal noise terms W_i contains the assumption that the variance of W_i is the same for all i. Quite often, however, the variance of W_i varies substantially over the data pairs. For example, the variance of W_i may be strongly affected by the value of x_i. (For a concrete example, suppose that x_i is yearly income and y_i is yearly consumption. It is natural to expect that the variance of the consumption of rich individuals is much larger than that of poorer individuals.) In this case, a few noise terms with large variance may end up having an undue influence on the parameter estimates. An appropriate remedy is to consider a weighted least squares criterion of the form $\sum_{i=1}^{n} \alpha_i (y_i - \theta_0 - \theta_1 x_i)^2$, where the weights α_i are smaller for those i for which the variance of W_i is large.

(b) **Nonlinearity.** Often a variable x can explain the values of a variable y, but the effect is nonlinear. As already discussed, a regression model based on data pairs of the form $\big(h(x_i), y_i\big)$ may be more appropriate, with a suitably chosen function h.

(c) **Multicollinearity.** Suppose that we use two explanatory variables x and z in a model that predicts another variable y. If the two variables x and z bear a strong relation, the estimation procedure may be unable to distinguish reliably the relative effects of each explanatory variable. For an extreme example, suppose that the true relation is of the form $y = 2x + 1$ and that the relation $z = 2x$ always holds. Then, the model $y = z+1$ is equally valid, and no estimation procedure can discriminate between these two models.

(d) **Overfitting.** Multiple regression with a large number of explanatory variables and a correspondingly large number of parameters to be estimated runs the danger of producing a model that fits the data well, but is otherwise useless. For example, suppose that a linear model is valid but we choose to fit 10 given data pairs with a polynomial of degree 9. The resulting polynomial model will provide a perfect fit to the data, but will nevertheless be incorrect. As a rule of thumb, there should be at least five times (preferably ten times) more data points than there are parameters to be estimated.

(e) **Causality.** The discovery of a linear relation between two variables x and y should not be mistaken for a discovery of a causal relation. A tight fit may be due to the fact that variable x has a causal effect on y, but may be equally due to a causal effect of y on x. Alternatively, there may be some external effect, described by yet another variable z, that affects both x and y in similar ways. For a concrete example let x_i be the wealth of the first born child and y_i be the wealth of the second born child in the same family. We expect y_i to increase roughly linearly with x_i, but this can be traced on the effect of a common family and background rather than a causal effect of one child on the other.

9.3 BINARY HYPOTHESIS TESTING

In this section, we revisit the problem of choosing between two hypotheses, but unlike the Bayesian formulation of Section 8.2, we will assume no prior probabilities. We may view this as an inference problem where the parameter θ takes just two values, but consistent with historical usage, we will forgo the θ-notation and denote the two hypotheses as H_0 and H_1. In traditional statistical language, hypothesis H_0 is often called the **null** hypothesis and H_1 the **alternative** hypothesis. This indicates that H_0 plays the role of a default model, to be proved or disproved on the basis of available data.

The available observation is a vector $X = (X_1, \ldots, X_n)$ of random variables whose distribution depends on the hypothesis. We will use the notation $\mathbf{P}(X \in A; H_j)$ to denote the probability that the observation X belongs to a set A when hypothesis H_j is true. Note that consistent with the classical inference framework, these are not conditional probabilities, because the true hypothesis is not treated as a random variable. Similarly, we will use notation such as $p_X(x; H_j)$ or $f_X(x; H_j)$ to denote the PMF or PDF, respectively, of the vector X, under hypothesis H_j. We want to find a decision rule that maps the realized values x of the observation to one of the two hypotheses (see Fig. 9.7).

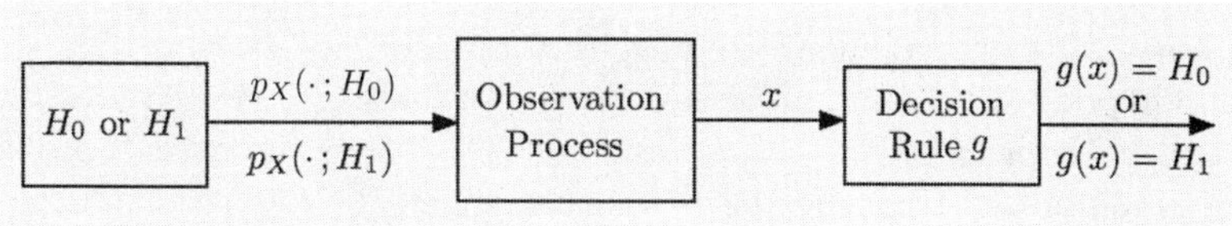

Figure 9.7: Classical inference framework for binary hypothesis testing.

Any decision rule can be represented by a partition of the set of all possible values of the observation vector $X = (X_1, \ldots, X_n)$ into two subsets: a set R, called the **rejection region**, and its complement, R^c, called the **acceptance region**. Hypothesis H_0 is **rejected** (declared to be false) when the observed data $X = (X_1, \ldots, X_n)$ happen to fall in the rejection region R and is **accepted** (declared to be true) otherwise; see Fig. 9.8. Thus, the choice of a decision rule is equivalent to choosing the rejection region.

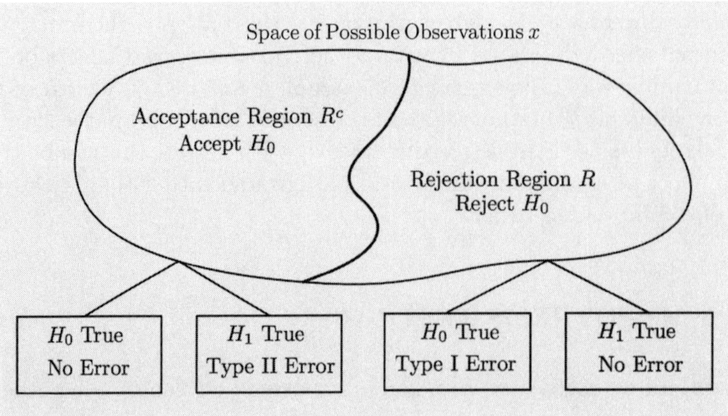

Figure 9.8: Structure of a decision rule for binary hypothesis testing. It is specified by a partition of the set of all possible observations into a set R and its complement R^c. The null hypothesis is rejected if the realized value of the observation falls in the rejection region.

For a particular choice of the rejection region R, there are two possible types of errors:

(a) Reject H_0 even though H_0 is true. This is called a **Type I** error, or a **false rejection**, and happens with probability

$$\alpha(R) = \mathbf{P}(X \in R; H_0).$$

(b) Accept H_0 even though H_0 is false. This is called a **Type II** error, or a **false acceptance**, and happens with probability

$$\beta(R) = \mathbf{P}(X \notin R; H_1).$$

To motivate a particular form of rejection region, we draw an analogy with Bayesian hypothesis testing, involving two hypotheses $\Theta = \theta_0$ and $\Theta = \theta_1$, with respective prior probabilities $p_\Theta(\theta_0)$ and $p_\Theta(\theta_1)$. Then, the overall probability of error is minimized by using the MAP rule: given the observed value x of X, declare $\Theta = \theta_1$ to be true if

$$p_\Theta(\theta_0)p_{X|\Theta}(x \,|\, \theta_0) < p_\Theta(\theta_1)p_{X|\Theta}(x \,|\, \theta_1)$$

(assuming that X is discrete).† This decision rule can be rewritten as follows: define the **likelihood ratio** $L(x)$ by

$$L(x) = \frac{p_{X|\Theta}(x \,|\, \theta_1)}{p_{X|\Theta}(x \,|\, \theta_0)},$$

and declare $\Theta = \theta_1$ to be true if the realized value x of the observation vector X satisfies

$$L(x) > \xi,$$

where the **critical value** ξ is

$$\xi = \frac{p_\Theta(\theta_0)}{p_\Theta(\theta_1)}.$$

If X is continuous, the approach is the same, except that the likelihood ratio is defined as a ratio of PDFs:

$$L(x) = \frac{f_{X|\Theta}(x \,|\, \theta_1)}{f_{X|\Theta}(x \,|\, \theta_0)}.$$

Motivated by the preceding form of the MAP rule, we are led to consider rejection regions of the form

$$R = \{x \,|\, L(x) > \xi\},$$

where the likelihood ratio $L(x)$ is defined similar to the Bayesian case:†

$$L(x) = \frac{p_X(x; H_1)}{p_X(x; H_0)}, \qquad \text{or} \qquad L(x) = \frac{f_X(x; H_1)}{f_X(x; H_0)}.$$

The critical value ξ remains free to be chosen on the basis of other considerations. The special case where $\xi = 1$ corresponds to the ML rule.

Example 9.10. We have a six-sided die that we want to test for fairness, and we formulate two hypotheses for the probabilities of the six faces:

$$H_0 \text{ (fair die):} \qquad p_X(x; H_0) = \frac{1}{6}, \qquad x = 1, \ldots, 6,$$

$$H_1 \text{ (loaded die):} \qquad p_X(x; H_1) = \begin{cases} \dfrac{1}{4}, & \text{if } x = 1, 2, \\[2mm] \dfrac{1}{8}, & \text{if } x = 3, 4, 5, 6. \end{cases}$$

---

† In this paragraph, we use conditional probability notation since we are dealing with a Bayesian framework.

† Note that we use $L(x)$ to denote the value of the likelihood ratio based on the observed value x of the random observation X. On the other hand, before the experiment is carried out, the likelihood ratio is best viewed as a random variable, a function of the observation X, in which case it is denoted by $L(X)$. The probability distribution of $L(X)$ depends, of course, on which hypothesis is true.

The likelihood ratio for a single roll x of the die is

$$L(x) = \begin{cases} \dfrac{1/4}{1/6} = \dfrac{3}{2}, & \text{if } x = 1, 2, \\ \dfrac{1/8}{1/6} = \dfrac{3}{4}, & \text{if } x = 3, 4, 5, 6. \end{cases}$$

Since the likelihood ratio takes only two distinct values, there are three possibilities to consider for the critical value ξ, with three corresponding rejection regions:

$$\begin{aligned} \xi < \frac{3}{4}: &\quad \text{reject } H_0 \text{ for all } x; \\ \frac{3}{4} < \xi < \frac{3}{2}: &\quad \text{accept } H_0 \text{ if } x = 3, 4, 5, 6; \text{ reject } H_0 \text{ if } x = 1, 2; \\ \frac{3}{2} < \xi: &\quad \text{accept } H_0 \text{ for all } x. \end{aligned}$$

Intuitively, a roll of 1 or 2 provides evidence that favors H_1, and we tend to reject H_0. On the other hand, if we set the critical value too high ($\xi > 3/2$), we never reject H_0. In fact, for a single roll of the die, the test makes sense only in the case $3/4 < \xi < 3/2$, since for other values of ξ, the decision does not depend on the observation.

The error probabilities can be calculated from the problem data for each critical value. In particular, the probability of false rejection $\mathbf{P}(\text{Reject } H_0; H_0)$ is

$$\alpha(\xi) = \begin{cases} 1, & \text{if } \xi < \dfrac{3}{4}, \\ \mathbf{P}(X = 1, 2; H_0) = \dfrac{1}{3}, & \text{if } \dfrac{3}{4} < \xi < \dfrac{3}{2}, \\ 0, & \text{if } \dfrac{3}{2} < \xi, \end{cases}$$

and the probability of false acceptance $\mathbf{P}(\text{Accept } H_0; H_1)$ is

$$\beta(\xi) = \begin{cases} 0, & \text{if } \xi < \dfrac{3}{4}, \\ \mathbf{P}(X = 3, 4, 5, 6; H_1) = \dfrac{1}{2}, & \text{if } \dfrac{3}{4} < \xi < \dfrac{3}{2}, \\ 1, & \text{if } \dfrac{3}{2} < \xi. \end{cases}$$

Note that choosing ξ trades off the probabilities of the two types of errors, as illustrated by the preceding example. Indeed, as ξ increases, the rejection region becomes smaller. As a result, the false rejection probability $\alpha(R)$ decreases, while the false acceptance probability $\beta(R)$ increases (see Fig. 9.9). Because of this tradeoff, there is no single best way of choosing the critical value. The most popular approach is as follows.

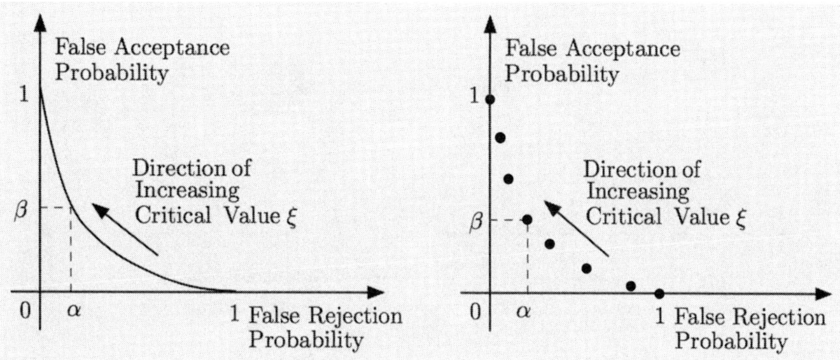

Figure 9.9: Error probabilities in a likelihood ratio test. As the critical value ξ increases, the rejection region becomes smaller. As a result, the false rejection probability α decreases, while the false acceptance probability β increases. When the dependence of α on ξ is continuous and strictly decreasing, there is a unique value of ξ that corresponds to a given α (see the figure on the left). However, the dependence of α on ξ may not be continuous, e.g., if the likelihood ratio $L(x)$ can only take finitely many different values (see the figure on the right).

Likelihood Ratio Test (LRT)

- Start with a target value α for the false rejection probability.
- Choose a value for ξ such that the false rejection probability is equal to α:
$$\mathbf{P}\big(L(X) > \xi; H_0\big) = \alpha.$$
- Once the value x of X is observed, reject H_0 if $L(x) > \xi$.

Typical choices for α are $\alpha = 0.1$, $\alpha = 0.05$, or $\alpha = 0.01$, depending on the degree of undesirability of false rejection. Note that to be able to apply the LRT to a given problem, the following are required:

(a) We must be able to compute $L(x)$ for any given observation value x, so that we can compare it with the critical value ξ. Fortunately, this is always the case when the underlying PMFs or PDFs are given in closed form.

(b) We must either have a closed form expression for the distribution of $L(X)$ [or of a related random variable such as $\log L(X)$] or we must be able to approximate it analytically, computationally, or through simulation. This is needed to determine the critical value ξ that corresponds to a given false rejection probability α.

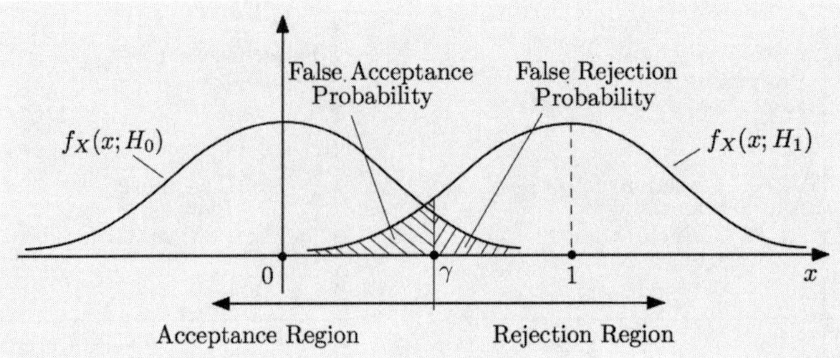

Figure 9.10: Rejection and acceptance regions in Example 9.11, and corresponding false rejection and false acceptance probabilities.

Example 9.11. A surveillance camera periodically checks a certain area and records a signal $X = W$ or $X = 1 + W$ depending on whether an intruder is absent or present (hypotheses H_0 or H_1, respectively). We assume that W is a normal random variable with mean 0 and known variance v. Since

$$f_X(x; H_0) = \frac{1}{\sqrt{2\pi v}} \exp\left\{-\frac{x^2}{2v}\right\}, \qquad f_X(x; H_1) = \frac{1}{\sqrt{2\pi v}} \exp\left\{-\frac{(x-1)^2}{2v}\right\},$$

the likelihood ratio is

$$L(x) = \frac{f_X(x; H_1)}{f_X(x; H_0)} = \exp\left\{\frac{x^2 - (x-1)^2}{2v}\right\} = \exp\left\{\frac{2x-1}{2v}\right\}.$$

For a given critical value ξ, the LRT rejects H_0 if $L(x) > \xi$, or equivalently, after a straightforward calculation, if

$$x > v \log \xi + \frac{1}{2}.$$

Thus, the rejection region is of the form

$$R = \{x \,|\, x > \gamma\}$$

for some γ, which corresponds to ξ via the relation

$$\gamma = v \log \xi + \frac{1}{2};$$

see Fig. 9.10. We set a target value α for the false rejection probability, and we proceed to determine γ from the relation

$$\alpha = \mathbf{P}(X > \gamma; H_0) = \mathbf{P}(W > \gamma),$$

and the normal tables. For example, if $\alpha = 0.025$, then $\gamma = 1.96\sqrt{v}$. We may also calculate the false acceptance probability,

$$\beta = \mathbf{P}\big(X \le \gamma; H_1\big) = \mathbf{P}\left(1 + W \le \gamma\right) = \mathbf{P}\left(W \le \gamma - 1\right),$$

by using again the normal tables.

When $L(X)$ is a continuous random variable, as in the preceding example, the probability $\mathbf{P}\big(L(X) > \xi; H_0\big)$ moves continuously from 1 to 0 as ξ increases. Thus, we can find a value of ξ for which the requirement $\mathbf{P}\big(L(X) > \xi; H_0\big) = \alpha$ is satisfied. If, however, $L(X)$ is a discrete random variable, it may be impossible to satisfy the equality $\mathbf{P}\big(L(X) > \xi; H_0\big) = \alpha$ exactly, no matter how ξ is chosen; cf. Example 9.10. In such cases, there are several possibilities:

(a) Strive for approximate equality.

(b) Choose the smallest value of ξ that satisfies $\mathbf{P}\big(L(X) > \xi; H_0\big) \leq \alpha$.

(b) Use an exogenous source of randomness to choose between two alternative candidate critical values. This variant (known as a "randomized likelihood ratio test") is of some theoretical interest. However, it is not sufficiently important in practice to deserve further discussion in this book.

We have motivated so far the use of a LRT through an analogy with Bayesian inference. However, we will now provide a stronger justification: for a given false rejection probability, the LRT offers the smallest possible false acceptance probability.

Neyman-Pearson Lemma

Consider a particular choice of ξ in the LRT, which results in error probabilities

$$\mathbf{P}\big(L(X) > \xi; H_0\big) = \alpha, \qquad \mathbf{P}\big(L(X) \leq \xi; H_1\big) = \beta.$$

Suppose that some other test, with rejection region R, achieves a smaller or equal false rejection probability:

$$\mathbf{P}(X \in R; H_0) \leq \alpha.$$

Then,

$$\mathbf{P}(X \notin R; H_1) \geq \beta,$$

with strict inequality $\mathbf{P}(X \notin R; H_1) > \beta$ when $\mathbf{P}(X \in R; H_0) < \alpha$.

For a justification of the Neyman-Pearson Lemma, consider a hypothetical Bayesian decision problem where the prior probabilities of H_0 and H_1 satisfy

$$\frac{p_\Theta(\theta_0)}{p_\Theta(\theta_1)} = \xi,$$

so that

$$p_\Theta(\theta_0) = \frac{\xi}{1+\xi}, \qquad p_\Theta(\theta_1) = \frac{1}{1+\xi}.$$

Then, the threshold used by the MAP rule is equal to ξ, as discussed in the beginning of this section, and the MAP rule is identical to the LRT rule. The probability of error with the MAP rule is

$$e_{\text{MAP}} = \frac{\xi}{1+\xi}\alpha + \frac{1}{1+\xi}\beta,$$

and from Section 8.2, we know that it is smaller than or equal to the probability of error of any other Bayesian decision rule. This implies that for any choice of rejection region R, we have

$$e_{\text{MAP}} \le \frac{\xi}{1+\xi}\mathbf{P}(X \in R; H_0) + \frac{1}{1+\xi}\mathbf{P}(X \notin R; H_1).$$

Comparing the preceding two relations, we see that if $\mathbf{P}(X \in R; H_0) \le \alpha$, we must have $\mathbf{P}(X \notin R; H_1) \ge \beta$, and that if $\mathbf{P}(X \in R; H_0) < \alpha$, we must have $\mathbf{P}(X \notin R; H_1) > \beta$, which is the conclusion of the Neyman-Pearson Lemma.

The Neyman-Pearson Lemma can be interpreted geometrically as shown in Fig. 9.11. We illustrate the lemma with a few examples.

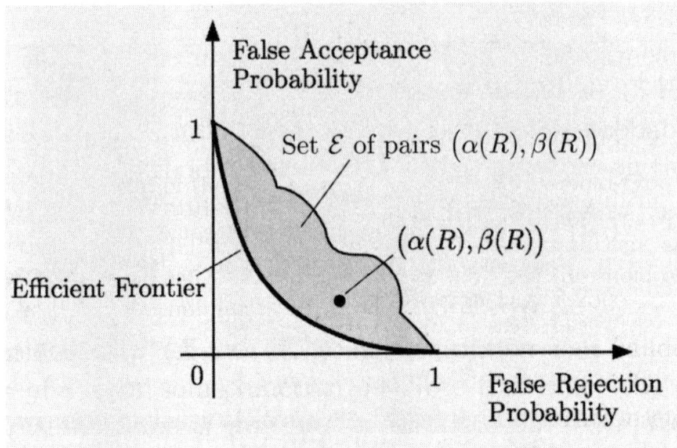

Figure 9.11: Interpretation of the Neyman-Pearson Lemma. Consider the set $\mathcal{E}$ of all error probability pairs $(\alpha(R), \beta(R))$, as R ranges over all possible rejection regions (subsets of the observation space). The **efficient frontier** of $\mathcal{E}$ is the set of all $(\alpha(R), \beta(R)) \in \mathcal{E}$ such that there is no $(\alpha, \beta) \in \mathcal{E}$ with $\alpha \le \alpha(R)$ and $\beta < \beta(R)$, or $\alpha < \alpha(R)$ and $\beta \le \beta(R)$. The Neyman-Pearson Lemma states that all pairs $(\alpha(\xi), \beta(\xi))$ corresponding to LRTs lie on the efficient frontier.

Example 9.12. Consider Example 9.10, where we roll a six-sided die once and test it for fairness. We consider the set $\mathcal{E}$ of all error probability pairs $(\alpha(R), \beta(R))$ as R ranges over all possible rejection regions (all subsets of the observation space $\{1, \ldots, 6\}$). The set $\mathcal{E}$ is shown in Fig. 9.12 and it can be seen that the error probability pairs $(1, 0)$, $(1/3, 1/2)$, and $(0, 1)$ associated with the LRTs have the property given by the Neyman-Pearson Lemma (i.e., lie on the efficient frontier, in the terminology of Fig. 9.11).

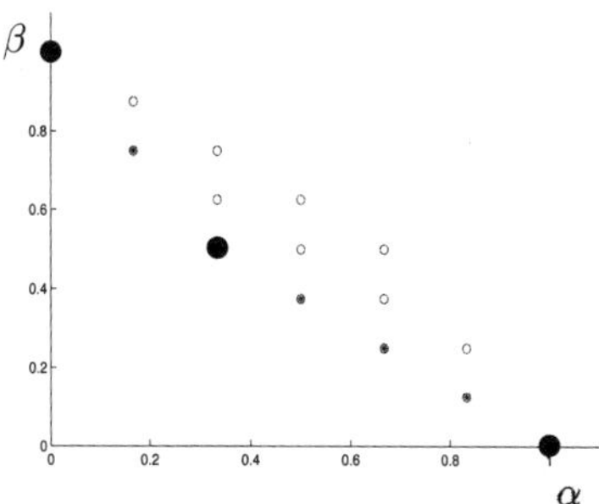

Figure 9.12: Set of pairs $\big(\alpha(R), \beta(R)\big)$ as the rejection region R ranges over all subsets of the observation space $\{1, \ldots, 6\}$ in Examples 9.10 and 9.12. The pairs $(1, 0)$, $(0, 1)$, and $(1/3, 1/2)$ are the ones that correspond to LRTs.

Example 9.13. Comparison of Different Rejection Regions. We observe two i.i.d. normal random variables X_1 and X_2, with unit variance. Under H_0 their common mean is 0; under H_1, their common mean is 2. We fix the false rejection probability to $\alpha = 0.05$.

We first derive the form of the LRT, and then calculate the resulting value of β. The likelihood ratio is of the form

$$L(x) = \frac{\dfrac{1}{\sqrt{2\pi}} \exp\Big\{ - \big((x_1 - 2)^2 + (x_2 - 2)^2\big)/2\Big\}}{\dfrac{1}{\sqrt{2\pi}} \exp\Big\{ - (x_1^2 + x_2^2)/2\Big\}} = \exp\big\{2(x_1 + x_2) - 4\big\}.$$

Comparing $L(x)$ to a critical value ξ is equivalent to comparing $x_1 + x_2$ to $\gamma = (4 + \log \xi)/2$. Thus, under the LRT, we decide in favor of H_1 if $x_1 + x_2 > \gamma$, for some particular choice of γ. This determines the shape of the rejection region.

To determine the exact form of the rejection region, we need to find γ so that the false rejection probability $\mathbf{P}(X_1 + X_2 > \gamma; H_0)$ is equal to 0.05. We note that under H_0, $Z = (X_1 + X_2)/\sqrt{2}$ is a standard normal random variable. We have

$$0.05 = \mathbf{P}(X_1 + X_2 > \gamma; H_0) = \mathbf{P}\left(\frac{X_1 + X_2}{\sqrt{2}} > \frac{\gamma}{\sqrt{2}}; H_0\right) = \mathbf{P}\left(Z > \frac{\gamma}{\sqrt{2}}\right).$$

From the normal tables, we obtain $\mathbf{P}(Z > 1.645) = 0.05$, so we choose

$$\gamma = 1.645 \cdot \sqrt{2} = 2.33,$$

resulting in the rejection region

$$R = \big\{(x_1, x_2) \,|\, x_1 + x_2 > 2.33\big\}.$$

To evaluate the performance of this test, we calculate the resulting false acceptance probability. Note that under H_1, $X_1 + X_2$ is normal with mean equal to 4 and variance equal to 2, so that $Z = (X_1 + X_2 - 4)/\sqrt{2}$ is a standard normal random variable. Thus, using the normal tables, the false acceptance probability is given by

$$\begin{aligned}\beta(R) &= \mathbf{P}(X_1 + X_2 \leq 2.33; H_1)\\ &= \mathbf{P}\left(\frac{X_1 + X_2 - 4}{\sqrt{2}} \leq \frac{2.33 - 4}{\sqrt{2}}; H_1\right)\\ &= \mathbf{P}(Z \leq -1.18)\\ &= \mathbf{P}(Z \geq 1.18)\\ &= 1 - \mathbf{P}(Z \leq 1.18)\\ &= 1 - 0.88\\ &= 0.12.\end{aligned}$$

We now compare the performance of the LRT with that resulting from a different rejection region R'. For example, let us consider a rejection region of the form

$$R' = \big\{(x_1, x_2) \mid \max\{x_1, x_2\} > \zeta\big\},$$

where ζ is chosen so that the false rejection probability is again 0.05. To determine the value of ζ, we write

$$\begin{aligned}0.05 &= \mathbf{P}\big(\max\{X_1, X_2\} > \zeta; H_0\big)\\ &= 1 - \mathbf{P}\big(\max\{X_1, X_2\} \leq \zeta; H_0\big)\\ &= 1 - \mathbf{P}(X_1 \leq \zeta; H_0)\,\mathbf{P}(X_2 \leq \zeta; H_0)\\ &= 1 - \big(\mathbf{P}(Z \leq \zeta; H_0)\big)^2,\end{aligned}$$

where Z is a standard normal. This yields $\mathbf{P}(Z \leq \zeta; H_0) = \sqrt{1 - 0.05} \approx 0.975$. Using the normal tables, we conclude that $\zeta = 1.96$.

Let us now calculate the resulting false acceptance probability. Letting Z be again a standard normal, we have

$$\begin{aligned}\beta(R') &= \mathbf{P}\big(\max\{X_1, X_2\} \leq 1.96; H_1\big)\\ &= \big(\mathbf{P}(X_1 \leq 1.96; H_1)\big)^2\\ &= \big(\mathbf{P}(X_1 - 2 \leq -0.04; H_1)\big)^2\\ &= \big(\mathbf{P}(Z \leq -0.04)\big)^2\\ &= (0.49)^2\\ &= 0.24.\end{aligned}$$

We see that the false acceptance probability $\beta(R) = 0.12$ of the LRT is much better than the false acceptance probability $\beta(R') = 0.24$ of the alternative test.

Example 9.14. A Discrete Example. Consider $n = 25$ independent tosses of a coin. Under hypothesis H_0 (respectively, H_1), the probability of a head at each toss is equal to $\theta_0 = 1/2$ (respectively, $\theta_1 = 2/3$). Let X be the number of heads observed. If we set the false rejection probability to 0.1, what is the rejection region associated with the LRT?

We observe that when $X = k$, the likelihood ratio is of the form

$$L(k) = \frac{\binom{n}{k}\theta_1^k(1-\theta_1)^{n-k}}{\binom{n}{k}\theta_0^k(1-\theta_0)^{n-k}} = \left(\frac{\theta_1}{\theta_0}\cdot\frac{1-\theta_0}{1-\theta_1}\right)^k \cdot \left(\frac{1-\theta_1}{1-\theta_0}\right)^n = 2^k\left(\frac{2}{3}\right)^{25}.$$

Note that $L(k)$ is a monotonically increasing function of k. Thus, the rejection condition $L(k) > \xi$ is equivalent to a condition $k > \gamma$, for a suitable value of γ. We conclude that the LRT is of the form

$$\text{reject } H_0 \text{ if } X > \gamma.$$

To guarantee the requirement on the false rejection probability, we need to find the smallest possible value of γ for which $\mathbf{P}(X > \gamma; H_0) \leq 0.1$, or

$$\sum_{i=\gamma+1}^{25}\binom{25}{i}2^{-25} \leq 0.1.$$

By evaluating numerically the right-hand side above for different choices of γ, we find that the required value is $\gamma = 16$.

An alternative method for choosing γ involves an approximation based on the central limit theorem. Under H_0,

$$Z = \frac{X - n\theta_0}{\sqrt{n\theta_0(1-\theta_0)}} = \frac{X - 12.5}{\sqrt{25/4}}$$

is approximately a standard normal random variable. Therefore, we need

$$0.1 = \mathbf{P}(X > \gamma; H_0) = \mathbf{P}\left(\frac{X-12.5}{\sqrt{25/4}} > \frac{\gamma - 12.5}{\sqrt{25/4}}; H_0\right) = \mathbf{P}\left(Z > \frac{2\gamma}{5} - 5\right).$$

From the normal tables, we have $\Phi(1.28) = 0.9$, and therefore, we should choose γ so that $(2\gamma/5) - 5 = 1.28$, or $\gamma = 15.7$. Since X is integer-valued, we find that the LRT should reject H_0 whenever $X > 15$.

9.4 SIGNIFICANCE TESTING

Hypothesis testing problems encountered in realistic settings do not always involve two well-specified alternatives, so the methodology in the preceding section cannot be applied. The purpose of this section is to introduce an approach to this more general class of problems. We caution, however, that a unique or universal methodology is not available, and that there is a significant element of judgment and art that comes into play.

For some motivation, consider problems such as the following:

(i) A coin is tossed repeatedly and independently. Is the coin fair?

(ii) A die is tossed repeatedly and independently. Is the die fair?

(iii) We observe a sequence of i.i.d. normal random variables $X_1, \ldots, X_n$. Are they standard normal?

(iv) Two different drug treatments are delivered to two different groups of patients with the same disease. Is the first treatment more effective than the second?

(v) On the basis of historical data (say, based on the last year), is the daily change of the Dow Jones Industrial Average normally distributed?

(vi) On the basis of several sample pairs (x_i, y_i) of two random variables X and Y, can we determine whether the two random variables are independent?

In all of the above cases, we are dealing with a phenomenon that involves uncertainty, presumably governed by a probabilistic model. We have a default hypothesis, usually called the **null hypothesis**, denoted by H_0, and we wish to determine on the basis of the observations $X = (X_1, \ldots, X_n)$ whether the null hypothesis should be rejected or not.

In order to avoid obscuring the key ideas, we will mostly restrict the scope of our discussion to situations with the following characteristics.

(a) **Parametric models:** We assume that the observations $X_1, \ldots, X_n$ have a distribution governed by a joint PMF (discrete case) or a joint PDF (continuous case), which is completely determined by an unknown parameter θ (scalar or vector), belonging to a given set $\mathcal{M}$ of possible parameters.

(b) **Simple null hypothesis:** The null hypothesis asserts that the true value of θ is equal to a given element θ_0 of $\mathcal{M}$.

(c) **Alternative hypothesis:** The alternative hypothesis, denoted by H_1, is just the statement that H_0 is not true, i.e., that $\theta \neq \theta_0$.

In reference to the motivating examples introduced earlier, notice that examples (i)-(ii) satisfy conditions (a)-(c) above. On the other hand, in examples (iv)-(vi), the null hypothesis is not simple, violating condition (b).

The General Approach

We introduce the general approach through a concrete example. We then summarize and comment on the various steps involved. Finally, we consider a few more examples that conform to the general approach.

Example 9.15. Is My Coin Fair? A coin is tossed independently $n = 1000$ times. Let θ be the unknown probability of heads at each toss. The set of all possible parameters is $\mathcal{M} = [0, 1]$. The null hypothesis H_0 ("the coin is fair") is of the form $\theta = 1/2$. The alternative hypothesis is that $\theta \neq 1/2$.

The observed data is a sequence $X_1, \ldots, X_n$, where X_i equals 1 or 0, depending on whether the ith toss resulted in heads or tails. We choose to address the problem by considering the value of $S = X_1 + \cdots + X_n$, the number of heads observed, and using a decision rule of the form:

$$\text{reject } H_0 \text{ if } \left| S - \frac{n}{2} \right| > \xi,$$

where ξ is a suitable **critical value**, to be determined. We have so far defined the shape of the **rejection region** R (the set of data vectors that lead to rejection of the null hypothesis). We finally choose the critical value ξ so that the probability of false rejection is equal to a given value α:

$$\mathbf{P}(\text{reject } H_0; H_0) = \alpha,$$

Typically, α, called the **significance level**, is a small number; in this example, we use $\alpha = 0.05$.

The discussion so far involved only a sequence of intuitive choices. Some probabilistic calculations are now needed to determine the critical value ξ. Under the null hypothesis, the random variable S is binomial with parameters $n = 1000$ and $p = 1/2$. Using the normal approximation to the binomial and the normal tables, we find that an appropriate choice is $\xi = 31$. If, for example, the observed value of S turns out to be $s = 472$, we have

$$|s - 500| = |472 - 500| = 28 \leq 31$$

and the hypothesis H_0 **is not rejected at the 5% significance level.**

Our use of the language "not rejected" as opposed to "accepted," at the end of the preceding example is deliberate. We do not have any firm grounds to assert that θ equals $1/2$, as opposed to, say, 0.51. We can only assert that the observed value of S does not provide strong evidence against hypothesis H_0.

We can now summarize and generalize the essence of the preceding example, to obtain a generic methodology.

Significance Testing Methodology

A statistical test of a hypothesis "$H_0 : \theta = \theta^*$" is to be performed, based on the observations $X_1, \ldots, X_n$.

- The following steps are carried out before the data are observed.
  - (a) Choose a **statistic** S, that is, a scalar random variable that will summarize the data to be obtained. Mathematically, this involves the choice of a function $h : \Re^n \to \Re$, resulting in the statistic $S = h(X_1 \ldots, X_n)$.

(b) Determine the **shape of the rejection region** by specifying the set of values of S for which H_0 will be rejected as a function of a yet undetermined critical value ξ.

(c) Choose the **significance level,** i.e., the desired probability α of a false rejection of H_0.

(d) Choose the **critical value** ξ so that the probability of false rejection is equal (or approximately equal) to α. At this point, the rejection region is completely determined.

- Once the values $x_1, \ldots, x_n$ of $X_1, \ldots, X_n$ are observed:

  (i) Calculate the value $s = h(x_1, \ldots, x_n)$ of the statistic S.

  (ii) Reject the hypothesis H_0 if s belongs to the rejection region.

Let us add some comments and interpretation for the various elements of the above methodology.

(i) There is no universal method for choosing the "right" statistic S. In some cases, as in Example 9.15, the choice is natural and can also be justified mathematically. In other cases, a meaningful choice of S involves a certain generalization of the likelihood ratio, to be touched upon later in this section. Finally, in many situations, the primary consideration is whether S is simple enough to enable the calculations needed in step (d) of the above methodology.

(ii) The set of values of S under which H_0 is not rejected is usually an interval surrounding the peak of the distribution of S under H_0 (see Fig. 9.13). In the limit of a large sample size n, the central limit theorem often applies to S, and the symmetry of the normal distribution suggests an interval which is symmetric around the mean value of S. Similarly, the symmetry of the rejection region in Example 9.15 is well-motivated by the fact that, under H_0, the distribution of S (binomial with parameter $1/2$) is symmetric around its mean. In other cases, however, nonsymmetric rejection regions are more appropriate. For example, if we are certain that the coin in Example 9.15 satisfies $\theta \geq 1/2$, a one-sided rejection region is natural:

$$\text{reject } H_0 \text{ if } S - \frac{n}{2} > \xi.$$

(iii) Typical choices for the false rejection probability α range between $\alpha = .10$ and $\alpha = 0.01$. Of course, one wishes false rejections to be rare, but in light of the tradeoff discussed in the context of simple binary hypotheses, a smaller value of α makes it more difficult to reject a false hypothesis, i.e., increases the probability of false acceptance.

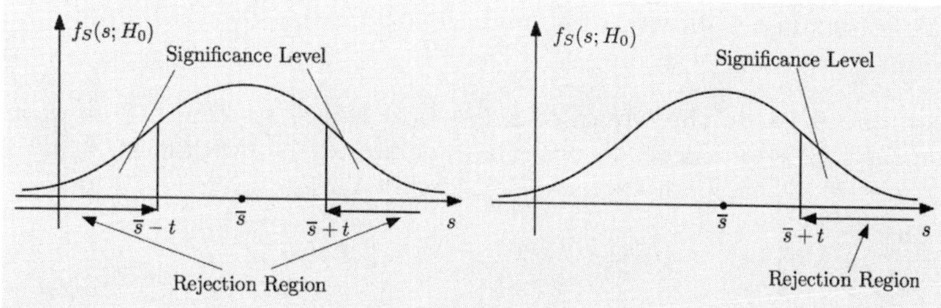

Figure 9.13: Two-sided and one-sided rejection regions for significance testing, based on a statistic S with mean $\overline{s}$ under the null hypothesis. The significance level is the probability of false rejection, i.e., the probability, under H_0, that the statistic S takes a value within the rejection region.

(iv) Step (d) is the only place where probabilistic calculations are used. It requires that the distribution of $L(X)$ [or of a related random variable such as $\log L(X)$] under the hypothesis H_0 be available, possibly approximately. In special cases, this is straightforward or involves an exercise in derived distributions. However, except for relatively simple situations, the distribution of S cannot be found in closed form. If n is large, one can often use well-justified approximations, e.g., based on the central limit theorem. On the other hand, if n is moderate, useful approximations may be difficult to obtain. For this reason, the choice of the statistic S is sometimes guided by the desire to obtain a tractable expression or approximation for the distribution of S. Alternatively, the distribution of S may be estimated by simulation, e.g., by generating many independent samples of X, and by using the resulting samples of $L(X)$ to build a histogram/estimated distribution.

Given the value of α, if the hypothesis H_0 ends up being rejected, one says that H_0 is **rejected at the α significance level.** This statement needs to be interpreted properly. It does not mean that the probability of H_0 being true is less than α. Instead, it means that when this particular methodology is used, we will have false rejections a fraction α of the time. Rejecting a hypothesis at the 1% significance level means that the observed data are highly unusual under the model associated with H_0; such data would arise only 1% of the time, and thus provide strong evidence that H_0 may be false.

Quite often, statisticians skip steps (c) and (d) in the above described methodology. Instead, once they calculate the realized value s of S, they determine and report an associated p-**value** defined by

$$p\text{-value} = \min\{\alpha \,|\, H_0 \text{ would be rejected at the } \alpha \text{ significance level}\}.$$

Equivalently, the p-value is the value of α for which s would be exactly at the threshold between rejection and non-rejection. Thus, for example, the null hypothesis would be rejected at the 5% significance level if and only if the p-value is smaller than 0.05.

A few examples illustrate the main ideas.

Example 9.16. Is the Mean of a Normal Equal to Zero? Here we assume that each X_i is an independent normal random variable, with mean θ and known variance σ^2. The hypotheses under consideration are:

$$H_0: \ \theta = 0, \qquad H_1: \theta \neq 0.$$

A reasonable statistic here is the sample mean $(X_1 + \cdots + X_n)/n$ or its scaled version

$$S = \frac{X_1 + \cdots + X_n}{\sigma \sqrt{n}}.$$

A natural choice for the shape of the rejection region is to reject H_0 if and only if $|S| > \xi$. Because S has a standard normal distribution, the value of ξ corresponding to any particular value of α is easily found from the normal tables. For example, if $\alpha = 0.05$, we use the fact that $\mathbf{P}(S \leq 1.96) = 0.975$ to obtain a rejection region of the form

$$\text{reject } H_0 \text{ if } \ |S| > 1.96,$$

or equivalently,

$$\text{reject } H_0 \text{ if } \ |X_1 + \cdots + X_n| > 1.96\,\sigma\sqrt{n}.$$

In a one-sided version of this problem, the alternative hypothesis is of the form $H_1: \ \theta > 0$. In this case, the same statistic S can be used, but we will reject H_0 if $S > \xi$, where ξ is chosen so that $\mathbf{P}(S > \xi) = \alpha$. Once more, since S has a standard normal distribution, the value of ξ corresponding to any particular value of α is easily found from the normal tables.

Finally, if the variance σ^2 is unknown, we may replace it by an estimate such as

$$\hat{S}_n^2 = \frac{1}{n-1} \sum_{i=1}^{n} \left(X_i - \frac{X_1 + \cdots + X_n}{n} \right)^2.$$

In this case, the resulting statistic has a t-distribution (as opposed to normal). If n is relatively small, the t-tables should be used instead of the normal tables (cf. Section 9.1).

Our next example involves a composite null hypothesis H_0, in the sense that there are multiple parameter choices that are compatible with H_0.

Example 9.17. Are the Means of Two Populations Equal? We want to test whether a certain medication is equally effective for two different population groups. We draw independent samples $X_1, \ldots, X_{n_1}$ and $Y_1, \ldots, Y_{n_2}$ from the two populations, where $X_i = 1$ (or $Y_i = 1$) if the medication is effective for the ith person in the first (respectively, the second) group, and $X_i = 0$ (or $Y_i = 0$) otherwise. We view each X_i (or Y_i) as a Bernoulli random variable with unknown mean θ_X (respectively, θ_Y), and we consider the hypotheses

$$H_0: \ \theta_X = \theta_Y, \qquad H_1: \theta_X \neq \theta_Y.$$

Note that there are multiple pairs (θ_X, θ_Y) that are compatible with H_0, which makes H_0 a composite hypothesis.

The sample means for the two populations are

$$\hat{\Theta}_X = \frac{1}{n_1}\sum_{i=1}^{n_1} X_i, \qquad \hat{\Theta}_Y = \frac{1}{n_2}\sum_{i=1}^{n_2} Y_i.$$

A reasonable estimator of $\theta_X - \theta_Y$ is $\hat{\Theta}_X - \hat{\Theta}_Y$. A plausible choice is to reject H_0 if and only if

$$|\hat{\Theta}_X - \hat{\Theta}_Y| > t,$$

for a suitable threshold t to be determined on the basis of the given false rejection probability α. However, an appropriate choice of t is made difficult by the fact that the distribution of $\hat{\Theta}_X - \hat{\Theta}_Y$ under H_0 depends on the unspecified parameters θ_X and θ_Y. This motivates a somewhat different statistic, as we discuss next.

For large n_1 and n_2, the sample means $\hat{\Theta}_X$ and $\hat{\Theta}_Y$ are approximately normal, and because they are independent, $\hat{\Theta}_X - \hat{\Theta}_Y$ is also approximately normal with mean $\theta_X - \theta_Y$ and variance

$$\text{var}(\hat{\Theta}_X - \hat{\Theta}_Y) = \text{var}(\hat{\Theta}_X) + \text{var}(\hat{\Theta}_Y) = \frac{\theta_X(1-\theta_X)}{n_1} + \frac{\theta_Y(1-\theta_Y)}{n_2}.$$

Under hypothesis H_0, the mean of $\hat{\Theta}_X - \hat{\Theta}_Y$ is known (equal to zero), but its variance is not, because the common value of θ_X and θ_Y is not known. On the other hand, under H_0, the common value of θ_X and θ_Y can be estimated by the overall sample mean

$$\hat{\Theta} = \frac{\sum_{i=1}^{n_1} X_i + \sum_{i=1}^{n_2} Y_i}{n_1 + n_2},$$

the variance $\text{var}(\hat{\Theta}_X - \hat{\Theta}_Y) = \text{var}(\hat{\Theta}_X) + \text{var}(\hat{\Theta}_Y)$ can be approximated by

$$\hat{\sigma}^2 = \left(\frac{1}{n_1} + \frac{1}{n_2}\right)\hat{\Theta}(1 - \hat{\Theta}),$$

and $(\hat{\Theta}_X - \hat{\Theta}_Y)/\hat{\sigma}$ is approximately a standard normal random variable. This leads us to consider a rejection region of the form

$$\text{reject } H_0 \text{ if } \frac{|\hat{\Theta}_X - \hat{\Theta}_Y|}{\hat{\sigma}} > \xi,$$

and to choose ξ so that $\Phi(\xi) = 1 - \alpha/2$, where Φ is the standard normal CDF. For example, if $\alpha = 0.05$, we obtain a rejection region of the form

$$\text{reject } H_0 \text{ if } \frac{|\hat{\Theta}_X - \hat{\Theta}_Y|}{\hat{\sigma}} > 1.96.$$

In a variant of the methodology in this example, we may consider the hypotheses

$$H_0: \ \theta_X = \theta_Y, \qquad H_1: \theta_X > \theta_Y,$$

which would be appropriate if we had reason to exclude the possibility $\theta_X < \theta_Y$. Then, the corresponding rejection region should be one-sided, of the form

$$\text{reject } H_0 \text{ if } \frac{\hat{\Theta}_X - \hat{\Theta}_Y}{\hat{\sigma}} > \xi,$$

where ξ is chosen so that $\Phi(\xi) = 1 - \alpha$.

The preceding example illustrates a generic issue that arises whenever the null hypothesis is composite. In order to be able to set the critical value appropriately, it is preferable to work with a statistic whose approximate distribution is available and is the same for all parameter values compatible with the null hypothesis, as was the case for the statistic $(\hat{\Theta}_X - \hat{\Theta}_Y)/\hat{\sigma}$ in Example 9.17.

Generalized Likelihood Ratio and Goodness of Fit Tests

Our last topic involves testing whether a given PMF conforms with observed data. This an important problem, known as testing for **goodness of fit**. We will also use it as an introduction to a general methodology for significance testing in the face of a composite alternative hypothesis.

Consider a random variable that takes values in the finite set $\{1, \ldots, m\}$, and let θ_k be the probability of outcome k. Thus, the distribution (PMF) of this random variable is described by the vector parameter $\theta = (\theta_1, \ldots, \theta_m)$. We consider the hypotheses

$$H_0 : \theta = (\theta_1^*, \ldots, \theta_m^*), \qquad H_1 : \theta \neq (\theta_1^*, \ldots, \theta_m^*),$$

where the θ_k^* are given nonnegative numbers that sum to 1. We draw n independent samples of the random variable of interest, and let N_k be the number of samples that result in outcome k. Thus, our observation is $X = (N_1, \ldots, N_m)$ and we denote its realized value by $x = (n_1, \ldots, n_m)$. Note that $N_1 + \cdots + N_m = n_1 + \cdots + n_m = n$.

As a concrete example, consider n independent rolls of a die and the hypothesis H_0 that the die is fair. In this case, $\theta_k^* = 1/6$, for $k = 1, \ldots, 6$, and N_k is the number of rolls whose result was equal to k. Note that the alternative hypothesis H_1 is composite, as it is compatible with multiple choices of θ.

The approach that we will follow is known as a **generalized likelihood ratio test** and involves two steps:

(a) Estimate a model by ML, i.e., determine a parameter vector $\hat{\theta} = (\hat{\theta}_1, \ldots, \hat{\theta}_m)$ that maximizes the likelihood function $p_X(x; \theta)$ over all vectors θ.

(b) Carry out a LRT that compares the likelihood $p_X(x; \theta^*)$ under H_0 to the likelihood $p_X(x; \hat{\theta})$ corresponding to the estimated model. More concretely, form the generalized likelihood ratio

$$\frac{p_X(x; \hat{\theta})}{p_X(x; \theta^*)},$$

and if it exceeds a critical value ξ, reject H_0. As in binary hypothesis testing, we choose ξ so that the probability of false rejection is (approximately) equal to a given significance level α.

In essence, this approach asks the following question: is there a model compatible with H_1 that provides a better explanation for the observed data than that provided by the model corresponding to H_0? To answer this question, we compare the likelihood under H_0 to the largest possible likelihood under models compatible with H_1.

The first step (ML estimation) involves a maximization over the set of probability distributions $(\theta_1, \ldots, \theta_m)$. The PMF of the observation vector X is multinomial (cf. Problem 27 in Chapter 2), and the likelihood function is

$$p_X(x;\, \theta) = c\,\theta_1^{n_1} \cdots \theta_m^{n_m},$$

where c is a normalizing constant. It is easier to work with the log-likelihood function, which takes the form

$$\log p_X(x;\, \theta) = \log c + n_1 \log \theta_1 + \cdots + n_{m-1} \log \theta_{m-1} + n_m \log(1 - \theta_1 - \cdots - \theta_{m-1}),$$

where we have also used the fact $\theta_1 + \cdots + \theta_m = 1$ to eliminate θ_m. Assuming that the vector $\hat{\theta}$ that maximizes the log-likelihood has positive components, it can be found by setting the derivatives with respect to $\theta_1, \ldots, \theta_{m-1}$ of the above expression to zero, which yields

$$\frac{n_k}{\hat{\theta}_k} = \frac{n_m}{1 - \hat{\theta}_1 - \cdots - \hat{\theta}_{m-1}}, \qquad \text{for } k = 1, \ldots, m-1.$$

Since the term on the right-hand side is equal to $n_m/\hat{\theta}_m$, we conclude that all ratios $n_k/\hat{\theta}_k$ must be equal. Using also the fact $n_1 + \cdots + n_m = n$, it follows that

$$\hat{\theta}_k = \frac{n_k}{n}, \qquad k = 1, \ldots, m.$$

It can be shown that these are the correct ML estimates even if some of the n_k happen to be zero, in which case the corresponding $\hat{\theta}_k$ are also zero.

The resulting generalized likelihood ratio test is of the form†

$$\text{reject } H_0 \text{ if } \quad \frac{p_X(x;\, \hat{\theta})}{p_X(x;\, \theta^*)} = \prod_{k=1}^{m} \frac{(n_k/n)^{n_k}}{(\theta_k^*)^{n_k}} > \xi,$$

where ξ is the critical value. By taking logarithms, the test reduces to

$$\text{reject } H_0 \text{ if } \quad \sum_{k=1}^{m} n_k \log\left(\frac{n_k}{n\theta_k^*}\right) > \log \xi.$$

† We adopt the convention that $0^0 = 1$ and $0 \cdot \log 0 = 0$

We need to determine ξ by taking into account the required significance level, that is,

$$\mathbf{P}(S > \log \xi;\, H_0) = \alpha,$$

where

$$S = \sum_{k=1}^{m} N_k \log\left(\frac{N_k}{n\theta_k^*}\right).$$

This may be problematic because the distribution of S under H_0 is not readily available and can only be simulated.

Fortunately, major simplifications are possible when n is large. In this case, the observed frequencies $\hat{\theta}_k = n_k/n$ will be close to θ_k^* under H_0, with high probability. Then, a second order Taylor series expansion shows that our statistic S can be approximated well by $T/2$, where T is given by†

$$T = \sum_{k=1}^{m} \frac{(N_k - n\theta_k^*)^2}{n\theta_k^*}.$$

Furthermore, when n is large, it is known that under the hypothesis H_0, the distribution of T (and consequently the distribution of $2S$) approaches a so-called "χ^2 distribution with $m-1$ degrees of freedom."‡ The CDF of this distribution

---

† We note that the second order Taylor series expansion of the function $y\log(y/y^*)$ around any $y^* > 0$ is of the form

$$y\log\left(\frac{y}{y^*}\right) \approx y - y^* + \frac{1}{2}\frac{(y-y^*)^2}{y^*},$$

and is valid when $y/y^* \approx 1$. Thus,

$$\sum_{k=1}^{m} N_k \log\left(\frac{N_k}{n\theta_k^*}\right) \approx \sum_{k=1}^{m}(N_k - n\theta_k^*) + \frac{1}{2}\sum_{k=1}^{m}\frac{(N_k - n\theta_k^*)^2}{n\theta_k^*} = \frac{T}{2}.$$

‡ The χ^2 distribution with ℓ degrees of freedom is defined as the distribution of the random variable

$$\sum_{i=1}^{\ell} Z_i^2,$$

where $Z_1, \ldots, Z_\ell$ are independent standard normal random variables (zero mean and unit variance). Some intuition for why T is approximately χ^2 can be gained from the fact that as $n \to \infty$, N_k/n not only converges to θ_k^* but is also asymptotically normal. Thus, T is equal to the sum of the squares of m zero mean normal random variables, namely $(N_k - n\theta_k^*)/\sqrt{n\theta_k^*}$. The reason that T has $m-1$, instead of m, degrees of freedom is related to the fact that $\sum_{k=1}^{m} N_k = n$, so that these m random variables are actually dependent.

is available in tables, similar to the normal tables. Thus, approximately correct values of $\mathbf{P}(T > \gamma; H_0)$ or $\mathbf{P}(2S > \gamma; H_0)$ can be obtained from the χ^2 tables and can be used to determine a suitable critical value that corresponds to the given significance level α. Putting everything together, we have the following test for large values of n.

The Chi-Square Test:

- Use the statistic
$$S = \sum_{k=1}^{m} N_k \log\left(\frac{N_k}{n\theta_k^*}\right)$$
(or possibly the related statistic T) and a rejection region of the form
$$\text{reject } H_0 \text{ if } 2S > \gamma$$
(or $T > \gamma$, respectively).
- The critical value γ is determined from the CDF tables for the χ^2 distribution with $m - 1$ degrees of freedom so that
$$\mathbf{P}(2S > \gamma; H_0) = \alpha,$$
where α is a given significance level.

Example 9.18. Is My Die Fair? A die is rolled independently 600 times and the number of times that the numbers $1, 2, 3, 4, 5, 6$ come up are

$$n_1 = 92, \quad n_2 = 120, \quad n_3 = 88, \quad n_4 = 98, \quad n_5 = 95, \quad n_6 = 107,$$

respectively. Let us test the hypothesis H_0 that the die is fair by using the chi-square test based on the statistic T, at a level of significance $\alpha = 0.05$. From the tables for the χ^2 with 5 degrees of freedom, we obtain that for $\mathbf{P}(T > \gamma; H_0) = 0.05$ we must have $\gamma = 11.1$.

With $\theta_1^* = \cdots = \theta_6^* = 1/6$, $n = 600$, $n\theta_k^* = 100$, and the given values n_k, the value of the statistic T is

$$\begin{aligned}\sum_{k=1}^{m} \frac{(n_k - n\theta_k^*)^2}{n\theta_k^*} &= \frac{(92-100)^2}{100} + \frac{(120-100)^2}{100} + \frac{(88-100)^2}{100} \\ &\quad + \frac{(98-100)^2}{100} + \frac{(95-100)^2}{100} + \frac{(107-100)^2}{100} \\ &= 6.86.\end{aligned}$$

Since $T = 6.86 < 11.1$, the hypothesis that the die is fair is not rejected. If we use instead the statistic S, then a calculation using the data yields $2S = 6.68$, which is both close to T and also well below the critical value $\gamma = 11.1$. If the level of significance were $\alpha = 0.25$, the corresponding value of γ would be 6.63. In this case, the hypothesis that the die is fair would be rejected since $T = 6.86 > 6.63$ and $2S = 6.68 > 6.63$.

9.5 SUMMARY AND DISCUSSION

Classical inference methods, in contrast with Bayesian methods, treat θ as an unknown constant. Classical parameter estimation aims at estimators with favorable properties such as a small bias and a satisfactory confidence interval, for all possible values of θ. We first focused on ML estimation, which is related to the (Bayesian) MAP method and selects an estimate of θ that maximizes the likelihood function given x. It is a general estimation method and has several desirable characteristics, particularly when the number of observations is large. Then, we discussed the special but practically important case of estimating an unknown mean and constructing confidence intervals. Much of the methodology here relies on the central limit theorem. We finally discussed the linear regression method that aims to match a linear model to the observations in a least squares sense. It requires no probabilistic assumptions for its application, but it is also related to ML and Bayesian LMS estimation under certain conditions.

Classical hypothesis testing methods aim at small error probabilities, combined with simplicity and convenience of calculation. We have focused on tests that reject the null hypothesis when the observations fall within a simple type of rejection region. The likelihood ratio test is the primary approach for the case of two competing simple hypotheses, and derives strong theoretical support from the Neyman-Pearson Lemma. We also addressed significance testing, which applies when one (or both) of the competing hypotheses is composite. The main approach here involves a suitably chosen statistic that summarizes the observations, and a rejection region whose probability under the null hypothesis is set to a desired significance level.

In our brief introduction to statistics, we aimed at illustrating the central concepts and the most common methodologies, but we have barely touched the surface of a very rich subject. For example, we have not discussed important topics such as estimation in time-varying environments (time series analysis, and filtering), nonparametric estimation (e.g., the problem of estimating an unknown PDF on the basis of empirical data), further developments in linear and nonlinear regression (e.g., testing whether the assumptions underlying a regression model are valid), methods for designing statistical experiments, methods for validating the conclusions of a statistical study, computational methods, and many others. Yet, we hope to have kindled the reader's interest in the subject and to have provided some general understanding of the conceptual framework.

PROBLEMS

SECTION 9.1. Classical Parameter Estimation

Problem 1. Alice models the time that she spends each week on homework as an exponentially distributed random variable with unknown parameter θ. Homework times in different weeks are independent. After spending $10, 14, 18, 8$, and 20 hours in the first 5 weeks of the semester, what is her ML estimate of θ?

Problem 2. Consider a sequence of independent coin tosses, and let θ be the probability of heads at each toss.

(a) Fix some k and let N be the number of tosses until the kth head occurs. Find the ML estimator of θ based on N.

(b) Fix some n and let K be the number of heads observed in n tosses. Find the ML estimator of θ based on K.

Problem 3. Sampling and estimation of sums. We have a box with k balls; $\overline{k}$ of them are white and $k - \overline{k}$ are red. Both k and $\overline{k}$ are assumed known. Each white ball has a nonzero number on it, and each red ball has zero on it. We want to calculate the sum of all the ball numbers, but because k is very large, we resort to estimating it by sampling. This problem aims to quantify the advantages of sampling only white balls/nonzero numbers and exploiting the knowledge of $\overline{k}$. In particular, we wish to compare the error variance when we sample n balls with the error variance when we sample a smaller number m of white balls.

(a) Suppose we draw balls sequentially and independently, according to a uniform distribution (with replacement). Denote by X_i the number on the ith ball drawn, and by Y_i the number on the ith white ball drawn. We fix two positive integers n and m, and denote

$$\hat{S} = \frac{k}{n}\sum_{i=1}^{n} X_i, \qquad \overline{S} = \frac{\overline{k}}{\overline{N}}\sum_{i=1}^{n} X_i, \qquad \tilde{S} = \frac{\overline{k}}{m}\sum_{i=1}^{m} Y_i,$$

where $\overline{N}$ is the (random) number of white balls drawn in the first n samples. Show that $\hat{S}$, $\overline{S}$, and $\tilde{S}$ are unbiased estimators of the sum of all the ball numbers.

(b) Calculate the variances of $\tilde{S}$ and $\hat{S}$, and show that in order for them to be approximately equal, we must have

$$m \approx \frac{np}{p + r(1-p)},$$

where $p = \overline{k}/k$ and $r = \mathbf{E}[Y_1^2]/\text{var}(Y_1)$. Show also that when $m = n$,

$$\frac{\text{var}(\tilde{S})}{\text{var}(\hat{S})} = \frac{p}{p + r(1-p)}.$$

(c) Calculate the variance of $\overline{S}$, and show that for large n,

$$\frac{\text{var}(\overline{S})}{\text{var}(\hat{S})} \approx \frac{1}{p + r(1-p)}.$$

Problem 4. Mixture models. Let the PDF of a random variable X be the mixture of m components:

$$f_X(x) = \sum_{j=1}^{m} p_j f_{Y_j}(x),$$

where

$$\sum_{j=1}^{m} p_j = 1, \qquad p_j \geq 0, \quad \text{for } j = 1, \ldots, m.$$

Thus, X can be viewed as being generated by a two-step process: first draw j randomly according to probabilities p_j, then draw randomly according to the distribution of Y_j. Assume that each Y_j is normal with mean μ_j and variance σ_j^2, and that we have a set of i.i.d. observations $X_1, \ldots, X_n$, each with PDF f_X.

(a) Write down the likelihood and log-likelihood functions.

(b) Consider the case $m = 2$ and $n = 1$, and assume that μ_1, μ_2, σ_1, and σ_2 are known. Find the ML estimates of p_1 and p_2.

(c) Consider the case $m = 2$ and $n = 1$, and assume that p_1, p_2, σ_1, and σ_2 are known. Find the ML estimates of μ_1 and μ_2.

(d) Consider the case $m \geq 2$ and general n, and assume that all parameters are unknown. Show that the likelihood function can be made arbitrarily large by choosing $\mu_1 = x_1$ and letting σ_1^2 decrease to zero. *Note:* This is an example where the ML approach is problematic.

Problem 5. Unstable particles are emitted from a source and decay at a distance X, which is exponentially distributed with unknown parameter θ. A special device is used to detect the first n decay events that occur in the interval $[m_1, m_2]$. Suppose that these events are recorded at distances $X = (X_1, \ldots, X_n)$.

(a) Give the form of the likelihood and log-likelihood functions.

(b) Assume that $m_1 = 1$, $m_2 = 20$, $n = 6$, and $x = (1.5, 2, 3, 4, 5, 12)$. Plot the likelihood and log-likelihood as functions of θ. Find approximately the ML estimate of θ based on your plot.

Problem 6. Consider a study of student heights in a middle school. Assume that the height of a female student is normally distributed with mean μ_1 and variance σ_1^2, and that the height of a male student is normally distributed with mean μ_2 and variance σ_2^2. Assume that a student is equally likely to be male or female. A sample of size $n = 10$ was collected, and the following values were recorded (in centimeters):

$$164, \quad 167, \quad 163, \quad 158, \quad 170, \quad 183, \quad 176, \quad 159, \quad 170, \quad 167.$$

(a) Assume that μ_1, μ_2, σ_1, and σ_2 are unknown. Write down the likelihood function.

(b) Assume we know that $\sigma_1^2 = 9$ and $\mu_1 = 164$. Find numerically the ML estimates of σ_2 and μ_2.

(c) Assume we know that $\sigma_1^2 = \sigma_2^2 = 9$. Find numerically the ML estimates of μ_1 and μ_2.

(d) Treating the estimates obtained in part (c) as exact values, describe the MAP rule for deciding a student's gender based on the student's height.

Problem 7. Estimating the parameter of a Poisson random variable. Derive the ML estimator of the parameter of a Poisson random variable based on i.i.d. observations $X_1, \ldots, X_n$. Is the estimator unbiased and consistent?

Problem 8. Estimating the parameter of a uniform random variable I. We are given i.i.d. observations $X_1, \ldots, X_n$ that are uniformly distributed over the interval $[0, \theta]$. What is the ML estimator of θ? Is it consistent? Is it unbiased or asymptotically unbiased? Can you construct alternative estimators that are unbiased?

Problem 9. Estimating the parameter of a uniform random variable II. We are given i.i.d. observations $X_1, \ldots, X_n$ that are uniformly distributed over the interval $[\theta, \theta + 1]$. Find a ML estimator of θ. Is it consistent? Is it unbiased or asymptotically unbiased?

Problem 10. A source emits a random number of photons K each time that it is triggered. We assume that the PMF of K is

$$p_K(k; \theta) = c(\theta)e^{-\theta k}, \qquad k = 0, 1, 2, \ldots,$$

where θ is the inverse of the temperature of the source and $c(\theta)$ is a normalization factor. We also assume that the photon emissions each time that the source is triggered are independent. We want to estimate the temperature of the source by triggering it repeatedly and counting the number of emitted photons.

(a) Determine the normalization factor $c(\theta)$.

(b) Find the expected value and the variance of the number K of photons emitted if the source is triggered once.

(c) Derive the ML estimator for the temperature $\psi = 1/\theta$, based on $K_1, \ldots, K_n$, the numbers of photons emitted when the source is triggered n times.

(d) Show that the ML estimator is consistent.

Problem 11.* Sufficient statistics – factorization criterion. Consider an observation model of the following type. Assuming for simplicity that all random variables are discrete, an initial observation T is generated according to a PMF $p_T(t; \theta)$. Having observed T, an additional observation Y is generated according to a conditional PMF $p_{Y|T}(y \mid t)$ that does not involve the unknown parameter θ. Intuition suggests that out of the overall observation vector $X = (T, Y)$, only T is useful for estimating θ. This problem formalizes this idea.

Given observations $X = (X_1, \ldots, X_n)$, we say that a (scalar or vector) function $q(X)$ is a **sufficient statistic** for the parameter θ if the conditional distribution of X

given the random variable $T = q(X)$ does not depend on θ, i.e., for every event D and possible value t of the random variable T,

$$\mathbf{P}_\theta\big(X \in D \,|\, T = t\big)$$

is the same for all θ for which the above conditional probability is well-defined [i.e., for all θ for which the PMF $p_T(t;\theta)$ or the PDF $f_T(t;\theta)$ is positive]. Assume that either X is discrete (in which case, T is also discrete), or that both X and T are continuous random variables.

(a) Show that $T = q(X)$ is a sufficient statistic for θ if and only if it satisfies the following **factorization criterion**: the likelihood function $p_X(x;\theta)$ (discrete case) or $f_X(x;\theta)$ (continuous case) can be written as $r\big(q(x),\theta\big)s(x)$ for some functions r and s.

(b) Show that if $q(X)$ is a sufficient statistic for θ, then for any function h of θ, $q(X)$ is a sufficient statistic for the parameter $\zeta = h(\theta)$.

(c) Show that if $q(X)$ is a sufficient statistic for θ, a ML estimate of θ can be written as $\hat{\Theta}_n = \phi\big(q(X)\big)$ for some function ϕ. *Note*: This supports the idea that a sufficient statistic captures all essential information about θ provided by X.

Solution. (a) We consider only the discrete case; the proof for the continuous case is similar. Assume that the likelihood function can be written as $r\big(q(x),\theta\big)s(x)$. We will show that $T = q(X)$ is a sufficient statistic.

Fix some t and consider some θ for which $\mathbf{P}_\theta(T = t) > 0$. For any x for which $q(x) \neq t$, we have $\mathbf{P}_\theta(X = x \,|\, T = t) = 0$, which is trivially the same for all θ. Consider now any x for which $q(x) = t$. Using the fact $\mathbf{P}_\theta\big(X = x, T = t\big) = \mathbf{P}_\theta\big(X = x, q(X) = q(x)\big) = \mathbf{P}_\theta(X = x)$, we have

$$\begin{aligned}
\mathbf{P}_\theta(X = x \,|\, T = t) &= \frac{\mathbf{P}_\theta(X = x, T = t)}{\mathbf{P}_\theta(T = t)} = \frac{\mathbf{P}_\theta(X = x)}{\mathbf{P}_\theta(T = t)} \\
&= \frac{r(t,\theta)s(x)}{\displaystyle\sum_{\{z \,|\, q(z)=t\}} r\big(q(z),\theta\big)s(z)} = \frac{r(t,\theta)s(x)}{r(t,\theta)\displaystyle\sum_{\{z \,|\, q(z)=t\}} s(z)} \\
&= \frac{s(x)}{\displaystyle\sum_{\{z \,|\, q(z)=t\}} s(z)},
\end{aligned}$$

so $\mathbf{P}_\theta(X = x \,|\, T = t)$ does not depend on θ. This implies that for any event D, the conditional probability $\mathbf{P}_\theta(X \in D \,|\, T = t)$ is the same for all θ for which $\mathbf{P}_\theta(T = t) > 0$, so T is a sufficient statistic.

Conversely, assume that $T = q(X)$ is a sufficient statistic. For any x with $p_X(x;\theta) > 0$, the likelihood function is

$$p_X(x;\theta) = \mathbf{P}_\theta\big(X = x \,|\, q(X) = q(x)\big)\, \mathbf{P}_\theta\big(q(X) = q(x)\big).$$

Since T is a sufficient statistic, the first term on the right-hand side does not depend on θ, and is of the form $s(x)$. The second term depends on x through $q(x)$, and is of the form $r\bigl(q(x),\theta\bigr)$. This establishes that the likelihood function can be factored as claimed.

(b) This is evident from the definition of a sufficient statistic, since for $\zeta = h(\theta)$, we have

$$\mathbf{P}_\zeta\bigl(X \in D \,|\, T = t\bigr) = \mathbf{P}_\theta\bigl(X \in D \,|\, T = t\bigr),$$

so $\mathbf{P}_\zeta\bigl(X \in D \,|\, T = t\bigr)$ is the same for all ζ.

(c) By part (a), the likelihood function can be factored as $r\bigl(q(x),\theta\bigr)s(x)$. Thus, a ML estimate maximizes $r\bigl(q(x),\theta\bigr)$ over θ [if $s(x) > 0$] or minimizes $r\bigl(q(x),\theta\bigr)$ over θ [if $s(x) < 0$], and therefore depends on x only through $q(x)$.

Problem 12.* Examples of a sufficient statistic I. Show that $q(X) = \sum_{i=1}^n X_i$ is a sufficient statistic in the following cases:

(a) $X_1, \ldots, X_n$ are i.i.d. Bernoulli random variables with parameter θ.

(b) $X_1, \ldots, X_n$ are i.i.d. Poisson random variables with parameter θ.

Solution. (a) The likelihood function is

$$p_X(x;\theta) = \theta^{q(x)}(1-\theta)^{n-q(x)},$$

so it can be factored as the product of the function $\theta^{q(x)}(1-\theta)^{n-q(x)}$, which depends on x only through $q(x)$, and the constant function $s(x) \equiv 1$. The result follows from the factorization criterion for a sufficient statistic.

(b) The likelihood function is

$$p_X(x;\theta) = \prod_{i=1}^n p_{X_i}(x_i) = e^{-\theta} \prod_{i=1}^n \frac{\theta^{x_i}}{x_i!} = e^{-\theta}\theta^{q(x)} \frac{1}{\prod_{i=1}^n x_i!},$$

so it can be factored as the product of the function $e^{-\theta}\theta^{q(x)}$, which depends on x only through $q(x)$, and the function $s(x) = 1/\prod_{i=1}^n x_i!$, which depends only on x. The result follows from the factorization criterion for a sufficient statistic.

Problem 13.* Examples of a sufficient statistic II. Let $X_1, \ldots, X_n$ be i.i.d. normal random variables with mean μ and variance σ^2. Show that:

(a) If σ^2 is known, $q(X) = \sum_{i=1}^n X_i$ is a sufficient statistic for μ.

(b) If μ is known, $q(X) = \sum_{i=1}^n (X_i - \mu)^2$ is a sufficient statistic for σ^2.

(a) If both μ and σ^2 are unknown, $q(X) = \bigl(\sum_{i=1}^n X_i, \sum_{i=1}^n X_i^2\bigr)$ is a sufficient statistic for (μ, σ^2).

Solution. Use the calculations in Example 9.4, and the factorization criterion.

Problem 14.* Rao-Blackwell theorem. This problem shows that a general estimator can be modified into one that only depends on a sufficient statistic, without loss of performance. Given observations $X = (X_1, \ldots, X_n)$, let $T = q(X)$ be a sufficient statistic for the parameter θ, and let $g(X)$ be an estimator for θ.

(a) Show that $\mathbf{E}_\theta\big[g(X)\,|\,T\big]$ is the same for all values of θ. Thus, we can suppress the subscript θ, and view

$$\hat{g}(X) = \mathbf{E}\big[g(X)\,|\,T\big]$$

as a new estimator of θ, which depends on X only through T.

(b) Show that the estimators $g(X)$ and $\hat{g}(X)$ have the same bias.

(c) Show that for any θ with $\text{var}_\theta\big(g(X)\big) < \infty$,

$$\mathbf{E}_\theta\Big[\big(\hat{g}(X)-\theta\big)^2\Big] \le \mathbf{E}_\theta\Big[\big(g(X)-\theta\big)^2\Big], \qquad \text{for all } \theta.$$

Furthermore, for a given θ, this inequality is strict if and only if

$$\mathbf{E}_\theta\Big[\text{var}\big(g(X)\,|\,T\big)\Big] > 0.$$

Solution. (a) Since $T = q(X)$ is a sufficient statistic, the conditional distribution $\mathbf{P}_\theta(X = x\,|\,T = t)$ does not depend on θ, so the same is true for $\mathbf{E}_\theta\big[g(X)\,|\,T\big]$.

(b) We have by the law of iterated expectations

$$\mathbf{E}_\theta\big[g(X)\big] = \mathbf{E}_\theta\Big[\mathbf{E}\big[g(X)\,|\,T\big]\Big] = \mathbf{E}_\theta\big[\hat{g}(X)\big],$$

so the biases of $g(X)$ and $\hat{g}(X)$ are equal.

(c) Fix some θ and let b_θ denote the common bias of $g(X)$ and $\hat{g}(X)$. We have, using the law of total variance,

$$\begin{aligned}
\mathbf{E}_\theta\Big[\big(g(X)-\theta\big)^2\Big] &= \text{var}_\theta\big(g(X)\big) + \mathrm{b}_\theta^2 \\
&= \mathbf{E}_\theta\Big[\text{var}\big(g(X)\,|\,T\big)\Big] + \text{var}_\theta\Big(\mathbf{E}\big[g(X)\,|\,T\big]\Big) + \mathrm{b}_\theta^2 \\
&= \mathbf{E}_\theta\Big[\text{var}\big(g(X)\,|\,T\big)\Big] + \text{var}_\theta\big(\hat{g}(X)\big) + \mathrm{b}_\theta^2 \\
&= \mathbf{E}_\theta\Big[\text{var}\big(g(X)\,|\,T\big)\Big] + \mathbf{E}_\theta\Big[\big(\hat{g}(X)-\theta\big)^2\Big] \\
&\ge \mathbf{E}_\theta\Big[\big(\hat{g}(X)-\theta\big)^2,
\end{aligned}$$

with the inequality being strict if and only if $\mathbf{E}_\theta\Big[\text{var}\big(g(X)\,|\,T\big)\Big] > 0$.

Problem 15.* Let $X_1, \ldots, X_n$ be i.i.d. random variables that are uniformly distributed over the interval $[0, \theta]$.

(a) Show that $T = \max_{i=1,\ldots,n} X_i$ is a sufficient statistic.

(b) Show that $g(X) = (2/n)\sum_{i=1}^n X_i$ is an unbiased estimator.

(c) Find the form of the estimator $\hat{g}(X) = \mathbf{E}\big[g(X)\,|\,T\big]$, and then calculate and compare $\mathbf{E}_\theta\Big[\big(\hat{g}(X)-\theta\big)^2\Big]$ with $\mathbf{E}_\theta\Big[\big(g(X)-\theta\big)^2\Big]$.

Solution. (a) The likelihood function is

$$f_X(x_1, \ldots, x_n; \theta) = f_{X_1}(x_1;\theta) \cdots f_{X_n}(x_n;\theta) = \begin{cases} 1/\theta^n, & \text{if } 0 \le \max_{i=1,\ldots,n} x_i \le \theta \le 1, \\ 0, & \text{otherwise,} \end{cases}$$

and depends on x only through $q(x) = \max_{i=1,\ldots,n} x_i$. The result follows from the factorization criterion for a sufficient statistic.

(b) We have

$$\mathbf{E}_\theta\big[g(X)\big] = \frac{2}{n}\sum_{i=1}^{n}\mathbf{E}_\theta[X_i] = \frac{2}{n}\sum_{i=1}^{n}\frac{\theta}{2} = \theta.$$

(c) Conditioned on the event $\{T = t\}$, one of the observations X_i is equal to t. The remaining $n-1$ observations are uniformly distributed over the interval $[0, t]$, and have a conditional expectation equal to $t/2$. Thus,

$$\mathbf{E}\big[g(X) \,|\, T = t\big] = \frac{2}{n}\mathbf{E}\Big[\sum_{i=1}^{n} X_i \,|\, T = t\Big] = \frac{2}{n}\Big(t + \frac{(n-1)t}{2}\Big) = \frac{n+1}{n}t,$$

and $\hat{g}(X) = \mathbf{E}\big[g(X)\,|\,T\big] = (n+1)T/n$.

We will now calculate the mean squared error of the two estimators $\hat{g}(X)$ and $g(X)$, as functions of θ. To this end, we evaluate the first and second moment of $\hat{g}(X)$. We have

$$\mathbf{E}_\theta\left[\hat{g}(X)\right] = \mathbf{E}_\theta\big[\mathbf{E}\left[g(X)\,|\,T\right]\big] = \mathbf{E}_\theta\big[g(X)\big] = \theta.$$

To find the second moment, we first determine the PDF of T. For $t \in [0, \theta]$, we have $\mathbf{P}_\theta(T \le t) = (t/\theta)^n$, and by differentiating, $f_T(t;\theta) = nt^{n-1}/\theta^n$. Thus,

$$\begin{aligned} \mathbf{E}_\theta\Big[\big(\hat{g}(X)\big)^2\Big] &= \Big(\frac{n+1}{n}\Big)^2 \mathbf{E}\left[T^2\right] = \Big(\frac{n+1}{n}\Big)^2 \int_0^\theta t^2 f_T(t;\theta)\,dt \\ &= \Big(\frac{n+1}{n}\Big)^2 \int_0^\theta t^2 \frac{nt^{n-1}}{\theta^n}\,dt = \frac{(n+1)^2}{n(n+2)}\theta^2. \end{aligned}$$

Since $\hat{g}(X)$ has mean θ, its mean squared error is equal to its variance, and

$$\mathbf{E}_\theta\Big[\big(\hat{g}(X) - \theta\big)^2\Big] = \mathbf{E}_\theta\Big[\big(\hat{g}(X)\big)^2\Big] - \theta^2 = \frac{(n+1)^2}{n(n+2)}\theta^2 - \theta^2 = \frac{1}{n(n+2)}\theta^2.$$

Similarly, the mean squared error of $g(X)$ is equal to its variance, and

$$\mathbf{E}_\theta\Big[\big(g(X) - \theta\big)^2\Big] = \frac{4}{n^2}\sum_{i=1}^{n}\mathrm{var}_\theta(X_i) = \frac{4}{n^2}\cdot n \cdot \frac{\theta^2}{12} = \frac{1}{3n}\theta^2.$$

It can be seen that $\dfrac{1}{3n} \ge \dfrac{1}{n(n+2)}$ for all positive integers n. It follows that

$$\mathbf{E}_\theta\Big[\big(\hat{g}(X) - \theta\big)^2\Big] \le \mathbf{E}_\theta\Big[\big(g(X) - \theta\big)^2\Big],$$

which is consistent with the Rao-Blackwell theorem.

SECTION 9.2. Linear Regression

Problem 16. An electric utility company tries to estimate the relation between the daily amount of electricity used by customers and the daily summer temperature. It has collected the data shown on the table below.

Temperature	96	89	81	86	83
Electricity	23.67	20.45	21.86	23.28	20.71
Temperature	73	78	74	76	78
Electricity	18.21	18.85	20.10	18.48	17.94

(a) Set up and estimate the parameters of a linear model that can be used to predict electricity consumption as a function of temperature.

(b) If the temperature on a given day is 90 degrees, predict the amount of electricity consumed on that day.

Problem 17. Given the five data pairs (x_i, y_i) in the table below,

x	0.798	2.546	5.005	7.261	9.131
y	−2.373	20.906	103.544	215.775	333.911

we want to construct a model relating x and y. We consider a linear model

$$Y_i = \theta_0 + \theta_1 x_i + W_i, \qquad i = 1, \ldots, 5,$$

and a quadratic model

$$Y_i = \beta_0 + \beta_1 x_i^2 + V_i, \qquad i = 1, \ldots, 5,$$

where W_i and V_i represent additive noise terms, modeled by independent normal random variables with mean zero and variance σ_1^2 and σ_2^2, respectively.

(a) Find the ML estimates of the linear model parameters.

(b) Find the ML estimates of the quadratic model parameters.

(c) Assume that the two estimated models are equally likely to be true, and that the noise terms W_i and V_i have the same variance: $\sigma_1^2 = \sigma_2^2$. Use the MAP rule to choose between the two models.

Problem 18.* Unbiasedness and consistency in linear regression. In a probabilistic framework for regression, let us assume that $Y_i = \theta_0 + \theta_1 x_i + W_i$, $i = 1, \ldots, n$, where $W_1, \ldots, W_n$ are i.i.d. normal random variables with mean zero and variance σ^2. Then, given x_i and the realized values y_i of Y_i, $i = 1, \ldots, n$, the ML estimates of θ_0 and θ_1 are given by the linear regression formulas, as discussed in Section 9.2.

(a) Show that the ML estimators $\hat{\Theta}_0$ and $\hat{\Theta}_1$ are unbiased.

(b) Show that the variances of the estimators $\hat{\Theta}_0$ and $\hat{\Theta}_1$ are

$$\text{var}(\hat{\Theta}_0) = \frac{\sigma^2 \sum_{i=1}^{n} x_i^2}{n \sum_{i=1}^{n} (x_i - \overline{x})^2}, \qquad \text{var}(\hat{\Theta}_1) = \frac{\sigma^2}{\sum_{i=1}^{n} (x_i - \overline{x})^2},$$

respectively, and their covariance is

$$\text{cov}(\hat{\Theta}_0, \hat{\Theta}_1) = -\frac{\sigma^2 \, \overline{x}}{\sum_{i=1}^{n} (x_i - \overline{x})^2}.$$

(c) Show that if $\sum_{i=1}^{n} (x_i - \overline{x})^2 \to \infty$ and $\overline{x}^2$ is bounded by a constant as $n \to \infty$, we have $\text{var}(\hat{\Theta}_0) \to 0$ and $\text{var}(\hat{\Theta}_1) \to 0$. (This, together with Chebyshev's inequality, implies that the estimators $\hat{\Theta}_0$ and $\hat{\Theta}_1$ are consistent.)

Note: Although the assumption that the W_i are normal is needed for our estimators to be ML estimators, the argument below shows that these estimators remain unbiased and consistent without this assumption.

Solution. (a) Let the true values of θ_0 and θ_1 be θ_0^* and θ_1^*, respectively. We have

$$\hat{\Theta}_1 = \frac{\sum_{i=1}^{n} (x_i - \overline{x})(Y_i - \overline{Y})}{\sum_{i=1}^{n} (x_i - \overline{x})^2}, \qquad \hat{\Theta}_0 = \overline{Y} - \hat{\Theta}_1 \overline{x},$$

where $\overline{Y} = \left(\sum_{i=1}^{n} Y_i \right)/n$, and where we treat $x_1, \ldots, x_n$ as constant. Denoting $\overline{W} = \left(\sum_{i=1}^{n} W_i \right)/n$, we have

$$Y_i = \theta_0^* + \theta_1^* x_i + W_i, \qquad \overline{Y} = \theta_0^* + \theta_1^* \overline{x} + \overline{W},$$

and

$$Y_i - \overline{Y} = \theta_1^* (x_i - \overline{x}) + (W_i - \overline{W}).$$

Thus,

$$\hat{\Theta}_1 = \frac{\sum_{i=1}^{n} (x_i - \overline{x}) \left(\theta_1^* (x_i - \overline{x}) + W_i - \overline{W} \right)}{\sum_{i=1}^{n} (x_i - \overline{x})^2} = \theta_1^* + \frac{\sum_{i=1}^{n} (x_i - \overline{x})(W_i - \overline{W})}{\sum_{i=1}^{n} (x_i - \overline{x})^2}$$

$$= \theta_1^* + \frac{\sum_{i=1}^{n} (x_i - \overline{x}) W_i}{\sum_{i=1}^{n} (x_i - \overline{x})^2},$$

where we have used the fact $\sum_{i=1}^{n}(x_i - \overline{x}) = 0$. Since $\mathbf{E}[W_i] = 0$, it follows that

$$\mathbf{E}[\hat{\Theta}_1] = \theta_1^*.$$

Also

$$\hat{\Theta}_0 = \overline{Y} - \hat{\Theta}_1\overline{x} = \theta_0^* + \theta_1^*\overline{x} + \overline{W} - \hat{\Theta}_1\overline{x} = \theta_0^* + (\theta_1^* - \hat{\Theta}_1)\overline{x} + \overline{W},$$

and using the facts $\mathbf{E}[\hat{\Theta}_1] = \theta_1^*$ and $\mathbf{E}[\overline{W}] = 0$, we obtain

$$\mathbf{E}[\hat{\Theta}_0] = \theta_0^*.$$

Thus, the estimators $\hat{\Theta}_0$ and $\hat{\Theta}_1$ are unbiased.

(b) We now calculate the variance of the estimators. Using the formula for $\hat{\Theta}_1$ derived in part (a) and the independence of the W_i, we have

$$\text{var}(\hat{\Theta}_1) = \frac{\displaystyle\sum_{i=1}^{n}(x_i - \overline{x})^2 \text{var}(W_i)}{\left(\displaystyle\sum_{i=1}^{n}(x_i - \overline{x})^2\right)^2} = \frac{\sigma^2}{\displaystyle\sum_{i=1}^{n}(x_i - \overline{x})^2}.$$

Similarly, using the formula for $\hat{\Theta}_0$ derived in part (a),

$$\text{var}(\hat{\Theta}_0) = \text{var}(\overline{W} - \hat{\Theta}_1\overline{x}) = \text{var}(\overline{W}) + \overline{x}^2\text{var}(\hat{\Theta}_1) - 2\overline{x}\,\text{cov}(\overline{W}, \hat{\Theta}_1).$$

Since $\sum_{i=1}^{n}(x_i - \overline{x}) = 0$ and $\mathbf{E}[\overline{W}W_i] = \sigma^2/n$ for all i, we obtain

$$\text{cov}(\overline{W}, \hat{\Theta}_1) = \frac{\mathbf{E}\left[\overline{W}\displaystyle\sum_{i=1}^{n}(x_i - \overline{x})W_i\right]}{\displaystyle\sum_{i=1}^{n}(x_i - \overline{x})^2} = \frac{\dfrac{\sigma^2}{n}\displaystyle\sum_{i=1}^{n}(x_i - \overline{x})}{\displaystyle\sum_{i=1}^{n}(x_i - \overline{x})^2} = 0.$$

Combining the last three equations, we obtain

$$\text{var}(\hat{\Theta}_0) = \text{var}(\overline{W}) + \overline{x}^2\text{var}(\hat{\Theta}_1) = \frac{\sigma^2}{n} + \frac{\overline{x}^2\sigma^2}{\displaystyle\sum_{i=1}^{n}(x_i - \overline{x})^2} = \frac{\sigma^2}{n} \cdot \frac{\displaystyle\sum_{i=1}^{n}(x_i - \overline{x})^2 + n\overline{x}^2}{\displaystyle\sum_{i=1}^{n}(x_i - \overline{x})^2}.$$

By expanding the quadratic forms $(x_i - \overline{x})^2$, we also have

$$\sum_{i=1}^{n}(x_i - \overline{x})^2 + n\overline{x}^2 = \sum_{i=1}^{n} x_i^2.$$

By combining the preceding two equations,

$$\mathrm{var}(\hat{\Theta}_0) = \frac{\sigma^2 \displaystyle\sum_{i=1}^{n} x_i^2}{n \displaystyle\sum_{i=1}^{n} (x_i - \overline{x})^2}.$$

We finally calculate the covariance of $\hat{\Theta}_0$ and $\hat{\Theta}_1$. We have

$$\mathrm{cov}(\hat{\Theta}_0, \hat{\Theta}_1) = \mathbf{E}\big[(\hat{\Theta}_0 - \theta_0^*)(\hat{\Theta}_1 - \theta_1^*)\big] = \mathbf{E}\big[\big((\theta_1^* - \hat{\Theta}_1)\overline{x} + \overline{W}\big)(\hat{\Theta}_1 - \theta_1^*)\big],$$

or

$$\mathrm{cov}(\hat{\Theta}_0, \hat{\Theta}_1) = -\overline{x}\,\mathrm{var}(\hat{\Theta}_1) + \mathrm{cov}(\overline{W}, \hat{\Theta}_1).$$

Since, as shown earlier, $\mathrm{cov}(\overline{W}, \hat{\Theta}_1) = 0$, we finally obtain

$$\mathrm{cov}(\hat{\Theta}_0, \hat{\Theta}_1) = -\frac{\overline{x}\,\sigma^2}{\displaystyle\sum_{i=1}^{n} (x_i - \overline{x})^2}.$$

(c) If $\sum_{i=1}^{n} (x_i - \overline{x})^2 \to \infty$, the expression for $\mathrm{var}(\hat{\Theta}_1) \to 0$ derived in part (b) goes to zero. Then, the formula

$$\mathrm{var}(\hat{\Theta}_0) = \mathrm{var}(\overline{W}) + \overline{x}^2 \mathrm{var}(\hat{\Theta}_1),$$

from part (b), together with the assumption that $\overline{x}^2$ is bounded by a constant, implies that $\mathrm{var}(\hat{\Theta}_0) \to 0$.

Problem 19.* Variance estimate in linear regression. Under the same assumptions as in the preceding problem, show that

$$\hat{S}_n^2 = \frac{1}{n-2} \sum_{i=1}^{n} (Y_i - \hat{\Theta}_0 - \hat{\Theta}_1 x_i)^2$$

is an unbiased estimator of σ^2.

Solution. Let $\hat{V}_n = \sum_{i=1}^{n} (Y_i - \hat{\Theta}_0 - \hat{\Theta}_1 x_i)^2$. Using the formula $\hat{\Theta}_0 = \overline{Y} - \hat{\Theta}_1 \overline{x}$ and the expression for $\hat{\Theta}_1$, we have

$$\begin{aligned}
\hat{V}_n &= \sum_{i=1}^{n} \big(Y_i - \overline{Y} - \hat{\Theta}_1 (x_i - \overline{x})\big)^2 \\
&= \sum_{i=1}^{n} (Y_i - \overline{Y})^2 - 2\hat{\Theta}_1 \sum_{i=1}^{n} (Y_i - \overline{Y})(x_i - \overline{x}) + \hat{\Theta}_1^2 \sum_{i=1}^{n} (x_i - \overline{x})^2 \\
&= \sum_{i=1}^{n} (Y_i - \overline{Y})^2 - \hat{\Theta}_1^2 \sum_{i=1}^{n} (x_i - \overline{x})^2 \\
&= \sum_{i=1}^{n} Y_i^2 - n\overline{Y}^2 - \hat{\Theta}_1^2 \sum_{i=1}^{n} (x_i - \overline{x})^2.
\end{aligned}$$

Taking expectation of both sides, we obtain

$$\mathbf{E}[\hat{V}_n] = \sum_{i=1}^{n} \mathbf{E}\big[Y_i^2\big] - n\mathbf{E}\big[\overline{Y}^2\big] - \sum_{i=1}^{n}(x_i - \overline{x})^2 \mathbf{E}\big[\hat{\Theta}_1^2\big].$$

We also have

$$\begin{aligned}
\mathbf{E}\big[Y_i^2\big] &= \text{var}(Y_i) + \big(\mathbf{E}[Y_i]\big)^2 = \sigma^2 + (\theta_0^* + \theta_1^* x_i)^2, \\
\mathbf{E}\big[\overline{Y}^2\big] &= \text{var}(\overline{Y}) + \big(\mathbf{E}[\overline{Y}]\big)^2 = \frac{\sigma^2}{n} + (\theta_0^* + \theta_1^* \overline{x})^2, \\
\mathbf{E}\big[\hat{\Theta}_1^2\big] &= \text{var}(\hat{\Theta}_1) + \big(\mathbf{E}[\hat{\Theta}_1]\big)^2 = \frac{\sigma^2}{\sum_{i=1}^{n}(x_i - \overline{x})^2} + (\theta_1^*)^2.
\end{aligned}$$

Combining the last four equations and simplifying, we obtain

$$\mathbf{E}[\hat{V}_n] = (n-2)\sigma^2.$$

SECTION 9.3. Binary Hypothesis Testing

Problem 20. A random variable X is characterized by a normal PDF with mean $\mu_0 = 20$, and a variance that is either $\sigma_0^2 = 16$ (hypothesis H_0) or $\sigma_1^2 = 25$ (hypothesis H_1). We want to test H_0 against H_1, using three sample values x_1, x_2, x_3, and a rejection region of the form

$$R = \{x \mid x_1 + x_2 + x_3 > \gamma\}$$

for some scalar γ. Determine the value of γ so that the probability of false rejection is 0.05. What is the corresponding probability of false acceptance?

Problem 21. A normal random variable X is known to have a mean of 60 and a standard deviation equal to 5 (hypothesis H_0) or 8 (hypothesis H_1).

(a) Consider a hypothesis test using a single sample x. Let the rejection region be of the form

$$R = \big\{ x \;\big|\; |x - 60| > \gamma \big\}$$

for some scalar γ. Determine γ so that the probability of false rejection of H_0 is 0.1. What is the corresponding false acceptance probability? Would the rejection region change if we were to use the LRT with the same false rejection probability?

(b) Consider a hypothesis test using n independent samples $x_1, \ldots, x_n$. Let the rejection region be of the form

$$R = \left\{ (x_1, \ldots, x_n) \;\middle|\; \left| \frac{x_1 + \cdots + x_n}{n} - 60 \right| > \gamma \right\},$$

where γ is chosen so that the probability of false rejection of H_0 is 0.1. How does the false acceptance probability change with n? What can you conclude about the appropriateness of this type of test?

(c) Derive the structure of the LRT using n independent samples $x_1, \ldots, x_n$.

Problem 22. There are two hypotheses about the probability of heads for a given coin: $\theta = 0.5$ (hypothesis H_0) and $\theta = 0.6$ (hypothesis H_1). Let X be the number of heads obtained in n tosses, where n is large enough so that normal approximations are appropriate. We test H_0 against H_1 by rejecting H_0 if X is greater than some suitably chosen threshold k_n.

(a) What should be the value of k_n so that the probability of false rejection is less than or equal to 0.05?

(b) What is the smallest value of n for which both probabilities of false rejection and false acceptance can be made less than or equal to 0.05?

(c) For the value of n found in part (b), what would be the probability of false acceptance if we were to use a LRT with the same probability of false rejection?

Problem 23. The number of phone calls received by a ticket agency on any one day is Poisson distributed. On an ordinary day, the expected value of the number of calls is λ_0, and on a day where there is a popular show in town, the expected value of the number of calls is λ_1, with $\lambda_1 > \lambda_0$. Describe the LRT for deciding whether there is a popular show in town based on the number of calls received. Assume a given probability of false rejection, and find an expression for the critical value ξ.

Problem 24. We have received a shipment of light bulbs whose lifetimes are modeled as independent, exponentially distributed random variables, with parameter equal to λ_0 (hypothesis H_0) or equal to λ_1 (hypothesis H_1). We measure the lifetimes of n light bulbs. Describe the LRT for selecting one of the two hypotheses. Assume a given probability of false rejection of H_0 and give an analytical expression for the critical value ξ.

SECTION 9.4. Significance Testing

Problem 25. Let X be a normal random variable with mean μ and unit variance. We want to test the hypothesis $\mu = 5$ at the 5% level of significance, using n independent samples of X.

(a) What is the range of values of the sample mean for which the hypothesis is accepted?

(b) Let $n = 10$. Calculate the probability of accepting the hypothesis $\mu = 5$ when the true value of μ is 4.

Problem 26. We have five observations drawn independently from a normal distribution with unknown mean μ and unknown variance σ^2.

(a) Estimate μ and σ^2 if the observation values are $8.47, 10.91, 10.87, 9.46, 10.40$.

(b) Use the t-distribution tables to test the hypothesis $\mu = 9$ at the 95% significance level, using the estimates of part (a).

Problem 27. A plant grows on two distant islands. Suppose that its life span (measured in days) on the first (or the second) island is normally distributed with unknown mean μ_X (or μ_Y) and known variance $\sigma_X^2 = 32$ (or $\sigma_Y^2 = 29$, respectively). We wish to test the hypothesis $\mu_X = \mu_Y$, based on 10 independent samples from each island. The corresponding sample means are $\overline{x} = 181$ and $\overline{y} = 177$. Do the data support the hypothesis at the 95% significance level?

Problem 28. A company considers buying a machine to manufacture a certain item. When tested, 28 out of 600 items produced by the machine were found defective. Do the data support the hypothesis that the defect rate of the machine is smaller than 3 percent, at the 5% significance level?

Problem 29. The values of five independent samples of a Poisson random variable turned out to be 34, 35, 29, 31, and 30. Test the hypothesis that the mean is equal to 35 at the 5% level of significance.

Problem 30. A surveillance camera periodically checks a certain area and records a signal $X = W$ if there is no intruder (this is the null hypothesis H_0). If there is an intruder the signal is $X = \theta + W$, where θ is unknown with $\theta > 0$. We assume that W is a normal random variable with mean 0 and known variance $v = 0.5$.

(a) We obtain a single signal value $X = 0.96$. Should H_0 be rejected at the 5% level of significance?

(b) We obtain five independent signal values $X = 0.96, -0.34, 0.85, 0.51, -0.24$. Should H_0 be rejected at the 5% level of significance?

(c) Repeat part (b), using the t-distribution, and assuming the variance v is unknown.

INDEX

Summary of Results for Special Discrete Random Variables

Discrete Uniform over $[a, b]$:

$$p_X(k) = \begin{cases} \dfrac{1}{b-a+1}, & \text{if } k = a, a+1, \ldots, b, \\ 0, & \text{otherwise,} \end{cases}$$

$$\mathbf{E}[X] = \frac{a+b}{2}, \quad \text{var}(X) = \frac{(b-a)(b-a+2)}{12}, \quad M_X(s) = \frac{e^{sa}\left(e^{s(b-a+1)} - 1\right)}{(b-a+1)(e^s - 1)}.$$

Bernoulli with Parameter p: (Describes the success or failure in a single trial.)

$$p_X(k) = \begin{cases} p, & \text{if } k = 1, \\ 1-p, & \text{if } k = 0, \end{cases}$$

$$\mathbf{E}[X] = p, \qquad \text{var}(X) = p(1-p), \qquad M_X(s) = 1 - p + pe^s.$$

Binomial with Parameters p and n: (Describes the number of successes in n independent Bernoulli trials.)

$$p_X(k) = \binom{n}{k} p^k (1-p)^{n-k}, \qquad k = 0, 1, \ldots, n,$$

$$\mathbf{E}[X] = np, \qquad \text{var}(X) = np(1-p), \qquad M_X(s) = (1 - p + pe^s)^n.$$

Geometric with Parameter p: (Describes the number of trials until the first success, in a sequence of independent Bernoulli trials.)

$$p_X(k) = (1-p)^{k-1}p, \qquad k = 1, 2, \ldots,$$

$$\mathbf{E}[X] = \frac{1}{p}, \qquad \text{var}(X) = \frac{1-p}{p^2}, \qquad M_X(s) = \frac{pe^s}{1-(1-p)e^s}.$$

Poisson with Parameter λ: (Approximates the binomial PMF when n is large, p is small, and $\lambda = np$.)

$$p_X(k) = e^{-\lambda}\frac{\lambda^k}{k!}, \qquad k = 0, 1, \ldots,$$

$$\mathbf{E}[X] = \lambda, \qquad \text{var}(X) = \lambda, \qquad M_X(s) = e^{\lambda(e^s - 1)}.$$

Summary of Results for Special Continuous Random Variables

Continuous Uniform Over $[a, b]$:

$$f_X(x) = \begin{cases} \dfrac{1}{b-a}, & \text{if } a \le x \le b, \\ 0, & \text{otherwise,} \end{cases}$$

$$\mathbf{E}[X] = \frac{a+b}{2}, \qquad \text{var}(X) = \frac{(b-a)^2}{12}, \qquad M_X(s) = \frac{e^{sb} - e^{sa}}{s(b-a)}.$$

Exponential with Parameter λ:

$$f_X(x) = \begin{cases} \lambda e^{-\lambda x}, & \text{if } x \ge 0, \\ 0, & \text{otherwise,} \end{cases} \qquad F_X(x) = \begin{cases} 1 - e^{-\lambda x}, & \text{if } x \ge 0, \\ 0, & \text{otherwise,} \end{cases}$$

$$\mathbf{E}[X] = \frac{1}{\lambda}, \qquad \text{var}(X) = \frac{1}{\lambda^2}, \qquad M_X(s) = \frac{\lambda}{\lambda - s}, \quad (s < \lambda).$$

Normal with Parameters μ and $\sigma^2 > 0$:

$$f_X(x) = \frac{1}{\sqrt{2\pi}\,\sigma} e^{-(x-\mu)^2/2\sigma^2},$$

$$\mathbf{E}[X] = \mu, \qquad \text{var}(X) = \sigma^2, \qquad M_X(s) = e^{(\sigma^2 s^2/2) + \mu s}.$$

	.00	.01	.02	.03	.04	.05	.06	.07	.08	.09
0.0	.5000	.5040	.5080	.5120	.5160	.5199	.5239	.5279	.5319	.5359
0.1	.5398	.5438	.5478	.5517	.5557	.5596	.5636	.5675	.5714	.5753
0.2	.5793	.5832	.5871	.5910	.5948	.5987	.6026	.6064	.6103	.6141
0.3	.6179	.6217	.6255	.6293	.6331	.6368	.6406	.6443	.6480	.6517
0.4	.6554	.6591	.6628	.6664	.6700	.6736	.6772	.6808	.6844	.6879
0.5	.6915	.6950	.6985	.7019	.7054	.7088	.7123	.7157	.7190	.7224
0.6	.7257	.7291	.7324	.7357	.7389	.7422	.7454	.7486	.7517	.7549
0.7	.7580	.7611	.7642	.7673	.7704	.7734	.7764	.7794	.7823	.7852
0.8	.7881	.7910	.7939	.7967	.7995	.8023	.8051	.8078	.8106	.8133
0.9	.8159	.8186	.8212	.8238	.8264	.8289	.8315	.8340	.8365	.8389
1.0	.8413	.8438	.8461	.8485	.8508	.8531	.8554	.8577	.8599	.8621
1.1	.8643	.8665	.8686	.8708	.8729	.8749	.8770	.8790	.8810	.8830
1.2	.8849	.8869	.8888	.8907	.8925	.8944	.8962	.8980	.8997	.9015
1.3	.9032	.9049	.9066	.9082	.9099	.9115	.9131	.9147	.9162	.9177
1.4	.9192	.9207	.9222	.9236	.9251	.9265	.9279	.9292	.9306	.9319
1.5	.9332	.9345	.9357	.9370	.9382	.9394	.9406	.9418	.9429	.9441
1.6	.9452	.9463	.9474	.9484	.9495	.9505	.9515	.9525	.9535	.9545
1.7	.9554	.9564	.9573	.9582	.9591	.9599	.9608	.9616	.9625	.9633
1.8	.9641	.9649	.9656	.9664	.9671	.9678	.9686	.9693	.9699	.9706
1.9	.9713	.9719	.9726	.9732	.9738	.9744	.9750	.9756	.9761	.9767
2.0	.9772	.9778	.9783	.9788	.9793	.9798	.9803	.9808	.9812	.9817
2.1	.9821	.9826	.9830	.9834	.9838	.9842	.9846	.9850	.9854	.9857
2.2	.9861	.9864	.9868	.9871	.9875	.9878	.9881	.9884	.9887	.9890
2.3	.9893	.9896	.9898	.9901	.9904	.9906	.9909	.9911	.9913	.9916
2.4	.9918	.9920	.9922	.9925	.9927	.9929	.9931	.9932	.9934	.9936
2.5	.9938	.9940	.9941	.9943	.9945	.9946	.9948	.9949	.9951	.9952
2.6	.9953	.9955	.9956	.9957	.9959	.9960	.9961	.9962	.9963	.9964
2.7	.9965	.9966	.9967	.9968	.9969	.9970	.9971	.9972	.9973	.9974
2.8	.9974	.9975	.9976	.9977	.9977	.9978	.9979	.9979	.9980	.9981
2.9	.9981	.9982	.9982	.9983	.9984	.9984	.9985	.9985	.9986	.9986
3.0	.9987	.9987	.9987	.9988	.9988	.9989	.9989	.9989	.9990	.9990
3.1	.9990	.9991	.9991	.9991	.9992	.9992	.9992	.9992	.9993	.9993
3.2	.9993	.9993	.9994	.9994	.9994	.9994	.9994	.9995	.9995	.9995
3.3	.9995	.9995	.9995	.9996	.9996	.9996	.9996	.9996	.9996	.9997
3.4	.9997	.9997	.9997	.9997	.9997	.9997	.9997	.9997	.9997	.9998

The standard normal table. The entries in this table provide the numerical values of $\Phi(y) = \mathbf{P}(Y \leq y)$, where Y is a standard normal random variable, for y between 0 and 3.49.

Example of use: To find $\Phi(1.71)$, we look at the row corresponding to 1.7 and the column corresponding to 0.01, so that $\Phi(1.71) = .9564$. When y is negative, the value of $\Phi(y)$ can be found using the formula $\Phi(y) = 1 - \Phi(-y)$.

ATHENA SCIENTIFIC BOOKS

OPTIMIZATION AND COMPUTATION SERIES

1. Introduction to Probability, 2nd Edition, by Dimitri P. Bertsekas and John N. Tsitsiklis, 2008, ISBN 978-1-886529-23-6, 544 pages
2. Dynamic Programming and Optimal Control, Two-Volume Set, by Dimitri P. Bertsekas, 2007, ISBN 1-886529-08-6, 1020 pages
3. Convex Analysis and Optimization, by Dimitri P. Bertsekas, with Angelia Nedić and Asuman E. Ozdaglar, 2003, ISBN 1-886529-45-0, 560 pages
4. Nonlinear Programming, 2nd Edition, by Dimitri P. Bertsekas, 1999, ISBN 1-886529-00-0, 791 pages
5. Network Optimization: Continuous and Discrete Models, by Dimitri P. Bertsekas, 1998, ISBN 1-886529-02-7, 608 pages
6. Network Flows and Monotropic Optimization, by R. Tyrrell Rockafellar, 1998, ISBN 1-886529-06-X, 634 pages
7. Introduction to Linear Optimization, by Dimitris Bertsimas and John N. Tsitsiklis, 1997, ISBN 1-886529-19-1, 608 pages
8. Parallel and Distributed Computation: Numerical Methods, by Dimitri P. Bertsekas and John N. Tsitsiklis, 1997, ISBN 1-886529-01-9, 718 pages
9. Neuro-Dynamic Programming, by Dimitri P. Bertsekas and John N. Tsitsiklis, 1996, ISBN 1-886529-10-8, 512 pages
10. Constrained Optimization and Lagrange Multiplier Methods, by Dimitri P. Bertsekas, 1996, ISBN 1-886529-04-3, 410 pages
11. Stochastic Optimal Control: The Discrete-Time Case, by Dimitri P. Bertsekas and Steven E. Shreve, 1996, ISBN 1-886529-03-5, 330 pages